The Amateur Radio Public Service Handbook

First Edition

A Guide to Radio Communications for Community Events, Emergencies, and Disasters

Editors
Michael Corey, KI1U
Becky Rodia Schoenfeld, W1BXY

Cover
Sue Fagan, KB1OKW

ARRL — The national association for AMATEUR RADIO
225 Main Street, Newington, CT 06111
www.arrl.org

Copyright © 2012 by

The American Radio Relay League, Inc.

Copyright secured under the Pan-American Convention

International Copyright secured

All rights reserved. No part of this work may be reproduced in any form except by written permission of the publisher. All rights of translation are reserved.

Printed in the USA

Quedan reservados todos los derechos

ISBN: 978-0-87259-484-5

First edition
First printing, 2012

Contents

Part 1: The ARRL and ARES

Chapter 1: The ARRL and Amateur Radio Public Service and Emergency Communications 7

Chapter 2: The Role of the Emergency Coordinator Within the ARES Section 10

Part 2: Served Agencies

Chapter 3: The American Red Cross and Amateur Radio Operators 20

Chapter 4: The Salvation Army and Emergency Communications — The SATERN Program 25

Chapter 5: Weather Events — The National Weather Service and the National Hurricane Center 30

Chapter 6: Working with Local Emergency Management Agencies 42

Part 3: Training and Readiness

Chapter 7: Leading and Training Volunteers 48

Chapter 8: Personal Safety, Survival, and Health 58

Chapter 9: Contesting and Field Day as Supplements to EmComm Training 64

Chapter 10: The Science of Radio — Basics for the EmComm Operator 72

Chapter 11: A Compact, Versatile Multi-Mode Kit for Emergency Communications 85

Chapter 12: Going Portable: Necessary Equipment 96

Part 4: Nets and the National Traffic System

Chapter 13: The National Traffic System 106

Chapter 14: Nets and Message Handling 115

Chapter 15: Network Theory and the Design of Emergency Communication Systems 126

Part 5: The Response

Chapter 16: Managing the Response 133

Chapter 17: ARES PIO: The Right Stuff 142

Part 6: Public Service

Chapter 18: Organizing Amateur Radio Communications for a Public Service Event 147

Chapter 19: Planning and Organizing Ham Communications for the Boston Marathon 154

Part 7: Digital Modes

Chapter 20: NBEMS Best Practices 160

Chapter 21: Ham Radio on the Internet: Hints and Tips for EchoLink Users 176

Chapter 22: The *Winlink 2000* System: Its Use in Emergency Communications 181

Chapter 23: D-STAR for Emergency Communications 186

Part 8: Other Relevant Organizations

Chapter 24: Hams and Military Working Together: How MARS Serves National Security 190

Chapter 25: Handihams and Emergency Communications: A Winning Team 198

Chapter 26: An International Perspective on Emergency Communications 204

Appendices 209

Index 302

About the ARRL

The seed for Amateur Radio was planted in the 1890s, when Guglielmo Marconi began his experiments in wireless telegraphy. Soon he was joined by dozens, then hundreds, of others who were enthusiastic about sending and receiving messages through the air — some with a commercial interest, but others solely out of a love for this new communications medium. The United States government began licensing Amateur Radio operators in 1912.

By 1914, there were thousands of Amateur Radio operators — hams — in the United States. Hiram Percy Maxim, a leading Hartford, Connecticut inventor and industrialist, saw the need for an organization to band together this fledgling group of radio experimenters. In May 1914 he founded the American Radio Relay League (ARRL) to meet that need.

Today ARRL, with approximately 155,000 members, is the largest organization of radio amateurs in the United States. The ARRL is a not-for-profit organization that:
- promotes interest in Amateur Radio communications and experimentation
- represents US radio amateurs in legislative matters, and
- maintains fraternalism and a high standard of conduct among Amateur Radio operators.

At ARRL headquarters in the Hartford suburb of Newington, the staff helps serve the needs of members. ARRL is also International Secretariat for the International Amateur Radio Union, which is made up of similar societies in 150 countries around the world.

ARRL publishes the monthly journal *QST*, as well as newsletters and many publications covering all aspects of Amateur Radio. Its headquarters station, W1AW, transmits bulletins of interest to radio amateurs, as well as Morse code practice sessions. The ARRL also coordinates an extensive field organization, which includes volunteers who provide technical information and other support services for radio amateurs as well as communications for public service activities. In addition, ARRL represents US amateurs with the Federal Communications Commission and other government agencies in the US and abroad.

Membership in ARRL means much more than receiving *QST* each month. In addition to the services already described, ARRL offers membership services on a personal level, such as the Technical Information Service — where members can get answers to all their technical and operating questions by phone, e-mail, or on the ARRL website.

Full ARRL membership (available only to licensed radio amateurs) gives you a voice in how the affairs of the organization are governed. ARRL policy is set by a Board of Directors (one from each of 15 Divisions). Each year, one-third of the ARRL Board of Directors stands for election by the full members they represent. The day-to-day operation of ARRL HQ is managed by an Executive Vice President and his staff.

No matter what aspect of Amateur Radio attracts you, ARRL membership is relevant and important. There would be no Amateur Radio as we know it today were it not for the ARRL. We would be happy to welcome you as a member! (An Amateur Radio license is not required for Associate Membership.) For more information about ARRL and answers to any questions you may have about Amateur Radio, write or call:

ARRL — The national association for Amateur Radio
225 Main Street
Newington CT 06111-1494
Voice: 860-594-0200
    Fax: 860-594-0259
    E-mail: **hq@arrl.org**
    Internet: **www.arrl.org/**

Prospective new amateurs call (toll-free):
800-32-NEW HAM (800-326-3942)
You can also contact us via e-mail at **newham@arrl.org**
or check out *ARRLWeb* at **www.arrl.org/**

Foreword

The mandate to serve the public has been at the core of Amateur Radio since its earliest days. As hams, we have the unique ability to assist those in need in times of trouble. We're experienced communicators. We know how to make radios work. We have the skill to efficiently communicate helpful and even life-saving information when other communications systems fail. In the first decade of the 21st century, Amateur Radio has answered the call to serve in response to disasters such as the terrorist attacks of 9/11, hurricanes Katrina and Ike, massive tornado outbreaks in the South and the Midwest, and the 2010 Haiti earthquake.

To provide emergency communications training to as many amateurs as possible, the ARRL continues to offer courses for emergency communications volunteers and field leaders. The courses are offered online as well as in person. Thousands of amateurs have honed their emergency skills through these courses. Additionally, the ARRL has promoted further training for amateurs through courses offered by served agencies such as the American Red Cross, the National Weather Service, and the Federal Emergency Management Agency.

The Amateur Radio Public Service Handbook provides information distilled from courses and learned through experience in the field, presented by a variety of subject matter experts. I hope you will find it to be a valuable resource.

Of course, this book, like our courses, would not have been possible without the contributions of many amateurs. The acknowledgments on the following page are by no means a complete listing of all of the amateurs who contributed to, or assisted with, the contents of this book. Whether all or part of their material was used in *The Amateur Radio Public Service Handbook*, or if we simply used their material as review or reference for what needed to be covered, we are sincerely grateful to each and every amateur involved in the production of this book for their dedication and willingness to participate.

73,
David Sumner, K1ZZ
ARRL Chief Executive Officer
May 2012

Acknowledgments

The ARRL extends sincere thanks to the following people and organizations for their contributions to this book:

Matthew Beckler, KB3VDJ; Harry Bloomberg, W3YJ; John Brodie, VA7XB; Lloyd Colston, KC5FM; Mark Conklin, N7XYO; John Davis, WB4QDX; Betsey Doane, K1EIC; Tim Duffy, K3LR; Steve Ford, WB8IMY; Ira Green, KB3RXA; Jeffrey Green, KB3RTG; Wayne Gronlund, N1CLV; Gordon Grove, WA7LNC; Matthew Hackman, KB1FUP/AAT1CD/AAA1SN; Scott Hartlage, KF4PWI; Grant Hays, WB6OTS; Pete Kemp, KZ1Z; Dave Kleber, KB3FXI; Sean Kutzko, KX9X; Edith Lennon, N2ZRW; Rob Macedo, KD1CY; Cris McBride, KB7QXQ; Major Patrick McPherson, WW9E; Dan Miller, K3UFG; Greg Mossop, G0DUB; Rick Palm, K1CE; Allen Pitts, W1AGP; G. Mike Raymond, K5HUM; Chuck Rexroad, AB1CR; Julio Ripoll, WD4R; Keith Robertory, KG4UIR; Santa Clara County ARES® / RACES; Steve Schwarm, W3EVE; Bill Sexton, N1IN / AAR1FP / AAM1RD; Mark Spencer, WA8SME; Marc Stern, WA1R; Howard "Skip" Teller, KH6TY; Phil Temples, K9HI; Patrick Tice, WA0TDA; Dave Trachtenberg, N4WWL / AFA3TR; Chuck Verdon, W5KAV; Steve Waterman, K4CJX; Tom Whiteside, N5TW; Duane Whittingham, N9SSN.

This is by no means an exhaustive list. Thanks to all who contributed their time, knowledge, and skills to this project.

Part 1: The ARRL and ARES

Chapter 1
The ARRL and Amateur Radio Public Service and Emergency Communications

From the early days of our service, Amateur Radio operators have regularly answered the call when their communities, states, and nation have needed them. Amateurs have responded to natural and manmade disasters, search and rescue operations, and power outages resulting in lost communication capability, and they have provided communications at marathons, festivals, and races. Over the last two decades, events such as the 9/11 attacks, hurricane Katrina, and earthquakes in Haiti, Chile, Japan, and New Zealand have cast a new light on how we view emergency preparedness and the role Amateur Radio plays in it. Many people and organizations take emergency preparedness more seriously now, and traditional thinking on the subject has begun to take new shapes.

Amateur Radio has enjoyed an increased level of attention from the role we play in responding to emergency events. We have also found ourselves challenged; there is now high value put on training, more people are getting licensed for emergency preparedness reasons, and traditional Amateur Radio emergency communications groups are examining how to meet the communications needs of the 21st century. All of these elements, combined with the continued progression of technology, means that the amateur who wants to be involved with emergency communications must continually improve his or her skills and knowledge base.

For amateurs wondering where to start, however, the fundamental elements of our role in providing emergency and public service communications are found at the very heart of the Amateur Radio Service: Part 97 of the FCC rules.

FCC PART 97.1 BASIS AND PURPOSE

The rules and regulations in this Part are designed to provide an Amateur Radio service having a fundamental purpose as expressed in the following principles:
(a) Recognition and enhancement of the value of the amateur service to the public as a voluntary noncommercial communication service, particularly with respect to providing emergency communications.
(b) Continuation and extension of the amateur's proven ability to contribute to the advancement of the radio art.
(c) Encouragement and improvement of the amateur service through rules which provide for advancing skills in both the communications and technical phases of the art.
(d) Expansion of the existing reservoir within the Amateur Radio service of trained operators, technicians, and electronics experts.
(e) Continuation and extension of the amateur's unique ability to enhance international goodwill.

One may look at Part 97.1 and assume that the section that addresses public service and emergency communications is (a). This assumption would be wrong; in fact, all of Part 97.1 is related to public service and emergency communications. Our ability to provide emergency communications is due to our work for technological advancement (b), our skill and technical ability (c), our recruitment and growth of the service (d), and our recognition that our service is international and that disasters do not discriminate based on nationality or location (e).

ARRL ORGANIZATION, HEADQUARTERS, AND THE FIELD ORGANIZATION

At the national level in the United States, Amateur Radio enjoys a strong advocate in the American Radio Relay League (ARRL). For nearly 100 years, the ARRL has represented Amateur Radio operators and the Amateur Radio Service in the US. The ARRL has also played a key role in Amateur Radio emergency communications. Most amateurs are familiar with the Amateur Radio Emergency Service (ARES), the National Traffic Service (NTS), Radiograms, and Field Day, but the ARRL supports our role in emergency communications in many other ways as well. Let's take a look at how the ARRL is organized, the role of ARRL HQ, the field organization, and various emergency communications programs that exist through the ARRL.

The League's leadership emanates from its membership. Members are the driving force of the ARRL. Through the ballot, members have control over the policies and direction of the League. Nationally, the ARRL has 15 Divisions; each represented by a Director and Vice Director elected by the ARRL members in their Division. The Directors and Vice-

Directors represent the members on the ARRL's Board of Directors. The Board of Directors determines the policies that are implemented by ARRL staff and elect the officers.

ARRL HQ AND EMERGENCY COMMUNICATIONS

At ARRL Headquarters in Newington, Connecticut. Amateur Radio emergency communications is supported through several departments and programs.

Emergency Preparedness Program

The ARRL's Emergency Preparedness Program (EPP) was established in 2007 in the Membership and Volunteer Programs department at HQ. It is managed by the Emergency Preparedness Manager. There are several key duties of the ARRL EPP, some of which are:
- Represent the ARRL with national served agencies
- Provide support and guidance to the field organization
- Represent ARRL at approved conventions and hamfests
- Maintain and report situational awareness during large-scale disasters and emergencies
- MoU compliance
- Monitor ARES and nationwide emergency communications status
- Administer the ARRL HQ Emergency Response Team and Ham Aid Program

While not a part of the Field Organization, the EPP is closely linked with the activities of ARES and other Amateur Radio groups and organizations that assist during emergencies.

Field Services

The Field Organization Team (within the ARRL Membership and Volunteer Programs Department at Headquarters) provides administrative support and guidance to all the members of the Field Organization, and that includes Section Managers, other Section Leaders, station appointees, and leaders and participants in the ARRL National Traffic System.

In an emergency, the Field Organization Team may assist the Emergency Preparedness Team at Headquarters in establishing and maintaining contact with ARRL Section Leaders and NTS Leaders who are pertinent to the emergency situation and response. Team members may also assist the overall Headquarters response by providing backup monitoring duty and / or handling other tasks as needed.

Additional ARRL HQ Resources

Other departments at ARRL Headquarters also play a role in Amateur Radio emergency communications. The Education Services Department coordinates the ARRL training courses Introduction to Emergency Communications (EC-001) and Public Service and Emergency Communications Management for Radio Amateurs (EC-016). The ARRL's Media and Public Relations Manager helps promote to the media and public the valuable contributions made to their communities by Amateur Radio operators. And the Membership and Volunteer Programs department supports Amateur Radio clubs.

AMATEUR RADIO EMERGENCY AND PUBLIC SERVICE COMMUNICATIONS

Whether you call it EmComm, public service communications, or auxiliary communications one thing is true: Amateur Radio operators have been some of the first to step forward when their communities and their country are in need. For over 100 years, the Amateur Radio Service has answered the call when a communications emergency occurs. We have done so under many different banners and names; ARES, RACES, MARS, ARECC, and SKYWARN, to name a few.

You will learn more about several of these organizations in this book. The Amateur Radio Emergency Service® (ARES®) in particular, affiliated with the ARRL, consists of licensed amateurs who have voluntarily registered their qualifications and equipment for communications duty in the public service when disaster strikes. Only licensed amateurs are eligible for membership, though membership in the ARRL is not necessary for membership. There are four levels of ARES organization: national, section, district, and local, with national emergency coordination under the supervision of the ARRL Membership and Volunteer Programs Manager at ARRL Headquarters.

So why is the Amateur Radio Service so valuable during a communications emergency? The answer isn't as straightforward as it may seem. It is found in who we are and what we do as a service.

Most Amateur Radio operators did not get their license because it was required of them; they got it because something about Amateur Radio appealed to them. In other words, they volunteered. They volunteered to learn more about communications, radio, technology, and the world in which they live. The FCC describes the Amateur Radio Service as a "voluntary noncommercial communication service." Think of all the ways in which amateurs volunteer; Elmering, assisting another amateur with station building, pitching in at Field Day, assisting a new ham who's building his or her first kit, volunteering at a served agency. During the first and second World Wars, amateurs volunteered en masse to serve their country as radio operators in the armed forces. Volunteerism is the key to Amateur Radio.

Since the events of September 11, 2001, "emergency communications" have become buzzwords in Amateur Radio. Many people have gotten their Amateur Radio licenses with emergency and disaster preparedness in mind. However, 99% of the time, an emergency doesn't exist. During that time, we should not and cannot sit around waiting for an emergency to occur, at which point we spring into action. We must keep our communication skills sharp by being active Amateur Radio operators; something achieved first and foremost by getting on the air. Amateurs are active every day in many ways;

ragchewing, DXing, experimenting, nets, contests. When an emergency does occur, we are valuable because we have been honing our skills the other 99% of the time.

We are also experimenters and tinkerers. We try new things and look for new innovative ways to communicate. True, our bread-and-butter modes of communication are traditional modes such as voice and CW, but we have a wide range of other communications tools in the toolbox.

THE RIGHT STUFF

In order to be an effective emergency communications volunteer, you'll need good Amateur Radio operating skills. Just having a license does not automatically make us emergency communicators or public service volunteers. Just as you don't get a driver's license so you can compete in the Indy 500, you don't get an Amateur Radio license to be an emergency communicator. Once you have your license, you must put it to good use and become a known quantity as an amateur.

This means getting on the air and exercising your privileges. If you have a Technician license, explore the bands above 50 MHz and 10 meters. As you upgrade your license, explore even further on the HF bands. Try working towards an award such as Worked All States or Worked All Continents. Get on the air for your state's QSO party. Try checking into a net.

It's important to become active in off-air activities as well. Start by joining a club. There are clubs that cater to about every interest in Amateur Radio; DX, contests, general interest, schools, and members that share a common profession or faith. Clubs are excellent ways to network and learn. Becoming a volunteer examiner or license class instructor is another great way to become active in the amateur community.

You can't get on the air without a station, and station building is a skill you must have for emergency communications. You can also help other Amateurs build their stations and hone your technical skills that way. Kit building is also one of the best ways to enhance your technical skills and learn more about what happens underneath your radio's cover.

An effective volunteer needs to exhibit certain personality traits: the ability to be proactive, ability to work outside his or her comfort zone, and a willingness to serve. It must be mentioned that there isn't any room for ego when it comes to being an emergency communications volunteer. You must understand that you will be told what to do, when to do it, and perhaps how to do it. You probably will have to be trained, and it's possible that your training may be a class you've already taken. And there's a good chance that praise and recognition may not come your way, even if you think you deserve it.

Being a good volunteer starts with your approach. When you volunteer to help you are answering the ultimate CQ: "We need help!" To answer that call effectively requires humility, ability, compassion, and adaptability.

Outstanding people skills are also of paramount importance for amateurs who want to be involved in emergency communications. We've all heard the types that like to complain, belittle those new to Amateur Radio, present a bad public image, or believe the only way something should be done is their way. These are not the qualities of a good emergency communications volunteer.

The emergency communications volunteer must take a genuine interest in his or her community and the served agencies they may work with. This interest is not lost on the public or served agency representatives. This interest will be seen in how you interact with people and how you perform your assigned tasks.

When you work with a served agency, you must keep their goals and mission in mind. This is, after all, what you are there to assist them with. When you understand what they are trying to accomplish, you can develop a clear view of how your abilities will fit in.

Not enough can be said about our image. Professional appearance gives an air of credibility. We are judged by the served agency before we even open our mouths. If you are working in an office environment, such as a National Weather Service forecast office or emergency management EOC, wearing a hard hat, a bright yellow safety vest, and a belt with a dozen HTs on it would not be appropriate. Likewise, when assisting with a search and rescue operation in a heavily wooded area it would not be appropriate to wear shorts and sandals and not have an HT or water with you.

We cannot lose sight of what makes us an asset, which is our regular use of the spectrum we have been granted. Our license opens all the doors of Amateur Radio for us. What makes us excel at the path we choose is how good of an Amateur Radio operator each of us becomes.

Chapter 2
The Role of the Emergency Coordinator Within the ARES Section

Though the title "Emergency Coordinator" is given to one person, the truth is, all the duties assigned to an EC really require a team effort. The EC is the team's coach, quarterback, and cheerleader all in one. In this chapter, we'll take a look at the Emergency Coordinator position, first from a general standpoint in terms of duties and, later in the chapter, using the specific context of ARES in the state of Washington.

FOR THE NEW EC

The minimum requirements for becoming a county EC include full ARRL membership and an FCC Technician Class (or higher) Amateur Radio license. Having an EmComm or leadership background is very helpful, but not required.

Because any good plan has at least one backup, the first thing a new Emergency Coordinator must do is locate and/or recruit his or her replacement, someone who would be willing to step into the role of EC if needed. Let's say you're the EC and when a disaster strikes, you're out of town on vacation — or you're ill, or your boss will not let you leave work, or you and your family are victims of the disaster. Any of those situations can and do happen. Your backup person doesn't have to be someone who is being groomed to take over as EC when the time is right for you to step aside, but he or she can be.

Next read, learn and understand the Section's and your county's ARES Emergency Communication plan. Take note that any good plan is always evolving and being revised. Don't make any changes to your county's plan right away. Build your team, get their input, and then make any needed changes. More importantly, your overall understanding of the plan is crucial. ARES volunteers in your county will be looking to you to find out what's going on and what they should be doing. Being able to put a written plan in their hands, and being confident in that plan, will make any event go much more smoothly.

DUTIES OF THE EMERGENCY COORDINATOR

The ARRL Emergency Coordinator (EC) is a key team player in ARES on the local emergency scene. Working with the Section Emergency Coordinator (SEC), the Zone Emergency Coordinator (ZEC), the District Emergency Coordinator (DEC), and Official Emergency Stations (OES), the EC prepares for, and engages in management of communications needs in disasters. The word "local" is the most important word in the previous sentences. All emergencies or disasters are local. The EC is the primary manager of ARES volunteers in a communication emergency. According to The ARRL Emergency Coordinator's Manual, an EC is expected to:

1. Promote and enhance the activities of the Amateur Radio Emergency Service (ARES) for the benefit of the public as a voluntary, non-commercial communications service.

Notice that this is number one on the list. It's your biggest duty. Promote ARES, enhance/grow/build your team, then recruit, recruit, and recruit some more. Never miss an opportunity to meet, greet, and recruit volunteers and leaders to your team.

2. Manage and coordinate the training, organization, and emergency participation of interested amateurs working in support of the communities, agencies, or functions designated by the Section Emergency Coordinator/Section Manager.

This is a catch-all paragraph, and it also should remind you that flexibility is a big part of any successful ARES team. You should also keep a "Manageable Span of Control" in mind as your team grows (remember your ICS/NIMS training); you will need to assign a leader, or in most cases several leaders (A-EC), to oversee many operational functions such as training, exercises, and logistics, to name a few. This may also include the training of new Amateur Radio licensees to help grow your ARES team.

3. Establish viable working relationships with federal, state, county, and city governmental and private agencies in the ARES jurisdictional area that need the services of ARES in emergencies.

Determine what agencies are active in your area, evaluate each of their needs in terms of which ones you are capable of meeting, then prioritize these agencies and needs. Discuss the planning with your SEC and then with your counterparts in each of the agencies to ensure they are all aware of your ARES group's capabilities and, perhaps more importantly, your limitations. The EC is a key part of the overall plan and partnership with each served agency. However, your best leaders (AEC) can help you with this matter.

As just one example, in Oklahoma, under the management of the county EC, an AEC leads a small team (Rapid Response Team or RRT) focusing on the day-to-day interaction and communication with a specific served agency. There is one RRT per each served agency. The size of each RRT is one AEC and three to seven ARES emergency communications volunteers — the size of the RRT follows NIMS best practice of "Manageable Span of Control." The EC cannot do it all. In a communications emergency, one person can't be everywhere.

Develop detailed local operational plans with "served" agency officials in your jurisdiction that set forth precisely what each of your expectations are during a disaster operation. Work jointly to establish protocols for mutual trust and respect. All matters involving recruitment and utilization of ARES volunteers are directed by you, in response to the needs assessed by the agency officials. Technical issues involving message format, security of message transmission, Disaster Welfare Inquiry policies, and others, should be reviewed and expounded upon in your detailed local operations plans, a Memorandum of Understanding (MoU), and in your county's ARES communication plan. A key point to remember is that ARES volunteers work under the management and direction of ARES leadership. ARES volunteers do not become part of a served agency's volunteer pool. ARES is a self-standing Non-Governmental Organization (NGO) with national partnerships and MoUs in place with may national disaster response providers. See this link for more information on MOUs with current national served agency partners: **www.arrl.org/served-agencies-and-partners**

4. Establish local communications networks run on a regular basis and periodically test those networks by conducting realistic drills.

A weekly local ARES net is a good place to start. This can be simplex, or via local repeater — you and your leadership team can decide. It's a good idea to make this NET part of your county's ARES communication plan. The net can be an on-air meeting place any time your ARES team is on standby or is active during a communications emergency.

5. Establish an emergency traffic plan, with Welfare traffic inclusive, utilizing the National Traffic System as one active component for traffic handling.

Establish an operational liaison with local and section nets, particularly for handling Welfare traffic in an emergency situation. As EC, it is important for you to have your county's ARES communication plan (your roadmap) worked out before any emergency arises. If your county's ARES leaders are knowledgeable about the ARES communication plan, then you will find that communications and message traffic will flow smoothly. This also applies to your core ARES emergency communicators — make sure they have working knowledge of your county's ARES communication plan.

6. In times of disaster, evaluate the communications needs of the jurisdiction and respond quickly to those needs. The EC will assume authority and responsibility for emergency response and performance by ARES personnel under his jurisdiction.

The EC is in charge, but he or she doesn't have to "do it all." During a communications emergency, it's a good idea to have each of your RRT leaders check in with each served agency and evaluate communication needs. The leaders then report their findings to you, the EC. This first evaluation of the communications needs will allow you and your leaders to make decisions and shift resources where needed.

7. Work with other non-ARES Amateur provider groups to establish mutual respect and understanding, and a coordination mechanism for the good of the public and Amateur Radio. The goal is to foster an efficient and effective Amateur Radio response overall.

This amounts to "play well with others." If your leaders and core volunteers are trained and ready to go, you will notice that other non-ARES Amateur Radio groups will look for you to lead the way. Any planning and MoUs you can have in place before any communications emergency arises will make working with any group a smoother process.

8. Work for growth in your ARES program, making it a stronger, more valuable resource and hence able to meet more of the agencies' local needs. There are thousands of new Technicians coming into the Amateur Radio service that would make ideal additions to your ARES roster. A stronger ARES means a better ability to serve your communities in times of need and a greater sense of pride for Amateur Radio by both amateurs and the public.

It bears mentioning again: Never miss an opportunity to meet, greet and recruit volunteers and leaders to your team. Growing your team is very important and recruiting really never stops.

9. *Report regularly to the SEC, as required.*

Every month on the first day of the month, send your report to your DEC. DECs will send your report to the SEC.

TRAINING FOR ECS

Emergency Coordinators are encouraged to earn certification in Level 1 of the ARRL "Intro to Emergency Communication" course (course EC-001; see **www.arrl.org/online-course-catalog** for details). ARES leaders should also take the "Public Service and Emergency Communications Management for Radio Amateurs" (EC-016) course available online from the ARRL.

All ARES officers are also expected to complete ICS-700.A, ICS-100.B, ICS-200.B, and ICS-800.B FEMA training. All ARES members must complete this training within one year from the time of application. Just like recruiting, the process training and remaining ready never stops. Just when you think you and your team are well trained, there's another course to take or more training to do. If you know that going in, you'll never be surprised by it.

ARES AND AMATEUR RADIO IN THE STATE OF WASHINGTON

Teamwork, recruiting, planning, clear communication, and training will make an ARES team function smoothly and grow, as the following example from Washington illustrates. To begin, we need to gain an understanding of Washington's unique geography and the Amateur Radio cultures associated with that geography.

Washington, part of the ARRL's Northwestern Division, is a state with two separate ARRL sections, Western Washington (WWA) and Eastern Washington (EWA). The Cascade Mountains cut a line through the state that affects all aspects of life, including the emergency resource of Amateur Radio. Like ham populations everywhere, the Radio Amateurs in these two sections have a lot in common, but like citizens of the two areas the mountains separate, they march to the beat of somewhat different drummers. However, when it comes to the emergency resource of Amateur Radio in Washington, these differences create very few challenges to cross-mountain cooperation. Of course, the mutual appreciation that "we are all in this together, mountains or not" makes for a positive start. After all, WWA and EWA share the same state Office of Emergency Management and enjoy common ties and mutual interests. Beyond that, and quite fortunately, Amateur Radio leaders on both sides of the state recognize the dependence each side has on the strengths of the other, strengths that emanate from different environmental influences (and the styles of life that follow), and a willingness to share with each other the most helpful innovations they have to offer. It's all done with the intense desire that the Amateur Radio Service may be a service to the public.

THE HAMS OF WWA AND EWA: WHO THEY ARE

The contrasting geography of Washington's west and east creates an interesting mix of people on which to base its Amateur Radio resource. To the west side of the Cascades is the emerald "rainy side" or "wet side," dominated by the economic, industrial, technological, cultural, and political powerhouse of the Puget Sound cities of Seattle and Tacoma. Despite the fact that most of WWA is rural, those cities serve as the pumping heart muscle of the busy I-5 corridor where the business of society is big and the bulk of the state's general population is concentrated. The seat of state government is there, in Olympia, as well as the administration of state services. The ham population of western Washington tends to be a direct reflection of this environment, influenced greatly by an abundance of highly motivated, educated, and experienced professionals in the world of high technology, education, business, and international trade.

East from the Cascades is the "dry side" with the Columbia Basin desert at its core and the first pine-covered hints of the Rocky Mountain foothills at its eastern-most extremes. Eastern Washington is a place where the land itself is the dominant focus of the economy (agriculture, mining, timber), and, therefore, where open space is one of the most important resources. Between WWA and EWA, EWA has the larger land mass and the considerably smaller general population. Big western vistas serve as insulation between a sprinkling of urban centers, the largest being Spokane, centered on the state's eastern edge. Spokane serves as an industrial and cultural hub for much of eastern Washington, portions of Idaho, Montana, and even British Columbia, extending the east side's influence well beyond its borders. Eastern Washington hams are not without their own high-technology influences, but such influence is not nearly so concentrated as on the west side, and its scale is decidedly smaller, existing in little pockets here and there.

Extending a great example of what has always been common in ham radio culture, the hams of WWA and EWA mix easily on the air and in person. The differences between them are dictated mostly by the needs and character of the places they have chosen to live, as well as the reasons they have made that choice — and they all know that. So, to be successfully involved in emergency public service activity at the inter-section level is to appreciate what you have in common with your neighbors and to appreciate your differences without frustration. To be a successful leader of such activity is to understand the strength and opportunities that diversity offers and to be a willing partner with one's neighbor in doing well in an accountable and professional way. What often evolves from this is agreement and cooperation. That is certainly the situation in the state of Washington where the most fortuitous agreement about the Amateur Radio resource in emergency service is about the ties that bind hams together, no matter where they live.

A COMMON BOND: THE ARES MODEL OF SERVICE

Washington has been, for some time, an ARES-friendly state, which is a very good thing indeed. This is due to years of dedicated service by several individuals in both sections who have been willing to patiently promote good programs at local, section, and state levels, carefully building agency respect and proving that ARES can do what they say they it can do. In the last decade, both independent and collaborative work in both sections has dovetailed nicely and converged in unprecedented progress.

ARES is really the point where the best efforts of mainstream Amateur Radio culture and the needs of public service agencies meet. As part of the ARRL's Field Services, ARES provides a structure by and for radio amateurs that can operate locally and be linked and supported nationally. The ARRL's vision further capitalizes on the fact that nowhere else but in Amateur Radio can the common citizen use and experiment with radio frequency communications technique and technology at such a level. Because of this, active and well-rounded radio amateurs generally end up knowing a lot about making radio work for them in a variety of situations and environments, depending little or not at all on commercial infrastructure and more on their wits and familiarity with their spectrum and their equipment to make effective communication happen. The ARRL's vision and support of the well-rounded citizen ham as the greatest component of the Amateur Radio emergency resource is what ARES is all about.

Therefore, as it continues to promote Amateur Radio in general and the advancement of its licensees in knowledge, ability and service, the League continues to create a better emergency resource to serve the public. More and more, ARES becomes Amateur Radio's gold standard of emergency service, the kind of resource the state of Washington needs and the kind of resource to which Amateur Radio is best suited.

THE BASIC ARES BUILDING BLOCK: THE COUNTY ORGANIZATION

In both Washington sections, hams become active in ARES by becoming members at the county level, registering their qualifications and equipment capabilities with the county organization where they live. Most ARES organizations in Washington are directly associated with counties, though WWA does have some ARES groups that are organized within particular cities or towns or around particular ham radio clubs. ARRL membership is not required for ARES membership, though it is encouraged as a means of achieving personal enrichment and democratic representation in the Amateur Radio culture. Becoming a member is also a way to help the League do its work on behalf of the Amateur Radio resource.

ARRL membership is required for the person in the county group appointed by Section ARRL officials to be the leader, known as the Emergency Coordinator (EC). The EC is the direct ARES point of contact with individual members and local served agencies at the ground level. It's the EC who leads the county organization and is responsible for recruiting, credentialing, training and retaining members; for planning and overseeing operational activities; for testing and evaluating service readiness; for creating and maintaining served agency relationships; and for keeping records and making monthly and annual reports to the League. An active and effective EC can make all the difference in an ARES group's quest to serve the public.

It's common for an EC to appoint one or more Assistant Emergency Coordinators (AEC) to assist with leadership responsibilities. AECs are not required to be ARRL members, making this leadership role available to anyone in the group. Different ECs appoint their AECs according to the needs of the group, often to be the lead for a particular group function such as membership or training, to be the group's representative to a particular served agency such as the American Red Cross or an office of the National Weather Service or to lead the group in exploration of a specific band, mode or technical activity.

ARES groups are responsible for nearly 2000 members scattered throughout the 39 counties of the state of Washington. WWA consistently maintains a total of around 1500 ARES members in 19 counties that make up Washington Homeland Security Regions 1 through 6. EWA consistently maintains a total of around 400 ARES members in 20 counties that make up Washington Homeland Security Regions 7, 8, and 9. **Figure 2.1** shows the Washington Homeland Security Regions. Borders of the ARES districts in both WWA and EWA correspond directly to those of the state's Homeland Security regions and are purposely numbered to match.

Each of these county groups has a story that is unique and documents the joys and challenges of service and leadership in both WWA and EWA. Most county groups are the result of initiatives made by radio amateurs long ago and have been in existence for years while some are new or reorganized since the 9/11 and Hurricane Katrina disasters awoke agencies anew to the potential of Amateur Radio in emergency service. Some groups enjoy great acceptance from their served agencies while some, for a variety of reasons, struggle to have their resource recognized or taken seriously. Some groups enjoy the benefits of having talented leaders while other members are reluctant to even take on the mantle of leadership. Some newer groups are the direct result of a non-ham county sheriff or emergency management director stirring up a sleepy ham community to unprecedented action (it's amazing what a little official acceptance and support can do). Some organizations are very large while some are very small, and some counties have such a small ham population that they can't support an ARES group and must look to their neighbors for coverage. The best chance for maximizing the successes of county ARES organizations and mitigating threats to their effectiveness is found, once again, in the fact that there is strength in numbers. That is why organization and cooperation beyond the

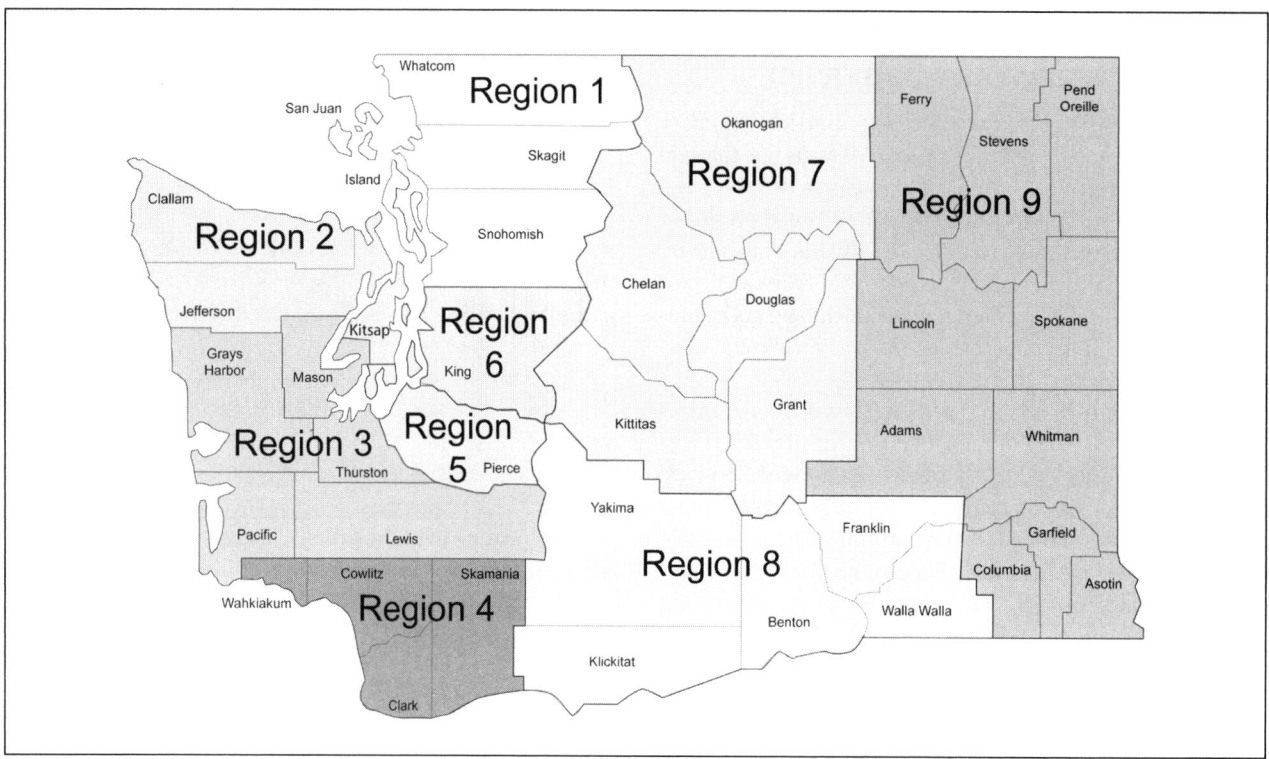

Figure 2.1 — This map of the state of Washington displays Homeland Security Regions and the counties included within them. ARES Districts in the state correspond directly to these regions and are numbered identically. Western Washington Section is made up of Districts 1 through 6. Eastern Washington Section is made up of Districts 7 through 9.

scope of county groups at the district and section levels is essential to ARES across the state.

SECTION AND DISTRICT LEADERSHIP

The elected Section Managers (SM) of the ARRL hold office to serve ARRL member constituents with the good of all of Amateur Radio in mind. In doing so, they are also responsible to serve the public interest, in harmony with the call to public service set forth in Part 97 of the FCC Rules and Regulations. As chief administrators of their sections, SMs have authority over ARES leadership appointments, from Emergency Coordinator appointees on up through the ranks. Washington's two Section Managers appoint Section Emergency Coordinators (SEC) to directly oversee the ARES organizations in their respective sections. The SECs also represent WWA and EWA in ARES matters pertaining to Washington as a whole and to served agencies that are administered at the state or Northwest level. In turn, the SECs appoint District Emergency Coordinators to serve as their advisors and regional assistants, providing leadership dedicated to the county groups in their particular districts. Served agencies that often operate at the regional level (such as regional health facilities and EMS resources) can rely on DECs for coordination of ARES assistance during emergencies and exercises. SECs and DECs have a major advisory role in the recruitment and appointment of county ECs as well as in the appointment of Official Emergency Stations (OES). OES appointees (often recommended by ECs) are members possessing special expertise or resources considered particularly valuable to ARES (Examples: skilled net controllers; owners of well-equipped stations with emergency power; members of particular technical ability). While OES appointees do not have the leadership responsibilities or authority of the SM, SEC, DEC, or EC, they are often invaluable mentors and become leaders by example.

The intention in both WWA and EWA is that leaders should meet regularly, just as ECs do with their group members. Since 2000, the EWA SEC has tried to conduct leadership summits twice a year to which all the DECs, ECs, and OES appointees are invited. ECs may invite AECs and other guests they choose. EWA has only recently added periodic meetings between the SEC and DECs only. WWA has a longer tradition of periodic meetings between the SEC and DECs and between DECs and their ECs, between two and four times a year. The business transacted in these meetings, along with the training and information offered, is all important, but face-to-face meetings between leaders yield a natural cooperation through the sharing of ideas and the building of working relationships and trust. The result is a better organization, bottom to top.

The purpose of all this attention to ARES leadership in the section is not to create an impossible bureaucratic hierarchy of bosses that keeps everybody under a controlling thumb. The leaders of WWA and EWA ARES want to facilitate the workings of a broader network of the section's county

Chapter 2

organizations, creating and maintaining lasting connections between them. This includes encouraging the county groups to get to know one another; to grow in proficiency and competence through standardized training and the testing of each other in special exercises; to participate in public service operating event activities that cross county, section, state, and possibly international borders. These things are essential in fostering cooperation and making the total section organization greater than merely the sum of its individual parts. Leadership in ARES is all about creating the opportunities for this to happen and making the resources of the ARRL readily available to assist that process. The importance of this concept cannot be stressed enough.

Other ARRL section appointees (non-ARES appointees) are great resources for Washington's ARES leaders. The Section Traffic Managers (STM) of both WWA and EWA has long been heavily involved in ARES activities. In addition to their work with the ARRL's National Traffic System, they have influenced, advised and trained ARES groups in traffic-handling modes and techniques. The STM in EWA has designed section ARES exercises and has been an invaluable mentor regarding *Winlink* and digital peer-to-peer modes for handling emergency information. Technical Advisors (TA) makes their expertise in technical matters available to help ARES members understand and enhance their gear's performance, expand their technical horizons and solve operational problems. The work of Public Information Officers (PIO) helps the public know and understand the work of ARES volunteers.

ARES AND RACES:
SERVICE TO WASHINGTON AND MORE

In service to government, neither WWA nor EWA fall victim to the familiar confusion surrounding the Radio Amateur Civil Emergency Service (RACES). Both see ARES as the solution to what a lot of hams have turned, through misunderstanding, into what might be called "the RACES enigma."

First of all, the RACES provision in Part 97 of the FCC Rules and Regulations should be seen as codified acceptance, conditional regulation and fierce protection of the Amateur Radio service rather than the creation of an actual organizational entity. Failure to grasp this very point is the main cause of "the RACES enigma," the misguided notion that RACES is a super-exclusive organization created and run by government to serve government. In reality, language of this provision officially embraces Amateur Radio as a resource to serve local or state civil preparedness agencies during emergencies while, at the same time, applying strict controls on this activity to maintain maximum order and to protect the Amateur Radio service from abuse at the hands of those same agencies. It does all of this while providing for Amateur Radio to remain conditionally functional to serve the public during a national communications emergency. In short, RACES is the intelligent concoction of intelligent people who, for the good of Amateur Radio, didn't ever want to revisit the total suspension of Amateur Radio activity that was imposed by the government during World War I and World War II. Therefore, RACES is a wonderful thing to keep in reserve for when it's needed, but forcing it upon a situation that doesn't benefit from its conditions or necessitate its restrictions only serves to cripple the emergency flexibility of the Amateur Radio resource.

Washington's RACES solution begins with registration in ARES. In most WWA and EWA counties, the process of ARES registration for a radio amateur includes completing an additional form for registration as a volunteer state emergency worker. Registrants may expect the county to conduct a rudimentary background check. This satisfies the RACES registration requirement (registering with the county department of emergency services) and ensures that the membership will have the flexibility of dual affiliation. If an emergency occurs that requires implementation of RACES (it has never happened yet) or a national communications emergency is declared (which hasn't happened since World War II) the membership will be covered seamlessly. What the public gets through this arrangement is a large number of willing ARES volunteers that can serve government and non-government agencies alike, simultaneously if necessary, any time they must, for as long as necessary, uninhibited by RACES restrictions, exclusions or limitations. The League has often called this the process of "switching hats" from ARES to RACES and back, to fit the situation at hand. In Washington, with little exception, it is the standard operating practice.

The cooperative nature of dual affiliation is reflected in ARES/RACES leadership throughout the state. Many county ARES Emergency Coordinators serve with official recognition as their county RACES Officers (RO), while some ARES organizations opt for a separate member to serve as the RACES Officer and share these duties with the EC. In either case, county RACES officers are responsible for registering their ARES volunteers in RACES and seeing to their preparedness as emergency communicators. ARES District Emergency Coordinators serve as Regional RACES Officers, responsible for preparedness at the regional level. The WWA and EWA Section Emergency Coordinators serve as Assistant State RACES Officers for their respective portions of the state. There is a State RACES Officer (SRO) who is Amateur Radio's liaison to Washington's Office of Emergency Management (which is part of the Military Department of the State of Washington, headquartered at Camp Murray, south of Tacoma). The State RACES Officer maintains the State RACES Plan and oversees operation of the state Emergency Operations Center RACES station at Camp Murray.

At this writing, RACES in Washington is in a state of total evolution. In a new state planning document, a more flexible view is being constructed with regard to the role of Amateur Radio in emergency service. RACES, a component of the FCC's Part 97 Rules and Regulations, will retain its role as an important tool to be used in the extraordinary circumstances for which it was designed. Otherwise, the plan will finally recognize a practice of long standing in which ARES organizations have provided the organizational and operational structure for RACES in most of Washington's counties. ARES will be listed as "…the primary or lead organization, in the State of Washington, for Amateur Radio emergency communications…" while updated language will rescue the state plan from a Cold War-era imprint that begs to be replaced by post-9/11 expectations and a post-Hurricane Katrina necessity for flexibility. The plan will be known broadly as the state's Amateur Radio emergency communications support plan, and the state Emergency Management Division's club station, W7EMD, will continue to serve as the Amateur Radio station at the state's Camp Murray EOC.

For an ARES leader in Washington, the benefits of this evolution will be mostly transparent in ordinary practice, though officially liberating in a way that will further encourage the building of a far-reaching Amateur Radio network in the state and beyond during an emergency. For instance, with the state EMD (and its link to FEMA) as the focus of response and recovery, the network may also include liaison with the many Amateur Radio groups serving relief organizations that would likely be involved. Another example would be liaison to services affiliated with Amateur Radio such as MARS.

PLANNING TO TRAIN TO THE PLAN

An ARES plan should be a document that ARES leaders and emergency managers can reference to clearly understand and implement the Amateur Radio resource in service to the county. The plan is also a focal point for training (training to the plan) as well as the starting point for designing exercises (testing the plan). No doubt, a new state plan will have a lot of influence in both Washington sections as plans at the county level are reviewed and rewritten, but one common message that leaders in both WWA and EWA will continue to stress is that a county's ARES plan should be an appendix to Emergency Support Function (ESF) #2 of the county's comprehensive emergency plan.

Much of a county comprehensive emergency plan is based on hazards to life and property identified through a special process of analysis. ARES in Washington has responded to incidents and/or exercises involving nearly all of the hazards about to be mentioned here. WWA has its winter storms with winds, rain, floods, and landslides. Mindful of its place on the Pacific Ring of Fire, WWA also has concerns about the Cascade volcanoes (nobody forgets Mt. St. Helens), and the impending "Big One," a projected magnitude 9 earthquake and the tsunami threat that isn't limited to only that potential incident. EWA shares the volcanic threat with WWA (WWA gets the blast destruction and the lahars; EWA gets most of the ash fallout). EWA has some flooding from runoff as well as flash flooding from severe thunderstorms which can also produce an occasional tornado but most often produce lightning-caused wild land fires (grass and timber). EWA gets big wind storms with blowing dust that also increase the fire threat. There are severe winter storms with freezing rain and deep and/or drifting snows. Chemical, biological, and nuclear threats are always present in both sections (EWA also has a nuclear power plant). Both

---

WWA TRAINING RECOMMENDATIONS

The Western Washington ARES leadership has developed the following minimum recommendations for all ARES personnel, in order to deploy efficient and well trained teams in times of emergencies and for public service events.

They recommend that Section Staff take these classes:
- IS-100a: Introduction to Incident Command System
- IS-200a: ICS for Single Resource and Initial Incidents
- IS-288: The Role of Voluntary Agencies in Emergency Management
- IS-700: National Incident Management System (NIMS), an Introduction
- IS-800b: National Response Framework, an Introduction
- Basic First Aid/CPR
- EmComm Level 1, 2, & 3 provided by ARRL

They recommend that District Staff take these classes:
- IS-100a: Introduction to Incident Command System
- IS-288: The Role of Voluntary Agencies in Emergency Management
- Basic First Aid/CPR
- EmComm Level 1, 2, & 3 provided by ARRL

They recommend that All ARES Personnel take these classes:
- IS-100a: Introduction to Incident Command System
- Basic First Aid/CPR
- EmComm Level 1 & 2 provided by ARRL

sections share terrorism as a hazard. WWA has notable experience with international terrorism threats while EWA has notable experience with terrorism incidents of the domestic kind. With a good knowledge of the kind of incidents that are probable as well as the added knowledge of what incidents are likely to warrant an ARES response, an ARES leader can write a plan covering each appropriate hazard, train to the plan and then exercise the plan.

Whatever the hazard, an ARES group in Washington responds with a set of fundamental skills and tools that are usually in demand from incident to incident. Set up a radio link where there is none. Enhance or back up a communications link that is overburdened. Send a written message with accountability across town, across the county, across the state, across the nation. Send that message via e-mail with no local Internet service provider operational. Do it without commercial power. Set up a voice network of operators to shadow local officials. Set up another for shelter operations. For that network provide liaison to a state network of relief workers. Provide spotter reports for the National Weather Service during and after severe local storms. Back up the 911 system. The list goes on. ARES leaders in Washington plan for satisfying these basic communications requests with competence and professionalism and then train to the plan.

It's important for ARES leaders to appreciate and remember that each county has its own emergency management culture that may have a lot to do with determining the planned mission of an ARES group. This includes attitudes held by served agency officials about Amateur Radio in general and ARES in particular, attitudes that probably vary greatly, even within county agencies. County emergency management officials will also see the value of ARES volunteers in different ways. In some counties, ARES is seen as a communications backup service only, while in other counties, ARES members play additional roles. In many counties, Radio Amateurs are trained as auxiliary communicators on public service bands for service in EOCs or portable or mobile command posts. Amateur Radio is an integral part of several county search and rescue programs. One county ARES group in WWA leads an Emergency Position Indication Radio Beacon (EPIRB) locating team. This diversity of mission extends to geography (and its associated human resources) which can also be an important factor in expectations concerning amateur service. The State of Washington Department of Health, with HRSA/ASPR grant funding, installed amateur stations in all of the hospitals in the state, as well as in several support agencies such as blood centers and poison centers. In WWA, these stations are operated by the well-established, well-staffed, and mission specific Medical Service Team. In EWA, where the pool of amateurs is small (and finding a group of hams to dedicate their service to a singular function is difficult), most of these hospital stations are operated by generalists from the county ARES groups where the health facilities are located.

With the ARES mission naturally varying from county to county, it follows that minimum mandatory training requirements, set by county emergency managers for ARES members, and may differ. This is a fact of life in both Washington sections. ARES leaders must remember, though, that members work on the air with volunteers from many different areas and agencies, and would benefit from training that might exceed the minimum expected by their own county OEMs. It is becoming clearer all the time that a list of common training classes, available from FEMA, ARRL, and county trainers, is beginning to take form across the state. WWA maintains a graduated schedule of basic training recommendations for ARES personnel, from regular members up through the ranks of leadership (see sidebar).

In WWA a non-profit coalition of volunteer communications teams, including ARES, presents the annual Communications Academy (www.commacademy.org) in the month of April. This successful, professional-grade training opportunity is held for the various emergency communications teams around the Pacific Northwest, but is open to anyone with an interest in emergency communications, volunteer or professional. The event attracts nationally-known speakers and exposes attendees to topics in emergency management, communications techniques and protocols, real-life emergency responses, and other pertinent subjects. The presentations are designed to promote the development of knowledgeable, skilled emergency communicators who will support their local communities during a disaster or emergency response.

WASHINGTON'S AMATEUR COMMUNICATION NETWORKS

The state's plan for its primary network of amateur resources is mostly in place and is tested ever more thoroughly as time progresses. HF, VHF, and UHF voice frequencies play a large role in state, section, district, and county and city communications. Increasingly, the *Winlink 2000* system and other peer-to-peer digital systems are becoming the preferred means of sending formal message traffic and documents in ARES with voice networks learning how to support that effort as a primary responsibility.

The state has long had a unifying HF net, the Washington State Emergency Net (WSEN), on 3985 kHz (7245 kHz secondary frequency), which has been maintained cooperatively by leaders of the WWA and EWA sections. WSEN is identified in every WWA and EWA county ARES plan as a designated HF meeting place for ARES resources in the state. It's sort of an on-air staging area from which larger state ARES operations grow. The purpose of its regular Monday evening and Saturday morning check-in sessions has been to increase ARES awareness in the state (showing the flag, as it were); to familiarize the county groups, leaders and members with one another as well as with other amateur support organizations in the state; to allow net control stations in all districts an opportunity to practice; to reaffirm the frequency's identity as a staging resource in the state when incidents of significant scale suggest an impending emergency. WSEN also brings WWA and EWA together during an emergency and prepares one group to support the other if neces-

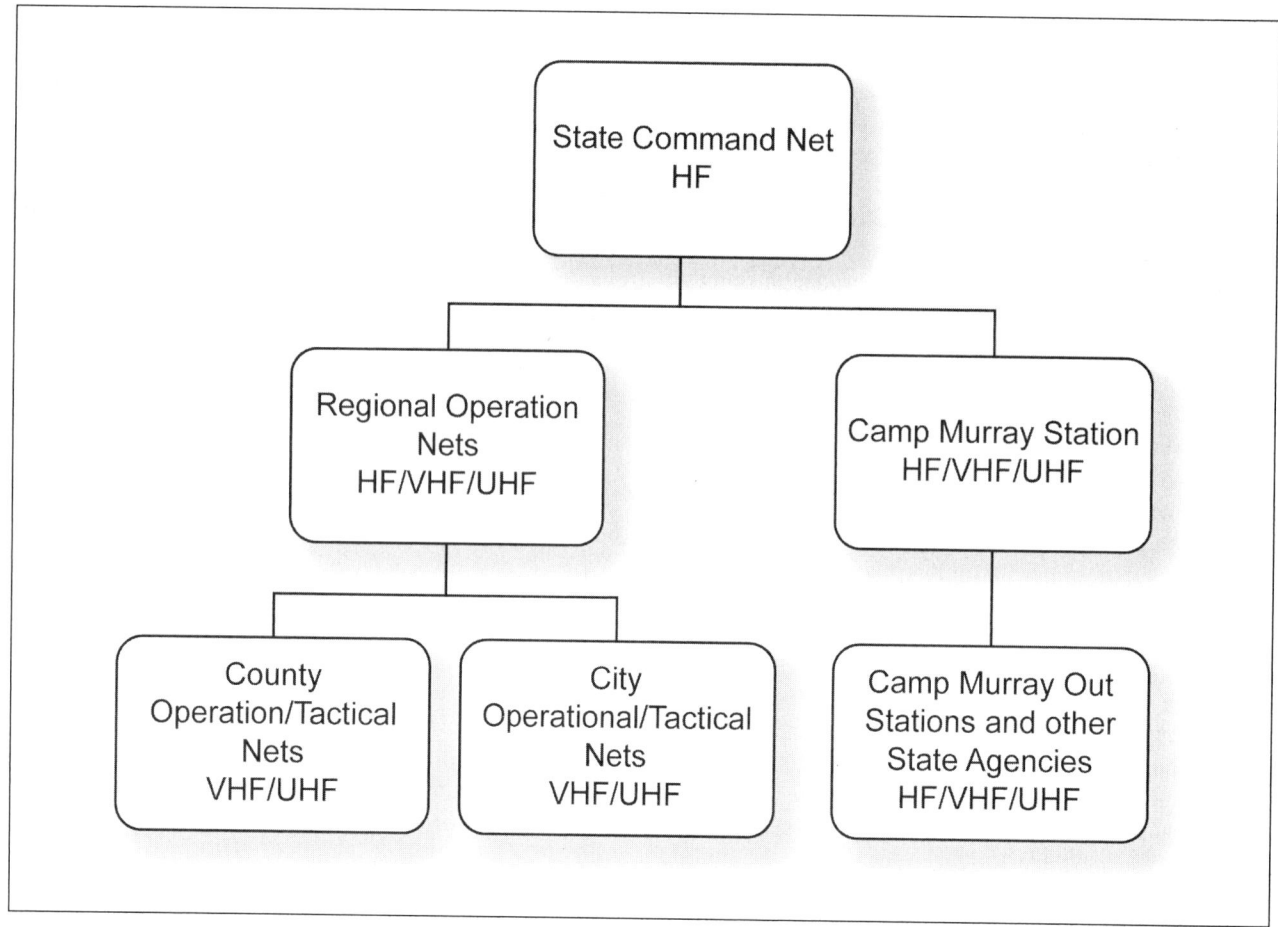

Figure 2.2 — This chart shows the flow of communication between the State Command Net, the state's Amateur Radio station at the Camp Murray EOC (with its various operational support stations and liaisons to state agencies), the District Nets, the County Nets, and the Local Nets. Liaison stations provide the connective links up the chain from Local Net to County Net, County Net to District Net, and District Net to the state Command Net. The station at the state's Camp Murray EOC functions at the Command Net level, but does not normally operate in a net control capacity.

sary. That has been the case at the beginning of more than one emergency in the state since the Mt. St. Helens disaster. It was also the case on the morning of September 11, 2001, and kept the state EMD in touch with at least one EWA county that, because of overloaded phones, could not otherwise contact the state. WSEN works because its identification with ARES is so strong. In an emergency, WSEN becomes the primary HF resource and information net for WWA and EWA ARES and other supporting groups.

When it's warranted, the state network can grow out of WSEN, as provided for in the state plan, and become several different nets of different focus linked by liaison stations. A State Command Net, would be established (in addition to WSEN), to link Regional (District) liaison stations with the Camp Murray State EOC station. Camp Murray would not be the Net Control Station for this net, but would be a part of the net as the link to state and federal resources through the state EOC. Regional Operations Nets would be established on an appropriate band/frequency for each region to link liaisons from County and City Operational/Tactical Nets involved. In EWA, with its smaller human resources, a single HF section net can usually serve the purpose of three Regional Operations Nets for Districts 7, 8, and 9 and provide necessary liaison to the State Command Net. County and, where appropriate, City Tactical/Operational Nets would be established to serve the affected areas or operate in support of them, as necessary. **Figure 2.2** illustrates this command structure.

TESTING THE NETWORKS = TESTING THE PLANS

In addition to the regular activities of the WSEN, operational tests of the state nets outlined in the state plan occur quarterly (actually, on the last Saturday of any calendar month holding five Saturdays) during the EOC-to-EOC exercises arranged and conducted by the Camp Murray station staff. It is the goal of the Camp Murray station manager for all of the ARES groups in the state to fully understand the network flow and be prepared to operate according to this plan as the network expands to meet the needs of an emergency response.

The local part of the network receives the most frequent tests. Nearly every county ARES group in Washington holds a net check-in on VHF or UHF at least once per week. These nets are the grass roots of ARES service in both sections and allow county groups to keep in touch, stay informed, give net controllers some experience on the air and allow groups to practice the procedures and techniques necessary to run an operational/tactical net at the county level.

To various degrees, the Simulated Emergency Test (SET) exercises of WWA and EWA ARES also test section nets and links to Camp Murray each year. Both sections invite each other's members and counties to be players, opening up the exercises beyond section borders. SET exercises, by tradition, are usually scheduled to take place in October, but EWA is known for occasionally scheduling the exercise in early February, creating a winter training experience for those counties that assemble field stations during the exercise. SET exercises are also great for testing digital mode operations. A winter 2011 SET exercise in EWA positioned voice operations at the county and section levels to serve almost totally as tactical support for the digital *Winlink* and peer-to-peer resources on VHF and HF which were loaded with experimental traffic and files of many kinds. All modes involved were very busy during the six-hour exercise.

Power users of the *Winlink 2000* system associated with ARES in Washington and the Northwest don't wait for exercises to test their equipment and knowledge. They are keenly aware of the system's emergency service potential and are eager to test themselves daily so as to boost their proficiency. They are also quite interested in promoting and mastering the performance of digital peer-to-peer modes that allow them to send and receive e-mail type messages without connection to the Internet. Their activities are some of the most exciting explorations of amateur technology going today.

County, regional, and state exercise opportunities sponsored by public service agencies just can't happen enough. The value of ARES groups training and exercising with their served agencies is huge, providing the best situational and contextual test of the membership and the state network this side of a real emergency. Short of that, some of the best practical experience ARES members can get in dealing with real-world situations is involvement in public service events. Washingtonians sponsor lots of events: parades, walkathons, fun runs, marathons, bicycle rides and races, and even motorcycle events (to name a few) include support from Amateur Radio groups. Even events in sparsely populated EWA can easily involve the coordination and service of 30 to 60 operators. The more an event involves the public and requires the special attention of public service agencies for safety and order, the more valuable the experience for hams can be. Planning and working these events with community leaders and agency personnel becomes a mutual trust-building exercise that pays off later for the ARES member or leader who meets a familiar and friendly face while working side-by-side in an emergency shelter, an EOC, a command post, or a field position.

MINING THE CULTURE

Perhaps the best experiential training that an ARES leader in Washington or anywhere can suggest to members is a no-brainer: total immersion in the Amateur Radio culture itself. There, a diverse world of activity and involvement in communications technology awaits the person who wants to become a smart, sharp, versatile ham radio operator. Active use of the operating privileges granted by the license (on some of the most valuable spectral real estate there is) can be a laboratory for learning and understanding the art of radio communication so that, in pursuing the aspects of the hobby they enjoy the most, hams acquire invaluable knowledge and experience, both theoretical and practical. All of this fascinating, challenging, and rewarding activity can help create a great emergency communicator for ARES and, in turn, an even greater asset for the Amateur Radio culture.

Finally, the culture of Amateur Radio provides countless additional layers of possible support, from the ARRL's National Traffic System nets to a host of other scheduled nets, operating events, and informal activity by many individuals of substance. It's tested every day and is always working in some way. Not every ham is going to naturally gravitate to emergency service activity. Not every ham can, nor is every ham qualified for it, or well-suited to it. In an ARES leader's world of registration, credentialing, and training mandates, it's still good to realize that there is real support of quality out there in the culture beyond the spotlighted ranks of the registered, willing, and able. At the very least, that there is almost always someone on the air on the Amateur Radio bands is one of the culture's particularly attractive assets and should not be discounted in official emergency circles, much less by the thoughtful and creative ARES leader.

Part 2: Served Agencies

Chapter 3
The American Red Cross and Amateur Radio Operators

The American Red Cross and Amateur Radio operators have enjoyed a long history of collaboration. For many decades, two-way radio provided the only communication connection in a disaster. The continual evolution of communication technologies used by the general public influences how the Red Cross establishes connectivity during a disaster. As new technologies emerge and are adopted, it is important to know that older technologies are not obsoleted. The old and new technologies stand together, at the ready, in an ever-growing toolbox of options for responders. The vital element in providing effective and efficient connectivity during a disaster is knowing the pros and cons of each.

In the case of Amateur Radio operators, their value is not in the radios they carry, but in their *"Semper Gumby"* ("always flexible") mindset, combined with a set of skills for adapting existing equipment in new ways to meet an objective. The operator can hack together solutions that take the best of established and emerging technologies, balanced with a "what's the simplest way to get this done" perspective.

MEMORANDUM OF UNDERSTANDING

The purpose of the Memorandum of Understanding (MoU) between the Red Cross and the ARRL is to offer a broad framework for how the two organizations will cooperate during a disaster. These two organizations have a shared desire to prepare for and respond to disaster, providing a service to survivors and to other organizations involved in the relief effort. It takes the combined efforts of the local Red Cross chapter, Amateur Radio Emergency Service® (ARES) unit, and radio club to weave the tapestry of collaboration within each community. Just as every chapter, unit, and club has different strengths and needs, each application of this MoU will be different.

The current MoU between the American Red Cross and ARRL, the national association for Amateur Radio, was signed on March 25, 2010, and is valid through 2015. The most current document can be found on various Internet sites, including **www.arrl.org/attachments/view/News/49455**. Amateur Radio operators who plan to participate with the Red Cross should take the time to read the MoU in its entirety. Although it contains standard language on the two organizations and their interaction, which most amateurs will already know, it also offers specific recommendations and a template for a local radio club/local Red Cross unit MoU under the umbrella of the larger agreement. For instance, Attachment C of the MoU provides a vehicle to formalize the relationship between the local Red Cross and ARRL field organizations. The local group just needs to complete the attachment to have a local formal agreement. While this is specifically written for Red Cross chapters and ARRL field organizations, it can also cover the relationship of the Red Cross and many local radio clubs.

It is the sincere hope of the ARRL and the Red Cross that all resources in a community will work together on a cooperative team and efficiently leverage local resources to best meet the needs of the survivors in that community. There is nothing in the MoU that requires an activation of the MoU so the ARRL and Red Cross can operate during a disaster. It is always in effect and can be used by any units that have agreed to it. All that is required is a shared mutual agreement of what needs to be done.

SHARED MEMBERSHIP

The best advocate for the ARRL and Red Cross relationship is someone who is a member of both organizations. The culture and politics of each will then be understood, enabling a volunteer to more effectively transition from the duties of one role to the other. For example, being a member of both organizations would give an ARRL member a deeper understanding of the need for a good disaster assessment, and would enable a Red Cross volunteer to communicate as he or she performs an assessment task. This is true for many organizational relationships that are activated during disasters, such as those between the Salvation Army and ARES; the Southern Baptist Convention and a local radio club; and the local Emergency Operations Center (EOC) and *Radio Amateur Civil Emergency Service* (RACES) unit. A full member of both organizations can quickly switch hats as needed.

As stated within the MoU:

Each organization will encourage interested volunteers to become members and participate in the activities of the other organization. Such volunteers shall meet the standards, have the responsibilities and be entitled to the privileges of each organization.

One caveat for the volunteer is to not overcommit to all the possible organizations he or she can serve with. Volunteers should make it clear during the planning stages what organization they will participate with primarily. This will prevent situations where an individual has been counted on to be in multiple places at the same phase in the disaster.

BACKGROUND CHECKS

The Red Cross policy and standards require a background check on anyone who is volunteering for the organization, a common practice with large volunteer organizations and many employers. This provides a measure of security for the disaster survivors, other workers, and the response organizations interacting with a Red Cross-vetted person. Prior to the signing of the MoU, there were some issues regarding the background check requirement, but these were worked out to the satisfaction of the ARRL and the Red Cross. The Red Cross still requires a background check for volunteers, which can be done through the Red Cross chapter, or the individual can get an equivalent background check at his or her own expense through a law enforcement entity. The volunteer coordinator at the local Red Cross unit can provide more details on the options available.

ASSUMPTIONS

The Red Cross respects that Amateur Radio operators are licensed by the FCC, and that there are laws regarding the operation of a radio station. When a Red Cross representative requests that an Amateur Radio operator use equipment in a way that violates FCC rules, the operator should inform the requester of the conflict. The Amateur Radio operator is expected to be the subject matter expert when it comes to what can and cannot be done under his or her license. People who are not Amateur Radio operators should not be expected to be intimately familiar with the FCC regulations and other guidelines that influence how Amateur Radio operators conduct themselves on the air.

The Red Cross is responsible for any licensing arrangements necessary for approved operations that occur outside Amateur Radio frequencies and license. A radio operator can request confirmation that a particular license and frequency has been properly arranged prior to use.

TRAINING AND EXERCISES

Volunteer communicators must train as they intend to respond. Planning followed by training and exercises based in that plan provide the starting point for the response. The ARRL is a leader in the field of Amateur Radio communications; the Red Cross is a leader in the field of non-governmental organization disaster response. Both entities have invested heavily in creating training materials for people interested in each pursuit. The Red Cross will recognize training certifications by the ARRL and vice versa. True local collaboration is evidenced by joint training exercises that assess and hone the developed plans. As the MoU states:

The Red Cross recognizes the leadership and expertise of the ARRL in the area of Amateur Radio communications. Where appropriate, the Red Cross will rely on materials created by the ARRL to train radio communicators. Additionally, the ARRL offers training in Amateur Radio emergency communications that is mutually beneficial to the ARRL and to the American Red Cross. Volunteers holding valid ARRL Emergency Communications certificates of completion will be recognized for this knowledge.

LOCAL RADIO CLUB AND ARC UNIT RELATIONSHIPS

The ARRL can only commit the resources of its organization and engage ARES units. Other local radio club entities may not be in the direct control of the ARRL. Red Cross chapters are in the middle of fundamental transformations in how they report to the national organization. Unique challenges and capacities exist in each locality and what works well in one area is not going to work well in another. Local politics and personalities will invariably affect organizational relationships, and it is up to the individuals in these organizations to determine if the impact will be positive or negative.

ROLES IN A DISASTER

Anyone interested in supporting any phase of a disaster can almost always find an appropriate role. An exact assignment may be obvious, or it may take a little time to flesh out the best role depending on a volunteer's skill set, attitude, and flexibility. All participants want a meaningful opportunity to use their skills, and many would rather be a little overworked than underused and underappreciated. A leader needs to recognize this and appropriately balance the workload and adjust roles as appropriate.

There are two ways to support the relief efforts of the Red Cross. One way is from within the organization and the other is from outside it.

Roles Internal to ARC: There are radio operators who are directly registered as Red Cross volunteers, many as part of the Disaster Services Technology (DST) group, which is responsible for all the technology related to disaster response that is employed on *Disaster Relief Operations (DROs)*. DST establishes and maintains *intraoperable* communications. The ARC must be able to "talk to itself" by connecting field units, disaster sites, and the national Disaster Operations Center. These communications could involve working on specific local Red Cross radio frequencies and on various types of equipment to establish the connections. The vast majority of information passed through Red Cross connections is internal communications passed through internal networks.

Serving Red Cross and Others as an External Partner: The Red Cross also welcomes radio operators who are established as independent radio resources, and who can be utilized by multiple organizations or individuals. When multiple organizations come together to form a multipurpose site, such as a joint field office, one radio station may be collocated to assist with any communication needs, instead of establishing a radio station for each organization. Staff would be able go to the radio station to request that messages be transmitted or to receive any incoming messages. The radio operators could be acting under the banner of their local radio club, RACES affiliation, ARES unit, or as individuals. There may be a verification, authorization, or permission needed from the agency that is the "landlord," or overall coordinator, for the site. Similarly, a radio station that is located apart from any official site, such as in a person's home, could support the communication needs of one or many organizations by passing messages, sharing information, and functioning as a net control station.

Whether the Amateur Radio operator is internal to the Red Cross or working with a partner organization, here are some examples of roles someone interested in Amateur Radio and other technologies may seek out.

Communicator: This very narrow focus may be necessary when a unique skill is needed to operate equipment, handle weak signals, use exotic modes, or when there are just so many messages that a dedicated traffic handler or net control is required. This person may be totally focused on the message and getting it somewhere. This person waits for someone to come with a message to transmit elsewhere, and thus performs a message handling function. Similarly, when there is a lot of radio traffic on the air, someone may be asked to participate in a net control function. This is common in the very early stages of a disaster response or during special events. However, dedicating a person to this role becomes overkill as the operational tempo of the response settles down, requests becomes more routine or sporadic, and cellular service is being restored.

Using Radio as a Tool: Some volunteers may be trained to perform a particular task and are also able to provide their own means of communicating their progress, such as via their personal HT. These embedded communicators may be distributing food supplies, setting up a shelter, performing a disaster assessment, or staffing the local EOC, and they use their radios to communicate updates to their supervisor as needed. Instead of requiring two people — one to perform the task and one to communicate it — the skills and abilities involved in this role are provided by the same person.

Technologist: A technologist is someone who has technical skills that extend beyond radio communications and uses radio as one option in a larger communications toolbox. Technologists' ability to shift from radio communications, to computers, to networking enables them to handle any technology. In a Red Cross response, they may be asked to set up an antenna, network computers, assemble a satellite dish, and help other workers resolve technical issues, all in the same day. This is the most common role sought by the Red Cross for DST volunteers.

HOW WILL AMATEUR RADIO OPERATORS BE USED?

There is a serious philosophical debate on how, when, and where Amateur Radio operators are best used. What follows is just this author's opinion and perspective; there are certainly others.

As stated earlier, every local Red Cross chapter and Amateur Radio field unit will have unique characteristics that need to be considered. There is no single right model for every situation. With a national perspective, disaster responses can be divided into three main categories: those that can be handled with resources available to the chapter; those that require resources of the national organization; and those that exceed resources of the national organization. Each category has its own set of operational needs.

Local Red Cross chapters collectively respond to nearly 70,000 disasters per year. That's about one disaster every eight minutes somewhere in the United States. The vast majority of these are single-family home fires, although a significant number of incidents are larger disasters within the community. The most common way an Amateur Radio operator will be involved in Red Cross disaster operations is through such localized events. It only takes a few radio operators to handle the needs of these operations from start to finish, and it gives the operators a real hands-on application of their emergency planning. While the national conversation is dominated by the need to plan for a catastrophic, wide-scale disaster; it is the success in local high-probability, low-impact events that will build trust between the Amateur Radio community and the organizations that need help.

Events that exceed the local chapter's ability to respond will activate the national organization to move people and resources into the area to meet the needs of those impacted by the disaster. In these situations, chapters are expected to hold their own for the first few days while reinforcements are mobilized and arrive. Amateur Radio operators who have planned and exercised regularly with the Red Cross chapter will act quickly to help fill communication gaps prior to the arrival of additional resources. Technologists, including Amateur Radio operators, would then adjust their roles to take advantage of the additional resources arriving to establish connectivity in the most efficient way to best meet specific needs.

A preexisting relationship is vital to smooth activation and maximum effectiveness during the "window of opportunity" that exists in the first few days of an event. Amateur Radio may be the only means of communication until additional resources arrive and while cellular and other network providers bring key elements of their systems back online. Delays in establishing or repairing communication systems to a pre-disaster state may lengthen the time frame during which Amateur Radio is critical. Regardless, it is important not to focus solely on the immediate aftermath and discount the longer-term value of Amateur Radio operators; they are vital during that initial window, and they can transition easily to other technical roles as pre-disaster communication systems come back online.

There are catastrophic scenarios that will exceed the national response capabilities of the Red Cross, FEMA, the Salvation Army, and other response agencies. These are easy to imagine: massive earthquakes in Southern California or the along the New Madrid fault line in the southern and midwestern United States; Category 5 hurricanes on the Gulf or East Coast; man-made disasters that might combine physical destruction with cyber-attacks. The list goes on.

According to disaster modeling, destruction of communication infrastructure will occur in concentric rings of severity — nothing will be heard from the very center of the damaged area, with increasing reports coming in as one moves away from the center of the impact. Initial communications will be sparse and through a variety of systems with significantly reduced capacity. There will be extreme congestion on any working communication systems around the periphery of the impacted area. Disasters of great magnitude also disrupt transportation networks, which will delay response resources moving into the area and evacuation out of the area. Amateur Radio operators will get on the air where they are and form ad hoc networks. People and organizations that have previously responded to the smaller, more common disasters will have practical experience to draw from when making decisions. Amateur Radio operators will be asked to do work based on prior successes, and this work will be high-visibility and momentous.

These scenarios represent low-probability, high-impact events that will get a lot of attention. The window of opportunity for Amateur Radio will be extended from days to weeks, and maybe months. Amateur Radio operators will be critical for passing information between sites. This visibility will bring out many well-meaning, yet inexperienced people whose actions may actually interfere with relief plans. It will be crucial for leaders in Amateur Radio to find ways to incorporate people with all levels of experience for inclusive planning.

In the end, everyone wants a meaningful volunteer experience where they use their skills and abilities to support the survivors' recovery. The partnership between the Red Cross and the Amateur Radio community enhances that experience and, more importantly, brings significant and crucial aid to those in need.

THE LOCAL ANGLE

Though an MoU for the relationship between the Red Cross and Amateur Radio exists at a national level, it is also possible for local Red Cross chapters working with Amateur Radio operators in their local areas to create a local MoU using the "tear-off" local MoU template, also known as Attachment C, found at the back of the MoU.

But even if that local understanding isn't formalized, it's important to keep in mind that effective emergency response begins within the community, enhanced by strong grass-roots alliances. Local Red Cross chapters and Amateur Radio operators can engage in many outreach activities to initiate and develop these alliances as they build their own "broad framework." For volunteers who aren't sure where to begin this process, here are some suggestions, based on real-life observations and experiences at the grassroots level of emergency preparedness.

INVOLVING HAMS AND RED CROSS VOLUNTEERS ALIKE

From a Red Cross perspective, there are several ways to establish a relationship with the local Amateur Radio community. One way is to invite local hams to tour the Red Cross chapter and see what it's all about. Not only might the chapter gain an Amateur Radio operator, it might gain a new Red Cross volunteer. Chapter leadership might consider making space available for the local ham radio clubs to have meetings, radio license classes, or exam sessions, which can lead to hams becoming interested in Red Cross activities, and Red Cross volunteers becoming interested in ham radio. Within a chapter organization, this relationship can also work to involve Red Cross Mass Care staff and possibly other disaster relief operations groups and activities such as logistics and client casework in a ham meeting or class in which each group can get to know the other's work and mission and how they can mutually benefit from a relationship.

Another good fit for hams can be with the Red Cross Damage Assessment staff. This function of the organization utilizes crews of people going into the field to provide damage assessments in an affected area. Sometimes these crews are in an area with poor cell phone or commercial radio coverage, in which case hams would be able to provide information back to the disaster headquarters (shadowing staff). In addition, if the hams are so trained, they would be able to help others better understand this process.

ESTABLISH A PRESENCE WITH A RED CROSS CLUB OR A CLUB CALL

Once Amateur Radio and/or DST starts gaining recognition within the local Red Cross chapter, consider setting up an Amateur Radio station. You can start with a simple station consisting of one or two VHF radios to include 2 meters and 70 cm and/or an HF radio. The station can comprise a few portable kits that can be set up, moved, and taken down as needed. A simple dipole antenna can be put up outside the chapter facility. Stations of the Greater Erie County (Pennsylvania) chapter and the Greater Chicago chapter started modestly and over time have been able to acquire equipment for 2 m/440 MHz, HF, packet, EchoLink, and more. If you locate it near disaster services staff, however, be mindful of the added noise the radios may create in an area that would be busy during disasters. Also, consider power requirements and paths for running coax, ground connections, and other cables to the outside. Work with the local disaster services director, building engineers, or whomever else will help make this project a success. You should stress to leadership that this would make a great location to set up Red Cross low-band VHF radios and other communications services to aid in their disaster services mission.

You might need to get creative for antennas. The Greater Erie County chapter was successful when Amateur Radio emergency communicators worked out a deal to allow Verizon to put a cell tower at the chapter office and, in exchange, were able to put ham antennas on the tower. Another possible approach is to work out a deal on an off-site commercial or amateur tower and link to that site from the chapter office. Speak to other chapters nationwide and see what they have been able to do.

DEVELOP COMMUNICATIONS EXERCISES

It is very important to develop a chapter communications exercise. This can be part of a chapter-wide activity or a standalone communications exercise, though the best strategy is to do both types of exercise over time. Begin with a communications exercise and see how many communicators you have and what equipment hams can bring (antennas, radios, masts, power supplies, etc.). You can start simply, conducting it at the chapter site, or perhaps between the chapter and another facility that might be used as a shelter or distribution site. Try to work communications into a larger, or even full-scale exercise in which you can deploy gear and personnel to other locations, such as shelter sites, service centers, and kitchens; the farther away the better.

Once the exercise is complete, hold an after-action meeting and generate a report to analyze out what worked and what did not. For example, was there adequate staffing, were the right types and quantities of equipment available, was there adequate power for communications equipment, what problems were encountered, what worked well?

As you can tell, it may take some trial and error initially, but it is very rewarding to combine Amateur Radio and the Red Cross, whether you sit at the chapter office working a radio, riding with someone doing disaster assessment out in the field, or providing communications in a shelter or other location in a disaster.

Chapter 4
The Salvation Army and Emergency Communications — The SATERN© Program

The Salvation Army Team Emergency Radio Network, or SATERN©, is an organization that provides critical disaster intelligence to Salvation Army national, regional, and local command units to assist Salvation Army disaster response. SATERN is further tasked to provide communication interoperability for Salvation Army personnel and response units on the disaster field. SATERN's mission is:

- To develop and maintain a corps of Amateur Radio operators skilled in emergency traffic and communications to assist the Salvation Army during times of disaster.
- To assist in training other Salvation Army personnel to access and use the resource of Amateur Radio for local, regional, national, and international disasters.
- To develop training materials and exercises designed to enhance the use of Amateur Radio within the Salvation Army Disaster Services programs.

Services and operations performed by SATERN volunteers may include:
- Providing communications support for mobile units and field operations sites during major emergencies.
- Providing emergency monitoring.
- Coordinating the delivery of food, cots, blankets, and other materials to shelters and other emergency centers.
- Fulfilling Health and Welfare inquiries.
- Representing the Salvation Army Disaster Services at emergency operations command and service centers.
- Participating in weekly radio nets for the development and continued readiness of the SATERN program.
- Conducting training programs for individuals interested in securing an Amateur Radio license.
- Providing other support as needed and appropriate.

THE TWO-LEVEL EMPHASIS OF THE SATERN PROGRAM

SATERN works on two levels:
1. It provides an at-the-ready, international, high-frequency network that trains six days a week to help mitigate the effects of a disaster by providing emergency communications when normal communications are down and by providing a conduit for Health and Welfare traffic when needed. When disaster strikes, SATERN is on the air for the duration of the event.

2. It provides strategic, on-the-ground communications for the Salvation Army disaster response units and personnel, and ensures communications interoperability for major disaster. SATERN field units are components of the Salvation Army Emergency Disaster Services (EDS) teams and take part in the regular training, exercise, qualification programs, and the day-to-day response of the Salvation Army EDS.

SATERN WITHIN THE SALVATION ARMY

SATERN was developed to function within the framework of the Salvation Army at all levels. It parallels the National, Territorial, Divisional, and Local or Corps levels.

At the national level is the National Director, who is appointed by National Command of the Salvation Army. The primary duty of the National Director for SATERN is the facilitation, coordination and maintenance of the network, including the maintenance of the SATERN organization's interface as a legitimate component of the Salvation Army's disaster response. The National Director relates to all national Amateur Radio and/or communications organizations, such as the ARRL, MARS, Radio Amateur Civil Emergency Service (RACES), and Radio Emergency Associated Communications Teams (REACT). The National Director (formerly called the National Disaster Coordinator) reports to the National Liaison for EDS at the National Headquarters.

The SATERN National Director has as his or her primary volunteer staff:
- SATERN National Net Director
- SATERN National 20 Meter Net Director
- SATERN National CW Net Director
- SATERN National IT Director
- SATERN National Health and Welfare Coordinator
- SATERN National Training Officer
- SATERN National Technical Services Officer
- SATERN *Ring* Editor (Newsletter)

THE HISTORY OF SATERN

by Major Patrick McPherson, WW9E

The beginnings of SATERN sprang from a passion for Salvation Army emergency response that I developed growing up in Kansas City, where the Salvation Army Division had a committed disaster response team, which included a Salvation Army Communications (SAC) Team that used Citizen's Band radios for communications. In 1972, I began building disaster response communications teams in the various corps I commanded as I transferred around the central states. In Dubuque, Iowa, for instance, the 75-person disaster team I formed included 18 Amateur Radio operators, among them Art Evans, N9KQ (former KA9KLZ), who eventually became an instrumental member of SATERN.

When I was transferred to Springfield, Illinois, in 1987 I was given the additional appointment of Heartland (Central Illinois and Eastern Iowa) Divisional Emergency Disaster Services (EDS) Coordinator. Over a decade earlier, the Rapid City Flood of 1972, during which no standard communications were available, showed that Amateur Radio could provide operability even after a catastrophe. The lessons learned from such events provided the impetus for starting SATERN. I garnered the support of Art, KA9KLZ, a skilled Navy MARS and NTS traffic handler, to help build the team. SATERN was born in June 1988.

A mere two months after its inception, SATERN responded to its first international disaster when Hurricane Gilbert battered the Caribbean for nine days, killing 341 people. SATERN principals worked with amateur operators in Atlanta and Jamaica to send Health and Welfare traffic back from the Salvation Army headquarters in Jamaica.

The organization continued to coalesce even as it faced new emergencies. In 1990, Patricia Duce, WZ9H, became the first Amateur Radio Liaison Officer (ARLO) for the Metro SATERN Division, covering the greater Chicago area, northern Illinois, and northwest Indiana. A week after she was appointed, a horrific F5 tornado hit the community of Plainfield, Illinois, killing 29 people, injuring 350, and causing $200 million worth of damage. Duce set up a resource net on the Chicago FM Club (CFMC) repeater on 146.76 MHz, and volunteers responded. The around-the-clock response operation required 64 hams each day for 11 days. Thirty-five vehicles, including mobile field kitchens (canteens), Salvation Army vans, volunteer vehicles, supply vehicles, and trucks, all had Amateur Radio communications. Every Salvation Army principal had an Amateur Radio shadow, and every command center had an Amateur Radio operator on site. The interoperability provided enabled a quick and efficient response operation.

The efficiency of the net operation during that response attracted other amateurs in the area to become part of SATERN, and the Metropolitan Division eventually boasted 211 SATERN members. Within months, the first SATERN VHF Net, using the CFMC repeater system, was started. Also in 1990, SATERN used the Chicago Area Packet Radio Association (CAPRA) packet system to help develop Western Territory SATERN.

In 1992, KA9KLZ, who was then National Net Director, opened the door for SATERN's evolution into a high-profile global response network by starting up the SATERN international net on 14.265 MHz. The Metropolitan Division began a series of Disaster Services Seminars, which amateurs from across the United States attended. When Hurricane Mitch struck in 1998, the FCC declared 14.265 MHz an emergency frequency for the duration of the event. This move greatly assisted the SATERN operation, and Honduras and Nicaragua reported casualties on the SATERN net. In responding to the massive need, SATERN began using the Internet to complement its traditional Amateur Radio operation for Health and Welfare operations.

SATERN emerged as one of the primary emergency networks in Amateur Radio. It developed relationships with the Hurricane Watch Net and the Maritime Mobile Service Network and was relied upon for help in Health and Welfare operations after major catastrophes from that point on. Also in 1998, SATERN was named an official program of the Salvation Army. Since that time, SATERN has grown throughout the United States and Canada and has also established memberships in the United Kingdom, Mexico, China, India, Europe, South America, the Middle East, New Zealand, South Africa, and Turkey.

SATERN activated on September 11, 2011 to coordinate emergency responses, including the communications between a blood bank in California and one near Ground Zero. A SATERN North

America Command was set up at Central Territorial Headquarters, and Australian and German stations checked in to help with "Stand by for America." The New York SATERN liaison coordinated the logistical/maintenance efforts for Salvation Army canteens throughout the area as well as the communications, with the ARRL and REACT teams coming in to assist.

That same year, SATERN helped facilitate divisional Emergency Disaster Services (EDS) responses for wildfires in the Western states. The SATERN network's response was commended by the Denver area EDS coordinator Michael Gelski, KBØPVD (SK). The ARRL called out ARES assistance for SATERN in the wildfire operation, and a video they produced about the events, *Amateur Radio Today*, illustrates not only Amateur Radio's potential, but also the Salvation Army's response using Amateur Radio volunteers. It was a great reflection of the Memorandum of Understanding (MoU) between the ARRL and the Salvation Army.

In 2003, Bermuda corps officer and SATERN radio operator Major Rick Shirran, VE3NUZ, relayed data from the midst of Hurricane Fabian via HF radio and the Internet Radio Linking Project (IRLP). Major Shirran was later appointed Territorial Disaster Director and SATERN Director for Canada and Bermuda, and is the current Director of SATERN. That year also saw the Metropolitan Division open a leading-edge EDS facility with a high-tech communications center for the SATERN North American Command.

In 2005, Hurricanes Katrina and Rita hit the Gulf Coast, leaving 1,800 dead and $81 billion in damages. International SATERN ran full force for 20 days, and its operators used all available modes of communications to save lives during the flooding. The State of Louisiana referred the Computer Science Corporation (CSC) to the SATERN Health and Welfare operation, and they devised a coordinated effort in which CSC set up a massive database and an 800 number and assigned CSC personnel to direct and control Health and Welfare messages from the SATERN server to SATERN personnel. SATERN eventually received over 61,000 Health and Welfare requests via its web page, at a rate of up to 20 applications per second at peak periods. Amazingly, SATERN was able to locate 25,508 people affected by Katrina, thanks to the huge mobilization of amateur operators on the national network. Also, for the first time, the Shared Resources High Frequency Radio Program (SHARES, for SHAred RESources) asked SATERN to take part in its network to provide logistical information during the disaster.

When a magnitude 7.0 earthquake hit Haiti on January 12, 2010, the SATERN 20 meter net on 14.265 MHz was immediately called up. As the event unfolded, a nighttime net was set up on the SATERN western nets frequency of 3.977 MHz to handle traffic from the disaster site. The ARRL met regularly with SATERN and other principals after the quake to plan continuing help for the island.

On May 22, 2011, an EF5 tornado hit Joplin, Missouri, a town of approximately 50,000 people, leveling nearly one third of it. The Salvation Army's Kansas and Western Missouri and Midland (Southern Illinois and Missouri) Divisions ran SATERN communications for 19 days following the tornado, and virtually all of the SATERN personnel responding to the emergency were also ARES members. The experiences in Haiti and Joplin demonstrate the value of the long-term relationship between the ARRL and the Salvation Army.

As disasters strike, SATERN stands ready to assist the Salvation Army.

Jim Andera, KØNK, checking in on the SATERN EmComm net. [KEN PANCZYK, W9KMP]

Each territorial level has a SATERN Coordinator for that region. The primary duty of the Territorial Coordinator for SATERN is the facilitation, coordination, and maintenance of the program within the given territory and the interface of that program with the territory. The Territorial Coordinator relates to all regional Amateur Radio or communications entities.

The Territorial Coordinator's primary volunteer staff is the SATERN Territorial Net Director who oversees the nets of that region. In some cases there is provision for an associate or assistant SATERN Coordinator for the territory. Each Territorial Coordinator relates to the Amateur Radio Liaison Officers of the divisions in the territory he or she coordinates.

At the divisional level is a SATERN Amateur Radio Liaison Officer (ARLO) whose primary job is the building, organization, and maintenance of an Amateur Radio unit to support and enhance Salvation Army emergency communication potential during disaster. The ARLO relates to Amateur Radio and/or communications entities at the regional level. The divisional ARLO also relates to the ARLOs at the local or corps level.

At the corps level, there is an ARLO for the local Salvation Army EDS team. The ARLO reports to the Corps Officer in charge of the area and relates to all Amateur Radio groups in the local area. The holder of this position is responsible for building the Amateur Radio component for the Salvation Army EDS response at the ground level. This local team typically consists of the ARLO, a VHF Net Manager, an HF Net Manager for the region (where necessary), and SATERN volunteers, but the team may be modified according to the needs of the region.

For instance, the local team in Chicago had an IT person develop the database, a photographer doing the PR, and a digital specialist helping to set up digital communications in the field. The SATERN volunteers performed tasks such as canteen or mobile field vehicle communications, or shadowing Salvation Army principals and communicating from various Salvation Army centers during the disaster incident.

SATERN OPERATIONAL NETS

SATERN uses 14.265 MHz as its international net frequency. This net meets at 1500 UTC Monday through Saturday throughout the year for training purposes. The Saturday net is considered part of the Salvation Army Radio Operators' Fellowship (SAROF), but is widely used by SATERN and in disaster response functions as a SATERN net. If the frequency is occupied, an adjacent frequency is used. When a national/international event occurs, a SATERN net is established on this frequency for the duration of the event specific to the needs of the disaster while there is propagation. An alternate nighttime frequency SATERN uses is 3.977.70.

There are several other nets held on a regular basis (see sidebar).

The SATERN callout for an international/national event comes from the National Director and is communicated on the air and through all means at hand to distribute the message (paging, cell phone, Internet, VoIP, IRLP, EchoLink, landline, etc.).

SATERN provides emergency communications support to the Salvation Army wherever needed on site, at the local level, and via VHF/UHF nets. Training for this occurs on a host of local nets, which are listed in the nets section on **www.satern.org**. All local SATERN teams are encouraged to develop a training network and participate in local ARRL/ARES networks in their area. Additionally, all regional SATERN leadership is encouraged to develop regional SATERN nets where needed and to participate in ARRL/ARES networks.

THE RELATIONSHIP OF SATERN TO AMATEUR RADIO

The element that makes SATERN work is Amateur Radio; that is to say all the hams who participate and stand by to help. I believe that SATERN contributes to the effectiveness of the Memorandum of Understanding (MoU) between

SATERN OPERATIONAL NETS

Net	Day	Frequency	Time
The SATERN International Net	Monday - Saturday	14.265 MHz	1500 UTC
The SATERN Eastern States Net	Saturday	7.265 MHz	1400 UTC
The SATERN Western States Net	Sunday	3.977.70 MHz	0400 UTC
The SATERN Southern States Net	Saturday	7.262 MHz	1500 UTC
The SATERN Central States Net	Saturday	7.265 MHz	1530 UTC

These territorial nets liaise with the international SATERN net on 14.265 MHz.

Other SATERN nets:

Net	Day	Frequency	Time
National SATERN CW Net	Thursday	10.115.0 MHz	2200 UTC
SATERN 60 Meter Net	Tuesday	5.330.50 MHz	0000 UTC
Kansas and Western Missouri Net	Tuesday	3.920 MHz	0130 UTC
Canadian and Bermuda Territories Net	Monday	3.740 MHz	0030 UTC
Oklahoma/Arkansas Network	Daily	3.903 MHz	1400 UTC

the ARRL and the Salvation Army, because in a very committed manner, the Salvation Army has embraced the dynamic of Amateur Radio and those who pursue it. It has incorporated their potential in its own operations, benefitting from the training, planning, and help offered by the ARRL through its membership.

Without the assistance of the wider Amateur Radio world, SATERN would not work. SATERN doesn't compete with other amateur entities; it embraces them to reap the advantages of the human and technological potential they represent. SATERN volunteers are frequently also members of the ARRL, ARES, RACES, MARS, REACT, SHARES, SKYWARN®, and a host of other organizations that are poised to help in catastrophe. This gives SATERN a high degree of resilience, ability, and experience. As well as its own personnel, SATERN encourages the training of ARRL, local Emergency Management Agencies (EMA), SKYWARN, and other communication entities to better respond to disasters.

In the Katrina response, for instance, there were glowing examples of pulling together, and publications such as *The ARRL Letter, The UK Radio Magazine,* and *World Radio* highlighted the effective coordination between SATERN, ARRL, MARS, SHARES, and other amateurs.

AMATEUR RADIO AS A RESOURCE FOR THE SALVATION ARMY'S EMERGENCY OPERATIONS

In addition to Amateur Radio expertise, every amateur operator has skills he or she has developed through life, which may benefit the program in areas other than communications. For instance, by adding Amateur Radio to SATERN's Metropolitan Division surrounding Chicago, the division gained someone trained in Informational Technology who provided the initial work on the SATERN personnel and EDS team database, as well as someone who managed a hydraulics plant and became the motor vehicle officer for the canteens and who later helped put together the division's 60,000-square-foot, state-of-the-art EDS center. Another amateur ran a food service for a local high school, and she became the food service officer for the disaster response team. Clearly, the most valuable resource Amateur Radio provides is not equipment, but people.

SATERN AS THE EYES AND EARS OF THE SALVATION ARMY DISASTER RESPONSE

SATERN's philosophy of being the "eyes and ears" of the Salvation Army's disaster response dictates that every member of an emergency team is singularly responsible to report the occurrence of disaster by developing and employing his or her listening and monitoring skills. The Salvation Army SATERN volunteer knows that he or she is individually responsible for reporting the occurrence of disaster. It is the volunteer's job to relay information to leadership to ensure a response is given when needed. Volunteers know that without their input, the word may not go out.

This aggregate commitment is essential to the Salvation Army's rapid, efficient response. SATERN's members represent a broad base of volunteers who have at their disposal uniquely effective monitoring tools. As they interface with other amateurs, their potential is further expanded. The result is a dynamic organization — at heart a listening post of thousands — that stands at the ready to help when disaster strikes.

Distinct from much cutting-edge technology, Amateur Radio provides a failsafe communication tool. Effective disaster response requires redundancy in every phase, and communications is no exception. Fragile, complex systems dependent on infrastructure too often fail, so there should a place on a well-equipped disaster team for every communication component, including Amateur Radio, etc. Interoperability is also critical, and SATERN volunteers offer the Salvation Army just that. A single team can go into an affected area, setting up a communications system with battery, generator, or solar power; they can hang antennas or mount them on their vehicles to be on the air in no time. With this unique ability, they connect officers, volunteers, distribution points, shelters, command centers, and liaison areas for the Salvation Army disaster response until normal communications are back in service.

SATERN GOING FORWARD

SATERN is now an integral component of the Salvation Army EDS response. It has a current membership of 4,500 volunteers who capably add to the skill set and resources of the Salvation Army Disaster program. Their ability to establish strategic communications and offer interoperability on the disaster scene have been demonstrated over and over, both nationally and internationally.

The Salvation Army now requires that all EDS members, including its SATERN communications component, undergo disaster training. This minimally includes an introduction to the Salvation Army and Incident Command System training. There are several courses that can be found offered on the Salvation Army EDS website at: **http://disaster.salvationarmyusa.org.**

The emphasis in the coming years will be to add to the list of divisional SATERN teams with ongoing training provided by the Salvation Army's disaster response.

Chapter 5
Weather Events — The National Weather Service and the National Hurricane Center

The SKYWARN program (see **Figure 5.1**) is run by the National Weather Service (NWS), which is an agency of the Federal government, and a division of the National Oceanic Atmospheric Administration (NOAA), United States Department of Commerce. The National Weather Service is charged with warning the public of severe or damaging weather and also issues timely aviation, marine, and land forecasts. There are 122 local NWS forecast offices and 11 NWS national prediction centers across the United States.

The mission of the NWS SKYWARN program is to provide a way for citizens to report severe weather conditions and damage reports to their local NWS office for the general protection of life and property. People trained in SKYWARN weather spotting don't have to be Amateur Radio operators, but having your Amateur Radio license does provide additional ways in which you can support the SKYWARN program.

Through the natural dispersion of Amateur Radio operators over an area, hams who conduct SKYWARN nets — typically on VHF/UHF repeaters, with other Amateur operators stationed at NWS Forecast offices — can provide real-time to near-real-time information when weather that meets National Weather Service reporting criteria occurs. This allows weather forecasters/meteorologists to know what is happening on the surface, and leads to greater confidence and lead time on weather warnings for a given area. The radar and satellite cannot see what is happening on the surface, but trained weather spotters and Amateur Radio operators can be the eyes and ears on the ground or at the surface to provide this critical information.

The National Weather Service typically conducts SKYWARN training classes so that citizens, both Amateur Radio operators and non-amateurs, can learn specific cloud formations and reporting criteria for the area that the National Weather Service office sponsoring the training covers. They also learn what defines "severe weather." Some areas require that people be trained in SKYWARN before participating in SKYWARN Amateur Radio nets.

Amateur operators involved with the ARRL and ARES (Amateur Radio Emergency Services) are encouraged to work with the Warning Coordination Meteorologist of the National Weather Service on coordinating the SKYWARN activity for a given NWS office. Each part of the country may have different ways to organize and accomplish the NWS SKYWARN mission of timely severe weather reporting for the protection of lives and property. Some areas have ARES oversee the SKYWARN program and may have Amateurs involved in ARES that are solely involved with ARES from a SKYWARN weather spotting perspective, while other areas have the SKYWARN program overseen separate from ARES, but offer collaboration between the two programs. This is contingent on what the NWS, as the served agency, wants and how the ARES and SKYWARN programs for a particular NWS office coverage area — known as a County Warning Area (CWA) — and ARRL section are organized. The National Weather Service has a Memorandum of Understand-

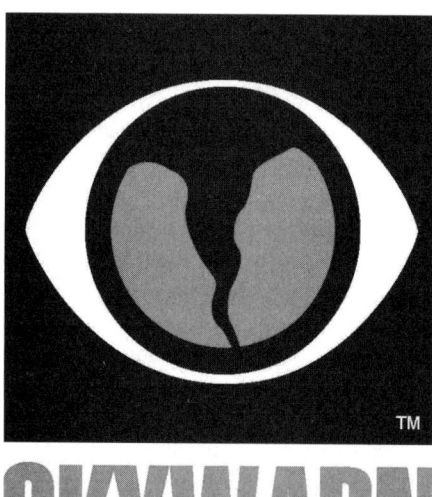

Figure 5.1 — SKYWARN is the National Weather Service's volunteer weather spotting program where timely severe weather reporting into the local National Weather Service Office protects life and property and supports timely severe weather warnings. [NWS/NOAA]

ing (MoU) with the American Radio Relay League (ARRL). In June 2011, the Memorandum of Understanding between the ARRL and NWS was renewed and updated (see **Figure 5.2**), and is available in the Appendix of this book.

Some of the best practices of Amateur Radio SKYWARN programs and of SKYWARN programs that are also connected to ARES feature consistent and continuous Amateur Radio operations at the local National Weather Service Forecast Office during severe weather events. Hams that go to the National Weather Service Forecast Office to operate an Amateur Radio station on site need to be aware that they could be subjected to a criminal background check and other security procedures based on the NWS Forecast Office's policy for allowing Amateur Operators and visitors into the facility. Amateur operators who operate from a local National Weather Service Forecast Office should be SKYWARN trained and trained in the National Incident Management System (NIMS)/Incident Command System (ICS). Other training requirements may be required based on the NWS Forecast Office and local Amateur Radio group supporting that NWS office's SKYWARN program.

Local/regional Amateur Radio SKYWARN programs across the country vary in use of EchoLink/IRLP/D-STAR and other Voice over Internet Protocol (VoIP) modes. In some areas, EchoLink and IRLP are cross-linked together to form a larger-scale network that can complement HF capability as needed, while other groups link repeaters together using EchoLink or IRLP, or utilize these modes for point-to-point connections that cannot be made purely over VHF/UHF radio means. *Winlink*, packet, and other digital modes such as *Fldigi* may also be utilized in various Amateur Radio SKYWARN programs depending on the capability of the Amateur Radio SKYWARN program and local National Weather Service Forecast Office. Other means of handling information are typically coordinated ahead of time in these situations.

In cases where Amateurs are not active at the local NWS Forecast Office, local SKYWARN groups will forward information to a specific NWS e-mail address for that Forecast Office; or via an NWS Forecast Office reporting form contained on the NWS office website; or an e-spotter reporting form; or a restricted phone spotter line that is given to trained SKYWARN Spotters only; and/or a system called NWS Chat. The e-spotter reporting form is at **http://espotter.weather.gov/** and covers approximately 100 of the 122 NWS Forecast offices. NWS Chat is an instant messaging program that is

Figure 5.2 — Kay Craigie, N3KN, and National Weather Service officials sign the updated MoU between the ARRL and the NWS. [ARRL]

utilized by a number of weather offices as another means of sending reports into the NWS Forecast Office. Its use is typically restricted to SKYWARN Net Controls and Coordinators, media, and other meteorology-based groups. The NWS Chat program link is **https://nwschat.weather.gov/** and access to the chat service can be requested via the "Request NWS Chat account — NWS Partners" link on the website.

Another means that many Amateur Radio SKYWARN programs use to gather information involves the Automatic Packet Reporting System, or APRS, and the Citizens Weather Observer Program, or CWOP. There is also the "Weather Underground" website, **www.wunderground.com**, which features many weather stations, including those that are on the APRS/CWOP network. Weatherbug, **www.weatherbug.com**, may also be utilized for reports from weather stations located at schools and other facilities that can also provide additional measured meteorological data. Many NWS Forecast Offices utilize this program to gather measured wind and rainfall information that can be critical in providing additional reports from across the region when wind and rainfall are concerns in a given weather situation. It also can give information on the rain-snow-ice line by reviewing temperature information given on these sites. The APRS/CWOP sites are often ingested to a system called Mesonet and made available online.

Many Amateur Radio SKYWARN programs maintain websites and e-mail lists that state when potential severe weather and major storm events could occur and state when SKYWARN activation is possible. These e-mail lists and websites are also utilized to publicize SKYWARN training sessions, and to publicize other topics related to weather spotting and Amateur Radio to the greater community, as well as to provide weather safety information. Some Amateur Radio SKYWARN programs are also using social media outlets such as Facebook® and Twitter as a notification network for activation and as a way of gathering reports. This is not a replacement for gathering reports via Amateur Radio, rather it is a means to enhance the level of reporting and notification of activation to a given area and can also be utilized to notify Amateur Radio operators who use social media about activation and where nets are activated.

Another best practice for Amateur Radio SKYWARN programs is a policy of self-activation. This does not mean that Amateur operators self-activate for any kind of weather-related situation. It means that they self-activate when a report of severe weather is received, or when specific warnings such

Figure 5.3 — Springfield, Massachusetts tornado damage, June 1, 2011. [RAYMOND WEBER, KA1JJM]

as a Flash Flood Warning, Severe Thunderstorm Warning, Tornado Warning (see **Figure 5.3**), or similar significant severe weather warning is issued by the NWS Office. Providing reports in these situations can be very critical, as it is likely a short duration or possibly non-forecasted event that causes the situation.

The meteorological and damage reports that are received from a SKYWARN program may not only be useful to the National Weather Service but may also be useful to Emergency Management Agencies and non-governmental organizations (NGOs) such as the American Red Cross, Salvation Army, and other faith-based NGOs that assist with the recovery phase of a weather-related disaster. This information can be shared over Amateur Radio if emergency management at the local or state level, NGOs, and the NWS SKYWARN program are coordinated to handle such needs. In addition, local media outlets can receive this information via scanners that monitor Amateur Radio bands or by e-mail or social media. This information can be used by Emergency Management, NGOs, and the media to prioritize response units in that phase of the operation to the hardest-hit areas as well as in the recovery phase for damage assessment, health and welfare, and shelter and relief operations. Rainfall, snowfall, flooding and weather-related damage reports gathered by the SKYWARN program and other sources are typically put together in the NWS Local Storm Reports and public information statements. This can include official NWS storm surveys that are conducted after a storm has passed, if damage is significant enough to warrant a survey based on preliminary damage reports from non-Amateur Radio and Amateur Radio SKYWARN spotters and other reports gathered from other agencies.

SKYWARN RECOGNITION DAY

The National Weather Service, in conjunction with the ARRL, holds a yearly event called SKYWARN Recognition Day (SRD) to recognize the strong partnership between the National Weather Service and Amateur Radio. The December event usually has over 100 NWS Forecast Offices on the air, contacting one another and their SKYWARN Spotters to recognize their significant contributions throughout the year. The event also provides an opportunity for neighboring Weather Service Forecast Offices to contact one another, which could be needed in times of a total communications failure. Participants may receive QSL cards from NWS offices that offer them, and a certificate is available for participating in this event as well. The SRD website located at http://hamradio.noaa.gov shows past SRD events, complete with pictures and statistics on contacts from all the participating NWS Forecast offices.

THE NATIONAL HURRICANE CENTER

Natural weather events such as hurricanes bring out Amateur Radio operators, nationwide and internationally. At the heart of the activation is the National Hurricane Center, located on the campus of Florida International University in Miami, Florida. The Amateur Radio station at the National Hurricane Center (NHC) is best known by its call sign, WX4NHC. According to the station's website at www.wx4nhc.org, its goals and objectives are:
1. Collect weather data "Surface Reports" from the Hurricane affected areas in real time for use by the Hurricane Forecaster;
2. Provide back-up emergency communications to and from the Hurricane Center during and after a direct hit on Miami;
3. Provide Hurricane Advisories over ham radio, when other sources are not available to the affected area; and
4. Enhance and promote the accuracy and availability of weather data surface reports.

Their mission: To Help Save Lives

WX4NHC has a pool of more than 30 specially trained volunteer operators who can be called upon to man the station in multiple-hour shifts for as long as needed. The efforts of these dedicated amateurs have had significant impact on hurricane forecasting. In 2010, surface reports gathered by the VoIP Hurricane Net and sent into WX4NHC during Hurricane Tomas' impact on St. Vincent and St. Lucia provided hurricane forecasters data that showed Tomas was stronger than satellite intensity estimates were giving. Another example is Hurricane Michelle in 2001. Satellite imagery made it apparent that Michelle was losing intensity as the storm impacted the Bahamas. Hurricane forecasters were going to reduce warnings for the Bahamas, however, an accurate report from an Amateur Radio operator on a sailboat docked on the island of Highbourn Cay reported sustained winds of 100 MPH with a gust to 115 MPH, using an anemometer mounted at 72 feet on the sailboat mast. This prompted hurricane forecasters to maintain the Hurricane Warnings for the region, despite the satellite imagery appearance of Hurricane Michelle. For this service, the coordinators of WX4NHC received a letter of commendation from Max Mayfield, who was the director of the NHC at that time.

Each year on a Saturday from 9 AM–5 PM ET (1300–2200 UTC) around the start of the Atlantic Hurricane Season on June 1, the Amateur Radio operators of WX4NHC put on a Communications Test in which they prepare the station to assure it is in proper working order before the start of the new season. They test equipment and make any necessary antenna and equipment repairs. They also assure that the station is not interfering with any equipment at the National Hurricane Center. The Communications Test is carried out by amateurs operating on various bands on HF throughout the day and by spending two hours on the EchoLink/IRLP VoIP Hurricane Net system. Often the Director of the National Hurricane Center is present for the Communications Test and makes contacts with amateur operators during the test. The event also serves as a test of equipment for the amateur operators who make contact with WX4NHC, as well as a way to promote Amateur Radio hurricane awareness.

Amateur Radio SKYWARN programs along the United States East and Gulf Coasts also deal with the threat of hurricanes that can cause a variety of weather impacts, including coastal flooding from storm surge; river, stream, and urban flooding from heavy rainfall; tornadoes; and damage to trees, wires, and structures from high winds. These Amateur Radio SKYWARN programs will activate for their local NWS Forecast offices as they do for severe thunderstorm and tornado events and other storm systems. WX4NHC will interact with these Amateur Radio SKYWARN programs via HF through the Hurricane Watch Net and using EchoLink/IRLP through the VoIP Hurricane Net. Local Amateur Radio SKYWARN programs will send liaisons to these nets or have WX4NHC monitor their local or regional SKYWARN programs EchoLink/IRLP and HF nets (where the Amateur Radio SKYWARN programs have this capability) to gather critical surface reports and maintain a critical communications link. There are also many other modes, including digital modes, to facilitate information flow from the local area up to the National Hurricane Center from the local and regional Amateur Radio SKYWARN programs as well. When a hurricane landfall occurs in southern Florida, WX4NHC ties into local VHF/UHF nets in their area, along with all the other modes they utilize, to gather surface reports. While local NWS Forecast Offices might get weather and damage information during hurricanes when other communication means are available, this may not happen on a continuous basis, therefore what Amateur Radio can provide in this area can be very timely. In addition, when other communication means fail or become overloaded, active Amateur Radio operators at the local National Weather Service Forecast Offices may be the only means of gathering meteorological information and damage reports that can be relayed to the National Hurricane Center.

This situation occurred during Hurricane Katrina in 2005, where the only means of getting weather and damage information out of the area was via an Amateur Radio link between the NWS Slidell, Louisiana Office and the National Hurricane Center. In addition, health and welfare traffic was passed from the NWS Slidell office to the National Hurricane Center to alert family members that meteorologists at the Slidell facility were well and working, but would be trapped at the Weather Office due to damage from the hurricane. A coordinated link between local NWS Forecast Office SKYWARN programs and the National Hurricane Center can be very critical in hurricane situations and this is something that should be practiced and exercised ahead of time if possible.

20 QUESTIONS FOR A STORM SPOTTER

Answered by Matthew Utley, KD5HAO, Wynne, Arkansas

Q1. How long have you been a storm spotter?
A. 10 years.

Q2. What types of severe weather are you activated for?
A. Any weather that has the potential of becoming severe that could directly affect property and life.

Q3. What type of training have you had?
A. Training is offered through the NWS to any persons interested in becoming a spotter for the NWS. Check with your local Amateur Radio group or NWS office to find out more about a class near you.

Q4. Approximately how many storm spotters are in your area?
A. 10 to 15.

Q5. How does information get from the spotter to the NWS?
A. The most common way is through the use of Amateur Radio. Amateur Radio plays a vital part in passing along severe reports directly to the NWS.

Q6. Does the local storm spotter group participate in any drills or exercises?
A. Yes. As in many cases, drills keep spotters in practice, and prepared for the real thing.

Q7. How does your local SKYWARN group include non-ham spotters?
A. While Amateur Radio has a big part in SKYWARN, anyone can participate, even if they don't have a ham license. Reports can be phoned into the NWS as well.

Q8. How often do you have local SKYWARN training?
A. Training classes are generally held from December through February, just before the storm season begins.

Q9. Describe the relationship between your SKYWARN group and local emergency management.
A. We assist local agencies in attempting to provide an early warning of approaching severe weather. It's all about saving lives!

Q10. What Amateur Radio modes and frequencies do you use for storm spotting?
A. Most areas have local SKYWARN nets that have a net control at the NWS. In some rural areas, however, a spotter may not be able to communicate with the net control. In those cases, the NWS is also monitoring an HF radio close by.

Q11. As a volunteer storm spotter, what are your primary safety concerns?
A. The main concern for any spotter is the effort to save lives. In order to do that, communication is necessary to relay reports of severe weather that could have an impact on human life. Amateur Radio plays a vital role in the community, providing an early warning when threatening weather approaches, thus allowing people a better chance getting to safety.

Q12. In your area are the evacuation centers, Red Cross, EOCs, etc equipped with Amateur Radio?
A. It is an ongoing effort to have a station wherever possible to assist in emergency communications. Most Amateur Radio operators can set up an emergency station very fast, if one is not already in place.

Q13. Describe how your local SKYWARN net is conducted.
A. All reports are communicated directly to the net control, who then passes along the report to the NWS. Scheduled nets are usually held weekly, to keep in practice for an actual net.

Q14. Does your local SKYWARN group conduct or participate in any training that is not SKYWARN?
A. Yes. ARES/RACES also play a big part in emergency communications. Contact your local Amateur Radio group to learn more.

Q15. What other Amateur Radio activities are you involved in?
A. Amateur Radio operators not only participate in emergency communications, they participate in other activities as well. Field Day is an excellent opportunity to witness and even get involved in the various operations of Amateur Radio.

Q16. What got you interested in storm spotting?
A. Participation in the local Amateur Radio club.

Q17. Do you have a go-kit? What's in it?
A. Yes. A portable ham radio, batteries, bottled water, first-aid kit, flashlight, etc. A go-kit is a good idea to have in place in emergencies.

Q18. What, in your opinion, is the most valuable tool a storm spotter can take along when spotting?
A. Radar. Utilizing radar allows spotters to know where a storm is, where it is headed, and what parts of the storm are severe.

Q19. How can Amateur Radio storm spotters improve?
A. It is the goal of every Amateur Radio spotter to keep room open for improvement. Any way a ham can better serve his or her community is certainly the desire for every amateur radio spotter.

Q20. When you are activated, how do you participate in SKYWARN? (ie mobile spotter, net control, relay, etc)
A. It is important that a spotter is able to rotate through each area when an activation occurs. All play a vital role in saving lives including participating as net control, relaying information, or even tracking a storm in the mobile. It is important to also have the proper training and participate in drills to have proper knowledge of how and what to do should an actual SKYWARN activation occur.

THE HURRICANE WATCH NETWORK AND THE VOIP HURRICANE NET

The WX4NHC station and its group of amateurs have worked with two organized hurricane nets: the Hurricane Watch Network (HWN) and the VoIP Hurricane Net. The HWN has been in existence since 1965, relaying reports from both land and sea stations for over 45 years. In 1965, Hurricane Betsy struck the Gulf Coast of the United States, first crossing the Florida Keys as a Category 3 before moving into the Gulf of Mexico where she intensified to a Category 4. Betsy then hit the Louisiana Gulf Coast as a Category 3 on September 9. Betsy was the first hurricane to hit the US that caused in excess of $1 billion in damages ($10 billion in 2005 dollars), earning the storm the nickname, "Billion-Dollar Betsy."

In the midst of Betsy's fury, Jerry Murphy, K8YUW, founded the Hurricane Watch Network. Hurricane nets were not a new thing. As early as the 1930s, Amateur Radio operators in hurricane-prone areas operated nets when hurricanes threatened. Since its beginning, the HWN has been a key part of the Amateur Radio response when hurricanes strike. Of primary concern for the HWN are hurricanes that form in the Atlantic and threaten populated areas in the Caribbean, Central America, Mexico, and the US. On rare occasions, the net has been activated for Pacific hurricanes. The general criteria for net activation are a named storm that is forecast to make landfall, and that is with 300 miles of doing so. Hurricanes or tropical storms that do not pose a threat to a populated area and stay out to sea would not require a net activation. Since the inception of the HWN, there have been 444 named storms in the Atlantic (as of 2009) and of those, an estimated 75% have threatened or actually made landfall.

Currently the HWN consists of about 40 members that serve as net control stations. They are strategically placed throughout the US, Canada, the Caribbean, Central America, and Mexico. Net members all serve as net control. Station that report weather conditions to the net are not formal members of the HWN per se, but play a critical role. Formal membership in the net is limited and highly selective. Participation in the net, however, is open and welcome to those reporting critical weather conditions related to the hurricane, particularly stations within 100 miles of the eye of the hurricane.

There are two primary purposes of the HWN; first, to disseminate the latest National Weather Service advisories on active named tropical storms and hurricanes on both the Atlantic and Pacific side of the Americas, including transmissions to any maritime Amateur Radio operators that may be in the affected area and; second, to gather real-time ground-level weather conditions from amateurs in the affected areas and get these reports to the National Hurricane Center via WX4NHC in a timely and accurate fashion. The HWN operates on 14.325 MHz and utilizes SSB and APRS for relaying weather information. When necessary, other bands such as 40 and 80 meters may be used. During net activation, stations checking in are directed by net control to report certain weather conditions: station and geographic location, time in UTC, wind speed (sustained and gusts), wind direction, and barometric pressure.

What happens to the information collected by HWN net control stations? This is where the National Hurricane Center (NHC) and WX4NHC come in. Since its beginning, the HWN has relayed information to the National Hurricane Center in Miami, Florida. In 1980 the NHC set up an Amateur Radio station (now WX4NHC) to which reports could be relayed. This was originally done by FAX and RTTY, and eventually via SSB. WX4NHC keeps a close working relationship with the HWN. Reports that come in via the HWN are relayed to WX4NHC and then on to the NHC. The net can also serve to help keep NWS offices in touch with the NHC and, if needed, relay important traffic to NWS offices.

HWN began working with WX4NHC in 1980 when the group at the National Hurricane Center was organized. Prior to that date, the HWN would call the National Hurricane Center directly with reports. HWN has worked closely with the WX4NHC group to coordinate reports during hurricanes from the Hurricane Net Watch frequency of 14.325 MHz (other HF bands may be utilized if propagation is not favorable on 20 meters). The HWN also works closely with the Maritime Mobile Net to warn maritime interests of hurricanes and gather surface reports from boats at sea.

The VoIP (Voice over Internet Protocol) Hurricane Net was organized in 2003 for the purpose of relaying reports from VHF and UHF repeaters that have either EchoLink or IRLP (Internet Radio Linking Project) nodes. The EchoLink and IRLP nets were combined in 2004, and are run on the EchoLink *WX_TALK* conference node: 7203/IRLP 9219 system. Despite the need for Internet infrastructure and power resources, the net has been very effective in getting reports directly from the areas affected by hurricanes and through relays just outside of the affected areas. The VoIP Hurricane Net works closely with the Caribbean Emergency Weather Net and local/regional Amateur Radio SKYWARN programs during hurricanes.

WEATHER DRILLS

Weather drills are an important part of keeping Amateur Radio SKYWARN programs sharp during periods of inactivity. They're also a way to try new practices in a scenario where such changes will not impact an important live operation in an actual severe weather event. Typically, weather drills are coordinated with multiple agencies that include local and state emergency management, NGOs, and the National Weather Service. Sometimes, the Federal Emergency Management Agency (FEMA) and the Department of Homeland Security (DHS) will be involved in these drill scenarios. Amateur Radio operators follow the drill scenario working with these agencies, and may also have input into the drill scenarios. Drills can be operational exercises, or they can be of a "tabletop" format in which the leaders of the various agencies work through their procedures without actually activating the various networks.

NWS WEATHER SAFETY INFORMATION

Tornado Safety
- Be on the lookout for other tornadoes that could form in the vicinity of the tornado you are watching.
- Never try to out-run a tornado in an urban or congested area, but immediately get into a sturdy structure after parking your car out of the traffic flow.
- Do not take shelter under bridges or overpasses. These structures do not offer ample protection and could increase the chance of injury or death.
- If you are caught outdoors, seek shelter in a basement, shelter or sturdy building. If you cannot quickly walk to a shelter, immediately get into a vehicle, buckle your seat belt and try to drive to the closest sturdy shelter. If flying debris occurs while you are driving, pull over and park. Now you have the following options as a last resort:
    - Stay in the car with the seat belt on. Put your head down below the windows, covering with your hands and a blanket if possible.
    - If you can safely get noticeably lower than the level of the roadway, exit your car, and lie in that area, covering your head with your hands.
    - Your choice should be driven by your specific circumstances.
- Flying and falling debris is the biggest hazard in a tornado. To be safe, you should get in, get down and cover up. Underground or in a safe room is best. If no underground shelter is available, get to the center of a sturdy building on the lowest level. Put as many walls between you and the tornado as possible. Stay away from windows and doors. Cover up to help minimize being injured by flying or falling debris.

Flash Flood Safety
"Turn Around, Don't Drown!"
- Do not attempt to drive or walk across a flooded road or low water crossing. You cannot be sure about the depth of the water or the condition of the roadway — it might be washed out.
- Two feet of moving water will carry away most vehicles.
- Six inches of fast-moving water can knock you off your feet.
- If your vehicle is suddenly caught in rising water, leave it immediately and get to higher ground.
- Be especially careful at night when flash floods are harder to recognize.

Lightning Safety
"When Thunder Roars, Go Indoors!"
- Remain in a hard-topped vehicle or an indoor location for at least 30 minutes after you hear the last thunder clap. If you use radio equipment, avoid contact with it or other metal inside your vehicle to minimize the impacts should lightning strike.
- If you are out on the water and skies are threatening, get back to land and find a fully enclosed building or hard-topped vehicle. Boats with cabins offer a safer, but not perfect, environment. Safety is increased further if the boat has a properly installed lightning protection system. If you are inside the cabin, stay away from metal and all electrical components.
- When reporting to the NWS, use a cordless or cell phone if available.
- Lightning victims do not carry an electrical charge, are safe to touch, and need urgent medical attention. If a person has stopped breathing, call 911 and begin CPR if the victim is not breathing.

Downburst Wind Safety
- Keep a firm grip on your vehicle's steering wheel to maintain control. Downbursts can occur suddenly with an abrupt change in wind speed and direction.
- If you can do so safely, point your vehicle into the wind to minimize the risk of the vehicle being blown over.
- Be prepared for sudden reductions of visibility due to blowing dust or heavy rain associated with downbursts.
- Point spotters observing from a substantial building should move away from windows as the downburst approaches.

Hail Safety
- Substantial structures, like a garage, offer the best protection from hail.
- Spotters in vehicles should avoid those parts of the storm where large hail is occurring.
- Hard-topped vehicles offer good protection from hail up to about golf ball size. Larger hail stones will damage windshields.

*Used with permission of the
National Weather Service.*

Figure 5.4 — Amateur Radio operators do a simulation of an activation of SKYWARN for a hurricane at the National Weather Service Taunton Forecast Office Amateur Radio station, WX1BOX, gathering mock reports from various locations using the Amateur Radio station. Pictured from left to right are Phil McLaughlin, KB1CYO, Steve Schwarm, W3EVE, Rob Macedo, KD1CY, and Jim Tynan, KC1JET. [NICK SNOW]

Weather drills are typically constructed with a timeline-based format in which messages instruct local or regional areas as to their actions in a given weather scenario (see **Figure 5.4**). Drills typically inject unexpected scenarios that agencies and Amateur Radio operators must react to, so that they can be reviewed when the drill is completed. Input is typically gathered from the National Weather Service and may include test warnings and advisories from that agency. It is important that the drill encompasses a realistic scenario that could occur over a given area, so NWS input in such an exercise is critical.

In the drill information, frequencies of operation should be stated, as should clear goals and objectives of what will be tested. Drills should be constructed to address the areas of a SKYWARN program that need work. This could range anywhere from participation, to how information funnels from local SKYWARN or ARES/SKYWARN groups to the National Weather Service and other agencies. It could focus on how neighboring SKYWARN groups interact with each other. Weather events see no boundaries; weather incidents that occur across an area can affect other areas. Having information and interaction ahead of time as to what may be coming heightens the awareness of spotters, and can assist the National Weather Service with additional lead time for critical warnings.

While "surprise" drills that are done with little advance publication may be good to get a sense of who is available at a given moment and to test activation procedures and plans, most drills that are coordinated with various served agencies are announced ahead of time through various means, including e-mail, websites, and social media outlets. In some drills and exercises, briefings and meetings are done with participants and controllers to ensure the objectives of the drill are understood and coordinated ahead of time. This should be treated similarly to most weather-related situations where there is advance warning of a potential severe weather scenario several days ahead of time, and the drill information and briefings therein tend to simulate what would occur if the severe weather situation was real.

During a weather drill, it is critical to say "This is a drill" often during exchanges in transmissions. While many agencies put out information on drills and exercises ahead of time, many people will be unaware that the drill is taking place. If our frequencies are monitored and people who are not "in the know" hear of the activity, it will likely cause them to call authorities to find out what's happening. To avoid this scenario, messages should contain something indicating that these are drill or test messages.

Moving through an exercise in an expedient manner is also important. Many drills only go on for several hours; the Net Control and the exercise controller should make it a priority to ensure the timeline is executed in a way where all messages and exchanges that are to be done as part of the exercise are completed. This means getting exercise participants to flow through the scenario and provide feedback understanding the time constraints that may occur during the drill. In drills, it can be difficult to formulate traffic that is imaginative and fits the scenario of the exercise. Net Controls and controllers of the exercise need to be sure that they can keep a solid traffic flow on the nets in an attempt to closely resemble the traffic flow of the real situation. One simple way that this has been done is to signify a color coding scheme for houses or structures and what each color symbolizes for degree of damage. Participants in the exercise can then look out over the area and, based on the color of the structure, indicate the damage based on the drill scenario. This should allow for participants to pass on exercise traffic in a way that best resembles what would occur in a real incident. You can also encourage participants to recall past scenarios of the weather situation that is being unfolded, and relay those past scenarios as part of the exercise. This will allow for traffic flow to best simulate what would occur in a real incident. Depending on the tasks of the exercise, whether the tasks cover multiple agencies, and if the exercise is statewide or regional in scope, liaisons between the various nets should be established. Where available, digital modes for both voice and data can be utilized, depending on the objectives of the exercise. Practicing moving traffic off frequency and having stations return to the main frequency may also be important, depending on the drill scenario that is being constructed.

After-Action Reports

When a drill is nearing completion, it's a good idea to have all participants file after-action reports (AAR). Whether we respond to severe weather as SKYWARN spotters, ARES members, or as part of some other group, a valuable part of the activation process involves assessing our performance, determining what went right or wrong, and taking corrective action. The after-action report makes all this possible.

Writing AARs is standard procedure in military, public safety, and emergency management. Disaster exercises and other training events also warrant an AAR to assess response and determine if the exercise objectives were met.

What is an After-Action Report?

The US Army provides a great explanation of what an after-action report is. In *A Leader's Guide to After-Action Reviews* (available online at www.au.af.mil/au/awc/awcgate/army/tc_25-20/tc25-20.pdf) they identify four key aspects of an AAR. First, it is a discussion of an event. While the Army may be concerned with an event such as combat, an event could be a hazardous materials incident, an earthquake, or a tornado. Second, an AAR focuses on performance standards. Third, an AAR explains to the reader what happened and why it happened. Fourth and finally, the AAR provides insight on how to sustain strengths and improve on weaknesses. By using a tool such as the AAR we can welcome mistakes in our response because the AAR gives us a way of recognizing and building off of them. There may be differences in what an after-action report contains depending on who wrote it and what event it covered, but the four basic goals of the report remain the same.

Who Writes the AAR and Who Reads it?

The AAR may be a collaborative effort. Early on in any critical incident, someone familiar with planning the response, the organizations involved, and the objectives of the response should be designated as the person responsible for compiling the AAR. Individuals involved in responding to the emergency can contribute by submitting reports on what happened, how they responded, problems encountered, and objectives met. In the end these account may be compiled and put into the final report either in their entirety or incorporated into a summary of reports received. However, for these reports to be accurate there must be a process in place to document events as they occur. Some ways of doing this are: logs and journals, written messages, action plans for specific events, and public information and media reports.

Let's use a major ice storm as an example. Such an event can cover a wide area, affecting several states. In this case the final AAR may be a collaborative effort between the Section Emergency Coordinators in the affected area, each one putting together an AAR for their area, with the final AAR being a compilation of each Section's AAR. Of course, accomplishing this would require collecting data from the local responders. This can come from reports written by local Emergency Coordinators, District Emergency Coordinators, SKYWARN storm spotters, SATERN members, and any other groups that may have responded. Just like any other event that may warrant an AAR, there must be some form of documenting events as they occur. Amateurs are familiar with logs and can use these to keep track of reported events, messages, and net activity. Field responders may keep a notebook journal to keep track of events and activities. Software is available to help keep track of resources and times. Needless to say, this can be a massive task. But the rewards of a good AAR are worth this critical follow-up work after the event. Later we will look at ways of streamlining the process of collecting this data.

Once the after-action report is written, who reads it? An AAR is designed to be a useful tool to those that plan response, respond, and assist in future planning. The AAR would definitely be of use to all parties involved in the emergency response. This could be the individual storm spotter, net control operators, local and section leaders, and any number of served agencies that may have been involved. Sharing our findings with other responders can benefit their response, as well as the Amateur Radio response. What an AAR is not, is a press release or any form of public information statement. It may contain information that is not suited for public release. This is not necessarily because there is anything to hide, but scrutiny by readers unfamiliar with all aspects of the emergency or response may not be entirely beneficial. If a public information statement is needed, that is a separate matter handled by a public information officer or spokesperson.

When Should the AAR be Written?

Ideally, the AAR should be written as soon after the event as possible. Depending on the size of the event, this may be difficult to do. Documentation from the event needs to be gathered, responders or leaders may need to be interviewed, surveys may need to be conducted, and perhaps workshops made up of responders may need to be held. A lot of work goes into gathering the information needed for a good AAR, and it must be done in a timely manner. By waiting too long after the event we may forget some details, neglect documentation, or not provide enough detail in the report. An Amateur Radio AAR is no different. Data must still be collected as soon as possible after the event, analyzed, and put into report form. The event is not over when the threat is gone. The event is over when we have included in our recovery effort a plan for better responding to the next emergency.

What Goes in the AAR?

There are different types of AARs. AARs for large-scale events such as hurricanes and earthquakes can contain massive amounts of information, while events such as an isolated tornado or hazardous materials spill may contain less. Likewise, the goals of an AAR for a disaster exercise may be different than for a real emergency, since the purpose of the exercise is training. Regardless of the size of the event or whether the AAR is formal or informal, the AAR contains certain key information. A sample after-action report for a localized event that required a relatively small response is shown opposite.

Lafayette County
Amateur Radio Emergency Service
After-Action Report

Date of activity: May 3, 2002

Description of activity: SKYWARN activation for severe weather

Duration of activity: 1500–2200 hours

Amateur radio groups participating: Lafayette County ARES/SKYWARN

Served agencies participating: Lafayette County Emergency Management, National Weather Service

Describe served agency participation: Emergency management assisted in gathering severe weather reports. All reports of severe weather were forwarded to the National Weather Service. National Weather Service issued weather watches, warnings, and statements.

Number of amateurs participating: 8

List of amateurs participating: K5DSG, KE5NQP, K5LMB, W5BJC, KD5JHE, N5RB, KE5TMY, WB5VYH

Person-hours of amateur service: 37

Describe goals of activity, both for served agency and serving group: Local SKYWARN members gathered severe weather reports and submitted them to local net control. Net control then relayed these reports in a timely manner to the National Weather Service via the Amateur Radio station at the National Weather Service Office. Requests to gather specific information were sent from the National Weather Service office via Amateur Radio nets to SKYWARN storm spotters in the local area.

Did the event fulfill the goals? Yes.

What went well? Call up procedures went well, the severe weather net was run effectively.

Areas needing improvement: Additional SKYWARN training is needed, one operator not involved with SKYWARN activities interrupted the net on several occasions, repeater coverage in some areas is not optimal.

Lessons learned: Study repeater coverage area and look at ways to improve, add a relay station to assist net control during times where traffic volume is high, coordinate efforts with neighboring county ARES/SKYWARN, look into future use of EchoLink.

General comments: Local SKYWARN members did a great job with severe weather reports.

Ideas for future exercises: A severe weather exercise coordinated with local emergency management, district ARES resources, and the National Weather Service would be useful in the near future.

END OF REPORT

THE IMPORTANCE OF BEING HONEST

Providing trustworthy reports is an important element of what makes volunteer storm spotters a valuable part of the warning system. Though, of course, we try to relay the correct information, sometimes mistakes are made. An isolated, honest error is tolerable, but routinely inaccurate or misleading information leads to negative consequences, including loss of credibility and the risk of conflicting information delaying an important warning or advisory.

Some individuals deliberately have made false or misleading reports. In 2007, the National Weather Service (NWS) office in Milwaukee, Wisconsin, received more than 25 false reports from the same computer in one weekend via the Weather Forecast Office's (WFO) web page for storm report submissions. The person making the reports, who appeared to have some knowledge of storm spotting and severe weather, falsely reported tornado damage and injuries. Local media interrupted broadcasts because of warnings generated from the NWS office as a result of these false reports. Complicating the situation, legitimate severe weather was present in the area and a tornado warning had been issued based on storm spotter reports. The Federal Bureau of Investigation (FBI) investigated and since then, new measures have been put into place at the WFO to reduce the incidence of false reports.

Unintentional false reports can result from inadequate training, poor communication, honest mistakes, or a combination of factors. Solutions may include more training, better coordination, or perhaps restricting an individual from acting as a storm spotter.

Knowingly submitting a false storm report is against the law and is a violation of the False Statements Accountability Act of 1996. Deliberately violating this federal statute can result in an individual being fined up to $250,000 and/or being sentenced to up to five years in prison.

The AAR begins with introductory information. This may include the type and location of the event. It should contain maps of affected areas (if available), a timeline of events, the dates and times of any proclamations or declarations (state of emergency, federal disaster area, etc.), and the duration of the event, as well as a general description of the event.

The second part covers a discussion of the response. This discussion will examine different levels in the response effort, likely covering field-level response, local government response, interactions within the operational area and the regional area, interaction with state-level agencies, and interaction with federal agencies. It will focus on planning, logistics, finance/administration and multi/inter-agency coordination. Granted, this may be beyond the scope of an Amateur Radio response, but this section of the report can help us focus on our response as part of a coordinated effort at local, section, division, and regional levels. There are times, though, when this may not be needed, such as when the event is isolated and does not involve participants at anything higher than a local or state level.

Next, the AAR covers the participants and systems involved in the response. Most times, Amateur Radio is only one part of a large-scale response. When Amateur Radio responds, we do not go alone. Other players may include public utilities, the American Red Cross, the National Weather Service, the Salvation Army, the media, and public safety, and an assessment of our interaction with them is critical. There may also be interaction with other levels of response, and assessment of these interactions is important as well. And, as you can probably guess, they are also assessing their interaction with us.

The AAR also addresses areas in our response that can be improved upon by further training. There is no such thing as a perfect response. If we get into the mindset that our actions during a disaster have no room for improvement, then we are setting ourselves up for an even bigger disaster in the future. Talk with all involved in the response and get feedback on what went wrong, as well as what went right. Problems both small and large should be addressed so they don't come back to haunt us in the future. This is also a good place to focus on training that the group may need. Does your group have problems with radioing in weather information that is not reportable? Identify that issue in the AAR, and discuss ways that the group can be trained in appropriate reports.

What to Do With the AAR Once it is Written

After the AAR has been properly distributed to those that need to read it, it should be kept on file. As a tool for future planning, it must be accessible for future use. One way to keep an AAR valuable is by conducting a future review. Let's say the AAR was written by the local Emergency Coordinator for a tornado event. Naturally all SKYWARN and ARES members will read it as soon as it is available, and it is likely it the local emergency management director will look at it as well. Now let's say a couple of months later, the SKYWARN and ARES groups are going to conduct an exercise to test their response to severe weather. The EC rereads the AAR and applies it to how the test is going to be conducted. This helps give some guidance to developing an exercise that encourages learning from real-life experience. Now let's say the same area is facing the threat of a tornado a year later. The EC and emergency manager can look over the AAR and find reminders of what worked, what didn't work, and eventually assess whether additional training and exercises paid off.

The AAR is a concept well-rooted in emergency management, the military, and public safety, but it is not necessarily exclusive to these fields. Amateur Radio can make valuable use of AARs as a tool for severe weather response.

Debriefing

A debriefing is another useful tool for assessing our response to severe weather. The debriefing typically involves more than just the Amateur Radio storm spotter group. It will likely involve local emergency management and/or public safety officials. It is also possible a debriefing may be done with representatives from the local NWS office.

The purpose of the debriefing is to review the effectiveness of the response and address issues of concern. Being present at the debriefing allows you to directly answer questions that may come up about the Amateur Radio response. It also gives you a forum in which to bring up issues of concerns for others involved and to ask questions. You will want to bring with you to the debriefing information from the event. Do not rely on memory and logbooks alone. Throughout the event, keep a separate diary of issues for the debriefing session. This "Debriefing Diary" should contain issues not appropriate for the station logbook, information you will need to retain if the logbook has to be handed over to someone else, as well as information about specific events, times, places, etc. that need to be mentioned. Here are some other items for a "Debriefing Diary:"

- What was accomplished?
- Is anything else still pending? Note unfinished items for follow-up.
- What worked well? Keep track of things that worked in your favor.
- What needed improvement?
- Ideas to solve known problems in the future.
- Key events.
- Conflicts and resolutions.

Remember to focus on constructive criticism during the debriefing, rather than engaging in attacks on actions taken, finger-pointing, or casting personal blame.

WEBSITE RESOURCES

National SKYWARN Home Page:	www.skywarn.org
Index of SKYWARN Web Pages on the Internet:	www.afn.org/~afn09444/weather/skywarn.html
NWS National SKYWARN Home Page:	www.weather.gov/skywarn/
Amateur Radio Station at the National Hurricane Center:	www.wx4nhc.org
Hurricane Watch Net:	www.hwn.org
VoIP Hurricane Net:	www.voipwx.net
SKYWARN Recognition Day:	http://hamradio.noaa.gov
Automatic Packet Reporting System APRS/ Citizens Weather Observing Program CWOP:	www.wxqa.com/
Weather Underground Personal Weather Stations:	www.wunderground.com/weatherstation/index.asp
Weatherbug:	http://weather.weatherbug.com/

Chapter 6
Working with Local Emergency Management Agencies

In order for a cooperative relationship between local emergency management agencies (EMAs) and local Amateur Radio Emergency Service (ARES) to work well during a disaster, the relationship must be built before a disaster ever takes place. For EMAs to appreciate the benefits of Amateur Radio, we must demonstrate our value to their mission. We will be most valuable if we understand their mission and their culture.

OPENING THE DOOR TO THE EMA

Contact between the local emergency management agency and the local Amateur Radio group can be made by either party. Typically though, it is the Amateur Radio club liaison or the local Emergency Coordinator (EC) of the local ARES group or in some cases the District Emergency Coordinator (DEC) — that reaches out to make the first contact with the EMA director. The purpose of this initial contact is to offer and explain the services that an ARES group can provide to the locality.

Where a relationship does not presently exist, it might be due to past volunteer relationships (ARES or other groups) that did not work well. Sometimes it is because the current EMA director had experiences wherein Amateur Radio proved to be "in the way," or of little value to the emergency management mission. Or, in some cases, a past EMA leader spoke poorly of the Amateur Radio contingent, and soured the present management's view of the value of the service. In some locales, there is also a concern about using volunteers in emergency operations for fear it will take away "paid positions" from other public safety first responders. In any event, the ARES Emergency Coordinator has to rebuild the EMA's trust and confidence in Amateur Radio.

There are also circumstances where the local communications infrastructure has not failed for the emergency management agency, or where a major disaster-related incident has not occurred in recent times. This can lead to an EMA that isn't very active or doesn't see the need for having a backup communications capability that will work when other means fail. In situations where the local emergency management agency doesn't solidify a relationship with the local Amateur Radio group, it is recommended that the Amateur Radio group make periodic contact to see if the interest changes with time.

The structure of a local emergency management office can vary from location to location. There may be areas in which the local emergency management director is also the police or fire chief, and therefore he or she may not be able to devote much time to emergency management, depending on his or her other responsibilities. The emergency management director may also be a volunteer and that may also impact the amount of time the director can spend on emergency management duties.

Sometimes, a "back door" approach can work nicely. The local Amateur Radio group could find out if other agencies (such as the department of public health or hospitals) or non-governmental organizations (NGOs, such as the American Red Cross or Salvation Army) are interested in Amateur Radio support. Once the ARES group is established with these served agencies or organizations, they can demonstrate their value in a way that may be meaningful to the EMA.

The appearance of the team is very important when in the field or in drill and practice. Understand that "looks do matter." Having a professional dress code for a local ARES group is important. Familiarize yourself with ARES vests and gear, and make a uniform dress code a requirement in your organization. Looking professional will go a long way toward the group being taken seriously.

To maintain a good relationship with the EMA, make sure operators understand that they are there to serve the EMA's staff — not to direct or intercede in the decision-making matters of that organization. Teach them how to "stand to the side" and remain out of the way when not needed for their communications role.

BECOMING FAMILIAR WITH THE EMA MILIEU

Understanding how the Incident Command System (ICS) operates in an emergency or disaster is essential if one is to succeed as a team player in the field. The most necessary requirement typically is National Incident Management System (NIMS) training, which involves the Incident Command System. Emergency management agencies want people who will be volunteering in their emergency operations center or

(EOC) or deploying to the field to have NIMS IS-700, IS-100, and IS-200 training. Emergency management agencies that accept Department of Homeland Security (DHS) money and grants have a requirement that all paid and volunteer personnel within the emergency management agency complete NIMS and ICS training. This training is offered for free in the classroom and can also be taken online via the FEMA NIMS training website at **http://training.fema.gov/IS/NIMS.asp** and the classes that need to be taken via their site are IS-700.a, IS-100.b, and IS-200.b. FEMA also offers a Communication and Information Management course, IS-704, which would be useful for any Amateur Radio group working with local emergency management. In addition, emergency communications classes are offered by the ARRL and may be offered by local Amateur Radio or ARES groups as well.

Some amateurs may be concerned about the time commitment involved in taking these training classes. In such cases, the local Amateur Radio group's leadership must get these amateurs to understand why the classes are an important part of understanding the working relationship between emergency management and ARES operators. If certain Amateur Radio operators cannot or will not take the classes, leadership should move them to positions that would not require them to be at the EOC or deployed in the field for emergency management. Home relay station and liaison support, as well as net control positions, may be better for these amateurs.

If your local EMA offers classes beyond the basic NIMS/ICS classes that are typically required for working with their agency, it is good for the local Amateur Radio group to participate in these classes. In addition, the local EMA may be willing to host other Amateur Radio and emergency communication related courses when there is sufficient interest to warrant them. The local Amateur Radio group, as they build a relationship with the local EMA, can request to hold these classes in conjunction with any regular training the local EMA does. This enhances the local Amateur Radio group's support of emergency communications for the local EMA, as well as supporting the EMA's operations.

CLEARANCES

Another item that local Amateur Radio groups need to consider when working with an EMA is that the agency's personnel security requirements will typically require a criminal background check. If a volunteer has any sort of serious infraction listed on their criminal record, the EMA may forbid that specific individual to work in their EOC. In these scenarios, it is important for the leadership of the local Amateur Radio group to handle the situation accordingly. While the individual with a criminal record cannot deploy to the EOC (or to the field) with the EMA, this person may be of use as a home station, as a relay or net control. If there still is a concern about that person based on the background check, the leadership needs to delicately handle this situation in a way that satisfies the EMA but also doesn't cause significant issues with the local Amateur Radio group.

PUBLIC SERVICE EVENTS TO PROMOTE AMATEUR RADIO

Another venue into the EMA and served agencies is public service events such as a local road race, parade, marathon, or many other kinds of functions. Demonstrating the usefulness of Amateur Radio as a backup communications tool through these events gains the attention of police, fire, and emergency management, and they are also likely to be involved in the event.

These events often happen every year and provide excellent practice and a way for the local Amateur Radio group to participate in and support the functions of the local EMA in a tangible way on a regular basis. It should be noted that during these events, much like in emergencies, Amateur Radio operators might be asked to perform communications via modes that are beyond Amateur Radio voice and digital communications. This can include communications via emergency management voice or digital modes. Having your local Amateur Radio group able to support all of these modes, along with typical Amateur Radio modes, provides more flexibility and support to their operation and will mean that your group will be more frequently utilized by your local EMA.

GAIN ACCESS TO PRACTICE DRILLS

Drills and exercises serve as good practice and are an opportunity for the local Amateur Radio group to show their value. They're also an opportunity for the Amateur Radio group to become familiar with the operations of the EOC and, if deployments in the field are done, operators may be deployed to facilities that could be staffed in a real disaster-related incident. Drills and exercises can be done several times during the course of the year and offer a great opportunity for the local Amateur Radio group and local EMA to strengthen their partnership and learn where to best utilize Amateur Radio.

Many drills and exercises involve damage to infrastructure. A local emergency management agency that can assess damage to its infrastructure and report that to NGOs, the National Weather Service if it is a weather related scenario, and to state and even federal emergency management allow for a better assessment for response and recovery efforts and can assist the NWS warning process if a weather-related scenario.

Using Amateur Radio to obtain that assessment and communicating that to the local EMA for distribution to other agencies can be a very important role for a local Amateur Radio group. In addition, assessing the extent of power outages, and access to stores and gas stations that are open can also be essential in a disaster scenario. Gaining practice in obtaining this information and making those assessments in a drill will allow for that information to be obtained that much more easily in a real disaster.

PUT YOUR EMERGENCY OPERATIONS CENTER ON THE AIR

How often is the Amateur Radio station at your local emergency operations center on the air? Only during emergencies? For nets? For training? In many cases, the EOC Amateur Radio station may only see activity when it's needed, but this really shouldn't be the case. The EOC Amateur Radio station plays two critical roles. First, it is the station you will rely on during an emergency. It is of significance not only to the operator at the EOC, but to all those communicating with your EOC via Amateur Radio. Second, it is an Amateur Radio showcase to present to your served agencies.

Putting this station on the air — not just regularly but as often as possible — accomplishes several things. Each time you are on the air, you are training. You are learning about propagation, band conditions, and improving your operating skills. In doing this, you also learn more about your station and its equipment — its strengths, weaknesses, and its capabilities. Additionally, you learn to identify problems in the station. Through regular activity, you will develop a baseline knowledge of how the station should perform. When something goes wrong, you will know it quickly and be better prepared to fix the problem.

This on-the-air activity also has other potential benefits. As you and your group spend more time exercising the station, your served agency will take notice. They will see that the station has value, and may even take interest in how well it is performing. And don't forget that through activity, you are given the chance to promote the Amateur Radio Service. Remember, there's more to it than emergencies and public service; don't miss an opportunity to show off the other facets of this great service.

The opportunities to get on the air are diverse. Your group could make it a goal to add an operating achievement to the wall, such as DXCC or Worked All States. Participating in a contest is a great way to hone your operating and traffic handling skills (a contest exchange is traffic!). It also provides a great way to test your station's capabilities. You can also design a friendly in-house competition between operators, perhaps to see who can make the most QSOs each month, or log check-ins to HF nets.

Never forget what truly makes Amateur Radio a great asset: the spectrum to provide communications. We have this spectrum because we use it, not because we talk about using it. As an Amateur Radio operator, you should get on the air as much as possible — and so should your EOC station.

Get your EOC station on the air as often as possible, via contesting, training, nets, and other activities. Here, Andy Zajac, N1ORK, is making a contact on 15 meters from Manchester CERT, K9OEM, in Manchester, Connecticut.

PARTNERING WITH CERT

Local emergency management may also have an active Community Emergency Response Teams or CERT. Many times, CERT conducts their own exercises, working closely with their local Emergency Management Agency. If a strong relationship is forged between the local EMA and the Amateur Radio group, it may extend to the local CERT, offering additional opportunities for practice drills and exercises.

In areas where CERT is active, the local Amateur Radio group has a unique opportunity to potentially recruit and train CERT members, and Amateur Radio Operators have an opportunity to learn from CERT members how to expand their skill set, making them more versatile in a disaster situation. Many successes with CERT involve the local Amateur Radio group offering a licensing class as well as training on how to use Amateur Radio and giving Amateur Radio emergency communications classes on net operations, Go-Kits, and the National Traffic System (NTS). Classes can also include digital modes such as packet, *Winlink*, *Fldigi*, and PSK31 as well as voice linking modes using Echolink, IRLP, and D-STAR. It is very important that the Amateur Radio group offers training above and beyond licensing. Involving CERT members in drills and exercises in a communications capacity, as well as involving them in other facets of Amateur Radio, including technical projects and other operations, can make CERT members who are Amateur Radio operators truly valuable during a disaster.

An additional way that a local Amateur Radio group can prove its worth to local emergency management is to offer a course on emergency communications for non-hams to the CERT members. Often, these groups receive two-way radios during deployments but have not had any training in their use or in proper radio procedures. These classes do not appear to be a regular part of the basic CERT Training and the material for these classes is not provided in the national CERT training despite it being offered in the national CERT Handbook. While providing the CERT members with valuable radio procedural training, it also offers the opportunity to introduce the advantages of moving on to Amateur Radio training.

COMMUNICATIONS CAPABILITIES OF THE EOC

For the local EMA, an EOC that is complete with various modes of Amateur Radio capability can help with the overall operation and connection between the local Amateur Radio group and the EMA. Having the key capabilities of VHF/UHF, HF, digital modes, and Internet voice linking modes can help maximize the EOC capability for Amateur Radio and can allow Amateur Radio to be there when all else fails. Quite often, funding is available for equipment, but operators have also donated equipment to their local EMA.

During a disaster-related incident, a local Amateur Radio group that has trained and has been involved in prior events and exercises with its local EMA will be able to perform many functions for the EMA. This can include interoperability between local agencies, communications to regional or state government, and communications with NGOs. In addition, the local Amateur Radio group can also provide situational awareness and disaster intelligence information by the natural dispersion of Amateur Radio Operators over a given area. This information can be very useful to the local EMA and gives the Amateur operators another role to fill if other communications means are still up and running.

Amateur Radio operators can be of more value to an EOC and local EMA deployment operation when they can support communications that don't involve Amateur Radio. This can include technical support and utilization of local EMA radios that may operate on frequencies outside of Amateur Radio, as well as web-based EOC programs that are used to facilitate communications between the local EMA and regional and state EMAs, as well as other agencies. There may also be additional digital modes that can be used outside of Amateur Radio. Knowing how to use these modes would make local amateurs versatile and valuable. Instead of being seen as Amateur Radio operators, they could be seen — and utilized — as emergency communication technicians who happen to use Amateur Radio.

The Amateur Radio Public Service Handbook

PUBLIC SERVICE
Emergency Communications

Putting Amateur Radio in Context in the EOC

RICK PALM, K1CE

In the October 2007 "Public Service" column, I took readers on a tour of a typical, modern county EOC to foster understanding of the EOC work environment.[1] The article appeared a scant 2 years after Hurricane Katrina, the costliest natural disaster of all time and one of the five deadliest storms in recorded history. Consequent changes in the emergency management and public safety arenas were inevitable and I decided to interview the two principals of the Flagler County, Florida, EOC team again to see how these changes have affected them and their EOC operations.

Troy Harper is emergency management chief and Bob Pickering, KB4RSY, is the emergency management technician. Both are veterans in their positions with vast experience in all aspects of the EOC and public safety. Pickering is a communications specialist and a former County Employee of the Year. If I were to pick two words to describe the pair, they would be "dedicated" and "enthusiastic."

Flagler County has a population of 90,000 with 19 miles of exposed coastline on the upper east coast of Florida. It has a western rural aspect with farms and forests. Although the Emergency Services Department was forced to cut back its staff from 18 to 12 as many state governments and agencies have slashed budgets since the 2008 economic crisis, Harper has managed to increase its funding through FEMA grants, which have allowed him and his team to effect the enhancements discussed below.

As far as telecommunications is concerned, two words were first off their lips: interoperability and redundancy. Since Katrina the mantra in the field has been "let's get to where we can talk to each other," which applies to both interagency communications and also to intra-EOC functioning. Their goal has been to "patch" communications systems together so that talk across system, function and agency is seamless, regardless of the radio or Internet service employed. For example, in the EOC's Public Safety Answering Point

Flagler Emergency Management Chief Troy Harper (right) and Emergency Management Technician Bob Pickering, KB4RSY, at the Flagler County, Florida Emergency Operations Center.

(PSAP) dispatch center "E-911," Harper had their system join the *Florida Interoperability Network* that uses voice over Internet protocol (VoIP) for instantaneous communications networking with other public safety agencies throughout the state.

Interconnecting Agencies

The EOC's primary workhorse and backbone for EOC operations remains their robust, hardened analog/digital 800 MHz trunking system, with many enhancements effected in the post-Katrina years. One of these enhancements is more physical antenna sites throughout the county, five in total, with another single site backup. The trunking system provides all communications between the EOC and those county government officials and workers involved with the various emergency support functions (ESFs). The system is also "hard-patched," that is, linked into the more old-fashioned but tried and proven VHF FM system for redundancy and also communications with Fire/Rescue and the Department of Forestry, which operate primarily using this mode.

The EOC's pagers are also on this VHF system, which has a single repeater and back-up simplex capability. Pickering picked up his beaten-up, heavy-duty Maxon handheld transceiver and beamed as he demonstrated instant communications across the entire Flagler EOC grid of radio and Internet systems. He can communicate with other EOCs, agencies and functions on his handheld transceiver.

For public alerting, the EOC employs the *CodeRed* notification system (like reverse 911) where citizens are called with warnings. The system is able to strip severe weather warnings from the NWS and call residents in the warning area immediately with advice like "Get under the bed now." The EOC also has AM and FM broadcasting facilities and is on the DHS/FEMA-sponsored National Warning System (NAWAS), an automated telephone party line to more than 2000 EOCs around the country.

The EOC also has access to the Shared Resources HF Radio Program (SHARES), an HF system sponsored by the National Communications System, with which the ARRL has a formal memorandum of understanding. It promotes interoperability between HF radio systems used by the Federal departments and agencies. "This role has taken on added importance with the widespread purchase and use of automatic link establishment (ALE) technology throughout the HF radio community," according to the NCS website.

For communications with the Florida state mega-EOC facility at Tallahassee, Harper and Pickering just pick up the phone (fancy name: the Public Switched Telephone Network or PSTN). If the landline is out or overloaded, the EOC relies on *EMnet*, (Emergency Management Network), a satellite-based emergency messaging system serving state and municipal government emergency operation centers, police, firefighters, broadcasters, hospitals and other organizations. It's a voice/data over IP system that is monitored at the Flagler EOC 24/7 and tested daily.

Formerly, the Auxiliary Communications room at the EOC featured a full HF Amateur Radio station, fixed mobile VHF/UHF FM radios and a bay of docked dual-band mobile radios on desk tops. Now, that equipment is sorted by type and kept and maintained in Pelican cases ready for instant deployment to the field to be operated by registered and certified amateur teams. More on this program later. The EOC also relies on the General Mobile Radio Service (GMRS) on UHF FM, CB

---

[1] R. Palm, K1CE, "A Tour of a Modern County EOC," *QST*, Oct 2007, pp 77-78.

Rick Palm, K1CE ♦ 31 Burning Ember Ln, Palm Coast, FL 32137 ♦ k1ce@arrl.net

Reprinted from the February 2012 issue of *QST*.

Local ham hero Bob Pickering, KB4RSY, veteran EOC communications specialist and ardent supporter of the local volunteer radio communications groups. The local Amateur Radio community has been fortunate to have Pickering, a former County Employee of the Year, as an advocate.

A photo of the operations room, with a desk for each of the Emergency Support Functions (ESFs).

REACT and others for communications with volunteers in the field in a disaster scenario. The EOC actively maintains dual-band, multimode radios that are monitored constantly for situational awareness with GMRS, the airport and the marine environment.

Turning the corner into the main operations room, where each ESF has a desk, computer and communications systems, the specialists pick up the phone first. If they are not working, they can pick up a deployable ground-based voice/data/Internet capable satellite phone (*TracStar*), similar to the old INMARSAT units, only with greater functionality and lower cost. Communications with the state EOC can also be achieved through the *ESATCOM* satellite system combined with land mobile system connectivity.

I asked the pair about any secret communications systems. They said they couldn't tell me.

The EOC runs *E Team* software for incident management and also *WebEOC*, another web-enabled crisis information management system that provides secure real-time information sharing with other EOCs to help managers make sound decisions quickly. Their *EMWIN* weather system described in 2007 has now been replaced with the weather function of the *EMnet* program. According to the *EMnet* website, "Emnet's most significant benefit is that it provides the EOC with a single, efficient and effective interface for all inbound hazard notices, and to all outbound warning systems. Over the last few years, EOCs have experienced an increase in the demands to continuously sort through e-mail, monitor numerous websites, radio and telephone networks, and to watch fax machines for urgent messages from many sources . . ."

New Systems Being Evaluated

Harper and Pickering are evaluating "new" communications options, modes that have been in traditional use by radio amateurs for a long time: SSTV, APRS and burst messaging systems (like packet). The pair is also looking to new cellular broadband networks, especially *LTE Advanced*, a new standard for wireless communication of high-speed data for mobile phones and data terminals, that will allow more interoperability with handheld radios and cell phones for communications with other EOC field operators in the region.

Amateur Radio in the EOC

In 2010, Flagler County Emergency Services, the governmental agency responsible for the management of the large EOC, elected to change the way it coordinates with volunteers, including several citizen-based emergency communications groups. Instead of having volunteer communicators and operators serve the EOC via liaison with leaders of the volunteer groups, emergency management now recruits, selects, registers and manages the volunteers directly.

Each volunteer applies and is trained for specific duties under the direct supervision of EM officials. The Flagler Emergency Management Volunteer (FEMV) program is open to all residents of Flagler County. All volunteers in this organization are trained, issued uniform shirts and an identification badge. FEMV members will be under the direction of Flagler County Emergency Management for preparedness, response, recovery and mitigation efforts.

The program now boasts 85 members, even before a public roll-out of the program expected soon. There are several units within the volunteer auxiliary: E-Comm, Training, Marketing/Recruitment, Logistics and Landing Zone. There are currently six members of the E-Comm unit, who are all radio amateurs, GMRS licensees and trained SKYWARN spotters. More E-Comm unit members are expected. The E-Comm unit is responsible for providing auxiliary communications support in the event of a disaster, under the direction of the EOC and under the umbrella of NIMS/ICS protocols. The EOC also trains CERT teams throughout the county, which use mostly GMRS and Family Radio Service (FRS) radios for communications.

Requirements for membership in FEMV include the FEMA ICS courses IS-100 and IS-700, on ICS and NIMS protocols. Volunteers are credentialed and "typed" or classified by their training certifications and experience and placed in a database so that as a situation develops, the EOC can alert the appropriate type of volunteers needed.

Harper and Pickering on Amateur Radio

I asked the pair about what they see as the most important elements necessary to keep Amateur Radio useful and relevant in today's continually evolving EOC environment: "Stick to Amateur Radio's core values of simplicity and on-the-fly innovation, while not losing sight of new technologies like D-STAR," they said. Harper and Pickering also said that Amateur Radio is their "When All Else Fails" system, but with the interoperability, redundancy and hardening of their own systems, the likelihood of all else failing is remote.

They said that radio amateurs can increase their value to emergency management by branching out and broadening their training and capabilities as volunteers into other areas besides just radio communications: "Gone are the days when a radio amateur just sits at a table with his handheld in front of him waiting for messages to be handed to him for relaying, and no other function," they said. "The bottom line is, the EOC wants people cross-trained for the fastest, most effective response to save lives and property as possible. The more hams can contribute to this effort, the more valuable they will be," Harper and Pickering concluded.

Reprinted from the February 2012 issue of *QST*.

Part 3: Training and Readiness

Chapter 7
Leading and Training Volunteers

How often have you heard people refer to someone as "a natural-born leader?" From the time you were a child playing in the park, groups of people have gravitated toward individuals who get things done and are fun to be with. Many people are innate leaders; in others, this skill must be nurtured. Your role as a member of a local Amateur Radio Emergency Communications volunteer team requires leadership skills as you provide training to your volunteers and guide them through both drills and actual response events to support the mission.

Among the most important assets individuals have to offer are time and effort. Amateur Radio thrives on people helping each other for the good of the community, domestically and internationally. Hams volunteer! Your volunteers will likely represent all ages, genders, skill levels, and backgrounds. You may have youths as well as adult retirees; people with limited educational skills to professionals including doctors and lawyers. The most effective leaders can meet the challenge of dealing with people of any age and background.

CHARACTERISTICS OF LEADERS

To understand leadership, one has to understand power. The old adage, "he who pays the piper calls the tune" applies. Whoever is in a position of power, such as the owner of a company, makes the final decisions and is the ultimate authority. The military uses the power of position to maintain order, as that organization deals with personnel numbering in the hundreds of thousands. Without an organized structure, their mission would not be accomplished in an orderly manner. Leadership by position is hierarchical, and is learned from an early age, as is the case when a parent says to a child, "Because I said so!"

A leader can also be selected by a group decision, perhaps by a board of directors electing the chairman of the board, a radio club electing its officers, or a nation electing a president. In each of these cases, the group identifies an individual who they believe can accomplish a task for a period of time. This type of leadership may be received through a formal or informal process. We are all familiar with the formal process, at a town meeting or the ballot box. The informal process is a more intuitive one, affecting us more directly in our everyday lives. Often, members of a group recognize an individual as someone who has traits they wish to emulate. This individual may or may not end up with a formal leadership position, but other members of the group go to that individual for guidance willingly.

Leadership by event may happen when unforeseen circumstances occur. In many cases a member of a group identifies a situation that calls for immediate attention and rises to the occasion. The tragic events of September 11 saw many ordinary people draw upon their innate leadership abilities to assist groups of people they did not know, who willingly took their direction. These individuals were not asked to do the job; they simply saw a need and wanted to do the right thing, which happened to involve taking command of the situation.

The factor that separates good leaders from poor ones is the respect of those with whom they interact. This respect is earned through diligence, clarity of vision, and the ability to effectively work with people. In this way, a leader will put together a group that will be more responsive, more creative, and more willing to go the extra mile to get the job accomplished according to high standards. Poor organizations are characterized by personnel who may do their jobs, but not a bit more. Their volunteers won't return if the environment feels cold and a smile is never seen.

What separates the leaders from the group is the ability to focus on the mission and the management of people. Leaders set realistic goals and build a team that works well together. As a leader, you are in charge of the overall success of a mission. Remember, the leader may be in charge, but the end result is always a team success. Leaders make decisions. They may take input from a variety of sources, but, ultimately, they lead. True leaders don't take a vote on every decision, as this would cause gridlock and nothing could be accomplished in a timely fashion.

Leaders set the example for others to follow. If a leader shows up looking disheveled and acting frazzled or seeming unprepared, he or she will not inspire confidence. Leaders demonstrate a calming, steady, positive influence through their presence and their actions.

WORKING WITH SERVED AGENCIES

Depending on the scope of the mission, you may be dealing with a handful of volunteers or several served agencies. The process of allocating volunteer resources needs to be flexible as events evolve. The motto "Semper Gumby"* — always flexible — applies.

You can't be fuzzy about the mission. Make sure that the goal is clearly understood. For example: Heavy rain is expected. You are tasked to set up, staff, and provide communications from Shelter A to the local EOC. This is more specific than just stating, "Go to a shelter," due to the many questions that arise. Which shelter? Will this assignment require VHF/UHF and/or HF capabilities? What is the problem to be addressed? This may affect which equipment you will need to take with you.

While you may be the leader of your group, you are one among many to the next step up in the chain of command. A leader must be able to deal professionally with those above and below his position. As a leader you represent your team, organization, and Amateur Radio to others.

To assist you in the initial stages of the leadership process, make yourself familiar with established protocols. Know the plan, and develop and maintain associations with other agencies in the system. A leader will do a much more effective job if he or she comes into a mission already knowing who does what. This also helps to avoid possible political conflicts that could develop if Memoranda of Understanding (MoUs) are not understood. National organizations such as the ARRL and American Red Cross are members of the National Voluntary Organizations Active in Disaster (VOAD). VOADs also exist at the state and local levels. ARRL section leadership should be involved with the state VOADs, and where local VOADs exist, DECs and ECs should be involved. More information about the VOAD movement can be found at www.nvoad.org.

Leadership requires practicality and analytical thinking. If amateurs are called out to provide communications for a shelter, consider what steps need to be covered in preparation for accomplishing the assignment. This might include:

- Numbers of volunteers needed
- Estimated time frame for effort
- Established call-up procedure
- Alternative methods of team member contact if primary resources are unavailable
- Equipment that will be needed
- Special provisions that must be provided
- Alternative transportation needed if a volunteer does not have a way to get to the shelter
- Shuttles to be set up
- Designated parking area for volunteers

As you think through the response, you focus on your served organization as well as others in the mix. For example, the Red Cross is in charge of shelter management and will provide food, cots, and first aid. The Red Cross may have an MoU with the Salvation Army for the feeding aspect of the mission. The shelter has a back-up generator with three days of fuel on hand. How many operators will you need to maintain communications between these two points? What operator rotation will you employ? Do all operators have to be at the shelter at the same time? If this shelter has been utilized before, are there any materials already on site? Once ops start arriving, who sets up the antenna? Who sets up the station? Which operators work well together? Who has experience? Remember, questions are constant and changing, and definite, accurate decisions are required. In the end, you do the best you can and strive to improve.

GETTING THE BEST FROM YOUR VOLUNTEERS

A crew rowing a boat does its best when all the personnel manning the oars are synchronized. You want every member of the group to feel like they are an integral part of the mission's success. In order for your team to be successful in any task, the environment must be set up to encourage positive group dynamics, starting with a practical working area. Try to create the most user-friendly environment as possible. Holding a meeting in the middle of a parking lot during a hot Florida summer isn't very productive. Participants would rather be anywhere else, so their contributions will be minimal. It's a better idea to meet under a shade tree or pavilion, or in an air-conditioned building, with cold refreshments available. In a relaxing environment, you will have the attention of volunteers. They will concentrate and contribute more when not distracted by outside influences.

Can you identify which people in your group have special knowledge or skills such as station assembly, maintenance, or operating proficiency?

Having amateurs cross-trained in a variety of special areas is a big benefit when human resources are limited. For example, an antenna party is planned. Those who arrive are ready and willing to tug on the rope. While getting organized, it is discovered that no one is comfortable with heights, there is only one safety belt, and there's no gin pole. This is poor planning. Conversely, some groups have a tower team (**Figure 7.1**). They have experience and all of the special tools to make the task go smoothly: tower jacks, gin pole, rope, hardhats, approved safety harnesses and tools.

Be aware of any special needs that volunteers may have. To a large extent, volunteers will normally take care of their personal needs such as medical or dietary requirements. There may be other needs to consider. Some amateurs have physical limitations that may affect their assignment. Be practical when making assignments, and you'll find that these operators can be a real asset to the team. Care must also be taken when making pair or small group assignments. Be aware of pairings that may cause a potential problem due to personality conflicts or other elements. Some amateurs naturally mix together based upon a mutual interest, lifestyle, station location, long-term friendships, or other commonalities. Take

*"semper flexibilis"

Figure 7.1 — The teamwork demonstrated in this picture was essential for interoperational capability. This portable telescoping antenna/trailer unit was utilized by an agency in Florida for use at in Gulfport, Mississippi, during Hurricane Katrina. [JIM GERHARDT, WA3DIT]

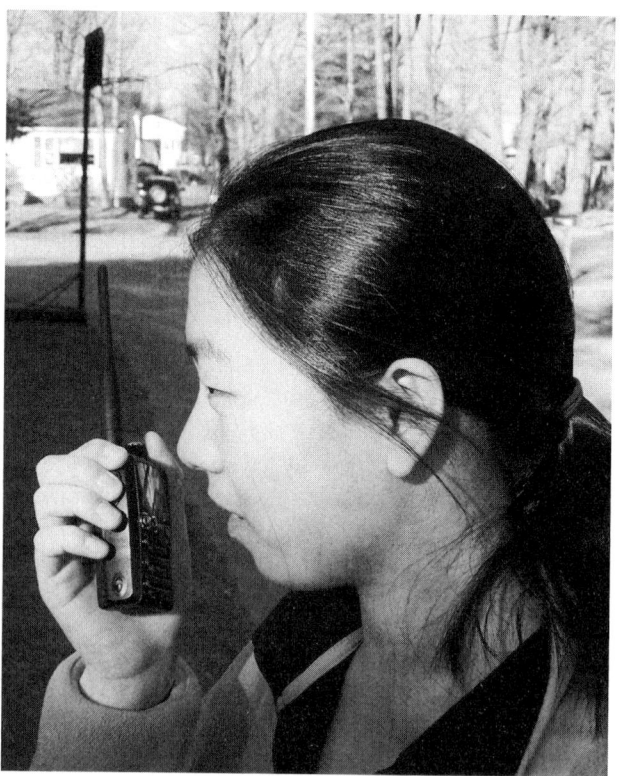

Figure 7.2 — Maximizing human resources is essential. The "many hands make light work" approach is definitely the way to go. An operator with less experience can handle traffic during a period of lower volume, with guidance from more experienced operators.

advantage of their natural chemistry. A little common sense goes a long way.

The matter of younger operators (**Figure 7.2**) requires more attention. Young operators may contribute in a productive manner. There are many examples of middle school students running NCS spots on NTS nets. Care must be taken to ensure that young operators are well supervised and never placed in any situation that may be even remotely dangerous. Certain work environments have age restrictions. Many EOCs require staff, even volunteers, to be 18 or 21 years of age, depending on policies or insurance requirements. In some organizations where youth are welcomed, it often helps if a parent or guardian accompanies the youth, participating in some way. Utilizing this approach will enlarge the pool of operators, support personnel, and create members for the future. Some groups require younger operators to have a permission-slip type of document, which may contain a waiver authorizing emergency medical care. Under no circumstances should alcohol be allowed at any event.

In any major situation, some volunteers seem to come out of the woodwork. They are well intentioned and may or may not have proper training and experience. How will you deal with spontaneous volunteers who are not trained or briefed? Though they may not be able to help out at the EOC or be given a shelter assignment, they may be able to help in other ways that are appropriate for a "beginner" volunteer. Everybody has to start somewhere.

Keep an eye out for volunteers among the group who demonstrate leadership qualities. Bring them over to the administrative side of the team slowly, sharing with them a variety of supervisory techniques so they may enhance their skills. As they feel more comfortable and demonstrate success, you may delegate to them ever-increasing responsibilities. During this mentoring process, you are building a team for the future, providing practical experience. In addition, this method provides a backup for key positions, should a person need to be replaced.

It's important to establish a warm rapport with volunteers, which can be achieved initially by calling people by their first names. Amateur Radio operators always use first names on the air, so this is a logical extension of your normal conversation. Never call upon an individual by their last name only. If using a last name, it should be preceded by Mr., Mrs., or Miss, to show respect for that individual.

If you need to address an issue with an individual, do so as privately and calmly as possible. Raising your voice initiates a no-win situation. It's likely that you'll lose that volunteer mentally, if not physically, and the other volunteers may lose respect for you. Treat your volunteers poorly and you'll be looking for other volunteers in the future. Speak to the person in a low voice, without stressful overtones, out of the sight and hearing range of others. No one appreciates being "spoken to" in front of others. Position yourself so your back is toward the door or wherever other people may be, so no one can hear what you are saying, read your lips, or interpret your facial expressions or body language.

Despite interacting on the air, some amateurs are uncomfortable in certain social or work situations. They may prefer contributing from their home stations, as an Official Relay, Emergency, or Bulletin station. Let them know it is okay if they prefer a more solitary assignment, but don't lose track of off-site hams, as that will really make them feel unappreciated. Remember, everyone can contribute to the mission. The object is to find the proper match of people, skills, and needs. Avoid micro-managing. Second-guessing every little thing done in the field will drive volunteers crazy and is unproductive. Give volunteers an opportunity to demonstrate their skills. Let them address issues in their immediate control independently, within the big picture of events.

Praise volunteers as much as possible. Be sincere, as a phony comment can be spotted a mile away, and sarcasm is always unacceptable. Why would any volunteer return in the future if they were made to feel incompetent? Even in a negative situation, you may be able to find a glimmer of light, so look for it!

Having a sense of humor goes a long way in making the time go by quickly. In a serious emergency, this may be dark humor, but it still reduces tension. Be mindful of the difference between humor and pranks, or practical jokes. The latter has no place in the process. Just a smile can help to decrease your stress level and that of others around you.

Don't be afraid to ask for assistance from your volunteers. The fact that you relied on their information will enable team members to feel like they are contributing to the mission. For example, if you have a computer issue and one of your volunteers has computer experience, take advantage of that knowledge. This action will make that volunteer feel important because he or she has been able to contribute to the mission in a special way.

THE IMPORTANCE OF FORMAL TRAINING

No single area of planning and preparation is as important as training of volunteers in emergency communications techniques and practices. Today, in addition to the technical aspects of operating sophisticated radio equipment, training in the various aspects of inter-agency and inter-organizational working relationships has become essential. Amateurs must understand the basics of emergency communications. Training comes in many forms. Formal classes sponsored by Amateur Radio clubs usually include radio operator licensing. The emergence of Community Emergency Response Teams (CERT — **www.citizencorps.gov/cert/**) in many communities has proven to be a natural fit for Amateur Radio and emergency training. Start with the ARRL's online courses in emergency communication at **www.arrl.org**.

Local/Area ARES/RACES groups often offer support for ARRL courses, such as their EC-001, or weekend seminars focused specifically on emergency communications. These rotating seminars provide information at various levels, from beginning volunteers to experienced management.

The new National Incident Management System (NIMS) is the language and operating plan of all prevention and response organizations. Knowledge of NIMS and training and certification in Incident Command System (ICS) practices is essential and, for some served agencies, required of volunteers. As discussed later in this chapter, ICS/NIMS courses, IS-100, 200, 700, and 800 are essential for Amateur Radio emergency communications work.

Demonstrations, such as building and erecting emergency antennas, soldering, displays of go-kits and mobile emergency vehicles and command centers often generate a lot of interest.

Types of Exercises

Local/Area offices of emergency management offices have very detailed Emergency Operating Plans (EOPs). As part of their planning, they conduct different types of drills. Not all drills are conducted annually, but rather on a rotational basis over a number of years. The annual ARES Simulated Emergency Test (SET) allows radio operators to hone their skills from portable, fixed, or mobile locations.

There are five main types of exercise activities that planners can use to validate policies and plans and train personnel.

Orientation is an overview or introduction. It is the least complex type of exercise, and its purpose is to familiarize participants with roles, plans, procedures, or equipment. It can also be used to resolve questions of coordination and assignment of responsibilities.

A *drill* is a coordinated, supervised exercise activity, normally used to test a single specific operation or function. During a drill, there is no attempt to coordinate organizations or fully activate a center. The purpose of a drill is to practice and perfect one small part of the plan and prepare for more extensive exercises, in which several functions will be coordinated and tested. A drill puts a tight focus on a potential problem area.

A *tabletop exercise* is a facilitated analysis of an emergency situation in an informal, stress-free environment. It is designed to elicit constructive discussion as participants examine and resolve problems based on existing operational plans and identify where those plans need to be refined. The success of the exercise is largely determined by group participation in the identification of problem areas.

A *functional exercise* is a fully simulated interactive exercise that tests the capability of an organization to respond to a simulated event. The exercise tests multiple functions of the organization's operational plan. It is a coordinated response to a situation in a time-pressured, realistic simulation.

A *full-scale exercise* is the most complex (and costly) exercise. Its purpose is to simulate a real event as closely as possible. It is an exercise designed to evaluate the operational capability of emergency management systems in a highly stressful environment that simulates actual response conditions. To accomplish this realism, it requires the mobilization and actual movement of emergency personnel, equipment, and resources. Ideally, the full-scale exercise should test and evaluate most functions of the emergency management plan or operational plan.

For further discussion of exercise planning, review FEMA Independent Study courses IS-120, IS-130, and IS-139.

Other community entities conduct drills periodically, as airports, hospitals, industries and schools all have emergency plans that have to be tested. Amateur Radio may be a part of the total plan. Keep in mind that many of these events occur during the work week, when volunteers aren't always available in large numbers.

Often amateurs are contacted by local civic organizations to provide primary or backup communications, depending on the event size. These activities not only provide a service to the community, but provide an opportunity for volunteers to use and develop their skills. Providing these services requires proper planning. Be sure to acquire all of the pertinent information from the agency to be served. Once the planning has begun, think not just about providing a service, but training less-experienced volunteers by placing them with veterans or at certain locations.

Other groups with whom Amateur Radio has had a long association, such as SKYWARN and the American Red Cross, conduct drills periodically.

Training Courses

It is imperative that volunteers actively participate in training. Consider having some sort of training meeting for your group at least once a month. These meetings may be formal or informal, depending on the size of the group. Schedule a rotating series of topics, designed to meet the needs of the group over time. This way at meetings, older members have a review and newer members are brought into the overall process. In larger groups, a common session may be held, followed by breaking into groups for specific discussions and study groups.

Training courses are available from three primary sources.

The ARRL

EC-001 Introduction to Emergency Communications. This revision of the former Emergency Communications Basic/Level 1 course is designed to provide basic knowledge and tools for any emergency communications volunteer. The course has six sections with 29 lesson topics. It includes required student activities and a 35-question final assessment, and is expected to take approximately 45 hours to complete over a nine-week period, with the student working at his or her own pace. Using this course, enrolled participants will be working with a mentor to monitor progress and provide feedback. While this course is intended for home learning, a local leader can provide motivation, be a resource, and provide local study group opportunities, if needed.

EC-016 Public Service and Emergency Communications Management for Radio Amateurs. This course is designed for Amateur Radio operators who will be in leadership and managerial roles organizing other volunteers in the support of public service activities and communications emergencies. The course is designed to provide a practical application of management theory, and teaches how to work in coordination with governmental and other emergency response organizations, deploying their services to provide communications in an emergency. After completing the final assessment for this course, the student earns a certificate of completion and should be ready for leadership roles in situations in which lives and property are at stake. Course prerequisites include *EC-001 — Introduction to Emergency Communications, ICS-100b — Introduction to ICS; ICS-200b — ICS for Single Incidents;* and *ICS-700 — Introduction to NIMS.*

Federal Emergency Management Agency (FEMA)

The Emergency Management Institute's (EMI) Distance Learning Branch offers an Independent Study Program, a distance learning program which offers free training to the nation's emergency management network and the general public. FEMA's Emergency Management Institute offers more than 100 training courses via the training site: **http://training.fema.gov/**. A certificate is issued to the student upon completion of each course. Courses with "IS" designators are offered online. These three courses are considered fundamental for volunteer development: *ICS-100 Introduction to Incident Command System, ICS-200 ICS for Single Incidents and Initial Action Incidents,* and *IS-700 NIMS: An Introduction.*

The following courses are important for volunteers pursuing *NIMS ICS-300,* as well as for anyone with specific functions within your local organization:

THE INCIDENT COMMAND SYSTEM

The Incident Command System (ICS) is an important organizational and management structure encompassed by the larger National Incident Management System (NIMS). Both are commonly used throughout the United States during national and local emergencies.

According to the website of the Federal Emergency Management Agency (FEMA),

The Incident Command System (ICS) is a standardized, on-scene, all-hazards incident management approach that
- *Allows for the integration of facilities, equipment, personnel, procedures, and communications operating within a common organizational structure.*
- *Enables a coordinated response among various jurisdictions and functional agencies, both public and private.*
- *Establishes common processes for planning and managing resources.*

Considered one of the most important best practices incorporated into the NIMS, the ICS brings multiple responding agencies, including those from different jurisdictions, under a single overall command structure to maximize efficiency.

ICS Beginnings

The ICS grew out of lessons learned from the chaotic, ineffective multi-agency response to wildfires that devastated Southern California, in the early 1970s. Today, every public safety organization, including fire, police, and other agencies, uses the ICS in responding to the call for their services. Most non-public organizations and served agencies use it as well, and it has been adapted for use in many other countries.

Prior to the widespread adoption of ICS principles, various agencies responding to a disaster often vied for control, duplicated efforts, and overlooked critical needs, reducing the potential effectiveness of the response. The ICS model, however, places the system's functional areas under the direction of a single Incident Commander (IC), allowing the response to be conducted in a coordinated manner, even when it involves multiple agencies and crosses jurisdictional lines. Within the ICS, each agency must recognize one "lead" coordinating person who handles one or more tasks that are part of a single overall plan and who interacts with other agencies in defined ways.

In large-scale responses, the IC may have the assistance of a Command Staff, consisting of Public Information, Safety, and Liaison Officers. In smaller incident responses, the IC may perform the duties of one, two, or all three of the Command Staff positions, if they are required at all.

Tasks within the ICS are subdivided into five major operating sections, or functional areas: Command, Operations, Planning, Logistics, and Finance/Administration (Intelligence/Investigations is an optional sixth). Again, what is activated is based on the particular incident needs.

Each of these operating sections, if activated, has its own Chief, who may also oversee various task forces. For instance, a Logistics Section Chief often handles the coordination of all interagency communication infrastructures involved in the response, including Amateur Radio when it is used in that capacity.

The Bigger Picture

The ICS is a subcomponent of the NIMS, which was established by Homeland Security Presidential Directive (HSPD)-5, Management of Domestic Incidents. According to FEMA, NIMS,

...provides a systematic, proactive approach to guide departments and agencies at all levels of government, nongovernmental organizations, and the private sector to work seamlessly to prevent, protect against, respond to, recover from, and mitigate the effects of incidents, regardless of cause, size, location, or complexity, in order to reduce the loss of life and property and harm to the environment.

NIMS comprises five major components: Preparedness, Communications and Information Management, Resource Management, Command and Management, and Ongoing Management and Maintenance. ICS, along with Multiagency Coordination Systems and Public Information, falls under the Command and Management component, which is designed to enable effective and efficient incident management and coordination by providing a flexible, standardized incident management structure.

Where Does Your EmComm Group Fit In?

Involvement in any incident where ICS is used is strictly by "invitation only," offering no role for "off-the-street volunteers." Should your group be invited to respond to an emergency, its place within the ICS will vary according to the specific situation. Depending on the situation and the degree of ICS structure being employed, the choice of using your group's services may be made by the served agency, the Communications Task Force leader, Logistics Chief, Incident Commander, or EmComm leadership itself. Your group might begin by supporting its own served agency, then end up supporting a new and unfamiliar one. It may find itself serving multiple agencies and see its mission change multiple times as the responsibility for incident management shifts from one agency to another.

> **AMATEUR RADIO EMERGENCY COMMUNICATION FIELD INSTRUCTORS**
>
> The ARRL's program of Amateur Radio Emergency Communication Field Instructors (formerly known as Certification Instructors, or CIs) certifies volunteers in offering instruction of the ARRL Amateur Radio Emergency Communications Course. Currently, Field Instructors may conduct classroom instruction covering the material in the Basic/Level 1 course.
>
> The League recommends that Field Instructors are appointees of ARRL, ARES, RACES, Red Cross, REACT, CERT, etc., with these organizations providing approval of the instruction and training program. Frequently ARRL Section Managers, Assistant Section Managers, Emergency Coordinators, and Assistant Emergency Coordinators conduct field instruction or are closely involved in the field instruction conducted within their section.
>
> Instructors should explain how the content of the course fits with the kinds of emergency responses one might be called upon to engage in their local community. Instructors should familiarize the students with the emergency responders in their community, include practice activities to illustrate the course content, and adapt the activities to the local situation to make them as practical and relevant as possible. The objective is for students to gain practical knowledge as well as the hands-on experience that is crucial to being prepared for local activation.
>
> Field Instructors must be registered with the ARRL Continuing Education Program and must meet certain requirements. All volunteers who wish to serve in the capacity as a Field Instructor with ARRL's Emergency Communications training program must renew their registration with ARRL. You'll find a form at **www.arrl.org/application-for-emcomm-mentor-or-field-instructor** where you can submit the necessary information. Please note: volunteers who have completed the *Public Service and Emergency Communications Management (EC-016)* course have met all of the requirements to serve as a Field Instructor.
>
> Field Instructors are asked to register their planned classes on the ARRL website so that interested hams in the local community may find out about them.

IS-800 National Response Framework
IS-240 Leadership & Influence
IS-241 Decision Making & Problem Solving
IS-250 Emergency Support Function 15 (ESF-15) External Affairs
IS-1 Emergency Manager: An Orientation to the Position
IS-288 The Role of Voluntary Agencies in Emergency Management
IS-244 Developing and Managing Volunteers
IS-120a An Introduction to Exercises
IS-130 Exercise Evaluation and Improvement Planning
IS-139 Exercise Design

In many cases, you can work on these courses as a study group. Local topics should include community issues such as MoUs, location of resources, and subjects that provide an overall general knowledge of the EOC operation, plans, and procedures.

Served Agency Training

If the served agency offers Emergency Communications volunteers job-specific training in areas related to communication, take advantage of that. Your EmComm managers should help you to learn how the served agency's organization works. Learn their needs and how you can best meet them. Work within your own organization to get any additional training or information you might need.

For instance, the American Red Cross offers self-study or classroom courses in mass care, damage assessment, and other areas that either directly involve or depend upon effective communication. The Red Cross offers a course, *Introduction to Disaster Services*, which is encouraged for those amateurs working with the Red Cross.

Many emergency management agencies offer additional training in areas such as radiological monitoring, sheltering, mass casualty response, and evacuation. Your own group may offer general or agency-specific training in message handling and net operations under emergency conditions. For example, the Eastern Washington State ARES and Western Washington State ARES organizations combine resources and forces for many ARES functions, including training. They sponsor a "communications academy" open to amateurs with an interest in emergency communications. At the time of this writing, a course being offered was *A Review of the Honshu 9.0 Earthquake and Tsunami, and Potential for a Similar Event in the Pacific Northwest*.

If your group has its own equipment, it should offer opportunities for members to become familiar with its setup and operation in the field. On your own, set up and test your personal equipment under field conditions to be sure it works as expected.

Experiential learning is the best way to gain and polish skills, so try to participate in any drills or exercises offered in your area. Some are designed to introduce or test specific

> **EMERGENCY TELECOMMUNICATIONS COURSE FOR YOUTH**
>
> The International Telecommunication Union and Telecommunication Development Bureau (ITU/BDT) has developed a basic online course on Emergency Telecommunications geared toward young men and women with no prior knowledge of emergency management or the technologies related to emergency warning and relief. The course is made available free of charge through the ITU's e-Learning Center, and gives students an overview of the concepts of emergency management and disaster response, the responsibilities of key institutions, and the role of volunteers. The material covers the following topics:
> - Introduction to Emergency Telecommunications and Disaster Management
> - Role of Emergency Telecommunications at various stages of Disaster Management
> - Building resilient telecommunication infrastructure for disaster mitigation
> - Multi-Hazard role of ICT in Disaster Management
> - Contribution of effective ICT for Disaster Mitigation to Economic and Social Development
> - Practical Measures to be taken when disasters strike to enhance response and relief.
>
> The course is available at **www.itu.int/ITU-D/youth/emergency_telecommunications/basic_course/**

skills or systems, others to test the entire response. ARRL's Field Day and Simulated Emergency Test are two good nationwide examples, but local organizations may have their own as well. See the Appendix of this book for a description of the TOPOFF 3 exercise, the largest single disaster drill ever to take place in the United States, with more than 10,000 participants from over 200 federal, state, local, tribal, private sector, and international agencies, organizations, and volunteer groups.

Independent public service projects may be a good exercise. Pre-installing antennas, such as the inexpensive roll-up J-poles, at shelters or other locations where volunteers are likely to be deployed, is a team-building project that could pay off in a later emergency situation — when an amateur shows up at that location, all they will have to do is hook up their gear. Training members can also be something as simple as encouraging hams to check into the National Traffic System (NTS) at least once a week to familiarize themselves with net procedures.

Guard against over-training. If you have full-blown activations every other week, volunteers will lose interest. Hams have families, jobs, and other obligations; don't overload them. Start on time and have a target end time. Ninety minutes is an adequate block of time allowing for sharing of information and a team-building activity.

During any post-event review, be sure to recognize your volunteers. Their input is valuable. They may see specific situations very differently. This approach will also make volunteers feel like an important part of the team.

Give credit to the volunteers, and associated groups, in all media interactions. Everyone likes to be appreciated. We all know stories of "glory hounds" that take credit for the work of others. Remember to share the satisfaction of a job well done with the volunteers who made it happen. Radio amateur volunteers don't get paid, but they are worth their weight in gold. If you treat them as such, with respect and care, they will be back — and bring their friends.

The Amateur Radio Public Service Handbook

Public Service

Rick Palm, K1CE, k1ce@arrl.org

Training and Certifications for ARES® Operators

If you don't know what you should know, you'll want to know about this Emcomm curriculum.

In the landmark report of the ARRL National Emergency Response Planning Committee filed with the League's Board of Directors in 2007, Chairman Kay Craigie, N3KN (now ARRL President) wrote for the panelists when she summed up the issue of training and certification for ARES® volunteers:

> For many years, Amateur Radio has longed to be taken seriously by governmental authorities as a professional-quality resource in disaster response. Although there are areas of the country where achieving and maintaining emergency management agencies' respect is still a struggle, Amateur Radio's service during 9/11 and the major hurricane disasters of the 21st century has brought us a new level of respect and new opportunities at the national level.
>
> Being taken seriously as a resource comes with a price, however. It is a price that must be paid by individual volunteers, not in dollars but in precious personal time. When the federal government instituted the National Incident Management System (NIMS), it imposed a set of requirements on state and local emergency management agencies and their personnel. Affected personnel included not only paid employees of emergency management and related agencies but also volunteers such as those in volunteer fire companies, ARES®, and RACES. If the emergency management agencies are to continue receiving federal funds, personnel must complete a number of FEMA training courses having to do with the Incident Command System (ICS) and NIMS. Individuals who do not complete the training will not be allowed to participate, even as volunteers.
>
> These FEMA courses are free of charge, available online or sometimes in person at emergency management offices, and not particularly difficult. The courses are useful in familiarizing volunteers with the specialized vocabulary and principles of the Incident Command System and showing where communications fits into the ICS structure. This is valuable knowledge, because if radio amateurs — particularly those in leadership positions — cannot "talk the talk," then authorities may well assume that we cannot "walk the walk."
>
> These formal requirements are here to stay and more may follow. At the national level, Amateur Radio has earned the respect we always wanted, bringing us closer to the emergency management establishment. The challenge now is persuading both casual ARES® volunteers and experienced volunteers to meet the requirements that follow from being part of the system. The national-level ARRL must be aware of that and develop ways to help local and section ARES® officials bring their volunteers, both old-timers and newcomers, into the new era.

Current Trends

Since the time of the report, the ARRL — along with the amateur community at large — has started to meet President Craigie's challenges and further embrace emergency communications and ARES®. This trend started with 9/11 and continued through Hurricane Katrina. There has been a concomitant rise in interest in the ARRL and FEMA courses by Emcomm operators, both serious and casual. Conversation on ICS/NIMS topics is now common on nets and in club meetings. The training scene has evolved rapidly in the past few years. ARRL HQ has ramped up its training resources and added a dedicated staff member for support.

In a recent survey by the ARRL Emergency Communications Advisory Committee (ECAC) of the ARRL Field Organization on ARES® topics, 55% of the ARRL sections require minimum training for active members. Of those sections requiring training, 38% require the ARRL EC-001, *Introduction to Emergency Communications* course; 75% require the FEMA IS-100, *Introduction to the Incident Command System*; 71% require IS-200, *ICS for Single Resources and Initial Action Incidents*; 67% require IS-700, *Introduction to the National Incident Management System* (NIMS), and 51% require IS-800, *National Response Framework* (NRF).

Of served agencies, it was reported that 78% require specific training of their volunteers. Seventy-five percent reported that most ARES® members are ICS trained.

Recommended Courses

I've run numerous recommendations in the *ARES® E-Letter* and have subsequently received more recommendations, which were then published, to enhance the value of our program and operators. In an effort to summarize these recommendations and give the ARES® operator some idea of useful courses to take, let's offer the following:

■ The ARRL EC-001 — This online course is designed to provide basic knowledge and tools for any emergency communications volunteer. Prerequisites include IS-100 and IS-700. The ARRL also recommends IS-250, *Emergency Support Function 15* and IS-288, *The Role of Voluntary Agencies in Emergency Management*.

■ American Red Cross or American Heart Association CPR and Automatic External Defibrillator (AED) courses — These courses are available at hospitals, colleges, and Red Cross offices and centers. Providing emergency communications in an actual emergency increases the likelihood of an ARES® volunteer having to assist someone needing CPR.

■ IS-100 — This course is a must-have not only because it is a requirement of most agencies, but because it imparts an understanding of the contemporary emergency management landscape. How can you function as a viable emergency communicator without a basic idea of what is going on around you on a disaster scene? Government agencies manage emergencies and disasters using the ICS as a standard playbook. You need to know how it works.

■ IS-700 — This course introduces NIMS, which serves as a "consistent nationwide

Reprinted from the March 2012 issue of *QST*.

template to enable all government, private-sector and nongovernmental organizations to work together during domestic incidents." The NIMS course is the other shoe for the government's emergency response framework and as such should be near the top of any ARES® volunteer's course list.

- IS-200 — According to FEMA, this course is "designed to enable personnel to operate efficiently during an incident or event within the ICS. IS-200 provides training and resources for ICS supervisory procedures."

- IS-230, *Fundamentals of Emergency Management* — Garth Kennedy, W9KJ, the emergency manager for the Naperville, Illinois EMA, recommends this course: "I manage a large emergency management agency. Most of our volunteers do not understand what constitutes 'Emergency Management.' As a result, we require IS-230 for any certification level in all of our specialties. I recommend adding this course to your list so ARES® operators will more fully understand the environment in which they work."

Mike Corey, KI1U, ARRL HQ's Emcomm planner and response manager, recommends the following core courses. For the rank-and-file ARES® field operator: ARRL EC-001, a basic SKYWARN class, IS-100, IS-200 and CPR/First Aid/AED.

For ARES® leaders including Emergency Coordinators, District ECs and Section Emergency Coordinators, ARRL HQ recommends ARRL EC-016, *Public Service and Emergency Communications Management for Radio Amateurs*, designed to train Amateur Radio operators for leadership and managerial roles organizing other volunteers. HQ also recommends an advanced SKYWARN class, IS-700, IS-800 and IS-802, *Emergency Support Function #2 – Communications*. And finally, Red Cross Disaster Services training is recommended, even if you do not work directly with the Red Cross. You should know how the Red Cross conducts field operations, an issue that was raised during the Hurricane Katrina mega-response.

Bottom Line

I recently toured a local EOC that is probably typical of most EOCs in size and functioning. The emergency manager turned on his projector to show me his database of volunteer resources: a large matrix of volunteers typed by their function, training in FEMA courses and others. It seemed to me, for the future of ARES®, the writing was, literally, on the wall.

OREGON ARES SHAKE EX 2011: AN EARTHQUAKE DISASTER SET

Vincent Van Der Hyde, K7VV, Oregon Section Emergency Coordinator, k7vv@arrl.net
John Core, KX7YT, Oregon Section ARES SET Coordinator, kx7yt@arrl.net

In the future a major earthquake will strike the Pacific Northwest. Within half an hour, a tsunami similar to the one that devastated Japan on March 11, 2011, will occur along the Oregon Coast.

On April 9, 2011, 1 month after the disastrous Japanese earthquake and tsunamis, Oregon ARES volunteers conducted a statewide simulated emergency test (SET) to determine their readiness to respond to just such a disaster. Although the SET was planned well before the events in Japan, the reality that similar events could happen in Oregon added to the realism of the exercise.

Geologists tell us that, historically, earthquakes in our region occur at 300 year intervals. Should Oregon be struck there would likely be catastrophic, widespread damage and an immediate need for ARES support.

The SHAKE EX 2011 SET was designed to test ARES ability to exchange very high volumes of written messages between the county Emergency Managers and the Oregon Emergency Management (OEM) office. Much of the radio traffic exchange occurred over the Oregon ARES Digital Network (OADN), which uses Winlink HF and VHF radio systems funded by the State of Oregon.

In addition to statewide communications activities, many counties held their own local drills in coordination with their local Emergency Managers, medical facilities and Community Emergency Response Teams (CERT). The local drills typically included HF radio systems at remote locations using portable antennas. Local drills included the transmission of photographs by radio to county and state EOCs and relaying simulated damage reports between stations.

About 130 members of Oregon ARES participated throughout the state sending and receiving about 2000 messages within the 6 hour SET period. Most of the traffic was sent by HF and VHF using the Winlink OADN system. During the height of the SET activity, OEM operators were receiving about one message per minute from ARES operators throughout Oregon.

The Oregon ARES Digital Network consists of HF radios equipped with Pactor 3 modems as well as V/UHF radios equipped with TNCs for local Winlink RMS gateways. Both radio systems are used with laptop computers loaded with Winlink *Airmail 3* software. With a few exceptions, each county EOC has an identical set of equipment. Training of ARES units receiving the equipment was completed in 2009 during installation. Since then, quarterly connectivity exercises and the twice-yearly statewide SETS have helped insure that the OADN system remains fully operational.

Lessons Learned

During a disaster of the scale anticipated during this SET, there will likely be an overwhelming volume of written and tactical traffic between emergency managers. At such times, it is essential that the flow of messages from ARES radio operators to and from these officials be accurate, efficient and timely. Although the technology used by ARES units to communicate worked well,

On the left, Don Kendall, N6VKW, Curry County (Oregon) Emergency Services Coordinator, is explaining to Bob Wilkinson, W7VN, ARES EC for Curry County the potential for transportation disruption from the probable collapse of the Patterson Bridge across the Rogue River in the predicted earthquake and tsunami.
[Lorraine Wilkinson W7RFC].

the flood of messages being received at many EOCs overwhelmed everyone's ability to log, manage and distribute them. Several options have been proposed to deal with this data management issue and they are discussed in the March issue of the *ARES E-Letter*.

The demonstrated ability of Oregon ARES volunteers to successfully support the large-volume written and tactical traffic demands of Oregon's emergency managers under the limited restrictions imposed during this SET were impressive. The tougher operating conditions imposed by a real disaster will prove to be the ultimate test, of course.

Chapter 8
Personal Safety, Survival, and Health

Disaster relief volunteers sometimes become so involved with helping others that they forget to take care of their own families and themselves. The needs of disaster victims seem so large when compared with their own, that volunteers sometimes feel guilty about taking even a moment for their own basic personal needs. However, if you are to continue to assist others, you need to keep yourself in good condition. If you do not, you risk becoming part of the problem. If your family is not safe and all their needs are not taken care of, worrying about them may prevent you from concentrating on your job.

HOME AND FAMILY FIRST

Before leaving on an assignment, be sure you have made all necessary arrangements for the security, safety, and general well-being of your home and family. Family members, and perhaps friends or neighbors, should know where you are going, when you plan to return, and a way to get a message to you in case of an emergency.

If you live in the disaster area or in the potential path of a storm, consider moving your family to a safe location before beginning your volunteer duties. Take whatever steps you can to protect your own property from damage or looting, and let a neighbor or even local police know where you are going, when you plan to return, and how to reach you or your family members in an emergency.

In addition to your deployment checklists, you might want to create a home and family checklist that covers everything your family will need while you are gone. Here are some ideas to get you started:

House
- Board up windows if you are in a storm's path
- Put lawn furniture and loose objects indoors if high winds are likely
- Move valuables from lower to upper levels of the home if flooding is possible
- Heating oil tanks should be filled
- Drain pipes if below-freezing temperatures and power loss are possible
- Shut off power and gas if practical and if structural damage is possible

Family
- Arrange for a safe place to stay if needed, preferably with friends or relatives
- Reliable transportation, with fuel tank filled
- Adequate cash for regular needs and emergencies (not ATM or credit cards)
- House, auto, life, and health insurance information to take along if evacuated
- Access to important legal documents such as wills, property deeds, etc.
- Emergency food and water supply
- AM/FM radio and extra batteries
- Flashlight with extra batteries and bulbs
- Generator, fuel, and safe operating knowledge
- Adequate supply of prescription medications on hand
- List of emergency phone numbers
- Pet supplies and arrangements (shelters will not take pets)
- List of people to call for assistance
- Maps and emergency escape routes
- A plan for contacting each other
- A plan for reuniting later

Should You Leave At All?

There are times when your family may need you as much or more than your emergency communications group does. Obviously, this is a situation that only you and your family can evaluate. If a family member is ill, if your spouse is unsure of his or her ability to cope without you, if evacuation will be difficult, or any similar concern arises, staying with your loved ones may be a better choice. If there is ever any doubt in your mind, your decision must be to stay with your family. This is also something you should discuss with your loved ones, and come to an agreement about well before any disaster, in order to avoid any last-minute problems.

SELF-CARE WHILE ON DUTY

Once you are working with your group, you will need to continue to take care of yourself. If you become over-tired, ill, or weak, you cannot do your job properly. If you do not take care of personal cleanliness, you could become unpleasant to be around. Whenever possible, each station should have at least two operators on duty so that one can take a break for sleep, food, and personal hygiene. If that is not possible, work out a schedule with the EmComm managers or your NCS to take periodic off-duty breaks.

Food

Most people need at least 2000 calories a day in order to function well. In a stressful situation, or one that involves a great deal of physical activity, you may need even more. Experienced EmComm managers and served agency personnel will usually be aware of this issue and take steps to see that volunteers' needs are met. If you are at a regular shelter, at least some of your nutritional needs may be taken care of. In other situations, you may be on your own, at least for a while. High-calorie, high-protein snacks such as cheese, jerky, or nuts will help keep you going, but you will also need food that is more substantial. You may need to bring along some freeze-dried camping food, a small pot, and a camp stove with fuel, or some self-heating, military-style "Meals, Ready to Eat" (MRE) packages.

Water

Safe water supplies can be difficult to find during and after many disasters. You will need at least a half-gallon each day just for drinking. In extremely hot or cold conditions, or with increased physical activity, your needs will increase significantly. FEMA recommends maintaining a three-day supply of water at all times in case of emergencies; plan for at least one gallon of water per person, per day.

Camping supply stores offer a range of water filters and purification tablets that can help make local water supplies safer. However, they all have limitations of which you should be aware. Filters may or may not remove all potentially harmful organisms or discoloration, depending on the type. Those with smaller pores (.3 microns is a very tight filter) will remove more foreign matter, but will also clog more quickly. Iodine-saturated filters will kill or remove most harmful germs and bacteria, but are more

---

SAMPLE PERSONAL SURVIVAL AND COMFORT NEEDS CHECKLIST
(Modify according to your own situation)

- ☐ Suitably sized backpack or duffel bag for clothing and personal gear
- ☐ Plastic storage tub for food and cooking gear
- ☐ Toiletry kit with soap, comb, deodorant, shampoo, toothbrush, toothpaste
- ☐ Toilet paper in zipper-lock freezer bag
- ☐ Small towel and washcloth
- ☐ Lip balm
- ☐ Facial tissues
- ☐ Sunscreen
- ☐ Insect repellent
- ☐ Prescription medications (one-week supply)
- ☐ Copies of medication and eyeglass/contact lens prescriptions
- ☐ Spare eyeglasses or contact lenses and supplies
- ☐ Hand lotion for dry skin
- ☐ Small first aid kit
- ☐ Non-prescription medications, including painkiller, antacids, anti-diarrheal, etc.
- ☐ Extra basic clothing: shirts, socks, underwear
- ☐ Gloves for protection or warmth
- ☐ Pocket flashlight
- ☐ Folding pocketknife
- ☐ Sleeping bag, closed-cell foam pad or air mattress, pillow
- ☐ Ear plugs (soft foam type in sealed package)
- ☐ Black eye mask
- ☐ Outer clothing for season and conditions (rain gear, parka, hat, face mask, etc.)
- ☐ Hardhat
- ☐ Reflective vest, hat
- ☐ Travel alarm clock
- ☐ Chemical light sticks
- ☐ Police or signal whistle
- ☐ Dust masks
- ☐ Phone/e-mail/address list for family, friends, neighbors, physician, pharmacy
- ☐ Emergency contact/medical information card in your wallet
- ☐ Spare car and house keys
- ☐ High energy or high protein snacks
- ☐ Food (freeze-dried or MREs)
- ☐ Coffee, tea, drink mixes
- ☐ Plate or bowl, knife, fork and spoon, insulated mug
- ☐ Camp stove, small pot, fuel and matches
- ☐ Battery or other lantern
- ☐ Water in heavy plastic jugs
- ☐ Water purification filter or tablets
- ☐ Magnetic compass, maps
- ☐ Duct tape, parachute cord
- ☐ Consider packing individual items or kits in zipper-lock freezer bags to keep the contents dry, clean, and neat.

expensive and impart a faint taste of iodine to the water. Most filters will remove Giardia cysts. All water filters require care in their use to avoid cross-contamination of purified water with dirty water.

Purification tablets, such as Halazone, have a limited shelf life that varies with the type, and give the water an unpleasant taste. Tablets will do nothing for particulate (dirt) or discoloration in the water. Be sure to read and understand the information that comes with any water purification device or tablet before purchasing or using it.

You can also use unscented laundry bleach (no perfumes, etc.) to purify water. After filtering out any particulate by pouring the water through cheesecloth, coffee filters, or paper towels, put 1/8 teaspoon of bleach in a gallon of water, mix well, and allow it to sit for thirty minutes. If it still smells slightly of chlorine, you can use it. If not, stir in another 1/8 teaspoon and wait another half hour. If it still does not smell of chlorine, discard the water and find a new supply. It will not taste great, nor will the chlorine bleach kill cysts like Giardia, but it may be enough.

If you have no other means, boiling for at least five minutes will kill any bacteria and other organisms, but will not remove any particulate matter or discoloration. Boiling will leave water with a "flat" taste that can be improved by pouring it back and forth between two containers several times to reintroduce some oxygen.

Sleep

Try to get at least six continuous hours of sleep in every 24-hour period, or at least four continuous hours with several shorter naps. Bring fresh soft foam earplugs and a black eye mask to reduce disturbances from light and noise around you. An appropriate sleeping bag, closed-cell foam pad or air mattress, and your own pillow will help give you the best chance of getting adequate rest. If caffeine keeps you awake, try to stop drinking coffee, tea, or other beverages containing caffeine at least four hours before going to bed. Allowing yourself to become over-tired can also make falling asleep difficult.

Top contesters will tell you that the key to successful contesting is understanding the sleep cycle and REM sleep, so you can manage when you sleep, and for how long. Participation in contests and Field Day can help you understand your sleep cycle better.

Personal Hygiene

If you pack only a few personal items, be sure they include toothpaste and toothbrush, a comb, and deodorant. If possible, bring a bar of soap or waterless hand cleaner, a small towel and washcloth, and a few extra shirts. Waterless shampoo is available from many camping stores. After two or three days, an unwashed person can be unpleasant to be around. Think of others and make an attempt to stay as clean and well-groomed as you can under the circumstances.

Medical Considerations

If you have a medical condition that could potentially interfere with your ability to do your job, it is a good idea to discuss this with your physician ahead of time. For instance, if you are a diabetic, you will need to avoid going for long periods without proper food or medication, and stress may affect your blood sugar level. Persons with heart problems may need to avoid stressful situations. Even if your doctor says you can participate safely, be sure you have an adequate supply of appropriate medications on hand, and copies of any prescriptions. Let your EmComm manager and any work partners know of your condition so that they can take appropriate actions if something goes wrong. Wear any medical ID jewelry you have. Keep a copy of any special medical information and emergency phone numbers in your wallet at all times.

Protect Your Eyes and Sight

If you wear eyeglasses or contact lenses, bring at least one spare pair. If you use disposable contact lenses, bring more changes than you think might be necessary, to avoid running out. Some contact lens wearers may want to switch to glasses to avoid having to deal with lens removal and cleaning under field conditions. If you have any doubts, consult your eye doctor ahead of time. Bringing a copy of your lens prescription along may also be a good idea, especially if you are likely to be some distance from home for a while.

Sunglasses may be a necessity in some situations. Working without them in bright sun can cause fatigue, and possibly eye damage. If you are in an area with large expanses of snow or white sand, prolonged periods of exposure can cause the retina to be burned, a very painful condition commonly known as "snow blindness." Use good-quality UV-blocking sunglasses at all times, and avoid prolonged exposure.

If you do not normally wear eyeglasses, consider a pair of industrial safety glasses or goggles to protect your eyes from wind-blown water, dust, and debris. Keep all spare eyeglasses or safety glasses/goggles in a felt-lined, hard-shell storage case to prevent scratching and breakage.

SAFETY IN AN UNSAFE SITUATION

Many disaster assignments are in unsafe places. Natural disasters can bring flying or falling debris, high or fast-moving water, fire, explosions, building collapse, polluted water, disease, toxic chemicals, and a variety of other dangers. Industrial buildings or facilities may contain toxic chemicals, which can be released in a disaster (there will be more on hazardous materials later in this chapter). Dams can break, bridges can wash out, and buildings can collapse. Areas can become inaccessible due to flooding, landslides, collapsed structures, advancing fires or storm surges.

You should always be aware of your surroundings and the dangers they hold. Never place yourself in a position

The Amateur Radio Public Service Handbook

where you might be trapped, injured, or killed. Try to anticipate what might happen and plan ahead. Always have an escape plan ready in the event that conditions suddenly become dangerous. Do not allow yourself to become "cornered" — always have more than one escape route from buildings and hazardous areas. If you can avoid being in harm's way, you can also prevent yourself from becoming part of the problem rather than part of the solution.

Be prepared to help others find or rescue you should you become trapped or isolated. Carry a police or signal whistle and a chemical light stick or small flashlight. Let others know where you are going if you must travel anywhere, even within a "safe" building. Try not to travel alone in dangerous conditions — bring a "buddy."

Wear appropriate clothing. Depending on the weather, your gear might include a hard hat, rain gear, warm non-cotton layers, work gloves, and waterproof boots. Always bring several pairs of non-cotton socks and change them often to keep your feet clean and dry. Create seasonal clothing lists suitable for your climate and the types of disasters you might encounter. As a volunteer communicator, you will not generally be expected to enter environments that require specialized protective clothing or equipment. Do not worry about purchasing these items unless they are required by your served agency.

SHELTER

In most cases, you will not need your own shelter for operating or sleeping. You may be able to stay or work in the emergency operations center, evacuation shelter or even your own vehicle. However, in some cases a tent, camp trailer, motor home or other suitable shelter may be necessary, if permissible. Your choice will depend on your needs and resources, as well as what is permitted at the location to which you are going.

Tents should be rated for high winds, and should be designed to be waterproof in heavy weather. Most inexpensive family camping tents will not survive difficult conditions. Dome tents will shed wind well, but look for published wind survival ratings since not all dome designs are equal. Your tent should have a full-coverage rain fly rather than a single waterproof fabric. The tent's bottom should be waterproof, extending up the sidewalls at least six inches in a "bathtub" design, but bring an extra sheet of plastic to line the inside, just in case. (Do not place a plastic ground cloth under a tent, as this will allow rain to run under and through any leaks in the tent floor.) Bring extra nylon cord and long ground stakes to help secure the tent in windy conditions. If you are not an experienced foul-weather camper, consult a reputable local outfitter or camping club for advice on selecting and using a tent.

HAZARDOUS MATERIALS AWARENESS

The term "hazardous materials" (hazmat) refers to any substances or materials that, if released in an uncontrolled manner (e.g., spilled), can be harmful to people, animals, crops, water systems, or other elements of the environment. The list of hazardous materials is long and includes explosives, gases, flammable and combustible liquids, flammable solids or substances, poisonous and infectious substances, radioactive materials, and corrosives. One of the major problems faced by emergency responders is determining which chemicals are involved and determining the potential hazards.

Amateur Radio operators may encounter hazmat incidents during their travels, or they may be asked to assist with emergency communications in such incidents. Proper training is required for your own safety (see **Figure 8.1**). Moreover, a wrong move by you during a hazmat operation can endanger not only your own safety, but also the safety of other responders as well as the entire local community.

Figure 8.1 — Hazmat personnel preparing for deployment.

Hazardous Chemicals On the Move

As the primary regulatory agency concerned with the safe transportation of such materials in interstate commerce, the US Department of Transportation (DOT) has established several systems to manage hazmat materials. These include definitions of various classes of hazardous materials, placards and other marking requirements for vehicles, containers and packages to aid in rapid identification of cargoes, and an international cargo commodity numbering system.

The DOT requires that all freight containers, trucks, and rail cars transporting these materials display placards (**Figure 8.2**) identifying the hazard class or classes of the materials they are carrying. The placards are diamond-shaped, 10 inches on a side, color-coded, and show an icon or graphic symbol depicting the hazard class (flammable, caustic, acid, radioactive, etc). They are displayed on the ends and sides of transport vehicles. A four-digit identification number may also be displayed on some placards or on an adjacent rectangular orange panel. If you have spent any time on the roads, you have undoubtedly seen these placards or panels displayed on trucks and railroad tank cars. You may recognize some of the more common ones, such as 1993, which covers a multitude of chemicals including road tar, cosmetics, diesel fuel, and home heating oil. You may have also seen placards with the number 1203 (gasoline) on tankers filling the underground tanks at your local gas station.

In addition to truck and rail car placards, warning labels must be displayed on most packages containing hazardous materials. The labels are smaller versions (four inches on a side) of the same placards used on vehicles. In some cases, more than one label must be displayed, in which case the labels must be placed next to each other. In addition to labels for each DOT hazard class, other labels with specific warning messages may be required. Individual containers also have to be accompanied by shipping papers that contain the proper product name, the four-digit ID number, and other important information about the hazards of the material.

Hazardous Chemicals in Buildings

The National Fire Protection Association (NFPA) has devised a marking system to alert firefighters to the characteristics of hazardous materials stored in stationary tanks and facilities. This system, known as NFPA 704M, can also assist citizens visiting a site in identifying the hazard presented by the stored substance. Use of the system is voluntary, unless specified by local codes.

The NFPA 704M label (**Figure 8.3**) is diamond-shaped and is divided into four parts, or quadrants. The left quadrant, colored blue, contains a numerical rating of the substance's health hazard. Ratings are made on a scale of 0 to 4, with a rating of 4 indicating a danger level so severe that a very short exposure could cause serious injury or death. A zero, or no code at all in this quarter, means that no unusual hazard would result from the exposure. The top quadrant of the NFPA symbol contains the substance's fire hazard rating. As you might expect, this quadrant is red. Again, number codes in this quadrant range from 0 to 4, with 4 representing the most serious hazard. The NFPA label's right quadrant, colored yellow, indicates the substance's likelihood to explode or react. As with the health and fire hazard quadrants, ratings from 0 to 4 are used to indicate the degree of danger. If a 4 appears in this section, the chemical is extremely unstable, and even under normal conditions may explode or react violently. A zero in this quadrant indicates the material is

Figure 8.2 — An example of the identification placards vehicles are required to display when transporting hazardous materials.

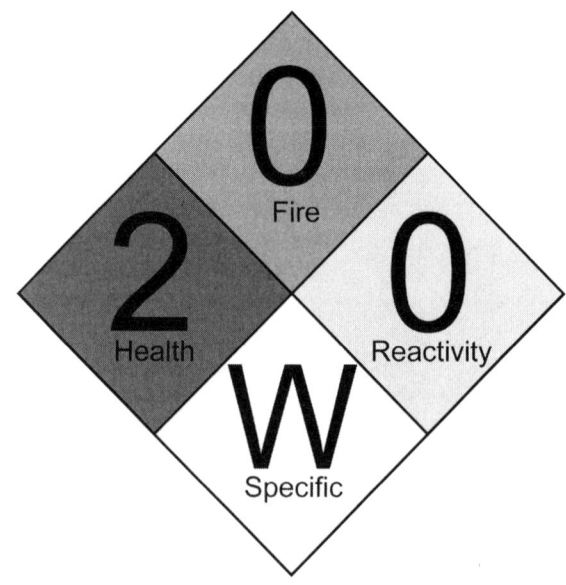

Figure 8.3 — An example of an NFPA 704M label.

considered stable even in the event of a fire. The bottom quadrant is white, and contains information about any special hazards that may apply. There are two possible codes for the bottom quarter of the NFPA symbol. "OXY" means the material is an oxidizer. It can easily release oxygen to create or worsen a fire or explosion hazard. The symbol "W" indicates a material that reacts with water to release a gas that is either flammable or hazardous to health. If the material is radioactive, the usual tri-blade "propeller" symbol for radioactivity will appear.

Guidelines for Handling Hazmat Incidents

Once you are in a safe position uphill and upwind, try to identify the material. However, it cannot be over-emphasized that you must stay well away from the site. Do not be tempted to get "just a little closer" so that you can read placards or other items. If you cannot read these items using a spotting scope or binoculars, simply report what you can see from a safe position. If you are able to see from a safe position, look for:

- The four-digit number on a placard or orange panel.
- The four-digit number preceded by the initials "UN/NA" on a shipping paper, package, or drum.
- The name of the material on the shipping papers, placard, or package.

Call for help immediately and let the experts handle the situation. Remember, even ordinary firefighters and police are prohibited by federal law from taking certain actions at some hazmat incidents. Do not attempt to take any action personally, beyond your report and preventing others from approaching. This is an instance when it is vitally important to know your limitations, not just for your own safety, but also for the safety of others.

When reporting a hazmat incident, include the following information:

a. Identify yourself.
b. Give your current location and the location of the incident, i.e. street address or cross streets, road and mile marker, distance from nearest town, etc.
c. Briefly describe what you see (from a distance), i.e. liquid spill, gaseous cloud, etc., and any placard numbers or other information you can safely see.
d. If a gaseous cloud or liquid spill exists, give the direction in which the contaminant is flowing or moving. Give any pertinent weather or other information you can observe from a safe distance that might help the experts in responding to the incident. Be concise.

To learn more, see the FEMA independent study course *An Introduction to Hazardous Materials* at **http://training.fema.gov/IS/crslist.asp.**

Chapter 9
Contesting and Field Day as Supplements to EmComm Training

Quality EmComm requires you to be on your toes and well-prepared for whatever the situation will throw at you. You've passed every FEMA course and have credentials that would impress a two-star General. But how much time have you actually spent on the air, passing traffic? Have you operated from numerous locales away from home, or from your car? Do you know propagation well enough to establish a regional link using RF and any one of numerous transmission modes at a moment's notice? Participating in the on-air activities of contesting and Field Day can give you these and other practical skills. In this chapter we'll take a look at contesting and Field Day with an eye toward the aspects that can enhance your emergency communications training.

WHAT IS A CONTEST?

Amateur Radio contesting is an on-air activity in which the object is to contact as many different stations as possible in as many different locations as possible. Contesting is all about passing traffic quickly and efficiently. Experienced contesters know how to use their gear to maximum capacity and efficiency, can exchange information incredibly quickly with minimal errors, and can set up a functional station virtually anywhere and make it work. Contesting will teach you how to operate and maintain any kind of station — from kilowatt transceivers and stacked monoband Yagi antennas to a QRP station with a makeshift antenna and battery power — effectively. You will learn more about propagation paths and transmission modes than you have ever known before, and, for most of the "major" events, you'll even get a report card showing you how well you did. If you're looking to improve your usefulness during an emergency, contesting offers a fun way to gain practical on-air experience that will make a positive difference in your operations if you're deployed.

Contests can run anywhere from one hour to 48 hours. Some contests are global in scope, while others are specific to a particular country or state. Contests cover every portion of the radio spectrum, from 160 meters all the way up to the microwave bands, and use every mode imaginable. There is some sort of contest every weekend, although some are larger than others.

Each month in *QST*, the ARRL publishes a column called "Contest Corral," which is a rundown of the contests for that month, with basic information about each of them. It also provides a link to the website of each contest listed so you can get more details on the event.

TYPES OF CONTESTS

Contests can be lumped into several different categories based on bands used, mode, or region.

HF contests occur on the HF bands from 160 through 10 meters (with no contest activity on 12, 17, or 30 meters). HF contests tend to be very large in scope, from worldwide events to events focused on a continent, country, or region.

VHF contests center on the six meter band (50 MHz) and higher. These events offer entry categories for fixed stations, mobile stations (known as "Rovers") and QRP portable stations (stations running very little power from a location other than where a current ham station exists), and also highlight forms of propagation found mostly on the VHF bands, such as sporadic-E, tropospheric ducting, and aurora. These unusual modes of propagation make it possible for signals on VHF frequencies (usually limited to "line-of-sight" distances) to travel hundreds or even thousands of miles, which can be very useful during an emergency on a local or even regional level.

DX contests are worldwide in focus, with the goal of trying to make QSOs with as many different countries as possible.

Domestic contests focus on a specific country, with the goal of talking to as many different states or provinces as possible. Many countries around the world sponsor a domestic event. The best-known domestic event in the US is the ARRL November Sweepstakes, or "SS."

QSO parties generally focus on states or provinces, with the goal of contacting as many counties in that state or province as possible. Most states in the US offer a QSO party, as do many of the Canadian provinces. While generally more casual than most domestic or DX contests, state QSO parties can still get quite intense and competitive. Some states combine their efforts into larger events, such as the 7-Land and New England QSO Parties.

CONTEST BASICS

With so many different contests to choose from, a basic understanding of what's required may be helpful to get you started.

All you need to participate is a rig and an antenna. Most contests have entry categories based on power; meaning you'll compete against others using the same amount of power as you. While most logging is done on computer, you can log on paper at first if you're not ready for computer logging. More on computer logging later.

It's a good idea to read the rules beforehand, to get a better understanding of entry categories and the required exchange. You can find links to contest websites at **www.arrl.org/contest-calendar** or at Bruce Horn, WA7BNM's, online contest calendar at **www.hornucopia.com/contestcal.**

Every contest has an "exchange." This is a swapping of information for contest credit. Some exchanges are very simple, like your state. Others, like the ARRL November Sweepstakes, have several pieces of data to exchange. Some of the most common elements of a contest exchange are a signal report, state, country, a sequential serial number, county, or in VHF contests, your grid square (see **www.arrl.org/grid-squares** for a description).

Working a contest is generally not the time to exchange pleasantries; it's a competition, so operators are trying to make the most of their time. An exchange between two contest stations takes only a few seconds. Brevity (and accuracy) is the name of the contesting game. When calling a station in a contest, give your call once. Always use phonetics, and use standard phonetics (especially if calling a DX station where English may not be the operator's first language). If you're acknowledged, copy the station's information. If you missed some of their contest exchange, ask for a fill by simply saying "again?" (or "?" on CW). When you have the station's exchange correct, simply reply with a "roger" (or "R" on CW), then give your exchange information. Avoid giving the station's exchange information in your reply, and avoid using the phrase "please copy." It is unnecessary and inefficient. Give your information once; if the other station needs a fill, they will ask for it. For most contests, the exchange of information between stations on SSB should take no more than 10 seconds, including call signs.

CONTEST OPERATING — "S & P" vs "CQ-CONTEST"

There are two ways to participate in a contest: Finding stations calling CQ and answering them ("search and pounce"), or calling CQ yourself and have stations find you ("running"). In terms of applying contest skills to emergency communications, calling CQ will give you more experience with rapid, efficient traffic handling, as it allows you to experience several stations calling you, under a bit of a time constraint, in order to exchange information quickly and accurately. During major HF contests, it's not uncommon for good operators to work stations at the rate of 100 an hour on CW or RTTY, or 200 an hour on phone. Think of it as being Net Control with 200 stations waiting to pass traffic, or being an air traffic controller at a major airport during the morning rush.

AN EFFICIENT CONTEST EXCHANGE

Here's an example of a perfect exchange of information during the ARRL DX SSB Contest, in which DX stations send a signal report and their transmit power, and US and Canadian stations send their state or province.

PJ2T: *CQ Contest, Papa Japan Two Tango.*
KX9X: *Kilo X-ray Nine X-ray.*
PJ2T: *Kilo X-ray Nine X-ray, 59 Kilowatt.*
KX9X: *Roger, 59 Connecticut*
PJ2T: *Thanks, Papa Japan Two Tango, contest.*

Notice how each station's information is sent only once and each station confirms they received the other station's information. This entire exchange, from PJ2T's CQ to the end of PJ2T's solicitation for another station to call him, takes about eight seconds.

RECORDING THE DATA: CONTEST LOGGING

Just as in traffic handling, contests require the information you gather to be logged. While some contests still allow the use of paper log sheets to record your information, the preferred method is to use a computer logging program. Numerous contest logging programs exist; however, it is generally better to use a logging program designed for contest operating, as opposed to a general-purpose logging program. Do an online search for "contest logging software" to explore the numerous options.

For almost all of the "major" contests, electronic logs must be submitted in an ASCII text format called *Cabrillo* (**Figure 9.1**). Complete information on the *Cabrillo* format and how to submit a log for ARRL contests can be found at **www.arrl.org/log-submission.**

AFTER THE CONTEST

As with every competitive event, we want to know how well we did. It takes a lot of time and computer processing to cross-check logs and publish results. Some smaller events have results available in a week or so, while global events with thousands of entries can take several months to publish the list of winners. ARRL publishes results online before the results are published in *QST*; many other contest sponsors do something similar.

```
START-OF-LOG: 3.0
CALLSIGN: W1AW
CLUB: NONE
CONTEST: ARRL-SS-CW
CLAIMED-SCORE: 208160
OPERATORS: W1AW
NAME: Maxim Memorial Station
ADDRESS: 225 Main Street
ADDRESS: Newington, CT 06111
ADDRESS: USA
CREATED-BY: N1MM Logger V10.10.2
LOCATION: CT
CATEGORY-OPERATOR: MULTI-OP
CATEGORY-TRANSMITTER: ONE
CATEGORY-BAND: ALL
CATEGORY-POWER: HIGH
CATEGORY-MODE: CW
CATEGORY-ASSISTED: ASSISTED
CATEGORY-STATION: FIXED
QSO: 07046 CW 2010-11-06 2101 W1AW    0001 M 38 CT    K9CT     0002 B 67 IL
QSO: 07045 CW 2010-11-06 2101 W1AW    0002 M 38 CT    NY3A     0002 U 73 EPA
QSO: 07044 CW 2010-11-06 2103 W1AW    0003 M 38 CT    WA0ACF   0002 A 61 WI
QSO: 07044 CW 2010-11-06 2104 W1AW    0004 M 38 CT    WI2E     0002 A 86 EPA
QSO: 07044 CW 2010-11-06 2105 W1AW    0005 M 38 CT    W4NZ     0008 U 60 TN
QSO: 07044 CW 2010-11-06 2105 W1AW    0006 M 38 CT    N3GGT    0004 A 88 EPA
QSO: 07044 CW 2010-11-06 2106 W1AW    0007 M 38 CT    AA4FU    0006 U 78 NC
QSO: 07044 CW 2010-11-06 2106 W1AW    0008 M 38 CT    N8TR     0011 B 79 OH
QSO: 07044 CW 2010-11-06 2106 W1AW    0009 M 38 CT    N3CW     0007 U 66 MDC
QSO: 07044 CW 2010-11-06 2107 W1AW    0010 M 38 CT    W3SD     0005 A 69 EPA
QSO: 07044 CW 2010-11-06 2107 W1AW    0011 M 38 CT    K3ZM     0007 B 69 VA
QSO: 07044 CW 2010-11-06 2108 W1AW    0012 M 38 CT    W9LO     0007 A 47 WI
QSO: 07044 CW 2010-11-06 2108 W1AW    0013 M 38 CT    W9SE     0005 B 52 IL
QSO: 07044 CW 2010-11-06 2110 W1AW    0014 M 38 CT    WA3KVN   0002 A 69 EPA
QSO: 07044 CW 2010-11-06 2111 W1AW    0015 M 38 CT    VO1BQ    0004 A 86 NL
QSO: 07044 CW 2010-11-06 2111 W1AW    0016 M 38 CT    N4UW     0011 Q 76 TN
QSO: 07044 CW 2010-11-06 2111 W1AW    0017 M 38 CT    W0HBH    0004 A 60 MO
QSO: 07044 CW 2010-11-06 2112 W1AW    0018 M 38 CT    K8AAX    0007 A 97 MI
QSO: 07044 CW 2010-11-06 2113 W1AW    0019 M 38 CT    K0PC     0014 Q 69 MN
QSO: 07044 CW 2010-11-06 2113 W1AW    0020 M 38 CT    K8GU     0017 A 93 MDC
QSO: 07044 CW 2010-11-06 2114 W1AW    0021 M 38 CT    AF9T     0018 A 77 WI
QSO: 07044 CW 2010-11-06 2116 W1AW    0022 M 38 CT    N9HDE    0012 A 85 IA
QSO: 07044 CW 2010-11-06 2118 W1AW    0023 M 38 CT    KG3V     0008 A 70 VA
QSO: 07044 CW 2010-11-06 2118 W1AW    0024 M 38 CT    W0UCE    0009 A 52 NC
QSO: 07044 CW 2010-11-06 2119 W1AW    0025 M 38 CT    W3BC     0009 U 66 WPA
QSO: 07044 CW 2010-11-06 2120 W1AW    0027 M 38 CT    VE2EZD   0005 A 77 QC
QSO: 21039 CW 2010-11-06 2120 W1AW    0026 M 38 CT    VE5ZX    0025 A 63 SK
QSO: 07044 CW 2010-11-06 2121 W1AW    0028 M 38 CT    K4BSK    0011 A 60 NC
QSO: 07044 CW 2010-11-06 2123 W1AW    0029 M 38 CT    K9OZ     0025 U 65 IL
QSO: 07044 CW 2010-11-06 2123 W1AW    0030 M 38 CT    W4PK     0010 U 54 VA
QSO: 07044 CW 2010-11-06 2124 W1AW    0031 M 38 CT    N4TZ     0017 A 61 IN
```

Figure 9.1 — An example of a *Cabrillo*-formatted log used in contesting. The "header" at the top contains information about the contest effort, such as call used, location, and entry category. After the header are lines of QSO information. Note the date, time, band, and both sent and received information on each QSO line.

If you don't want to wait for the results to come online (and who does?), you can get a rough idea of how you did by researching Claimed Scores on the ARRL's Logs Received page at **www.arrl.org/logs-received**. Another source of claimed scores is the 3830 e-mail list. You can find subscription info at **www.contesting.com**.

Once results have been published, major contest sponsors offer a Log-Checking Report (LCR; see **Figure 9.2**) for everyone who submits an electronic log. An LCR will compare your claimed score to your actual score, with a band-by-band breakdown of how many QSOs you made, what QSOs were removed from your score due to exchange information copied incorrectly, or a claimed QSO in your log that did not appear in the other station's log. It will also show you mistakes you made, such as logging a station's QTH in South Dakota when they were really in North Dakota. Your Log-Checking Report is an invaluable tool; it will show you your mistakes and help you see patterns of operating that need to be improved on, which results in improved accuracy. In emergency communications, accurately copying data is of paramount importance; reviewing your contesting LCRs can help you improve your accuracy by showing you what skills you may need to work on.

Almost all contests offer a certificate for a notable achievement, such as first place in your ARRL Section, state, province, or country. Some of the larger events offer plaques for exceptional performance, such as first place in your entry category, your call area, or your ARRL Division. It's nice to have an operating achievement recognized with an award!

NOVEMBER SWEEPSTAKES: A TRAFFIC HANDLER'S DREAM

If ever there was a contest that could help you learn how to pass traffic efficiently, the ARRL November Sweepstakes is it. "SS," as it is known, got its start in 1929, when hams around the country were encouraged to send in official traffic via wireless to ARRL Headquarters to wish ARRL founder Hiram Percy Maxim a happy birthday. So many people enjoyed the process, ARRL received requests for a competition amongst all ARRL Sections to see who could handle the most traffic.

Sweepstakes has evolved over time, but its ties to traffic handling remain. SS has one of the longest exchanges in all of contesting: six pieces of information, including the call of the station you're working, a sequential QSO number, a letter denoting your entry class, your own call, a two-digit number for the year you received your first Amateur Radio license, and your ARRL Section. That's a lot of information to get right! In addition to the exchange, there's a goal to work all ARRL Sections in the contest for a "Clean Sweep."

SS is a 30-hour event, but stations are permitted to operate only 24 out of the 30 hours. The best operators will make well over 1000 QSOs, with an error rate of less than 1%. Now that's some serious traffic handling!

INCORPORATING CONTESTING INTO YOUR TRAINING

With so many contests occurring on a yearly basis, it's easy to incorporate them into training with your group or just yourself. Most contests offer a multioperator category, in which several people operate from one location as a team, use one call, and contribute to a common score. This is one of the finest team-building opportunities Amateur Radio has to offer. If your state is holding a QSO party, use that weekend as an exercise. You could operate the event from a club member's house or your EOC, you could operate portable from a rare county in the state, or as a mobile from multiple counties (if the rules allow it). If you have a large group, split into two or more groups and operate from different portable locations. All of the participants will learn valuable lessons of deployment: packing and hauling gear, setup in a remote location, use and maintenance of alternative power sources (if you don't operate on commercial power), and keeping accurate records.

If you have a lot of Technicians in your group, consider sponsoring an FM simplex contest. It can be a very simple

---

```
          SWEEPSTAKES LOG REPORT FOR W1AW
CATEGORY-POWER: HIGH
CATEGORY-ASSISTED: ASSISTED
CATEGORY-OPERATOR: MULTI-OP
CATEGORY-STATION: FIXED

TIME ON CALCULATION
------------------
2101-0715 =  615
1016-1708 =  413
1844-2014 =   91
2102-2231 =   90
2316-0258 =  223
             ------
   On time = 1432 minutes (max of 1440).

CALLSIGN CHECK RESULTS
----------------------
WB4HLW is a unique call. Received QSO# = 1.
AC0W is a busted call. The correct call is KC0W.
K8CD is a busted call. The correct call is K8CN.
K3BVU is a unique call. Received QSO# = 1.
N6BW is a busted call. The correct call is N6DW.
W6RKP is a busted call. The correct call is W6RKC.
K1BG is a busted call. The correct call is K1DG.

You had 10 calls in your log which were not found in the database of good
callsigns. 7 of them have been judged to be incorrect. These will be
removed from your score – along with an additional penalty of one QSO
per per busted callsign.
Unique percentage = 0.2

EXCHANGE CHECK RESULTS
----------------------
QSO #43   AL1G   : B 89 Ak    should be B 98 Ak
QSO #176  AC2N   : U 01 NLi   should be Q 91 NFl
QSO #203  KT0F   : A 71 Co    should be A 81 Co
QSO #213  N2EY   : B 67 EPa   should be A 67 EPa
QSO #242  W3ABT  : Q 32 EPa   should be A 32 EMa
QSO #271  N1EU   : U 65 EMa   should be U 65 ENy
QSO #272  N9NE   : Q 58 Mi    should be Q 58 Wi
QSO #463  W7KF   : B 75 Mt    should be B 65 Mt
QSO #475  AJ4F   : U 69 STx   should be A 69 STx
QSO #538  W4UCZ  : A 57 Ga    should be A 58 Ga
QSO #753  KL7AF  : B 57 Ak    should be B 77 Ak
QSO #865  N5DM   : A 65 NTx   should be A 65 STx
QSO #1021 KD6WKY : A 92 Sb    should be A 92 Eb

99.1% of your non dupe QSOs had their exchanges checked.

There were 23 exchange errors found. These QSOs will be removed from
your score with no penalties.

CROSS CHECK RESULTS
-------------------
QSO #209: Received QSO#   2 should be  72 W2GG
QSO #248: Received QSO#  67 should be  77 N9IO
QSO #296: Received QSO#   3 should be 203 K0MPH
QSO #345: Received QSO# 192 should be 198 VA3EC
QSO #361: Received QSO# 137 should be 237 NX4N
QSO #540: Received QSO# 419 should be 519 W6PH
QSO #641: Received QSO#  23 should be 123 KE0G
QSO #838: Received QSO# 761 should be 771 W5RU
QSO #860: Received QSO#  81 should be 118 AE4CJ

79.8% of your remaining good QSOs were cross checked.
There were 15 bad cross check QSOs removed.
```

Figure 9.2 — After a major contest's results have been tabulated, you will get a Log-Checking Report (LCR). A cross between a report card and a post-deployment briefing, a contest LCR will show where you made mistakes and provide information on how to do better the next time.

WHERE DOES TRAINING BEGIN?

Regular training, whether a new course or a refresher, has several benefits. Of course, the learning experience benefits the individual ham, but training also lets served agencies know that we take their mission, and our relation to it, seriously.

Our focus on training, though, has missed something very important. Where do we start training? What is the first thing you should do if you want to help with emergency communications and public service? What is the most important training we can take part in? The answer was found in your Amateur Radio license study manual.

The most important thing we can do is simply to get on the air and make use of the spectrum we've been granted. Training starts by turning on your radio. Any time you get on the air is a training and learning experience. It doesn't matter what you do on the air: Field Day, ragchewing, nets, DXing, contests, etc. What matters is that you get on the air. Not just when there is an ARES event or an emergency, but at every opportunity. Get on the air!

Think about this: what would we be doing if we didn't have spectrum to use? We get to keep our spectrum not by spelling out all the "what ifs" and doomsday scenarios. We keep it by using it, and using it a lot. There are groups that wouldn't mind taking some of our spectrum from us. Saying we need it in an emergency doesn't always work; after all, even public safety feels the sting of spectrum grabs.

So how is getting on the air training? By being active on the Amateur Radio bands, you will increase your knowledge of propagation, rules and regulations, station building, antennas, and modes of communication, and build networks through the QSOs you make. The bands are a perpetual learning environment, where there is always something to be gained, no matter how many years you have been active. For proof, ask those in your EmComm group that have been there, done that, and have a closet full of T-shirts. Getting on the air is always a learning experience.

You may not get a certificate for getting on the bands (well, perhaps, if you try DXing or contesting) and your served agency may not understand your excitement for logging that new country on 10 meters, but you will be learning and growing as an amateur. So get on the air and start training!

---

event — have different entry categories based on power, and operate on 2 meter FM for just one hour, with cities as multipliers. The object: make as many contacts with as many different stations in as many different cities as possible. You can allow Rovers as well, so people can operate from different cities during the event. Use these rules as a jumping-off point and modify them based on your own local circumstances. It's fun, different, and an excellent team and skill builder.

Another good event for the newly licensed is the ARRL 10 Meter Contest, held during the second weekend in December. Have a class close to the contest time in which everybody learns to build their own 10-meter dipole, then have everyone deploy their antenna and operate the contest. Have an after-contest pizza party and discuss everybody's experiences.

The ARRL June VHF QSO Party has a "QRP Portable" category with a 10-watt PEP power limit, where operators set up in the field away from a pre-existing station. This event is perfect for a deployment exercise: antennas on VHF are generally smaller than HF, so it's a lot easier to transport gear. Any high-elevation location, such as a hilltop, mountain, roof, bleachers, public park, or recreation area will give you an excellent opportunity to work many stations, and give you experience using the VHF bands in a way you may not have used them before (see **Figure 9.3**).

One of the best things you can do to enhance your training is operate as part of a multioperator team from a seasoned

Figure 9.3 — Operating portable during any contest is a great way to test your go-kit's capabilities. Here ARRL Contest Manager Sean Kutzko, KX9X, operates from a hilltop during a VHF contest with a QRP rig, battery power, and small Yagis for 6 and 2 meters. [SEAN KUTZKO, KX9X]

Figure 9.4 — Phil Koch, K3UA, works a DX station at a high rate from multi-multi station K3LR using SSB. This is the 15-meter station at K3LR. K3UA is talking and turning the rotors at the same time, displaying efficiency that is necessary for both contesting and EmComm operating. [CLAUDIO FERNANDEZ, LU7DW]

contester's station (**Figure 9.4**). The odds are you'll get to use a piece of gear you're not familiar with. During a deployment, you may be exposed to several different pieces of unfamiliar equipment. Any opportunity you get to use a new piece of gear, grab that chance; spending time during an actual emergency learning how to use a piece of gear is wasting valuable time.

TROUBLESHOOTING DURING A CONTEST OPERATION

Murphy's Law states, "If anything can go wrong, it will." This has never been more true during a deployment. Operators are usually working under harsh conditions, using alternative power sources and sub-standard antennas. When you're in an environment where your operating conditions are not optimum, things will go wrong.

The amount of work it takes to maintain even a moderate contest station is significant. There are lots of antennas, radios, computer interfaces, antenna feed lines, various connectors and adapters — the list goes on. If a piece of equipment goes south during a contest, you need to be able to fix it on the fly or have a workaround, or your score suffers. This directly applies to an emergency communications deployment as well, except during a deployment, your "score" could be extremely critical information.

While it's impossible to plan for every scenario, pre-planning can go a long way toward helping manage the inevitable visit from Murphy. Do you have a basic toolkit in your go-kit? Can you bring spare batteries, connectors, etc.? Do you know how to use an ohmmeter to find shorts in your antenna or feed line? Do you know how to make various antennas (dipoles, J-poles, etc.) in case you find a need you didn't plan for?

In all cases, it's critical to have a flexible mind during a field deployment. During a recent tornado deployment in the south, the 47 MHz antenna on a served agency's Emergency Response Vehicle quit working. The amateurs assisting the served agency had a 6 meter (50 MHz) vertical with them, but were unaware how easily it could be modified to restore the served agency's communications on 47 MHz. It may not be possible to be prepared for every potential situation, but the more you educate yourself, know your resources, and are prepared to "think outside the box," the better off you'll be.

FIELD DAY: ALTERNATIVE SCENARIOS FOR EMCOMM APPLICATIONS

On every ARRL Field Day since the first one in 1933, amateurs make a point of taking gear away from their homes to put together a station running on non-commercial power and less-than-optimum antennas. Informally, this activity has been going on since the beginning of radio itself, even before Field Day existed. Why? ARRL Field Day serves several purposes: it's an emergency communications exercise, a contest, a public outreach event, and a reason to get the club together, but the bottom line is that it's fun to operate away from home. Field Day is also the most popular Amateur Radio on-air event of them all; in 2011, 39,287 participants made 1,434,363 QSOs in 24 hours. That's some serious operating.

Most amateurs that have participated in Field Day have done so with their local club. Those interested in emergency communications have probably operated from an EOC (Emergency Operations Center) that was given the "F" classification in Field Day several years ago. The EOC usually is associated with a served agency and provides shelter, generator power, and other support to amateurs. If you choose to use Field Day as an emergency communications exercise, this chapter lays out several scenarios to give you ideas about how to use the event to better test your skills at operating under sub-standard conditions. Not all of these scenarios will appeal to everybody, but at least consider what would be involved in operating under the conditions outlined here. Give some of them a try during your next Field Day — and remember to have fun doing them.

Scenario 1: From the EOC

This is what many EmComm operators are familiar with: activating a permanent or semi-permanent station from an EOC established by a served agency.

Pros: Quality shelter is usually available. Alternative power sources (generators, battery banks) have generally been kept in good working condition. Permanent antennas may already exist at the EOC, such as an HF beam on a tower, good dipoles for 40 meters and 80 meters, and beams or good verticals for VHF/UHF. Climate control (air conditioning) may be provided.

Available Bonus Points: Too many to list! Alternative power, getting elected officials to visit your site, media publicity…almost every possible bonus point that can be awarded for a Field Day operation is available to you with this kind of operation.

Scenario 2: From Your Home, Using Emergency Power

This scenario is most likely one with which amateurs can identify. Who hasn't been without power at their home for an extended time? Learning how to operate your home station with a generator or battery backup power source will give you practical experience in a fun environment, which will come in handy should you need to be active from home during a power outage caused by a flood, snowstorm, hurricane, or other natural disaster. Kill the power at your QTH if you want to add more realism to the exercise. Sleep with the windows open, think about how to cook and stay comfortable without power… just don't open your fridge or freezer.

Pros: It's your home! All your personal comforts are at your disposal.

Available Bonus Points: While it may not be practical to have an elected official visit your house, or for you to set up a table where the public can get information on Amateur Radio, you can still earn plenty of bonus points for your Field Day effort. Send an official message through the NTS system to your ARRL Section Manager, copy the W1AW Field Day Bulletin, try making QSOs via satellite, and, of course, use an alternative power source to bring in plenty of bonus points.

Scenario 3: Load Up the Car

You're eating breakfast when the phone rings: you're needed to help set up communications at an aid station in the wake of the previous evening's tornadoes in the neighboring county. You load up your car with whatever gear you can find, and hit the road.

Most guidelines for deployment to an affected area suggest you should be prepared to be completely self-supporting for 72 hours. You can use this scenario to truly operate Field Day "in the field" as a portable station, testing your skills and self-support capabilities as a one-man or two-man operation, entering the "B" class. Rent a nearby campsite and take all the gear you'd need to survive for a three-day deployment off the electrical grid.

This scenarios presents additional concerns. What will you use for a power source? A small generator? A deep-cycle battery? Will you have enough fuel? What do you want to use for antennas? How will you get them in the air? What kind of antenna will be most practical? Do you have proper shelter? Food and water?

Your vehicle can hold an incredible amount of equipment. If you don't have a go-kit ready, use this exercise as a way to prepare one.

Pros: A true Field Day operation from a field location. This will help you consider the logistics of setting up a portable station and making it work. You will learn about power consumption, and antenna design and requirements, and have to make some decisions on what is practical to bring and what isn't.

Available Bonus Points: Most available bonus points apply in this scenario.

Scenario 4: Pack It In

This scenario is not going to appeal to everybody. However, if you really want to test your self-sufficiency capabilities, consider a Field Day operation where you have to carry everything with you: radio, antenna, power source, food, water, clothing, and shelter.

You can have a friend or relative drop you off at a campsite and pick you up again three days later. Go hiking in a state or national forest, or maybe canoe down a river or around a large lake, all while carrying everything you'll need with you. This is the most grueling of all the scenarios presented here. Weight becomes a very real issue. Some considerations:

- A gallon of water weighs 8 pounds
- A 7 Ah sealed lead-acid (SLA) battery weighs about 6 pounds
- A standard compact dc-to-daylight rig weighs around 6 pounds

For this scenario to be truly successful, a battery that is capable of being recharged with solar power is your most likely power source. Numerous such packages exist. QRP power is the most efficient use of your limited energy resources.

QRP rigs come in many shapes and sizes. You can get several that are capable of transmitting on a band or two (like the Ten-Tec QRP series), or full-featured, highly portable QRP rigs like the Elecraft KX3 or the Yaesu FT-817ND. Special attention should be given to the amount of current your QRP rig consumes. Some are as high as 450 mA on receive, while other radios specifically designed for backpack operating have current consumption as low as 50 mA on receive. Dipole or vertical antennas can be made with very small gauge wire, since your transmit power is so low. A 40-meter dipole made with AWG 26 or 28 will weigh almost nothing. Lightweight rope or twine is available from any home supply store. To get your antenna in the air, you can use rope tied to a variety of weights to raise your antenna 30 feet or so over a nearby tree. Lightweight, collapsible fiberglass masts exist if your operating site doesn't have trees.

Pros: A true test of complete independence from the grid.

Available Bonus Points: With weight restrictions essentially forcing you to QRP power and a battery, you will be rewarded with a score multiplier of 5.

CONCLUSION

With participation in amateur radio contesting at an all-time high, there is little doubt that contesting can give you and your EmComm team solid experience in traffic handling, electronics troubleshooting, technical advancement, deployment skills through portable and mobile operations, and a lot of good times and enjoyment, too. Don't pass up the opportunity to expand your Amateur Radio world: a well-rounded, active amateur with experience in several on-air disciplines will be more useful and flexible during times of emergency.

Chapter 10
The Science of Radio —
Basics for the EmComm Operator

Too often among emergency organization-served agencies there's an attitude of, "Don't bother me with the details, all I need to know is how to turn the radio on and transmit." Anyone familiar with radio operations, however, knows this attitude is naive, particularly in the context of preparing for and operating during emergency situations. On the other hand, emergency managers and planners do not need to be technical experts in radio to be able to make informed decisions about equipment procurement, integration, and application in emergency management plans. There is a middle ground between being able to design, build, and repair a radio, and only knowing how to spell R-A-D-I-O.

This chapter will address the basics of the science of radio, with the goal of giving you some tools to put radio in an understandable context so you and your served agency can be more effective users of radio. We'll discuss how radio waves get from point A to point B. This is a complex topic, but some understanding of the mechanism of radio wave propagation and the capabilities and limitations of certain wavelengths used in emergency organizations is important for effective communications. Building on the discussion of wave propagation, we will next cover how to use radio waves as the medium to convey a message from one point to another through the processes of modulation and demodulation. Finally, we will demystify the "black box" known as a radio by looking inside a generic piece of equipment. We'll break down the complex circuitry of electronic devices into five basic building blocks that are configured to accomplish the task of communication via radio.

Before covering these topics, however, a review of human communications — and radio's role in it — is in order.

RADIO AS AN EXTENSION OF HUMAN COMMUNICATIONS

Why is the examination of human communications important in our context? To answer in question form: How often have you been misunderstood? How often has the idea or meaning you thought you were conveying been totally turned around during the communication process? It stands to reason that a periodic review of what actually happens during the natural act of human communications will help us become better radio communicators. This is particularly true while communicating in emergency or stressful situations.

In human communications, there are two major players: the sender, or person trying to convey some message to an intended audience; and the audience, the receiver of that intelligence. The receiver (here not to be confused with a radio receiver; the term is used in both contexts in this chapter) may be a single person or a collection of individual receivers, such as an audience of thousands, as human communications is simply an interaction between the sender and the receiver.

Again, the sender has some message he or she wants to convey to the receiver. This message may be as trivial as a pleasantry to nurture a relationship ("Hello, how are you today?") or as critical as a command to trigger a time-sensitive action ("FIRE! Evacuate the building through that exit NOW!"). The message content and purpose is as varied as the senders and receivers participating in the communication are numerous.

The mechanism used by the sender to convey the message is called the medium. The communications medium may be voice and vocalization, as in conversation; sight, as in reading these printed words; data for the sending of numerical information; visual light for projection of pictures or imagery; invisible light to control devices such as a TV set; or radio waves to connect the sender to the receiver via the medium of radio, which is the focus of this text. The methods can be combined, of course, such as illustrating a spoken lecture with images projected on a screen, melding auditory and visual media. This list of media is by no means complete, and it continues to grow as technology advances.

The choice of media is sometimes dictated by circumstance; sometimes it is chosen to expedite or facilitate the transfer of the message from the sender to the receiver. Wise communicators will choose carefully the appropriate medium at their disposal to accomplish the intended task. For instance, rapid action to evacuate a building during a fire does not require detailed explanations of what is happening and why action is required — a simple loud warning horn, simple directive lights, and a single word on a highly visible, recognizable sign (EXIT), plus verbal commands demanding action are most effective. Conversely, a detailed list of required supplies to restock a depleted inventory of shelter materials is best conveyed through digital or written text that provides a record of the requested supplies for filling the order accurately and to account for inventory used later. In emergency situations, radio may or may not be the most appropriate medium, but it may be the only choice available to communicate a message from the sender to the receiver.

Inherent in the use of any medium for communications is noise and distraction, the ubiquitous impediments that hinder the transfer of the message from the sender to the receiver. The good sender (as well as good receiver) will attempt to identify as many noises and distractions as possible and either lessen their effects on the communication process or make adjustments to accommodate for those distractions over which there is no control.

There are many noises and distractions present when using radio, and while they're not necessarily unique to the medium, they must be addressed in order to get the message through. For one thing, because radio can be used to connect the sender and receiver over great distances (which, of course, is also one of the reasons we use the medium), neither party in the communication may be familiar with the local area of the other. Unknowns may include the terrain, road networks, time zones, and the current safety and living conditions; for instance — have both areas been affected in the same way by an emergency incident? In other words, the distances bridged between participants in the communication may also prevent the participants from understanding the local context affecting the communication process.

There is also static and noise associated with using radio waves to transmit messages. Static crashes or signal fades may blank out words that are vital to the message. Single sideband (SSB) modes are more difficult to understand than frequency modulation (FM) modes; it takes a certain trained ear to listen to SSB effectively. Participants who are used to text transmissions via e-mail, with its near-instantaneous, near-perfect accuracy of seemingly limitless capacity, would find it virtually impossible to send the same message via PSK31 or packet — modes that necessitate an adjustment in the length of the message (quantity of data to be transmitted) as well as the content of the message (critical words may have to be repeated to allow for the possibility of lost characters).

Amateur Radio operators quickly learn to adapt and adjust for the noises and distractions inherent in using the medium of radio, and it is incumbent on the Amateur Radio participants in an EmComm situation to help the supported agency, the non-radio operator, and other involved communicators to make adjustments also. But before we address the necessary steps, let's look at possibly the most important part of human communications: feedback.

Feedback for Effective Communications

The final component of the human communications model recognizes that human communications is not a one-way process, but a two-way, simultaneous exchange of information. In radio jargon, effective human communication is full duplex, not simplex.

Perhaps most of the information exchange between humans in the communication process occurs non-verbally and via feedback. For instance, eye contact between the sender and receiver indicates a personal connection and conveys a sense of intimacy or personal attention. It also provides information about the level of interest, understanding, attention, and comprehension. Drooping eyes in an audience indicates that the topic may not be interesting to the receiver.

Of course, such feedback is open to interpretation. The types of feedback mechanisms are countless, and the effective sender will monitor the feedback from the receiver or audience and, in response, make spontaneous adjustments to the message or media to get the necessary information across through all the noise and distractions. With this model of human communications as the backdrop, let's now look at radio as the communications medium, with an eye toward some of things Amateur Radio operators must do to help our fellow communicators, as well as ourselves, use radio to its fullest potential.

The major limitation of the typical Amateur Radio system used for communications in an EmComm situation is the lack of instantaneous feedback between sender and receiver. Admittedly, there are full duplex radio connections — where the sender and receiver are sending their messages (feedback) simultaneously — as well as systems that use a combination of auditory, visual, and textual media simultaneously. But these systems are specialized, complex, expensive, and not very resilient and therefore not appropriate for the backup emergency communications that embody the "when all else fails" strategy of Amateur Radio.

The Amateur Radio station that supports emergency communications likely uses off-the-shelf transceivers, often the personal radios of the operators. The use of this typical equipment means the senders and receivers in the communications process need to take turns. The sender transmits while the receiver listens; once the push-to-talk (PTT) switch is released and the radio frequency is relinquished to other users of the channel, the receiver can send (or transmit) feedback to the sender. Because of the nature of basic radio technology, feedback is rarely immediate.

So what can we do to mitigate the lack of feedback when using radio? How can we best assist and guide the non-radio operators among our served agencies during an emergency communications situation? Here are some strategies you can employ.

First, and most importantly, slow down. When there is a lack of feedback between the sender and receiver in a communications event, the sender naturally assumes that the message being sent is getting through and is understood because there is no information to indicate otherwise. This may cause the sender to start talking even faster.

Second, choose words wisely. During communication, the sender knows what he or she wants to say and what is coming next during the process, while the receiver is constantly trying to keep up and can only anticipate what is coming next. The receiver is always one step behind, and if the sender throws unfamiliar or difficult words or complex details into the communication, the receiver must stop anticipating what is coming next and focus on the meaning of that unfamiliar word or unnecessary detail, and the communications virtually stops. An effective sender senses the difficulty and adjusts; without feedback, a sender may just keep going, unaware that effective communication is not occurring. Whenever possible, avoid unfamiliar words and jargon, but if you can't, slow down and define those words, or give clarifying details to allow the receiver to keep up. By doing so, you actually provide your own sort of feedback.

Third, insert breaks specifically for feedback. Because the typical radio system is simplex and the channel must be shared, the effective radio communicator will pause frequently and ask for confirmation that the message is being received. The periodic insertion of the plain English phrase, "How copy?" is very effective, despite being a bit of radio jargon.

Fourth, use standard phonetics to spell out complex or difficult words. "Alpha" for the letter "A," "Bravo" for the letter "B," and so forth. The noises and short duration disruptions inherent in radio can block out signals and, if timed just so, severely corrupt the reception of the intended message. Even verbalizing the individual letters might not get through during poor conditions. For example, trying to convey the location Detroit by verbalizing D-E-T-R-O-I-T may end up as D ? ? R ? I T. But if the sender uses the phonetic "Echo" for the letter "E," even though the receiver only catches "cho," a familiarity with the standardized phonetic alphabet helps the receiver infer that "cho" really is "Echo," or "E." So "Delta, Echo, Tango, Romeo, Oscar, India, Tango" will get the message through when "Detroit," or "D-E-T-R-O-I-T," might not. (Note, too, that the use of phonetics also slows the process down.)

The phonetic alphabet used should be a commonly accepted one — just as unfamiliar words slow down communications, unfamiliar phonetics can impede getting the message through. The use of a non-standard phonetic alphabet is common among emergency responding agencies and has been identified as one impediment to interoperability among responding agencies. With proper training, though, radio users learn that "Delta" is more effective than "Dog" in getting the letter "D" communicated through static.

There are, of course, other steps that will enhance communications, but these provide a good start. To recap, using a simple model for human communications we have looked at the communication process and learned how to evaluate what may prevent our message from getting from the sender to the receiver and how to take steps to make our communications more effective. In the remainder of this chapter, we will put the "magic" of radio into an understandable, layman-understandable context.

HOW DO RADIO WAVES GET FROM POINT A TO POINT B?

For many people, one of the mysterious things about radio communications is how radio waves get from one place to another. Radio waves are part of a wide chunk of wave-based energy within the electromagnetic spectrum, which ranges from radio frequencies, to visible and invisible light, and beyond. The radio frequencies of the spectrum are divided into bands — manageable blocks of frequencies, if you will — that behave in a similar way when traveling from one point to another. For instance, those in the high frequency (HF) band can travel very long distances by skipping or being reflected off the ionized upper reaches of the Earth's atmosphere, called the ionosphere, in a process known as sky wave propagation. The very high frequency (VHF) and the ultra high frequency (UHF) bands tend to travel locally only, in straight lines, through line of sight propagation, meaning that if there are no obstacles between stations using VHF or UHF, they will be able to communicate. Generally, public service agencies are assigned specific frequencies or channels within the VHF or UHF bands, because their areas of responsibility are more local (cities, towns, or county). A major advantage of using Amateur Radio for communications is that the Amateur Radio service is allocated a very large segment of the radio frequency spectrum, from HF through VHF, UHF, and beyond, allowing Amateur Radio operators to move around the spectrum and take advantage of changing propagation conditions. Understanding how the individual bands of radio frequencies propagate and travel lets the radio communicator choose the appropriate band, for which he or she is authorized, to facilitate the connection between the sender and the receiver. For instance, if great distances are involved, HF may be the answer. But signals in the HF band may skip right over an intended short-range receiver and never be heard. Moreover, the communications modes authorized in the HF bands (SSB and AM) tend to be vulnerable to static crashes from lightning strikes and propagation disruptions (or enhancements) caused by solar sunspot activity and solar flares.

On the other hand if the situation calls for a high volume of local-interest information in response to a localized emergency situation, then VHF or UHF is more appropriate. The communications modes generally authorized in the VHF and UHF bands (FM) tend to be less prone to naturally generated static and noise, and because the propagation on these bands is not dependent on ionosphere reflection, they are relatively immune to solar disruptions. However, because the wavelengths of VHF and UHF radio signals are very close to dimensions of common obstacles that might get in the way of the wave's travel, radio waves in these upper frequency bands can be reflected and/or absorbed, depending on the local terrain. Even the leaves on trees can be the right dimensions to serve as millions of antennas, absorbing VHF and UHF radio energy and preventing it from reaching the intended destination.

HOW RADIO WAVES TRAVEL: A HANDS-ON DEMONSTRATION

It may be helpful to perform this demonstration to further the understanding of how radio waves travel. For the following approach to be meaningful, a fundamental basis in the principles of the electron and magnetism is needed. Your audience may be familiar with magnetism, but the electron may be more difficult; if so, try relating it to the current supplied by a battery.

Radio waves are part of the electromagnetic spectrum of wave energy and consist of two components: an electrostatic one and a magnetic one, thus the name. How are these components related? Let's look at a simplified explanation that, while not absolutely correct technically, will help the layman radio user grasp the essence of a very complex topic.

Virtually all of electronics, including radio, can be boiled down to two fundamental principles: 1) moving electrons create magnetic fields; and 2) moving or changing magnetic fields cause electrons to move. In an electronic device, the circuits and components that make up the device manipulate these fundamental processes (moving electrons and moving magnetic fields) to accomplish some task. Understanding this relationship goes a long way toward understanding radio.

For the demonstration, you will need a bundle of three equal lengths of copper, aluminum, and PVC plastic pipe scraps around ¾ inches in diameter, and a very strong rare Earth magnet (neodymium magnet cylinder), which can be procured at **www.gaussboys.com/magnets/cylinders/**. Conduct the demonstrations as follows:

Step 1. Show the audience that the magnet will not stick to the pipes in your bundle and ask them why. *(Answer: Ferric materials [metals with iron] tend to be magnetic, but since there is no iron in your copper, aluminum, or plastic pipes, the magnet is not attracted to these and will not stick.)*

Step 2. Ask the audience what they think will happen if they drop the magnet into one of the pipes and let it fall through? *(Answer: Logically, it will fall right through, but you will get a variety of answers and theories.)*

Step 3. Choose an audience member and to come up and hold your bundle of pipe scraps so that he or she can sight down through the pipes while dropping the magnet into each one sequentially and observe what happens.

Step 4. Ask the demonstrator to drop the magnet into the plastic pipe and report what is observed. *(Answer: Most will anticipate that the magnet will fall straight through unimpeded, as it does; this provides a reference for what follows.)*

Step 5. Ask the audience member to drop the magnet through the copper pipe and report what is observed. *(Answer: I will not spoil the demonstration for you, you have to see it to believe it.)*

Step 6. Finally, ask the audience member to drop the magnet through the aluminum pipe and report what is observed.

Step 7. Discuss the demonstration to point out the connection between the fundamental principles.

Remind the audience of the two fundamental principles of electronics. Explain that as the magnet falls through the pipe, it creates a moving or changing magnetic field that causes the electrons in the pipe to move. The moving electrons in the pipe, in turn, create an opposing magnetic field that prevents the magnet from falling straight through the pipe. The magnet falls straight through the PVC pipe because one of these principles — moving electrons — is not involved (plastic is an insulator, after all). You can also discuss why the magnet appears to fall faster through the aluminum pipe than through the copper one. The answer is that copper is a better conductor of electricity, and a better conductor means more moving electrons, more moving electrons means a stronger opposing magnetic field and therefore a slower-moving magnet.

Now make the connection between the falling magnet demonstration and how radio waves travel from point A to point B. You may want to ask the following questions to help clarify the concept:

1. What is the purpose of transmitter equipment in a radio station? *(Answer: It moves electrons in the transmitting antenna back and forth at a rate equal to the radio frequency being used.)*

2. What do these moving electrons in the antenna create? *(Answer: Moving electrons create a magnetic field.)*

3. What does the induced magnetic field do to the electrons in the vicinity? *(Answer: This is where you take a little instructor license and stretch the truth a bit, explaining that moving or changing magnetic fields cause electrons to move.)*

4. What, in turn, is the result of these moving electrons? *(Answer: The result is alternating electrostatic and magnetic fields that radiate the radio energy out away from the transmitting antenna at approximately the speed of light, barring any impediments that might interfere [as discussed above].)*

5. As the final magnetic field washes across the receiver antenna at the destination, what happens to the electrons in the antenna? *(Answer: They move down the coax or antenna field line to the receiver. The receiver then takes the moving electrons and extracts the message attached to the radio wave. The final result is human communication via the medium of radio.)*

This brings us back to the original question: Why do we need to know how radio waves get from point A to point B? Radio waves are basically alternating electrostatic and magnetic fields that each generate the other as the radio wave energy radiates away from the originating point. Anything that gets in the way of these alternating fields potentially may attenuate, reflect, absorb, or otherwise affect the wave in a way that can degrade the effectiveness of the wave at carrying the message being sent. This is the medium's noise component in our human communications model. Effective communicators attempt to be as aware as possible of potential impediments that could prevent their message from getting through to the receiver, and they will take steps to mitigate or prevent the noise and distraction from interfering with the communications process.

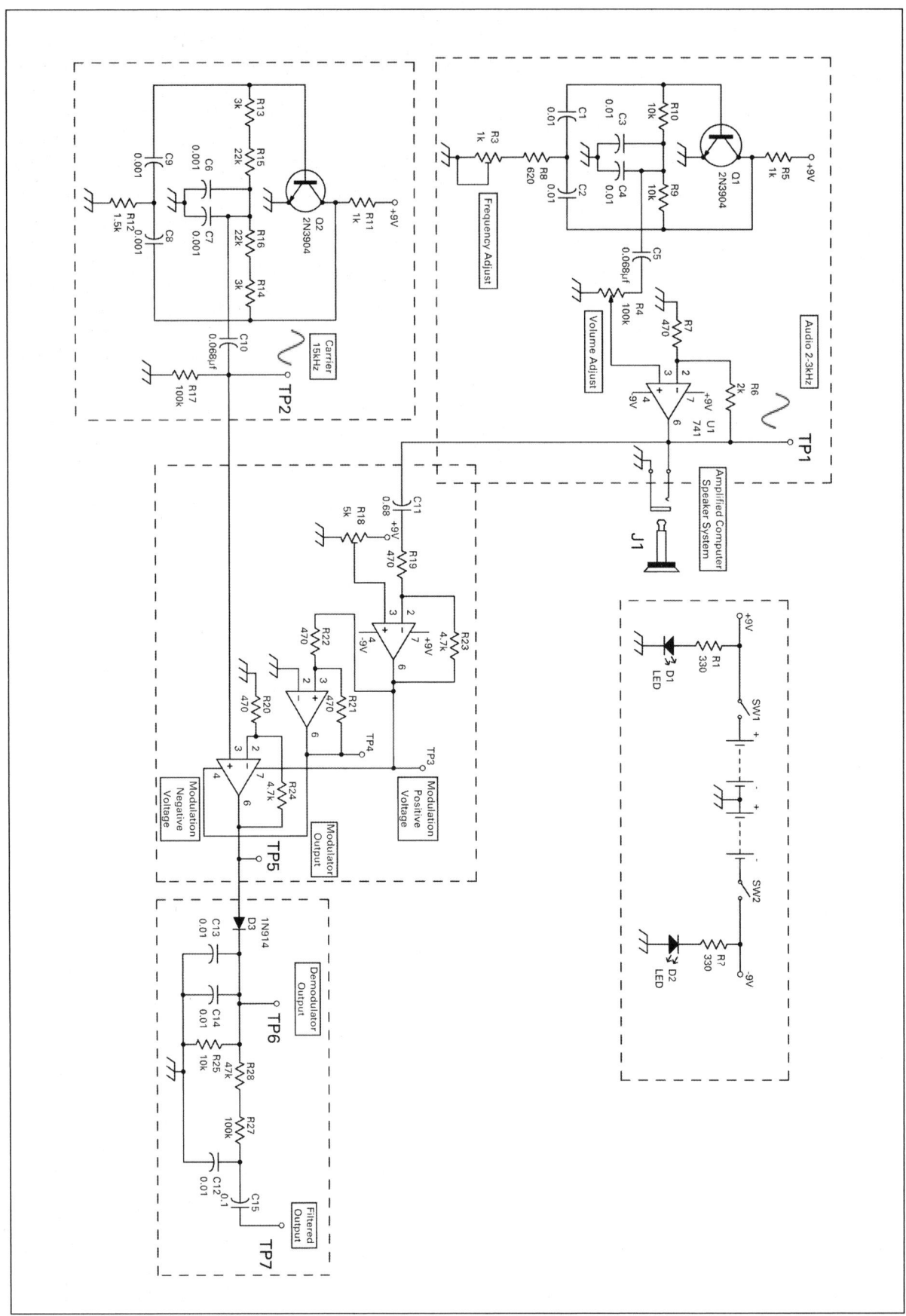

Figure 10.1 — Modulation/de-modulation demonstration circuit diagram. You can create one of these circuits and follow the steps in the text to demonstrate modulation and demodulation to personnel of a served agency.

The radio communicator, therefore, should anticipate some possible attenuation (reduction) of radio energy in the VHF and UHF bands in heavily forested areas; whereas an area with rugged mountains and exposed rock surfaces is full of good reflectors that will interfere with radio signals in the UHF bands. That reflected radio energy bounces all over the place, with each reflected portion of the radio signal traveling different distances. This process is called *multipathing* and it means that portions of the radio signal will arrive at the receiving station at different times. Although these time differences are very slight, they are significant enough to cause distortion (noise or corruption) of the received signal. In severe cases, the multipathing distortion can make the radio signal unreadable. Consequently, UHF may not be the most appropriate band to use for a canyon rescue operation.

Such a basic understanding of radio wave propagation is as important in the preparations that emergency managers make before an event happens as it is in the emergency situation itself. Some may argue that emergency management leaders should depend on "radio experts" to help them navigate through the maze of radio equipment and technology available, to purchase the right equipment, to operate on the right band, and to accomplish their community's emergency communications mission. Seeking expert guidance is wise, but it is wiser still for these emergency managers to have a fundamental understanding of radio wave propagation so they can make informed evaluations of the experts' recommendations and choose technology that is appropriate to the task. In an actual event, who is going to be operating the radio equipment to effect emergency communication? Probably not the radio consultant who gave the procurement advice — it will be the emergency manager and his or her staff members who employ the radio technology to effect emergency communication. Take a look at the sidebar "How Radio Waves Travel" for a brief practical demonstration you can do for anybody who might need a visual representation.

TECHNICAL ASPECTS

Referring to the model of human communications, the sender has a message he or she wants to convey to the receiver. If radio is chosen as the medium for relaying that message, the sender would then use the appropriate electronic technology to combine or embed the message to be sent with a radio wave that will carry the message to the receiver (by the process you explored with the magnet falling through the pipe).

This process of attaching the message to the radio carrier wave is called modulation. The radio wave leaves the transmitting antenna with the attached message and radiates out into space. At the receiving antenna, the final magnetic wave of the radio wave causes electrons in the antenna to move down to the receiver equipment where the radio carrier wave is removed, through a process called *demodulation*, to recover the original message. The recovered message then is presented to the receiver of the communication over a loudspeaker, video display monitor, printed text, or whatever else may appropriate for the format of the message.

This is basically how radio technology is used as the medium for human communications. There is, however, one important limitation in this process for our purposes: Any time you manipulate a waveform (in this case by attaching the message to the radio wave, transmitting that radio wave, and then recovering the attached message and eliminating the radio wave) there is going to be distortion or noise injected into the process that makes the use of radio as the medium less than perfect. Again, the effective communicator will know as much as possible about the noise and distractions present during the communication event and do whatever is possible to mitigate those impediments. Facilitating that is the purpose of these technical discussions.

The explanation of radio as a communications medium can be enhanced through the use of some training aids, including a circuit that simulates the radio communications process (message, radio wave, modulation, demodulation, noise) and an oscilloscope to display the radio wave at incremental steps though the process. A sample illustrative circuit that can be easily duplicated by hand-wiring a protoboard is shown in **Figure 10.1**. The board, used in conjunction with an oscilloscope, is an effective visual aid that helps your audience understand the modulation/demodulation process.

Here are some ideas on how you might employ these resources during your discussions with the served agency. This circuit offers a number of test points. Moving the oscilloscope probes from one test point to the next lets you sample and display the simulated message and the radio carrier wave through the modulation/de-modulation process to visually reinforce the following concepts.

Wave Fundamentals

Begin the lessons by presenting the vocabulary of waveforms so the audience will understand the subsequent lessons on modulation when that vocabulary is used. Connect the oscilloscope to TP1 to display the audio waveform, and connect a computer speaker amplifier to J1 to provide sufficient audio levels for the audience to hear. The projected display will look like that shown in **Figure 10.2**. Use this screen to illustrate the terms that describe a waveform (crest, trough, wavelength, frequency, period, amplitude). Then adjust the frequency through the range by adjusting R6. The frequency range is between approximately 2 and 3 kHz. During this portion of the demonstration, emphasize the relationship between frequency and pitch. Next, emphasize the inverse relationship between frequency and wavelength. Finally, vary the amplitude of the wave using R8. Emphasize also the relationship between amplitude and loudness (strength) of the wave while pointing out the independence of amplitude and frequency.

Modulation

Set up the board and oscilloscope to connect one channel of the scope to TP1 and the other channel to TP2. The

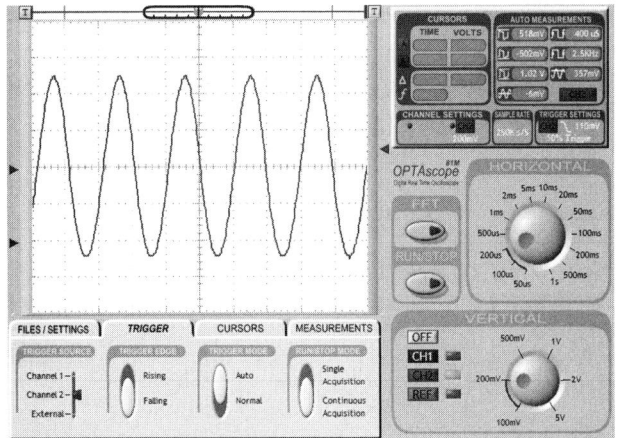

Figure 10.2 — Connecting an oscilloscope and a computer amplifier speaker to the circuit will result in an oscilloscope screen that looks like this. Use this screen to illustrate the terms that describe a waveform (crest, trough, wavelength, frequency, period, amplitude).

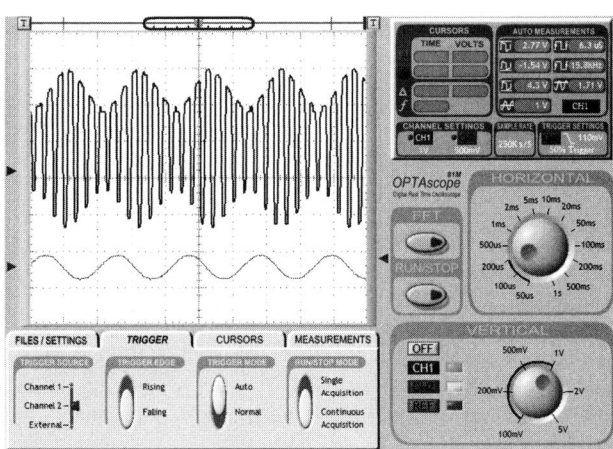

Figure 10.5a — Varying the amplitude of the audio wave. Fifty percent modulation.

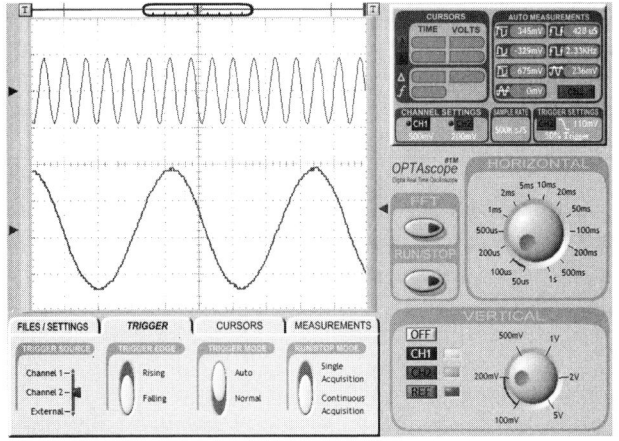

Figure 10.3 — Beginning the demonstration of modulation. The upper (higher frequency) trace is the simulated radio carrier wave, and the lower (lower frequency) trace is the simulated intelligence (the message to be sent).

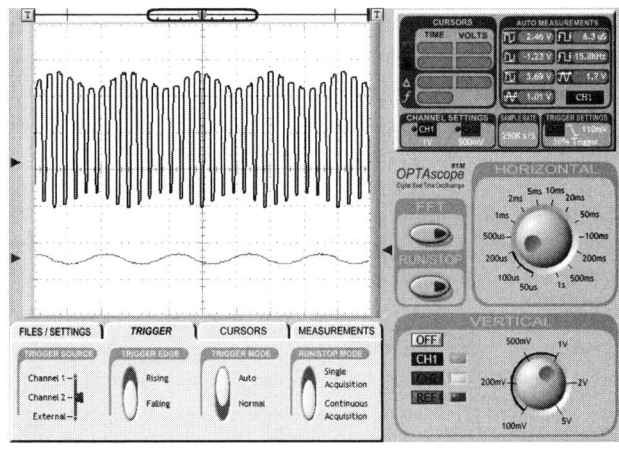

Figure 10.5b — Varying the amplitude of the audio wave. Twenty-five percent modulation.

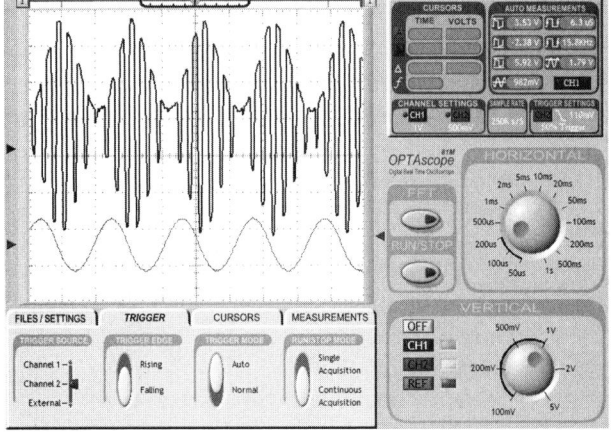

Figure 10.4 — The upper trace is the amplitude modulated radio carrier wave, and the bottom trace is the simulated intelligence (the message to be sent).

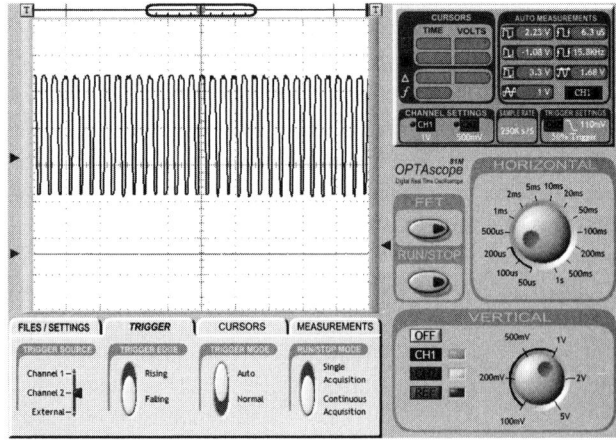

Figure 10.5c — Varying the amplitude of the audio wave. Zero percent modulation.

The Amateur Radio Public Service Handbook

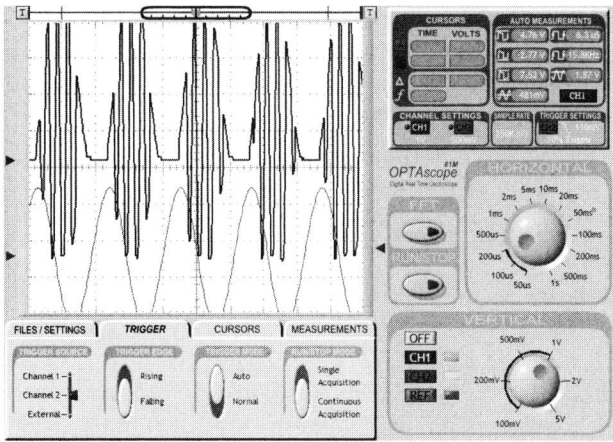

Figure 10.6 — Overmodulation causes distortion of the carrier waveform.

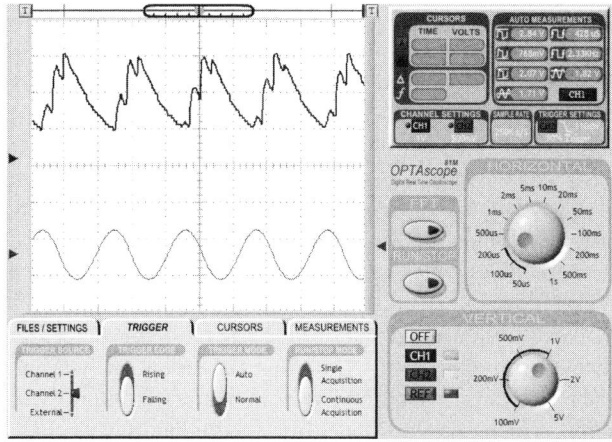

Figure 10.7 — Demodulation of the AM waveform. The alternating current (ac) component has been eliminated and a direct current (dc) wave results. Some remnants of the carrier wave are still present (upper trace) that corrupt the desired audio waveform.

scope display will look like that shown in **Figure 10.3**. The upper (higher frequency) trace is the simulated radio frequency carrier wave; the bottom (lower frequency) trace represents the audio to be embedded on the carrier wave during modulation.

Next, move one channel of the scope from the carrier on TP2 to the demodulator output on TP5. The modulator output is displayed in upper trace, as shown in **Figure 10.4**. Use this waveform to illustrate the concept of amplitude modulation, demonstrating how the amplitude of the carrier wave is varied in step with the amplitude of the audio wave being imprinted on the carrier.

The concept of amplitude modulation is further reinforced by varying the amplitude of the audio wave using the R8 control. The panels shown in **Figures 10.5a**, **b**, and **c** depict approximate levels of 50 percent, 25 percent, and 0 percent modulation.

The concept of overmodulation by increasing the audio wave amplitude beyond the 100 percent modulation level is illustrated in **Figure 10.6**. Varying amounts of distortion become evident as the modulation level goes beyond 100 percent. Explain that this is the kind of noise the sender must avoid during the communications process. As the audio wave increases beyond the 100 percent modulation certain point, the carrier waveform becomes distorted (flat topping) and no longer faithfully follows the audio waveform. At this point, the radio carrier wave with the embedded message is radiated out into space and is on its way to the receiver, which brings us to the next part of the demonstration.

Demodulation

Set up the demodulation demonstration by configuring the board for 100 percent modulation and connecting one channel of the scope to the audio waveform at TP1 and the other channel to the output of the demodulator at TP6. TP6 is connected to a simple diode rectifier that is used to demodulate the AM waveform. The display will look like that shown in **Figure 10.7**. Use this display to illustrate that the carrier

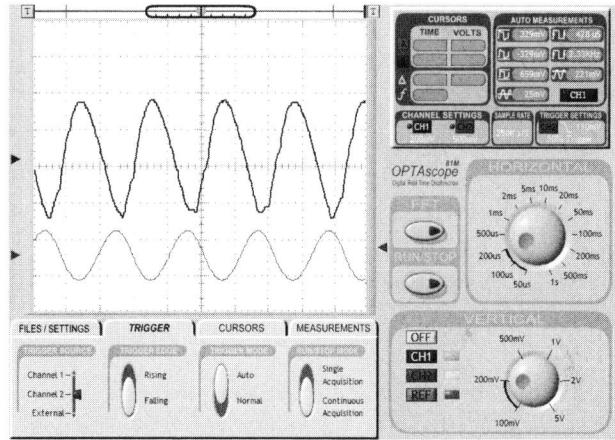

Figure 10.8 — Filtering of the demodulator output to eliminate the remnants of the radio carrier wave. At this point, the audio output is a fair duplication of the original audio waveform.

waveform has been modified by the demodulator; that is, the alternating current (ac) component has been eliminated and a direct current (dc) wave results. Point out that there are still some remnants of the carrier wave still present (upper trace) that corrupt the desired audio waveform.

Next configure the board with one scope channel connected to the filter output at TP7. The display will look like that shown in **Figure 10.8**.

Use this display to point out that a filter has removed virtually all of the remnants of the carrier that remained after demodulation and that the audio output is a fair duplication of the original audio waveform. This is a good opportunity to point out that the modulation/demodulation process is not perfect and that there is always going to be some distortion of the audio waveform during the process, due to the imperfections of electronic circuitry.

INSIDE A RADIO

If you dare to open the cover of a radio transmitter, all the tiny individual components that make up the circuitry can be very intimidating. But because the radio circuitry will inject noise and distortions that can impede communications, the effective communicator must have some level of knowledge and understanding of the fundamentals of how a radio works.

A radio, or any electronic device for that matter, can be broken down into five building blocks of circuit components that are assembled like puzzle pieces to build the radio. The remainder of this chapter will offer some tools that you can use to present these building blocks in a way that provides some foundation knowledge for your supported agency. These building blocks are deceptively simple, so a word of caution: don't try to make them more complicated than they really are. Real complications don't arise until you get to the individual electronic component level (the integrated circuits, transistors, resistors, capacitors, and inductors that make up the circuits or building blocks), and that degree of understanding warrants numerous dedicated courses in electronics.

The five building blocks are:

1. Oscillator
2. Amplifier
3. Rectifier
4. Mixer
5. Filter

That's it. Now let's take a look at each one.

Oscillator

The purpose of the oscillator is to generate ac from a dc supply. Oscillators can generate very low, sub-audible frequencies and very high frequencies in the microwave radio frequency range. The oscilloscope display shown in **Figure 10.9** illustrates the ac waveform produced by our oscillator circuit, which is powered by a dc battery. Connecting an oscilloscope to the output of an oscillator will demonstrate that the dc power from the battery is converted to an ac signal.

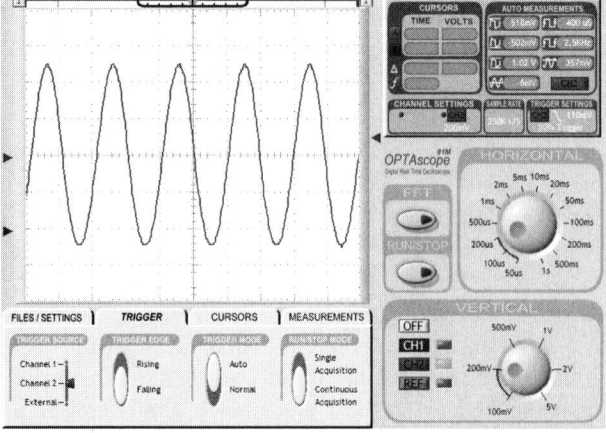

Figure 10.9 — This screen shows an ac waveform produced by the oscillator circuit, which is powered by a dc battery. Connecting an oscilloscope to the output of an oscillator will demonstrate that the dc power from the battery is converted to an ac signal.

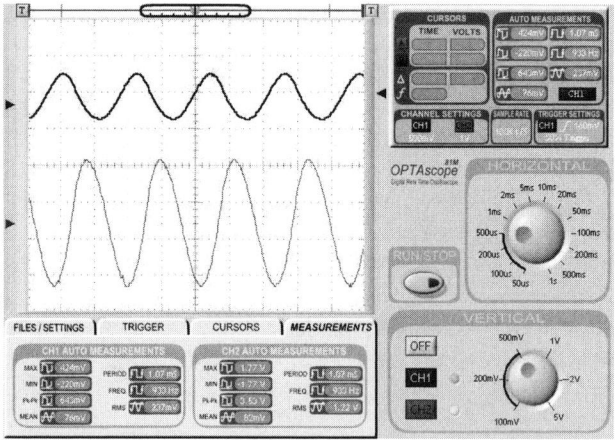

Figure 10.10 — Amplifier input (upper trace) and amplifier output (lower trace), with a significant increase in amplitude.

Amplifier

An amplifier increases the strength of a signal. The amount of amplification is called the gain of the amplifier. Voltages and currents used in wireless technology are incredibly small. Even very high-powered commercial broadcast stations create voltages on a receiver antenna that are in the millionths of a volt range. Amplifiers are, therefore, needed to increase the strength of weak currents so that they can be used in communications. The oscilloscope display shown in **Figure 10.10** depicts the input to the amplifier (upper trace) and the output (lower trace; both traces have the same vertical scale factor) with a significant increase in amplitude.

Rectifier

Rectifiers convert ac into dc, which is what all wireless technology depends on for the power source. When wireless devices are portable, such as a cell phone, handheld transceiver, or global positioning system (GPS), the power comes from batteries, which produce dc by chemical reaction. When these same devices are used in the home or other more permanent location where commercial power is available (from a wall outlet) a converter is used to change the current from the home (ac) into current the device can use (dc) to operate or recharge the batteries. A "wall-wart," the familiar little box that plugs into a wall outlet and has a connecting cord running to a wireless device, is simply a rectifier.

The primary component in a rectifier is the diode. These small devices are like one-way valves and allow current to only flow in one direction. If a current tries to flow in the opposite direction, it is stopped by the diode. As the name

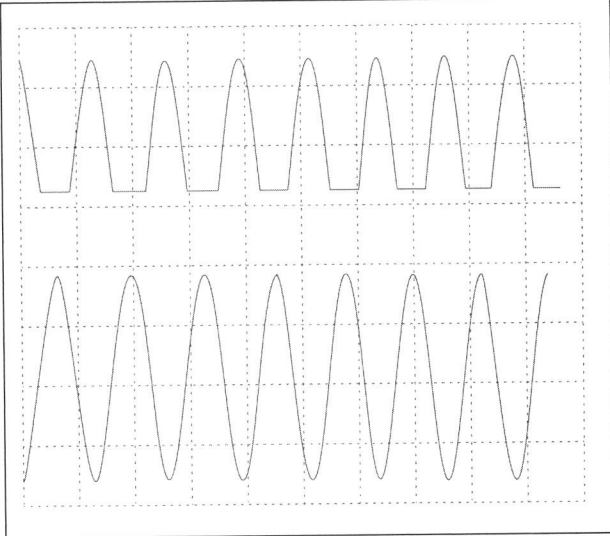

Figure 10.11 — The bottom trace is the ac current, and the upper trace is the rectified dc current, ready for further filtering.

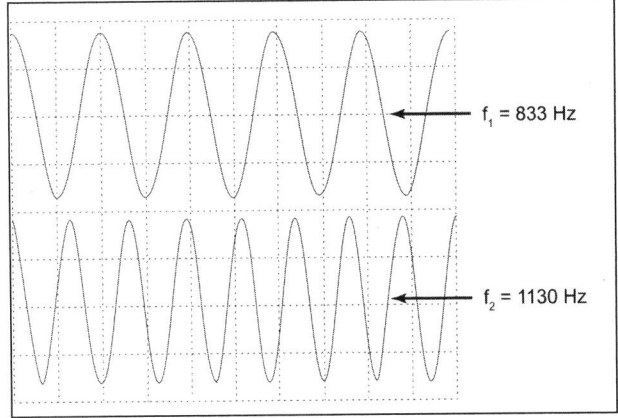

Figure 10.12 — Two frequencies prior to mixing.

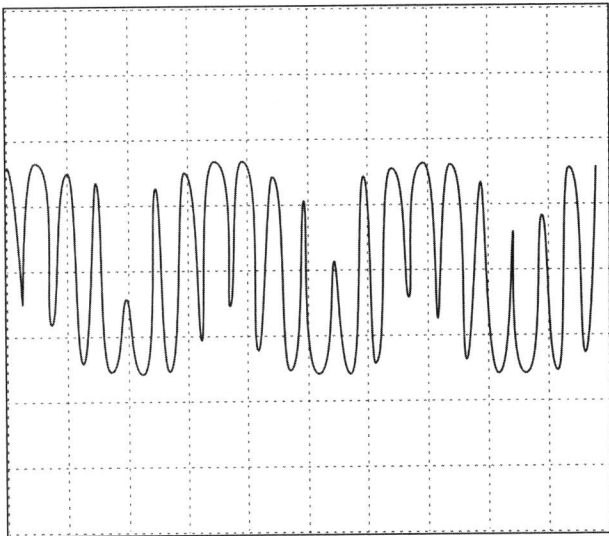

Figure 10.13 — The mixer combines the input frequencies.

implies, alternating currents reverse direction each cycle, while direct current moves in only one direction. If you look at an ac waveform, you see a sine-wave shape with one portion of the wave on the positive side of the y-axis and one side on the negative side. If this wave were applied to a diode, one half of the wave would be allowed to pass through (depending on the direction of the diode), and the other half would be blocked. The current allowed to pass will have only one polarity, and that means it only travels in one direction. In this way the diode rectifies the ac and converts it into dc.

The dc that comes out of the rectifier is not smooth, however, and will need some filtering to take off the rough edges before a wireless device can use it. You can tell when the filter in a rectifier has failed or is not working properly. These "ripples" in the rectified dc, if allowed to enter a wireless device, will produce a loud audible hum in the output speaker. The filters are not perfect, however, and some of the rough edges will get through to the wireless device, so you may hear a little hum even with a properly operating rectifier.

The oscilloscope display in **Figure 10.11** illustrates the function of the rectifier. The bottom trace is the ac waveform input, and the upper trace is the pulsed dc (positive component only) output of the rectifier, ready for further filtering.

Mixer

Again, as the name implies, a mixer combines two signals. In our discussion of modulation we spoke about the importance of the mixer in combining the message to be sent with the radio carrier wave. Another use of the mixer is to combine two frequencies, one collected by the antenna and one generated by a variable frequency oscillator controlled by the tuning dial. The combination of these two frequencies lets the radio user select the desired frequency, which is then passed to other parts of the radio for filtering and additional processing. Without the mixer, a dedicated radio would be required for each frequency of interest.

Mixers can be thought of as frequency adding and subtracting circuits, because the output of the mixer is actually two frequencies: the sum and the difference of the input frequencies. If the mixer input frequencies are 660 Hz and 2000 Hz, the output of the mixer would be 2660 Hz (660+2000) and 1340 Hz (2000–660). The output waveform of these combined frequencies appears to be complex, but with careful observation you can begin to see how the two frequencies depicted in **Figure 10.12** are mixed in **Figure 10.13** (the rapid change is the sum, the meandering change is the difference). In an actual wireless system, the output of the mixer is fed to a set of filters that would attenuate the undesirable frequencies and pass the desired frequency to the next stage of the system.

Filter

A filter reduces, or attenuates, unwanted frequencies whileactuallowing wanted frequencies to *pass*, which means

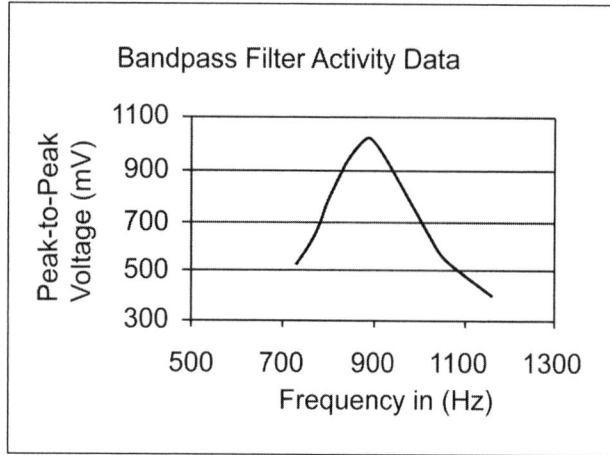

Figure 10.14 — Filters that pass a desired range of frequencies are called band-pass filters. In this case, the frequency of approximately 900 Hz is the desired output of this band-pass filter; frequencies above and below that are attenuated.

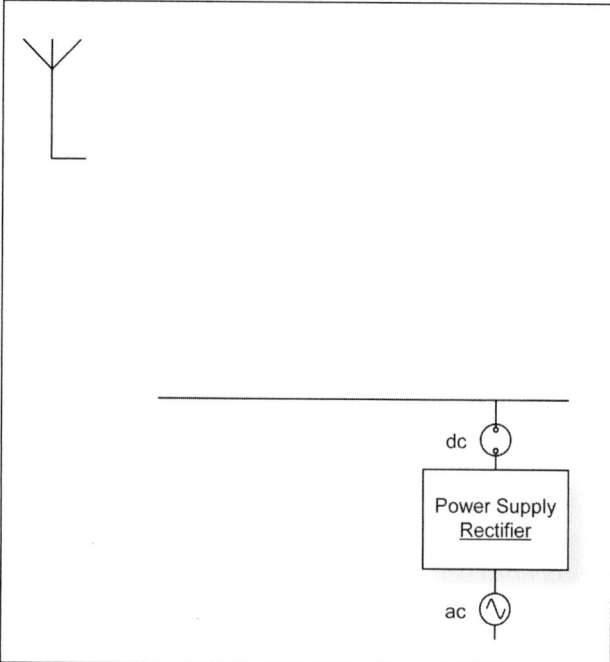

Figure 10.15 — Antenna and power source for a hypothetical amplifier.

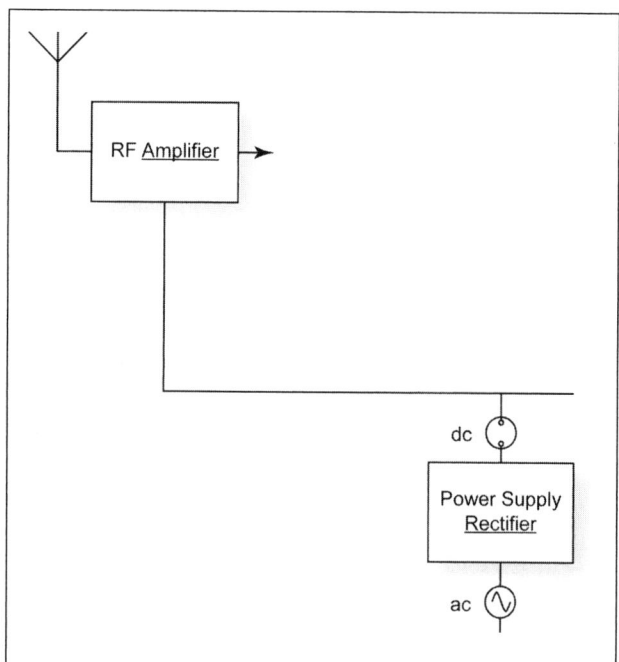

Figure 10.16 — Radio frequency amplifier.

that the desired frequency waves are allowed through the filter without being appreciably degraded or reduced in amplitude, while undesired frequencies are significantly reduced in strength. Just think about how many radio and television stations are in your local area. Now multiply that number by tens of thousands. You're still only approaching the number of radio stations and other signals that are on the air at any given moment. How do you eliminate some of those competing or interfering signals? You filter the unwanted signal out of the picture.

There are three basic types of filters: those that pass frequencies higher than a designated cutoff frequency (high-pass), those that pass frequencies lower than a designated cutoff frequency (low-pass), and those that pass a desired range of frequencies (band-pass). **Figure 10.14** illustrates the output of a band-pass filter, the frequency of approximately 900 Hz is the desired frequency; frequencies above and below that are attenuated.

PUTTING IT ALL TOGETHER

Now that we have broken down the complex concept of a radio into five basic building blocks (oscillator, amplifier, rectifier, mixer, and filter), it's time to assemble a radio receiver (in the abstract) from these components. The following sequence of figures will take us through the concepts one step at a time.

Figure 10.15 depicts the antenna and the power source for our hypothetical receiver. Electronic equipment needs some sort of power source, either a battery or a connection to the ac power supplied to our homes, business, and offices. Most of the electronic equipment we use today requires dc power to operate, so if the equipment's power source is the ac outlet, which of the building blocks converts ac into dc current? The rectifier, our third building block.

As we learned during the demonstration of the magnet falling through a pipe, radio waves are made of alternating electrostatic and magnetic fields. As a radio wave reaches the receiving destination, the final magnetic wave washes across the antenna and causes the electrons to move in the antenna and down the feed line. As we also learned, the voltages induced in the antenna by the millions of radio waves striking it are on the order of millionths of a volt (.000001

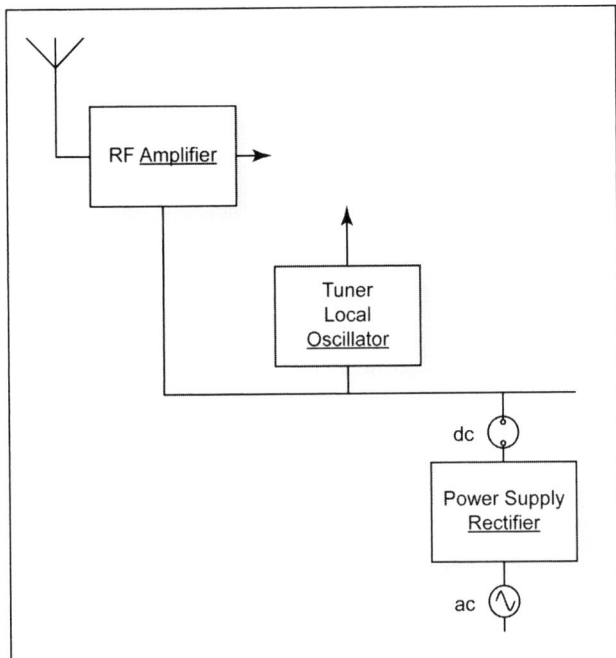

Figure 10.17 — The local oscillator (so called because the radio wave is confined to the inside of the receiver) converts the dc current from the power supply into an ac wave.

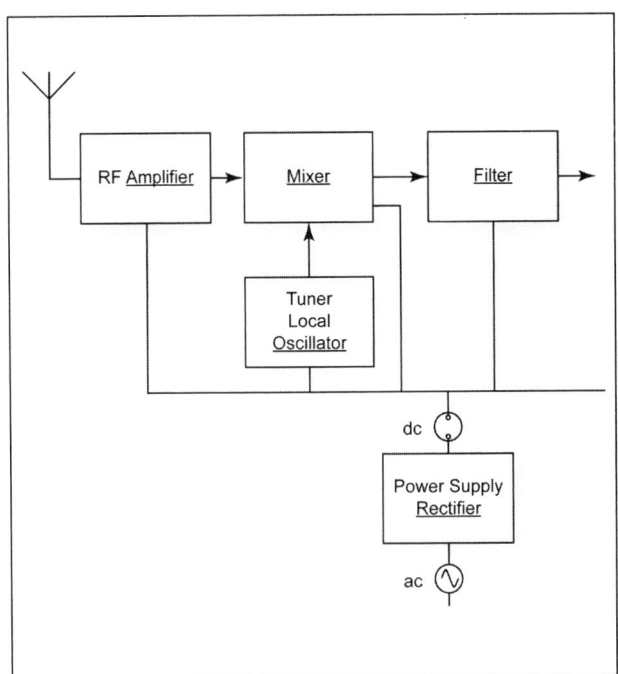

Figure 10.19 — The filter removes unwanted mixer output frequency.

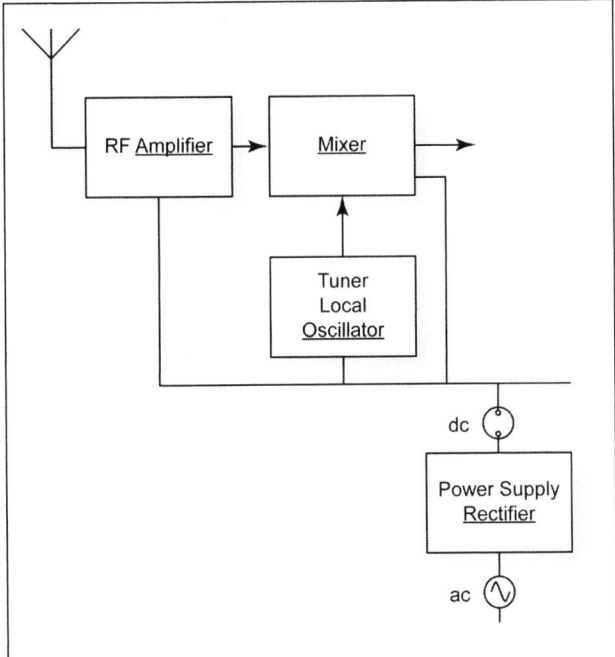

Figure 10.18 — The mixer combines the input frequencies.

volt) and the strength of these signal levels needs to be boosted to make them more usable. Which building block increases the amplitude of the input signal? The amplifier (see **Figure 10.16**). This figure shows an amplifier circuit specifically designed to amplify radio frequency signals.

It is impractical to have a radio for every channel we want to use for communications, so we need a way to select the single radio frequency we want from among the millions of signals coming down the antenna feed line and through the RF amplifier. Most radios have a way of tuning in the desired frequency with some sort of manual control. This control is connected to a circuit that generates a specific frequency within the radio receiver that will be used to isolate the desired frequency. Which of the building blocks will convert the dc current from the power supply into an ac wave, in this case the tuner frequency? The oscillator (see **Figure 10.17**). The oscillator depicted here is a local oscillator, because the radio wave is confined to the inside of the receiver.

Now that we now have the desired radio frequency (plus a million others that are unwanted) and a tuner frequency for tuning the radio, which of the building blocks do we need to combine those frequencies with a predictable result? The mixer (see **Figure 10.18**). The mixer's output, depending on the specific type of mixer circuit used, will have two components: the sum and the difference of the tuner frequency and the desired frequency.

The range of frequencies we have to deal with has become more manageable; all that remains to do is eliminate those we don't want. What building block attenuates, or reduces in amplitude, the unwanted output frequency from the mixer? The filter (see **Figure 10.19**).

So we now have the desired radio carrier wave with the attached message (modulated carrier), and all the extraneous, unwanted radio signals received by the antenna have been attenuated to insignificant levels. The carrier wave has done its job and is no longer needed. The next step in the receiver is to remove the radio carrier wave through demodulation.

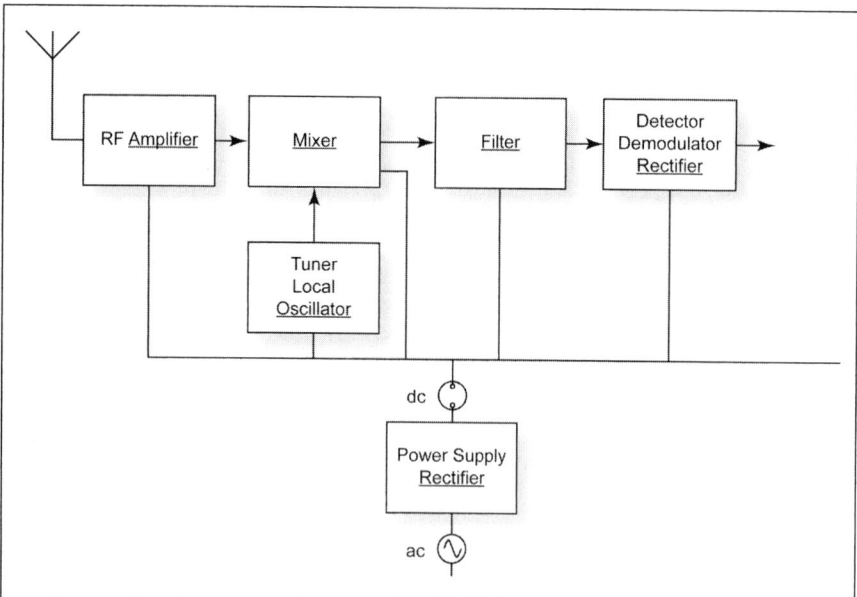

Figure 10.20 — The demodulator converts the radio carrier wave into dc.

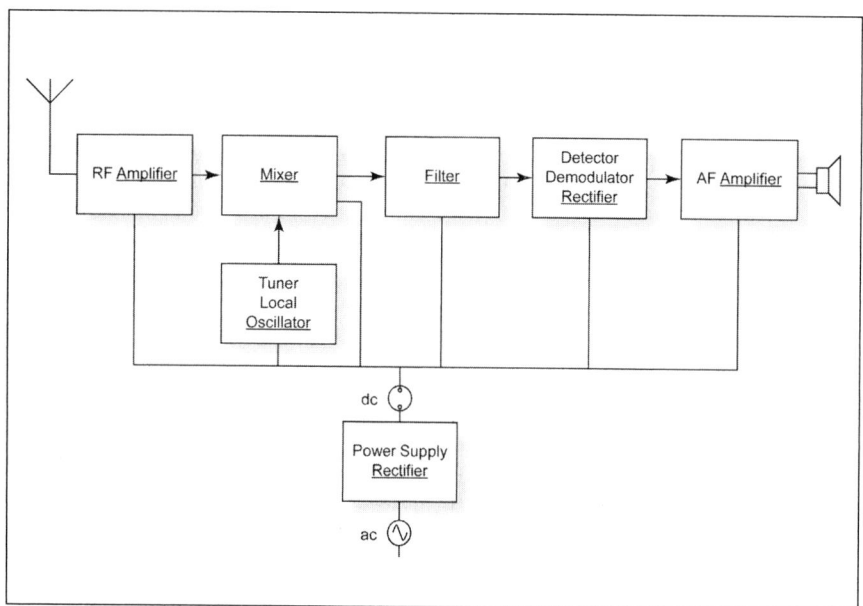

Figure 10.21 — The audio amplifier boosts the output of the demodulator to drive the speaker.

Which building block does that? The rectifier (**Figure 10.20**). The demodulator is basically just a rectifier circuit.

The output of the rectifier is now the original intelligence that was sent, with some noise and distortion injected along the way. However, the voltage levels of the signals coming out of the demodulator are still very, very low, on the order of tenths or thousandths of a volt, which is not much power to drive a loud speaker, or even earphones.

A final building block is required to increase the amplitude of the small voltage signals coming out of the demodulator. The last component of our receiver is, therefore, another amplifier (see **Figure 10.21**), in this case one specifically designed for audio frequencies. The result: the magic of radio, but now seen as not so mysterious.

SUMMARY

The value that Amateur Radio brings to local communities that must deal with emergency situations is not only a flexible backup communications infrastructure with trained operators, but also technical expertise in the use of radio as the medium for communications to help prepare and train the served agencies to use radios effectively. Radio can be intimidating, making the communicator hesitant to delve into the technical side of radio to the level needed to become an effective user of radio. However, radios — indeed virtually all electronic devices — can be simplified because they are made up of five basic circuit building blocks that perform specific tasks. A significant understanding of how a radio works can be gained by putting these five building blocks together to make a radio. Demystifying radio will help the communicator be a more effective user of radio as a communications medium.

The demodulation process is specific to the mode of transmission, and there are various kinds of demodulator circuits with different names, though all perform basically the same function: removal of the carrier wave to recover the attached intelligence. In most cases the radio carrier wave is a rapidly changing ac wave that has some parameter (for instance, the amplitude or frequency) modified in step with the embedded intelligence. If the ac radio wave is converted to a dc current, all that remains is a varying dc current with a magnitude proportional to the waveform of that of the attached intelligence. This is a complex topic, but the important point is that the demodulator converts the radio carrier wave into dc.

It takes more know-how about radio than just "how do I turn it on" and "how do I transmit" to use a radio effectively. On the other hand, you also don't need to be an electronics engineer either. Just some basic understanding and reflection of what happens during human communications and some basic understanding of radio technology will go a long way in the user being able to get the message through "When all else fails."

Chapter 11
A Compact, Versatile, Multi-Mode Kit for Emergency Communications

In 2007, a few dedicated members of the Surrey Emergency Program Amateur Radio Society (SEPARS) in British Columbia responded to a need for transportable emergency communication capability that had been identified by Surrey Fire Services. The lessons of some recent disasters, including Hurricane Katrina and a more local catastrophe, the 2003 Kelowna fire storm, brought home the serious implications of system failures of vulnerable communication and power networks that were prone to large-scale breakdown.

SEPARS was asked to conceive and construct a portable Amateur Radio package capable of supporting Surrey Emergency Program's communication function under a variety of disaster scenarios, whether local or regional in scope. The constructed kits were funded by the City of Surrey's Emergency Program and were to be capable of communication at HF, VHF, and UHF frequencies, including CW, SSB, and FM voice (including analog and D-STAR digital), PACTOR® III, and packet. Portability was essential, such that the equipment could be transported in an SUV or passenger van and be deployed by trained radio amateurs within 60 minutes of arrival at destination. This chapter describes the successful results of a year-long effort, which included conceptual planning; construction of a mock-up, a prototype, and two cloned copies; as well as trials and modifications.

FEATURES

Each kit is comprised of three units, containing radios (**Figure 11.1**), antennas, and HF antenna support poles (**Figure 11.2** and **Figure 11.3**). A complete list of components is provided at the end of this chapter.

The radios — two ICOM IC-2820H dual-band VHF/UHF, an IC-91AD handheld, and a Kenwood TS-480SAT HF — and ancillary items selected for the kits were chosen because of their appropriate band coverage and combination of features, compatibility with emergency gear utilized by other interacting emergency groups, small size, and proven reliability.

The radios and antennas are housed for storage and transport in hard-shelled Pelican cases with wheels, providing

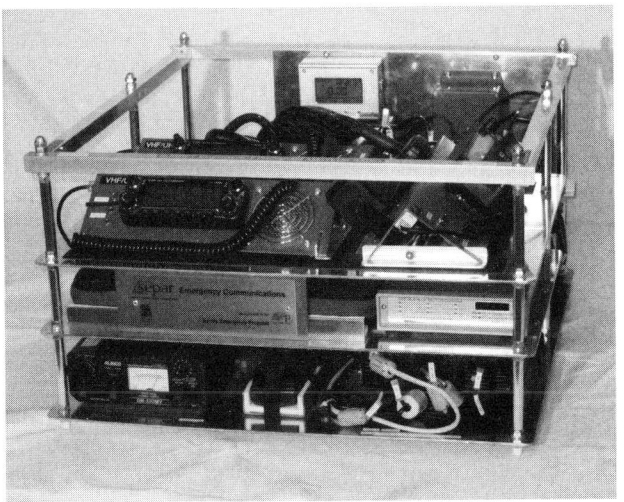

Figure 11.1 — Contents of Pelican Case #1, the prototype kit, with three control heads secured to the top shelf by hook-and-loop fasteners. Also visible are a laptop in its sleeve (behind the SEPARS plaque), an SCS PTCII-pro modem (at right on the middle shelf), and a DM-330MV power supply, IC-91AD handheld radio, and a Kenwood TS-480SAT (lower shelf).

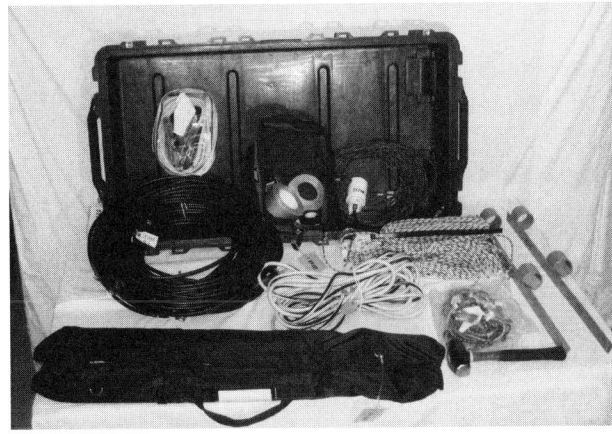

Figure 11.2 — Pelican Case #2 with contents displayed (lid not shown). The VHF/UHF antenna bag is in the lower foreground with coax, heavy-duty extension cord, rope, bungees, three-pound hammer, custom pole stakes, OCF dipole, battery jumper cables, and utility bag items behind.

The Amateur Radio Public Service Handbook

Figure 11.3 — Contents of the antenna mast bag: tripod, 12 four-foot fiberglass poles, guy ring, and stakes. Wire loops fastened to the base of the tripod allow it to be staked down. Note that two of the poles (lower right) have hose clamps at the lower ends to compensate for the absence of joint reinforcing rings, which are present on the other poles.

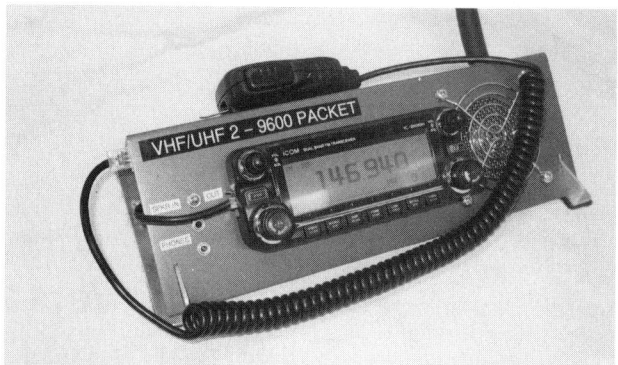

Figure 11.5 — A VHF/UHF panel with the IC-2820H control head attached by magnets to a mounting plate (not visible) supplied with the radio. The speaker grill is stainless steel mesh with a protective cover designed for cooling fans.

lockable, waterproof, robust protection. A heavy-duty, military-style duffel bag contains the collapsible pole sections for the HF antenna. A 90 amp-hr AGM sealed battery and a 1 kW Honda gasoline generator, packed separately, are essential accessories for each kit.

Each of the three radios is connected by an eight-foot-long cable bundle to its remote control head, an arrangement that permits three operators to be physically separated from each other when the kit is deployed (**Figure 11.4**). The con-

Figure 11.6 — The TS-480SAT control panel with the SGC noise cancelling DSP speaker fastened to the back. In this view, the VE7XDT power/SWR meter has been relocated from the base unit to the control head, and is held in place by hook-and-loop fasteners. Both a 12 V and signal cable must be provided at the control head for the meter to function at this location. When the meter is not in place, the spot on the back of the panel is occupied by the hand mic.

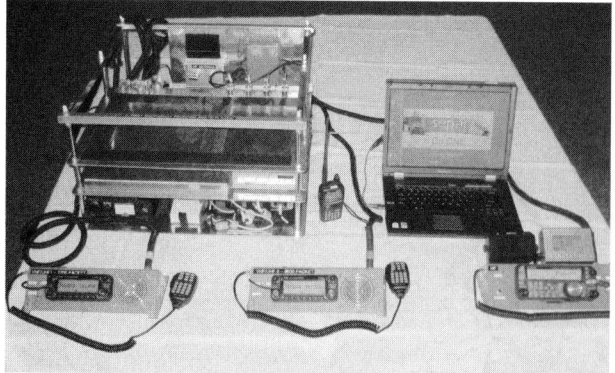

Figure 11.4 — Three control heads removed from the base are displayed close together, but each can be relocated a distance of eight feet from the base. The IC-91AD handheld radio has been removed from its storage bracket on the lower shelf. The laptop can also be moved the full length of its cabling.

trol heads, fastened to custom aluminum panels (**Figure 11.5** and **Figure 11.6**), are held firmly in place on the top shelf of the base unit by hook-and-loop fasteners during transport and until separated for operation.

The RF output from Ant 1 of the two IC-2820H radios and Ant 2 of the TS-480 is connected to its own dummy load when the radio is stowed, thus affording some backup protection against damage to the transmitter in the event the operator fails to connect a resonant antenna before powering up the next time. Since all three radios are provided with two antenna connectors each, the unused IC-2820H (receive only) connectors are capped and Ant 1 of the TS-480 is connected directly to the RF pickup for the external SWR/power meter. An SCS PTCII-pro modem does double duty by enabling PACTOR III on HF, and packet on VHF and UHF. One of the

IC-2820H radios is permanently assigned to 1200 baud and the other to 9600 baud, by separate connections to the SCS modem ports.

Power to the radios and accessories is provided by: a) a 32 amp Alinco DM-330MV switching power supply operating on 120 VAC from the mains or from a 120 VAC generator, or b) a 12 volt AGM sealed battery, depending on circumstances. All 12 volt power is routed through a West Mountain PG40 PWRgate and RIGrunner 4012 distribution panel. Modifications to this configuration (described later) were required to produce acceptable power output during battery operation.

Anderson Powerpoles® with 35 amp contacts are employed on all 12 volt connections in the interest of standardization. The 12 volt AGM battery is kept charged with a battery tender during storage but is also charged during connection to the power supply and the PWRgate when 120 VAC power is available.

Each of the three identical kits is identified by color-coded labels, which encourages components in the field to be returned to their respective kit without confusion. Where labeling is not feasible, colored tape is used. All interconnection wiring is also labeled to ensure correct connection when changes must be made and to aid in troubleshooting.

A separate RF power/SWR meter provides an accurate digital readout of HF power output and SWR, and can be relocated from the base unit to the TS-480SAT control panel for convenience of operating (**Figure 11.7**). An external SGC noise cancelling DSP speaker for the HF radio provides additional flexibility in the degree of noise reduction and the width of the DSP filter. Both the foregoing items may be considered optional because they duplicate features found in the HF radio and, in the interest of simplicity, could be omitted from the design.

ANTENNAS

Wide-base tripods designed to hold photographic screens are used to mount the VHF/UHF antennas to a height of up to 12 feet (**Figure 11.8**).

During storage and transport, the vertical element of each VHF/UHF antenna is affixed to its tripod upside-down. Ground plane elements are stored in a 3/4" diameter PVC tube with threaded cap attached by a short, flexible cable to a screwdriver used in assembly, and retained in the same bag as the antennas (**Figure 11.9**). The Allen-type setscrews for mounting the Larsen dual-band antenna ground radials have

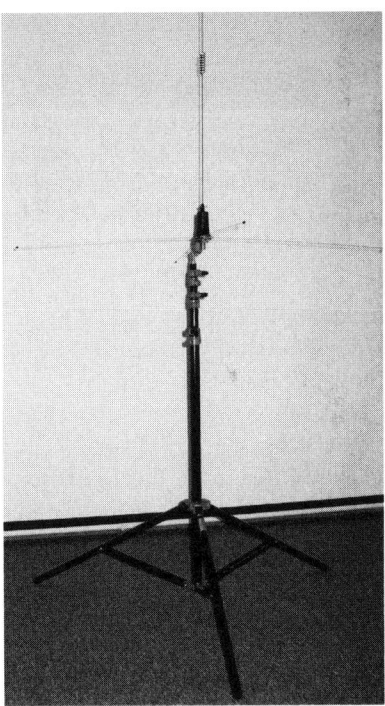

Figure 11.8 — The Larsen dual-band antenna (shown in fully lowered position) can be raised to a height of 12 feet.

Figure 11.7 — The VE7XDT power/SWR meter RF pickup (lower) is permanently affixed to the back panel of the base, as shown, but the digital readout (upper) can be relocated from its hook-and-loop fastening to the remote HF radio control head if a signal cable and 12 V power are provided.

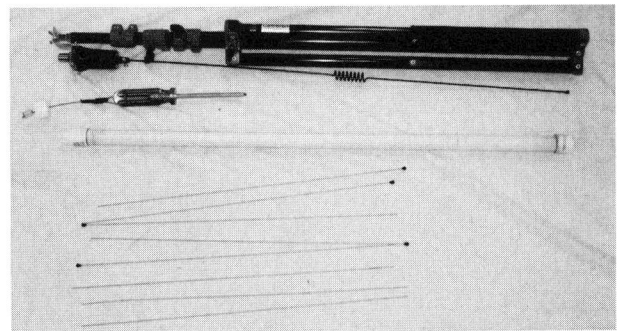

Figure 11.9 — Contents of the VHF/UHF antenna case, showing the antenna mounted upside-down for storage. A PVC tube with screwdriver cabled to a screw-on cap holds the ground plane elements.

Figure 11.10 — The OCF dipole in use, showing three guy ropes affixed to the guy ring, pulley and halyard for raising the balun, coax and antenna wire. Since this photo was taken, the top of the pole has been modified so that it now consists simply of an eye bolt through the top of the pole with a link and pulley.

been replaced with standard 8-32 Robertson head screws for ease of assembly.

Fiberglass poles for the HF antenna are military surplus "camouflage support poles," available from a number of sources on the web. Eight of the poles are telescoped together and inserted into a tripod base to form a mast 30 feet high for supporting an off-center-fed (OCF) dipole in an inverted V configuration (**Figure 11.10**). The OCF dipole is ideal for multiband operation using a single wire, as it is resonant on 75/80, 40, 20, 17, 12, 10, and 6 meters. A custom aluminum ring with links is inserted between the top and next-highest pole for attachment of three guy ropes. The remaining four poles in the set are used to support each end of the antenna legs at a safe height above ground.

COMPUTER, SOFTWARE, AND PROGRAMMING

A laptop computer in each kit runs Airmail and Paclink for *Winlink 2000*. Equipment manuals, reference material, and training documents are stored as PDF files in memory for easy reference.

The laptop is powered with a combination 120 Vac/12 Vdc adaptor, the latter of which can be connected to a 12 V cigarette jack on the rear panel of the unit.

The ICOM radios are programmed with local emergency frequencies using CS-2820 software, an OPC 1529R serial cable (alternatively OPC-478 or its USB equivalent), and an OPC-474 cable for radio-to-radio cloning. Programming of the IC-91AD requires RS-91 software, using the same cables as those for the IC-2820H.

CONSTRUCTION

A band saw equipped with metal-cutting blade is an essential construction tool for this project. Chassis punches in sizes of 5/8", 3/4", 7/8", and 1" are also useful, but not essential if holes are drilled then filed out to size. Careful attention to layout of components is critical in order to fit everything within the available space, including allowance for cable runs and access to front and rear jacks. A layered construction is utilized, with the equipment positioned on the three shelves as shown in **Figures 11.11a**, **b**, and **c**.

Outside dimensions of the radio assembly are 23" x 23" x 13 1/2" high, a size that makes for a snug but comfortable fit into the Pelican 1640 Case. The shelves are made from 1/8" aluminum sheet, joined together by 5/16" threaded steel rod, stiffened by nuts and lock washers separated by ferrules made from 3/8" ID round aluminum tubing. Front and rear top horizontal structural elements are made from 3/4" x 3/4" aluminum angle, while the adjacent top side elements consist of 3/4" x 3/4" hollow square aluminum stock.

The bottom shelf is set 3/4" up from the table top on extensions of the threaded rods to accommodate protruding fasteners. Spacing between the bottom and middle shelves is 4", and between middle and top shelves is 3 1/4". Acorn nuts and plastic tip protectors finish off the top and bottom (respectively) of the threaded rods.

The radio control panels are made from .090" thick aluminum cut to 7" x 12" and bent longitudinally into a 3" x 4" L-shape, as shown in **Figures 11.5** and **11.6**. Access to a metal shop bending brake is recommended for the larger bends, such as these. Additional .075" aluminum stock is required for the back utility panel and custom brackets.

After layout and drilling of the aluminum parts, the edges and corners were rounded and smoothed and the surfaces polished with buffing compound — with a view to a professional-looking final product.

Standard cable lengths, as provided, were seldom suitable without modification. The preferred procedure is to cut long cables to proper length and install new connectors. Where this is not feasible, cables can be shortened by doubling up inside shrink tubing or forming into coils, with the excess hidden under the radios.

The mounting bracket provided for the TS-480SAT is used without modification, with the radio raised to the high-

---

Opposite
Top: Figure 11.11a — Layout of top shelf, showing position of three removable radio control panels. [PHILIPP RUTISHAUSER]

Below: Figure 11.11b — Layout of middle shelf. [PHILIPP RUTISHAUSER]

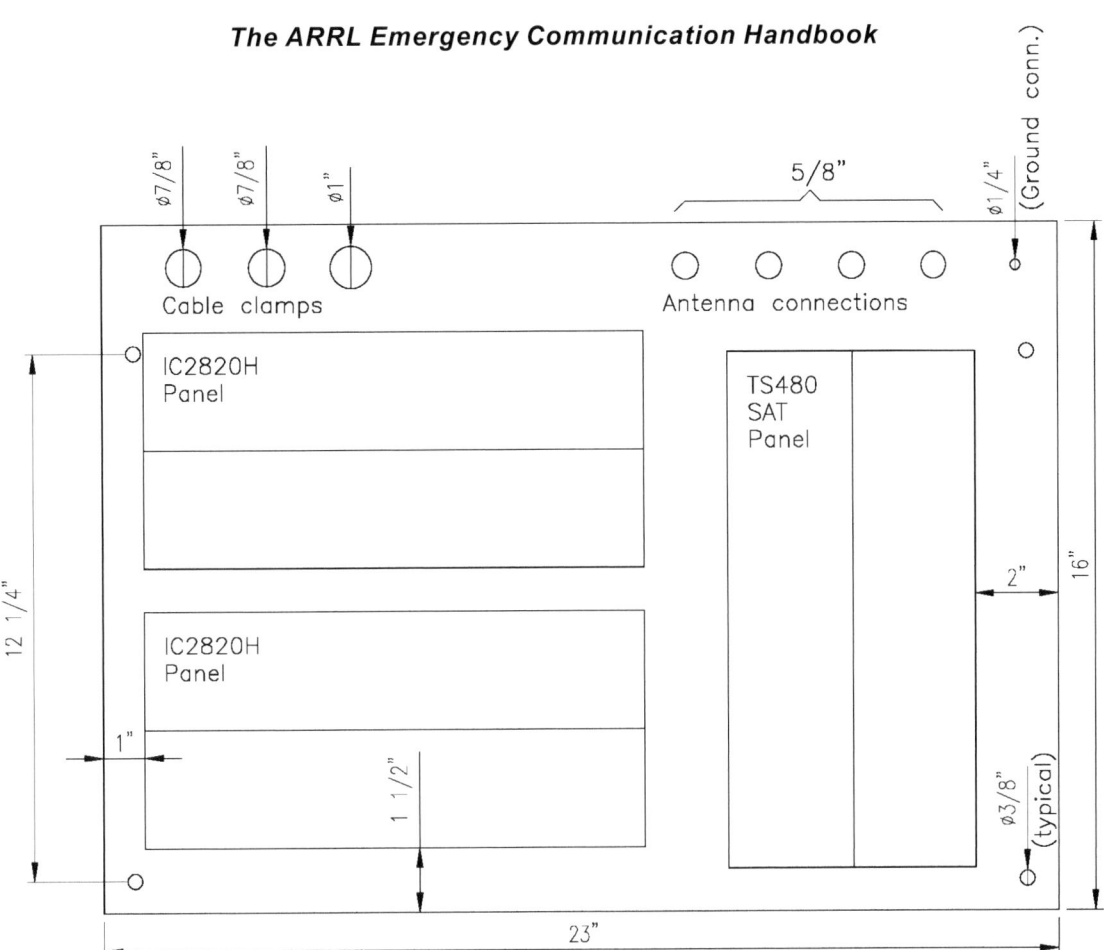

Figure 11.11a

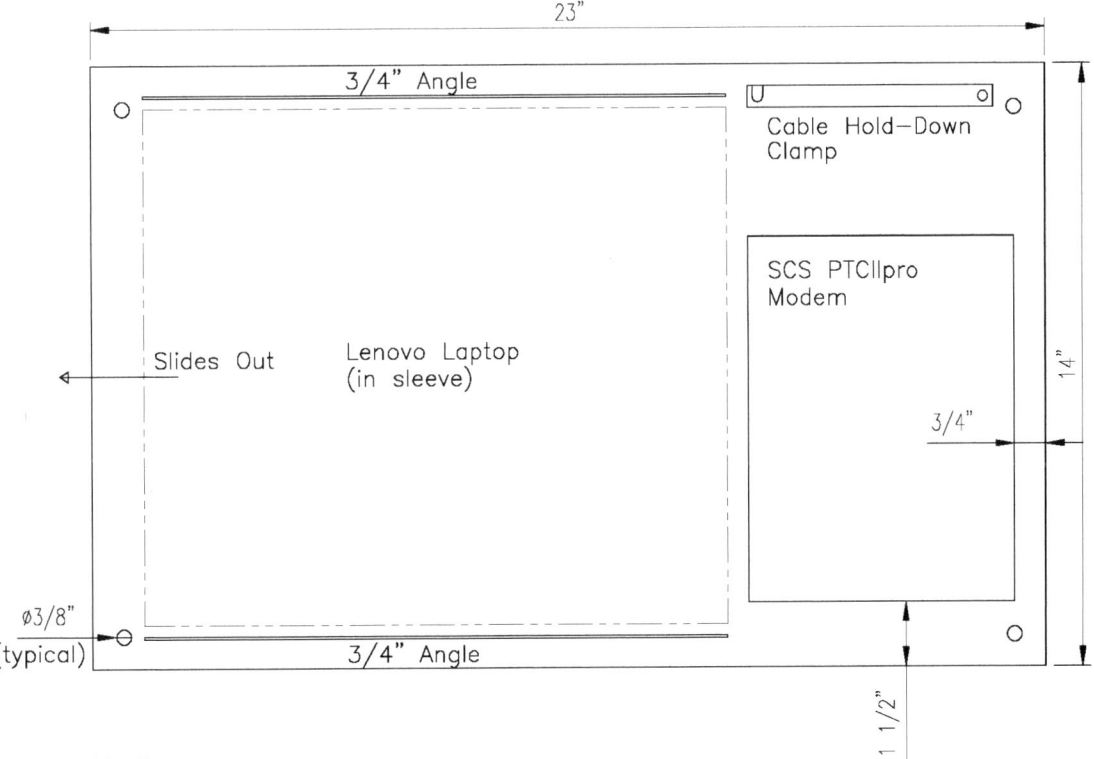

Figure 11.11b

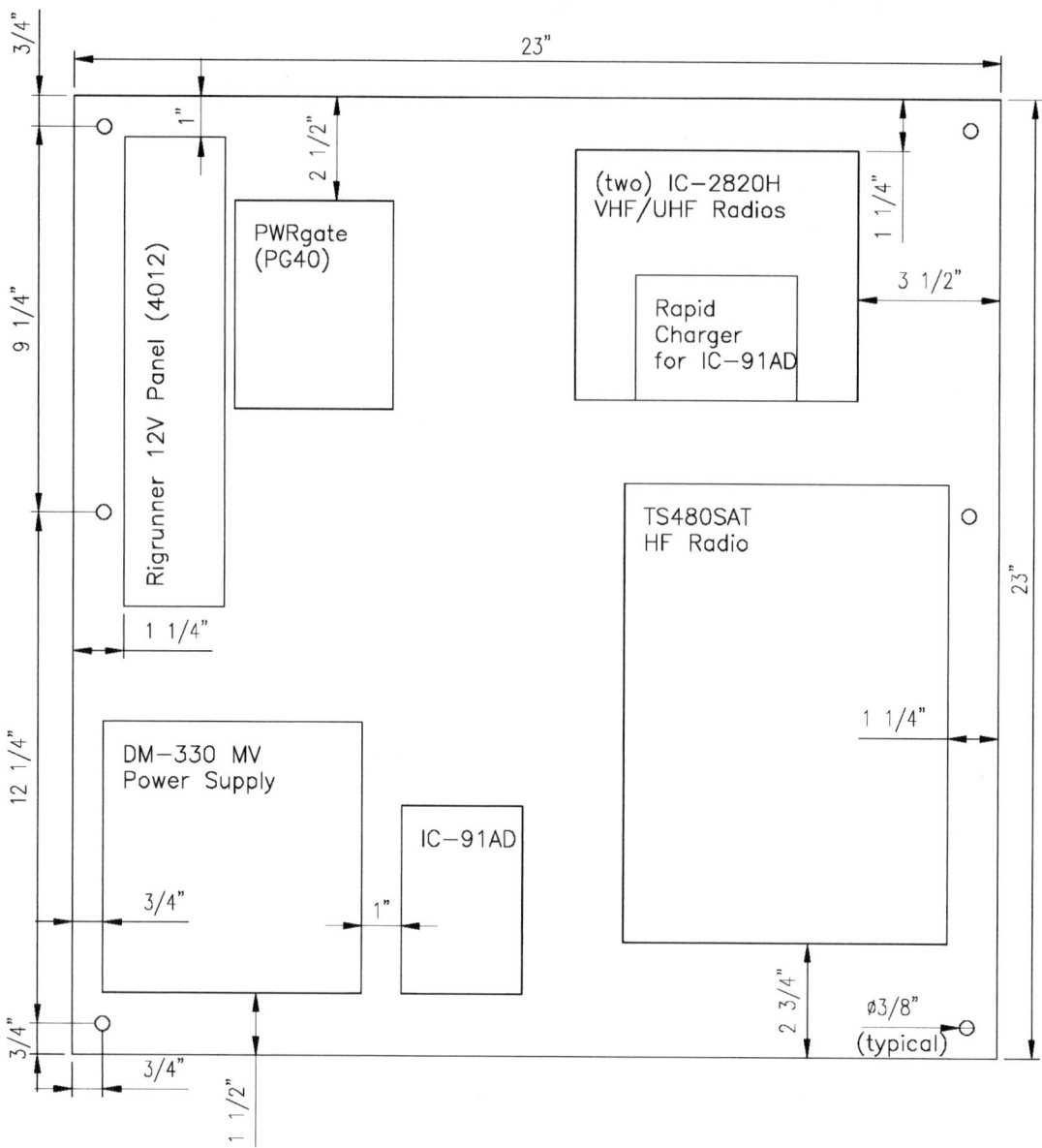

Figure 11.11c — Layout of bottom shelf. The rapid charger for the handheld radio is fastened to the top of the IC-2820H using hook-and-loop fasteners. [PHILIPP RUTISHAUSER]

est available position on the bracket to allow for cable placement beneath.

The IC-2820Hs, which are stacked two high at the rear of the unit, require a custom bracket made from .075" aluminum stock, also with allowance for excess cable length stored beneath the radios. The modem mounting bracket and the IC-91AD holder are fabricated to suit, using strips of hook-and-loop fastener. The DM-330MV power supply is secured to the lower shelf by attachment bolts installed after removal of its bottom panel and drilling of bolt holes.

The radio control panels and brackets, after bending and drilling, were roughened with emery cloth, then primed and painted with several coats of textured enamel and a top coat of clear acrylic.

Mounting feet on the control panels, made from 3/4" x 3/4" angle, are inserted into cut slots and bonded to the panels with "JB Weld." This bond is further strengthened by a tightly-bolted 3/16" threaded rod running longitudinally across the underside of the panel (**Figure 11.12**) providing a convenient anchor for cables, which are secured by zip straps so that they will not pull loose during transport and handling.

Cables joining the main unit and remote control head are bundled within 1/2" split sheathing and secured at each end with standard 1/2" or 3/4" electrical (Loomex-type) cable clamps, the larger size clamp being required for the TS-480SAT because of the greater number of cables contained within the bundle.

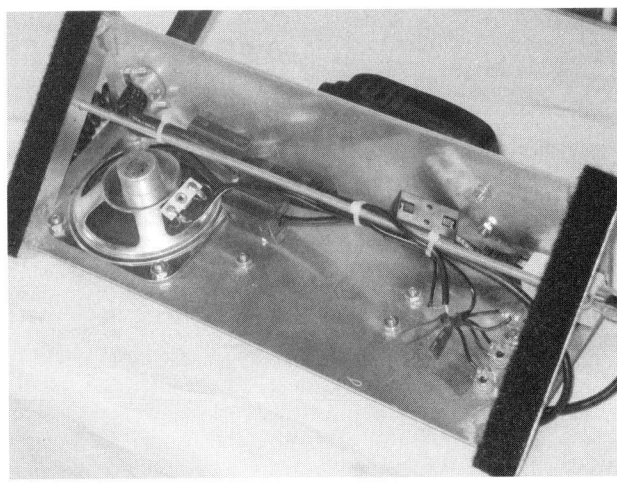

Figure 11.12 — The underside of a VHF/UHF radio control panel. Corners of the "feet" are bonded to the panel with "JB Weld" and are further strengthened by the bolted 3/16" threaded rod running across the full length of the panel. The rod provides a convenient anchor for cables. Hook-and-loop fasteners on the panel feet facilitate securing to the base unit. Note the liberal use of split ferrite cores here and elsewhere.

Dummy loads are made from 45 ohm 35 watt Ameritron non-inductive resistors mounted on acrylic sheet and soldered to coax via copper contacts made from standard 1/2" pipe clamps that have been cut in half (**Figure 11.13**). The dummy loads provide an SWR of 1.0 up to 28 MHz, 1.7 at 144

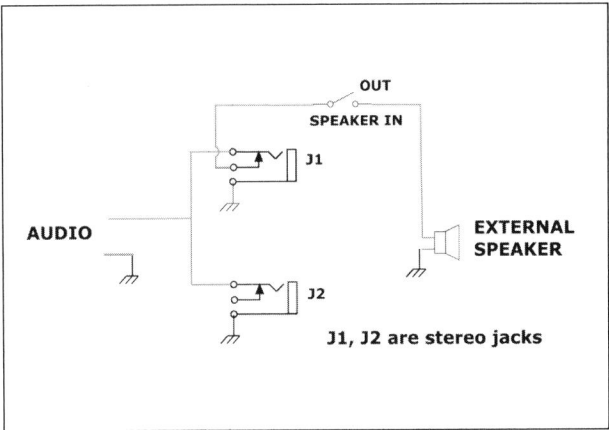

Figure 11.14 — Wiring of SGC external speaker audio allows use of a) speaker only, b) headphones only, or c) speaker plus headphones, depending on which jacks and switch position are used. Volume for both is acceptably controlled by the radio volume control.

MHz and 2.2 at 440 MHz. The resistors are mounted inside a Hammond 1590CFL flanged cast aluminum box bolted to the back panel. RG8X coax connects the resistors to their respective radios by way of double male PL-259 joiners connected to UHF-type bulkhead adaptors at the rear of the top shelf of the base unit.

External speakers, mounted in the control panels, are required for the IC-2820Hs, as their built-in speaker is located in the radio which may be some distance away from the control head during operation.

The TS-480SAT utilizes an SGC external speaker providing DSP filtering, which may be considered optional as this radio does have its own speaker built into the control head. Headphone and external speakers for all radios are wired to allow the operator to select: a) speaker only, b) headphones only, or c) speaker plus headphones, in accordance with the simple circuit shown in **Figure 11.14**, study of which will show that the selection is influenced by which of the two headphone jacks is being used.

Alternatively, if the headphone jack on the HF radio's external SGC speaker is used instead of one of the panel jacks, the headphone audio will be responsive to the SGC noise reduction and DSP filtering. Otherwise, these functions are controlled entirely by the radio during headphone use.

Relative loudness of the headphones and external speaker, operating together or independently, has been found to be acceptable when controlled by the radio volume control, without the need for independent volume controls.

The potential for undesired RF coupling associated with the long cables connecting physically separated units is addressed by the liberal use of split ferrite cores — approximately 45 per kit — probably more than necessary, but providing good insurance (**Figure 11.15**).

Two persons are required to move and lift each case comfortably and safely, considering that No.1 Pelican case containing the radio kit weighs approximately 120 lbs. and No. 2

Figure 11.13 — Dummy loads are 45 ohm Ameritron 103-9545 non-inductive resistors with clips, mounted on acrylic sheet and bolted to soldering lugs made from ½" copper pipe clamps.

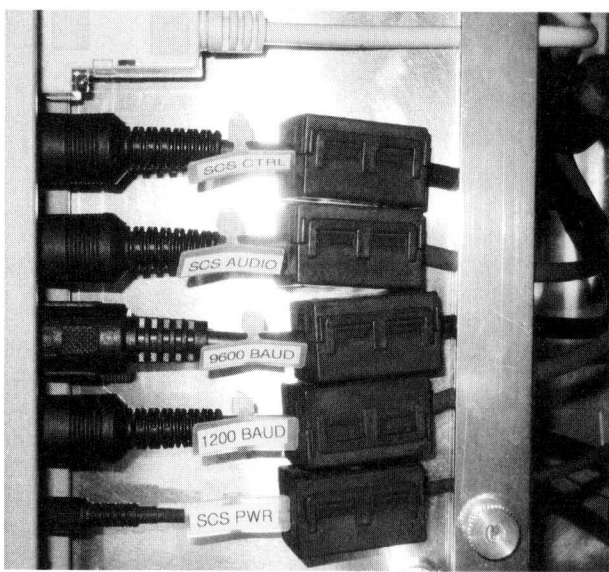

Figure 11.15 — Cable connections at the rear of the SCS modem, showing use of split ferrite cores and typical labeling. The removable hold-down bar prevents cables working loose during equipment transport.

Figure 11.16 — Top view of the completed base kit, showing three radio control panels held in place with hook-and-loop fasteners, with cables coiled and stored in the space at the rear.

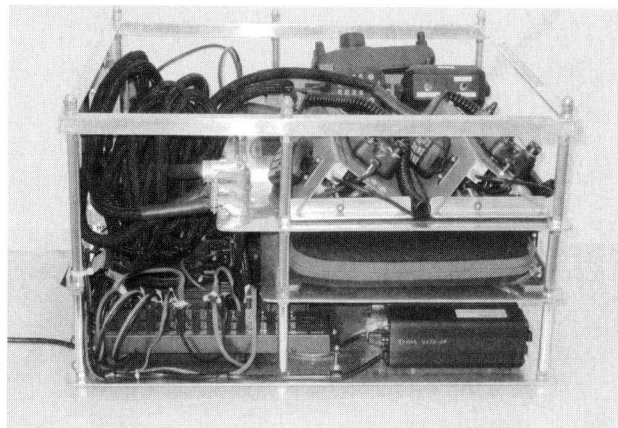

Figure 11.17 — Left-side view of the completed base kit, showing RIGrunner 12 V distribution panel and power supply on lower shelf, laptop in sleeve on middle shelf, and radio control heads on upper shelf, with cables coiled and stored in the space at the rear.

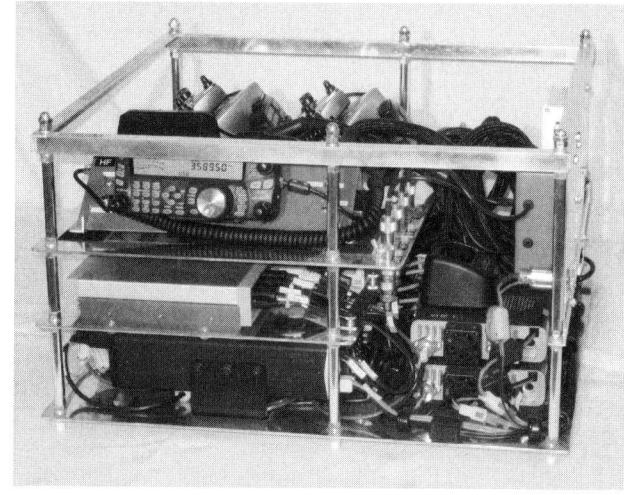

Figure 11.18 — Right-side view of the completed base kit, showing HF control panel, SCS modem, and HF radio on the left. The VHF/UHF radios are stacked two high on the lower right-hand shelf, with the rapid charger secured on top with hook-and-loop fasteners. The excess cable lengths are coiled and stored under the TS-480SAT radio.

Pelican case with the antennas, coax, and miscellaneous items weighs approximately 130 lbs. Handles and wheels on both cases make transport relatively easy, and the complete assembly of three units plus battery and generator fits in the rear compartment of a standard SUV or passenger van.

Other design features of the assemblies may be seen by examining top, left-side, and right-side views (**Figures 11.16, 11.17,** and **11.18**, respectively). **Figure 11.19** shows the base unit stowed in its Pelican case along with the document pouch and utility bag, ready for transport. Wiring connections for the power supply, radios, and modem are shown in **Figure 11.20** and **Figure 11.21**.

MODIFICATIONS

During field trials, when the radios were being powered only by the battery, the voltage drop across the PG40 PWRgate was found to be excessive. To remedy this, the unit was modified by shorting Schottky diode D1, as shown in **Figure 11.22**. This modification not only allows the power supply to charge the battery at a higher rate than the 1 amp dictated by D2/R1, but also facilitates the full battery voltage to be applied across the load when it is the source of power. Protection against power bleed-off from the battery to the power supply is still provided by D3. An alternative solution which does not necessitate any modifications of the PG40, is to leave the power supply and load connected to the PWRgate normally but to connect the battery directly to one of the fused load terminals on the RIGrunner panel. With this approach, only D3 of the PWRgate is operative, and the result is the same as shorting out D1.

During testing, it was observed that the external SGC noise canceling speaker used with the TS-480SAT did not function at initial power-up; the reason was never determined, but may be due to the sequence of powering up. This was resolved by installation of a momentary-off switch on the TS-480SAT panel, which allows interruption of the 12 V supplied to the SGC; once power is restored with the switch, the SGC functions normally.

A programmable external CW keyer is included in the kit even though the TS-480SAT has its own internal keyer. This was found necessary because many "non-iambic" CW operators found it difficult to adapt to the Mode B characteristics of the Kenwood built-in keyer.

FINAL COMMENTS

Each complete kit cost around $10,000 and required approximately 50 person-hours of construction time once the concepts, layout, and mock-up had been finalized. Though pricey, the kits provided the versatility that justified the cost to the sponsoring emergency agency.

Not all construction details can be provided in an article of this length. The construction team will be pleased to answer specific questions on request (see contact information at the end of the chapter). A CD containing high-resolution photographs and other information is available to those wanting to duplicate the construction or use it as a starting point for their own design.

Since construction was completed in 2007, the kits have seen service in numerous emergency exercises, including three Field Days, without exhibiting any serious deficiencies or problems.

However, since 2007 the following modifications have been made to enhance their functionality:
- Both IC-2820H radios are now equipped with D-STAR modules, and D-RATS is being increasingly used for digital messaging as an alternative to *Winlink*.
- Plans are underway to add an Alinco DR-235T 220 MHz radio to each kit, to be mounted on the lower shelf in the position occupied by the IC-91AD handheld radio, which will have to be relocated.
- One of the tripods in each kit now accommodates a Larsen BSA-220C antenna for 220 MHz in addition to the 2m/440 MHz dual band antenna, both of which are mounted on a single 1" angle aluminum cross bar (**Figure 11.23**).

A final note: The power demand from several radios operating simultaneously has been found to be sufficiently large that one or more 12 V batteries should always be

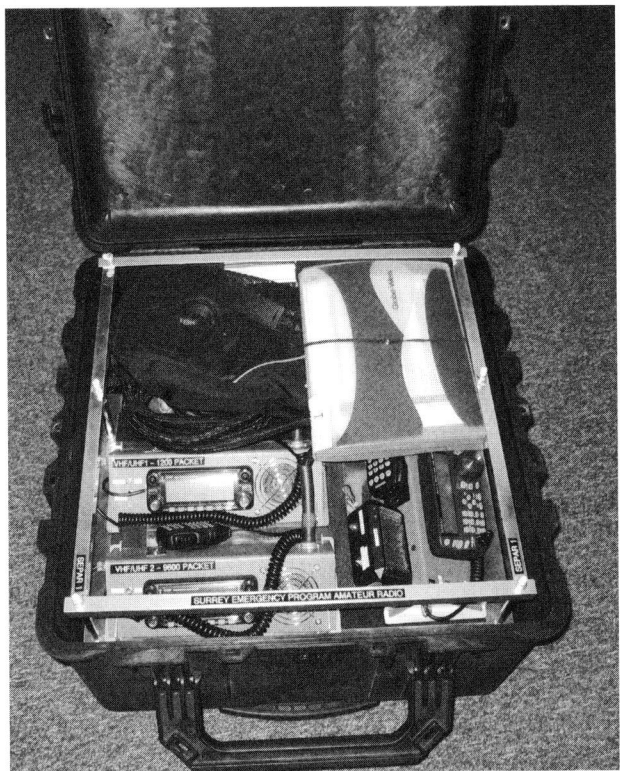

Figure 11.19 — The radio kit, plus documentation pouch and utility case (containing headphones, laptop power cables, and CW keyer) all fits snugly inside a Pelican 1640 case.

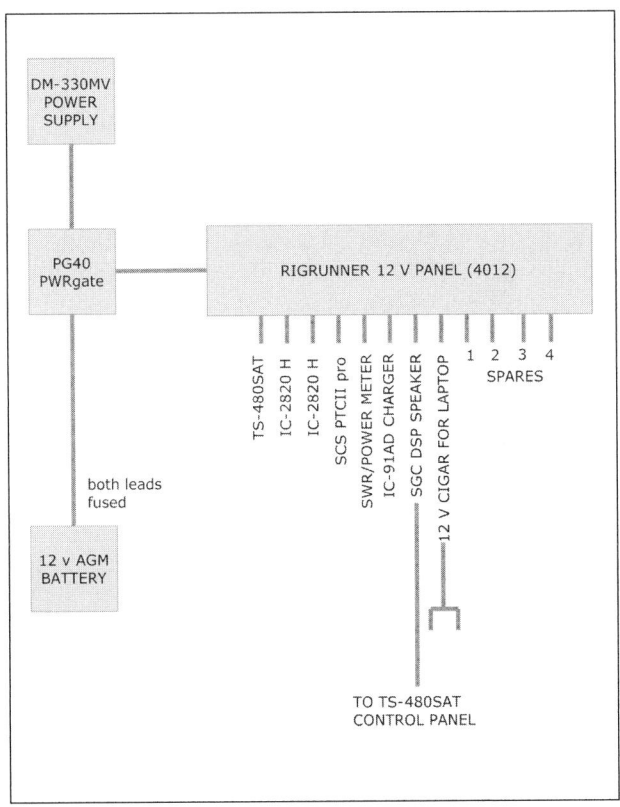

Figure 11.20 — Wiring of the 12 V system.

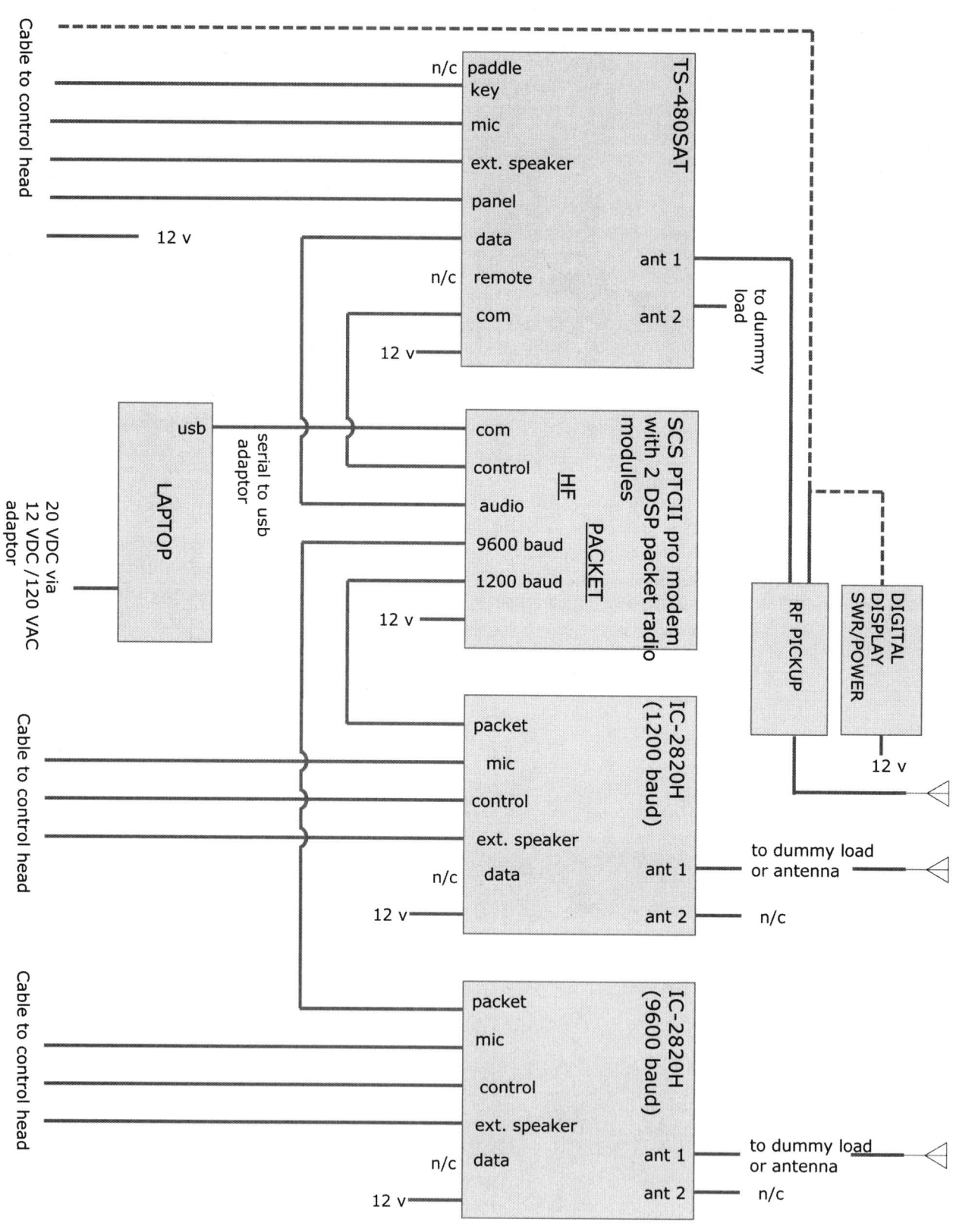

Figure 11.21 — Interconnection cabling of radios, modem, and computer.

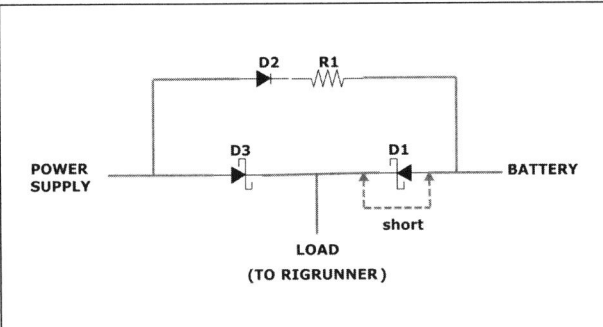

Figure 11.22 — Modification made to the PWRgate eliminates the large voltage drop across Schottky diode D1 when drawing power from the battery. Shorting of D1 also allows the power supply to charge the battery through D3 at whatever rate the battery requires, up to the current limit of the power supply. See the text for an alternative to this modification.

Figure 11.23 — A 45" length of 1" aluminum angle is drilled at both ends to accommodate the coax fitting for a 2 m/ 440 MHz dual band antenna at one end and a 220 MHz antenna at the other.

connected to the PG40 Powergate to supplement the power supply.

For more information on the fabrication of this kit, please visit the SEPAR website at **www.separ.comm.sfu.ca/** or feel free to contact the design and construction team at the following addresses: SEPARS Coordinator Fred Orsetti, ve7io@arrl.net; design and construction team members Bill Gipps, bill.gipps@ideasmcs.com, Drew Elvins, va7drw@shaw.ca, and John Brodie, va7xb@rac.ca; and designer of the digital power/SWR meter used with the HF radio, Dino Gueorguiev, ve7xdt@rac.ca.

KIT COMPONENTS

Unit #1 (Radios)
- Pelican case (model # 1640)
- Kenwood TS-480SAT HF radio; MC-58DM mic; PG4Z cable kit
- Two ICOM IC-2820H VHF/UHF radios (D-STAR added later) OPC-440 cable extension
- ICOM IC-91AD hand-held VHF/UHF radio with D-STAR; HM-75a mic
- Alinco DR-235T 220 MHz radio
- Alinco DM-330MV power supply
- SGC noise cancelling DSP speaker for the TS-480
- West Mountain 4012 RIGrunner fused power distribution panel
- West Mountain PG40 PWRgate
- SCS PTCII-pro modem
- Two SCS packet radio modules for modem, one for 1200 baud and other for 9600 baud
- SCS cables for above: (2) #8120, (1) #8080, and (1) #8050
- Lenovo laptop computer plus Lenovo 12 Vdc/120 Vac adaptor
- Custom VE7XDT Power/SWR meter for HF radio
- ICOM BC-139 rapid charger for handheld radio
- Three Heil Traveler single-side headphones & mic
- Headset/mic adaptors for ICOM (HSTA-IM) & Kenwood (HSTA-KM) radios
- Two Sony double-side headsets
- Unified Microsystems XT-4 CW memory keyer
- Document pouch containing manuals, programmed memories list, and radiogram forms

Unit #2 (Antennas & Misc.)
- Pelican case (model # 1780)
- Buckmaster 7 Band off-center-fed dipole
- Two Larsen NMO 2 meter / 70 cm dual band VHF/UHF antennas and base kits
- Larsen BSA-220C antenna and base kit
- Two Cameron collapsible tripods for above
- 333 feet of LMR400 coax in lengths of 150', 100', 50', and 33' c/w coax connectors and joiners
- 30 feet contractor's extension cord
- Battery jumper cables
- Ropes for guys and halyard
- Utility bag containing spare batteries for handheld radio, LEDs, spare fuses
- Electrician's tape, duct tape, assorted bungee cords, three-lb. hammer
- 12 V battery tender

Unit #3 (HF Antenna Mast Sections)
- Heavy duty duffel bag
- 12 four-foot lengths of telescoping fiberglass poles
- Tripod for above
- Guy ring and stakes

Chapter 12
Going Portable: Necessary Equipment

As an emergency communications volunteer, you never know which challenges an emergency situation will present. You might have ac power, or just the batteries you bring along. Safe drinking water may be available, or you may have only your canteen. Sometimes you can find out in advance what sort of conditions will be likely at your assignment, but many times no one will know — particularly during the early stages of an emergency. Therefore, you need to be prepared for every eventuality.

Being prepared for an emergency communication deployment involves a wide range of considerations, including radio equipment, power sources, clothing and personal gear, food and water, information, and specialized training. No two deployments are the same, and each region offers its own specific challenges. What is appropriate for rural Minnesota in January probably won't work for urban southern California in any season.

OPERATIONAL PRE-PLANNING

When the time comes to activate, you'll need to know where to go and what to do. Having this information already available will help you respond more quickly and effectively. It will not always be possible to know these things in advance, particularly if you do not have a specific assignment. Answering the following basic questions may help.

- On which frequency should you check in initially?
- Is there a "backup" frequency?
- If a repeater is out of service, which simplex frequency is used for the net?
- Which nets will be activated first?
- Should you report to a pre-determined location or will your assignment be made as needed?

Learn about any place to which you may be deployed, to familiarize yourself with its resources, requirements, and limitations. For instance, if you are assigned to a particular shelter, you might ask your superiors to schedule a visit, or talk to others who are familiar with the site to find out details such as the following:

- Will you need a long antenna cable to get from your operating position to the roof?
- Are antennas or cables permanently installed, or will you need to bring your own?
- Will you be in one room with everyone else, or in a separate room?
- Is there dependable emergency power to circuits at possible operating positions?
- Does the building have an independent and dependable water supply?
- Is there good cell phone or beeper coverage inside the building?
- Can you reach local repeaters reliably with only a rubber duck antenna, or do you need a more efficient antenna, or one with gain?
- If the repeaters are out of service, how far can you reach on a simplex channel?
- Will you need an HF radio to reach the net?

If you will be assigned to an EOC, school, hospital, or other facility with its own radio system in place, learn under what conditions you will be required or able to use it, where it is, and how it works. In addition to radios, consider copiers, computers, fax machines, phone systems, and other potentially useful equipment.

Consider escape routes. If you could be in the path of a storm surge or other dangerous condition, know all the possible routes out of the area. If you will be stationed in a large building such as a school or hospital, find the fire exits, and learn which parking areas will be the safest for your vehicle.

JUMP KITS / GO-KITS

Any experienced emergency responder knows the importance of keeping a kit of necessary items ready to go at a moment's notice. This is often called a "jump kit" or "go-kit." Without a jump kit, you will almost certainly leave something important at home, or bring items that will not do the job. Gathering and packing your equipment at the last moment also wastes precious time. It is important to think through, ahead of time, each probable deployment and the range of situations you might encounter. Here are a few basic questions you will need to answer:

- Which groups or served organizations will you need to join, and which equipment will you need to do so?

- Will you need to be able to relocate quickly, or can you bring a ton of gear?
- Will you be on foot, or near your vehicle?
- Is your assignment at a fixed location or will you be mobile?
- How long might you be deployed — fewer than 48 hours, or even a week or more?
- Will you be in a building with reliable power and working toilets, or in a tent away from civilization?
- What sort of weather or other conditions might be encountered?
- Where will food and water come from?
- Are sanitary facilities available?
- Will there be a place to sleep?
- Do you need to plan for a wide variety of possible scenarios, or only a few?
- Can some items do "double duty" to save space and weight?

Other questions may occur to you based on your own experience. If you are new to emergency communications or the geographical area, consult with other members of your group for their suggestions.

Most people seem to divide jump kits into two categories: one for deployments lasting under 24 hours, and one for deployments lasting up to 72 hours. For deployments longer than 72 hours, many people include more of the items that they will use up, such as clothing, food, water, and batteries. Others may add a greater range of communication options and backup equipment as well. See **Sidebar 12.1** at the end of this chapter for information on creating a handi-talkie go-kit, and the Appendix of this book for checklists of go-kit equipment suitable for deployments of varying durations.

EQUIPMENT SELECTION

This is, perhaps, the biggest question at hand: what sort of gear do you take into the field with you? Portability, versatility, and ease of use are all factors. Here are some possibilities to consider.

Transceivers

VHF/UHF. The most universal choice for EmComm is a dual-band FM 35-50 watt mobile transceiver. Radios in this class are usually rugged and reliable, and can operate at reasonably high duty cycles, although an external cooling fan is always a good idea if one is not built in. Handheld transceivers should be used only when extreme portability is needed, such as when "shadowing" an official or when adequate battery or other dc power is not available. Handheld radios should not be relied upon to operate with a high duty cycle at maximum power, since they can overheat and fail.

Both portable and mobile dual-band radios can be used to monitor more than one net, and some models allow simultaneous reception on more than one frequency on the same band (this is sometimes known as "dual watch" capability). Some mobiles have separate external speaker outputs for each band. For high-traffic locations, such as a Net Control or Emergency Operations Center, a separate radio for each net is a better choice, since it allows both to be used simultaneously by different operators. (Antennas must be adequately separated to avoid "de-sensing.")

Many dual-band transceivers also offer a "cross-band repeater" function, useful for linking local portables with distant repeaters, or as a quickly deployable hilltop repeater. True repeater operation is only possible if all other mobile and portable stations have true dual-band radios. Some so-called "dual" or "twin" band radios do not allow simultaneous or cross-band operation — read the specifications carefully before you purchase one.

HF. Operation from a generator-equipped Emergency Operations Center can be done with an ac-powered radio, but having both ac and dc capability ensures the ability to operate under all conditions. Most 12 V HF radios fall in either the 100-watt or QRP (less than 5 watts) categories. Unless power consumption is extremely important, use 100-watt variable output radios. This gives you the ability to overcome noise at the receiving station by using high power, or to turn it down to conserve battery power when necessary.

Do not use dc-to-ac inverters to power HF radios. Most use a high-frequency conversion process that generates significant broad-spectrum RF noise at HF frequencies that is difficult to suppress. Direct dc powering is more efficient in any case.

Voltage Tolerance and Current Drain

Some transceivers nominally powered using 12 volts dc actually have a rather narrow range of voltage (e.g., 13.0 to 13.8 volts) over which they will operate properly, and even a high-quality battery part way through its discharge cycle can easily fall below such a tolerable range. Transceivers with a wide range of acceptable input voltages (e.g., 11.5 to 15 volts) are preferable in limited-power situations; they will keep operating as the external battery discharges.

Similarly, some transceivers draw much more power than others during receive. If your chosen rig has a current drain on the high side, look for menu settings that will lower the overall drain, especially if you will be operating from a limited power source.

Radio Receiver Performance

For radios on all bands, several aspects of a radio receiver's performance can affect its suitability for EmComm. These include sensitivity (ability to receive weak signals), selectivity (ability to reject signals on adjacent frequencies) and intermodulation rejection (ability to prevent undesired signals from mixing within the receiver and causing interference). If you are inexperienced at comparing radio specifications, be sure to ask for guidance from a more experienced ham.

When operating near public service and business radio transmitters, an FM receiver's "intermodulation rejection" is important. Mobile radios generally have better intermodulation rejection than handheld radios, but you should review each individual radio's specifications. External intermodulation (band-pass) filters are available, but they add to the expense, complexity, size, and weight of the equipment. Band-pass filters will also prevent you from using a broadband radio to monitor public service frequencies. Some older "ham bands only" FM mobile radios have better front-end filtering than newer radios with broadband receive capability, making them more immune to intermodulation and adjacent channel interference. Receiver filters are important for effective HF operation. Choose appropriate filters for the types of operations you are most likely to use, including CW, RTTY, and phone. Digital Signal Processing (DSP) may be the single most important filtering feature available. Internal or external DSP circuits can allow clear reception of signals that might not otherwise be possible in situations with heavy interference. "Noise blankers" are used to reduce impulse noise from arcing power lines, vehicle and generator ignition systems, and various other sources. While most all HF radios have some form of noise blanker, some work better than others. Test your radio in suitably noisy environments before designating it for emergency communications use.

Antennas

VHF/UHF. A good antenna, mounted as high as possible without incurring large feed line losses, is more important than high transmitter power. Not only does it provide gain to both the transmitter and receiver, but a higher gain antenna may also allow output power to be reduced, thus prolonging battery life. In relatively flat terrain, use a mast-mounted single or dual-band antenna with at least 3dBd gain. If you are operating in a valley, the low angle of radiation offered by a gain antenna may actually make it difficult to get a signal out of the valley. Low or "unity" gain antennas have "fatter" radiation lobes and are better suited for this purpose. Unity gain J-poles are rugged, inexpensive, and easily built (see **Figure 12.1**). For directional 2-meter coverage with about 7-dBd gain, a three- or four-element Yagi can be used. Collapsible and compact antennas of this type are readily available. For permanent base station installations, consider a more rugged commercial two-way collinear antenna, such as the well-known Stationmaster™ series. Most 2-meter versions will also perform well on 70 cm. Commercial open dipole array antennas will work well for a single band, and are more rugged than a fiberglass radome encased collinear antenna.

A magnetic-mount mobile antenna is useful for operating in someone else's vehicle. They can also be used indoors by sticking them to any steel surface, such as filing cabinets, beams or ductwork, even upside down.

Handheld radio antennas, known as "rubber duckies," have negative gain. Use at least a 1/4-wave flexible antenna for most operations, and consider a telescoping 5/8-wave antenna for long-range use in open areas where the extra length and lack of flexibility will not be a problem. "Roll-up J-pole" antennas made from 300 ohm television twinlead wire

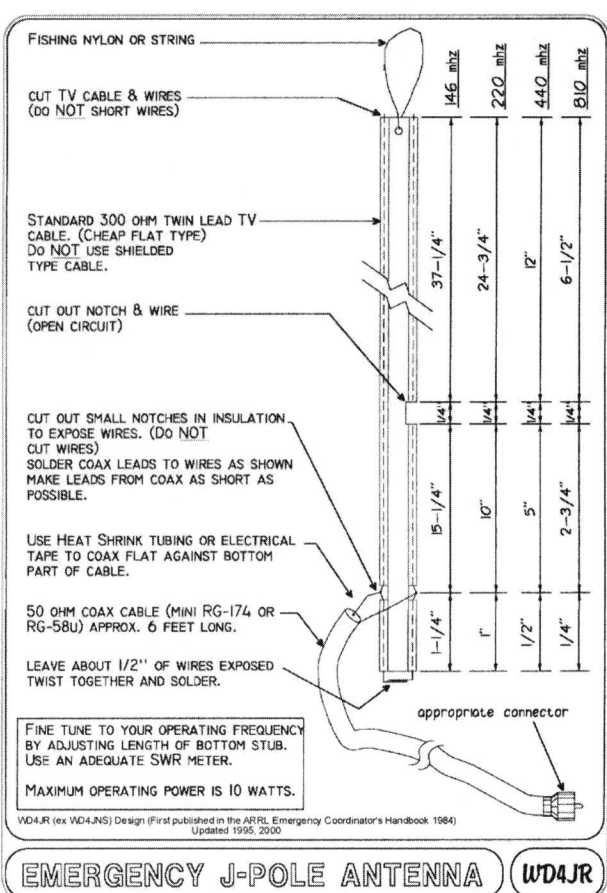

Figure 12.1 — Building a portable J-pole is relatively quick and simple.

can be tacked up on a wall or hoisted into a tree with heavy-duty string. In addition to unity gain, the extra height can make a big difference. Even a mobile 1/2-wave magnetic mount antenna can be used with hand-helds when necessary.

HF. There is no single perfect antenna for HF operation. Your choice will depend upon the size and terrain of the area you need to cover, and the conditions under which you must install and use the antenna.

For local operations (up to a few hundred miles), a simple random wire or dipole hung at less than 1/4 wavelength above the ground works well and is easy to deploy. This is known as a Near Vertical Incidence Skywave (NVIS) antenna. The signal is radiated almost straight up and then bounces off the ionosphere directly back downward. During periods of high solar activity, NVIS propagation works best on 40 meters during the day, switching to 80 meters around sunset. During low parts of the sunspot cycle, 80 meters may be the most usable daytime NVIS band, and 160 meters may be needed at night. The 60-meter band is also ideal for NVIS operation. See the article at the end of this chapter for information on building a portable NVIS antenna.

An antenna tuner is necessary for most portable wire antennas, (especially for NVIS antennas), and is a good idea for any HF antenna. The antenna's impedance will vary with its height above ground and proximity to nearby objects, which can be a real problem with expedient installations. An automatic tuner is desirable, since it is faster and easier to use, and many modern radios have one built in. Include a ground rod, clamps, and cable in your kit since almost all radios and tuners require a proper ground in order to work efficiently.

For communication beyond 200 miles, a commercial trapped vertical may work, although it has no ability to reject interfering signals from other directions. Mobile whip antennas will also work, but with greatly reduced efficiency. The benefits of a mobile antenna are its size and durability.

Directional (beam) antennas offer the best performance for very wide area nets on 10 to 20 meters, since they maximize desired signals and reduce interference from stations in other directions. This ability may be critical in poor conditions. Beam antennas also have a number of limitations that should be considered. They are usually expensive, large, and difficult to store and transport. In field installations, they can be difficult to erect at the optimum height, and may not survive storm conditions. One strategy is to rely on easily installed and repaired wire dipole antennas until conditions allow the safe installation of beam antennas (i.e., Moxon antennas or wire beams).

Feed Line

Feed line used at VHF and UHF should be low-loss foam dielectric coaxial cable. For short runs of 30 feet or less, RG-58 may be suitable. For longer runs, consider RG-8X or RG-213. RG-8X is an "in-between" size that offers less loss and greater power handling capability than RG-58 with far less bulk than RG-213. If you wish to carry only one type of cable, RG-8X is the best choice.

On HF, the choice between coaxial cable and commercial (insulated — not bare wire) "ladder" line will depend on your situation. Ladder line offers somewhat lower loss but more care must be taken in its routing, especially in proximity to metal objects, or where people might touch it. Coaxial cable is much less susceptible to problems induced by routing near metal objects or other cables.

Operating Accessories

Headphones are useful anywhere, and are mandatory in many locations. Operators in an Emergency Operations Center or a Command Post where multiple radios are in use must use headsets. They are also beneficial in locations such as Red Cross shelters, to avoid disturbing residents and other volunteers who are trying to rest.

Some radios and accessory headsets provide a VOX (voice operated transmit) capability. During EmComm operations, this should always be turned off, and manual "push-to-talk" buttons should be used instead. Accidental transmissions caused by background noise and conversations can interrupt critical communications on the net. As an alternative to VOX, consider using a desk or boom microphone and foot switch to key the transmitter. A microphone/headset combination and foot switch also works well, as does a speaker mic for HTs.

Batteries

Battery power is critical for emergency communications operations. AC power cannot usually be relied upon for any purpose, and portable operation for extended periods is common. Batteries must be chosen to match the maximum load of the equipment, and the length of time that operation must continue before they can be recharged.

NiCd, NiMH, and Li-ion

For handheld transceivers, the internal battery type is determined by the manufacturer. NiMH batteries store somewhat more energy than NiCd batteries for their size. Many smaller radios are using Lithium-ion (Li-ion) batteries, which have much higher power densities, without the so-called "memory effect" of NiCds. Many handhelds have optional AA alkaline battery cases, and are recommended EmComm accessories. Common alkaline batteries have a somewhat higher power density than NiCd batteries, are readily available in most stores, and may be all you have if you cannot recharge your other batteries. Most handheld radios will accept an external 13.8 V dc power connection for cigarette lighter or external battery use.

External batteries of any type can be used with a handheld, as long as the voltage and polarity are observed. Small 12–15 V gel cells and some battery packs intended for power tools and camcorders are all possibilities. For maximum flexibility, build a dc power cable for each of your radios, with suitable adapters for each battery type you might use. Molex™ plugs work well for power connections, but Anderson™ Powerpoles can withstand repeated plugging and unplugging without deterioration and have become the standard used by most ARES units. This standardization allows easier swapping and sharing of equipment if needed.

Lead-Acid

There are three common types of lead-acid batteries: flooded (wet), VRLA (Valve Regulated Lead-Acid), and SLA (Sealed Lead-Acid). Wet batteries can spill if tipped, but VRLA batteries use a gelled electrolyte or absorptive fiberglass mat (AGM technology) and cannot spill. SLA batteries are similar to VRLA batteries, but can be operated in any position — even upside down. All lead-acid batteries are quite heavy.

Lead-acid batteries are designed for a variety of applications. "Deep cycle" batteries are a better choice than common automotive (cranking) batteries, which are not designed to provide consistent power for prolonged periods, and will be damaged if allowed to drop below approximately 80% of their rated voltage. Deep cycle batteries are designed for specific applications and vary slightly in performance characteristics. For radio operation, the best choice would be one specified for UPS (uninterruptible power source) or recreational vehicle (RV) use. For lighting and other needs, a marine-type battery works well. For best results, consult the manufacturer before making a purchase.

Sealed lead-acid (SLA) or "gel cells," such as those used in alarm or emergency lighting systems, are available in smaller sizes that are somewhat lighter. These batteries are also the ones sold in a variety of portable power kits for Amateur Radio and consumer use. Typical small sizes are 2, 4, and 7 Ah, but many sizes of up to more than 100 Ah are available. SLA batteries should never be deeply discharged. For example, a 12 V SLA battery will be damaged if allowed to drop below 10.5 volts. Excessive heat or cold can damage SLA batteries. Storage and operating temperatures in excess of 75 degrees F or below 32 degrees F will reduce the battery's life by half. Your car's trunk is not a good place to store them. Storage temperatures between 40 and 60 degrees will provide maximum battery life.

Battery "Power Budgeting"

The number of ampere/hours (Ah — a rating of battery capacity) required, called a "power budget," can be roughly estimated by multiplying the radio's receive current by the number of hours of operation, and then adding the product of the transmit current multiplied by the estimated number of hours of transmission and by the duty cycle for that mode. For a busy net control station, the transmit current will be the determining factor because of the high percentage of transmit time. For low-activity stations, the receiver current will dominate. The value obtained from this calculation is only a rough estimate of the ampere/hours required. The Ah rating of the actual battery or combination of batteries should be up to 50% higher, due to variations in battery capacity and age.

Don't confuse the percent of time transmitting with duty cycle, which is mode-specific (e.g., 100% for FM and digital, 50% for CW, and 30% for uncompressed SSB).

Here's an example of an estimated 24-hour power budget:
Receive current: 1 amp x 24 hours = 24 Ah
Transmit current: 8 amps x 6 hours = 48 Ah (figuring 6 hours as the 25% transmit time)
Total Ah: 72 Ah estimated actual consumption
Actual battery choice 72 x 1.5 = 108 Ah figuring 50% higher due to variations.

Chargers, Generators, and Solar Power

You should have two or more batteries so that one can be charging while another is in use. The method of charging varies according to the type of battery you're using.

NiCd and NiMH batteries. The type of charger required depends on the battery — for instance, most NiCd chargers will also charge NiMH, but not Li-ion batteries. Several aftermarket "universal" chargers are available that can charge almost any battery available. A rapid-rate charger can ensure that you always have a fresh battery without waiting, although rapid charging can shorten a battery's overall lifespan.

Lead-acid batteries. Always consult the battery's manufacturer for precise charging and maintenance instructions, as they can vary from battery to battery. It is best to slow-charge all batteries, since this helps avoid over-heating and extends their overall life span.

In general, automotive and deep cycle batteries can be charged with an automobile and jumper cables, an automotive battery charger, or any constant-voltage source. If a proper battery charger is not available, any dc power supply of suitable voltage can be used, but a heavy-duty isolation diode must be connected between the power supply and the battery. (This is important, since some power supplies have a "crowbar" overvoltage circuit, which short-circuits the output if the voltage exceeds a certain limit. If a battery is connected, the crowbar could "short-circuit" the battery with disastrous results). The output voltage of the supply must be increased to compensate for the diode's voltage drop. Take a measurement at the battery to be sure.

Wet batteries. These should be charged at about 14.5 volts, and VRLA batteries at about 14.0 volts. The charging current should not exceed 20% of the battery's capacity. For example, a 20-amp charger is the largest that should be used for a battery rated at approximately 100 Ah. Consult the battery's manufacturer for the optimum charging voltage and current whenever possible.

Deep cycle batteries do not normally require special charging procedures. However, manufacturers do recommend that you use a charger designed specifically for deep cycle batteries to get the best results and ensure long life.

SLA or "gel-cell." Gel-cell batteries must be charged slowly and carefully to avoid damage. All batteries produce hydrogen gas while recharging. Non-sealed batteries vent it out. SLA batteries do something called "gas recombination," meaning that the gas generated is "recombined" into the cells. SLA batteries actually operate under pressure, usually about 3 psi. If the battery is charged too quickly, the battery generates gas faster than it can recombine it, and the battery over-pressurizes. This causes it to overheat, swell up, and vent, which can be dangerous and will permanently damage the battery. The charging voltage must be kept between 13.8 and 14.5 volts. Wherever possible, follow the battery manufacturer's instructions. Lacking these, a good rule of thumb is to keep the charging current level to no more than 1/3 the battery's rated capacity. For example, if you have a 7 Ah battery, you should charge it at no more than 2 amps.

The time it takes for a SLA battery to recharge completely will depend on the amount of charge remaining in the battery. If the battery is only 25% discharged, then it may recharge in a few hours. If the battery is discharged 50% or more, 18–24 hours may be required.

Solar panels and charge controllers. These are readily available at increasingly lower costs. These provide yet another option for powering equipment in the field when weather and site conditions permit their use. When choosing solar equipment, consult with the vendor regarding the required size of panels and controller for your specific application.

DC-to-ac inverters. While direct dc power is more efficient and should be used whenever possible, inverters can be used for equipment that cannot be directly powered with 12 V dc. Not all inverters are suitable for use with radios, computers, or certain types of battery chargers. The best inverters are those with a "true sine wave" output. Inverters with a "modified sine wave" output may not operate certain small battery chargers, and other waveform-sensitive equipment. In addition, all "high-frequency conversion" inverters generate significant RF noise if they are not filtered, both radiated and on the ac output. Test your inverter with your radios, power supplies and accessories (even those operating nearby on dc) and at varying loads before relying upon it for emergency communications use.

Effective filtering for VHF and UHF can be added rather simply (using capacitors on the dc input, and ferrite donuts on the ac output), but reducing HF noise is far more difficult. Inverters should be grounded when in operation, both for safety and to reduce radiated RF noise.

As an alternative to an inverter, consider a mid-sized 12 V computer UPS (uninterruptible power source). Smaller, square-wave UPS units are not designed for continuous duty applications, but larger true sine-wave units are. Most true sine-wave units use internal batteries, but with minor modifications can be used with external batteries. The larger commercial UPS units run on 24 or 48 volts, and require two or four external batteries in series. UPS units will have a limit on the number of depleted batteries they can re-charge, but there is no limit to the number of batteries that can be attached to extend operating time.

Generators are usually required at command posts and shelters for lighting, food preparation, and other equipment. Radio equipment can be operated from the same or a separate generator, but be sure that co-located multiple generators are bonded with a common ground system for safety. Not all generators have adequate voltage regulation, and shared generators can have widely varying loads to contend with. You should perform a test for regulation using a high-current power tool or similar rugged device before connecting sensitive equipment. A voltmeter should be part of your equipment any time auxiliary power sources are used.

Noise levels can be a concern with generators. Some are excessively noisy and can make radio operations difficult and increase fatigue. A noisy generator at a shelter can make it difficult for occupants to rest, and can result in increased levels of stress for already stressed people. Unfortunately, quieter generators also tend to be considerably more expensive. Consider other options, such as placing the generator at a greater distance and using heavier power cables to compensate. Placing a generator far from a building can also prevent fumes from entering the building and causing carbon monoxide poisoning, an all-too-common problem with emergency generators.

Several other devices may be helpful when dealing with generators or unstable ac power sources. High quality surge suppressors, line voltage regulators and power conditioners may help protect your equipment from defective generators. Variable voltage transformers ("Variacs") can be useful to compensate for varying power conditions.

EQUIPMENT FOR OTHER MODES

If you plan to operate one of the digital modes (packet, APRS, AMTOR, PSK31, etc.), then you will also need a computer and probably a TNC or computer sound card interface. Some newer radios have built-in TNCs. Be sure to identify all the accessories, including software and cables, needed for each mode. Include the power required to operate all of the radios and accessories when choosing your batteries and power supply. The internal battery in your laptop computer will probably not last long enough for you to complete your shift. Be prepared with an external dc power supply and cable, or a dc-to-ac inverter. If you need hard copy, then you will also need a printer, most of which are ac powered. In addition to your Amateur Radio equipment, you may find a few other items useful, such as:

- Multiband scanning radio (to monitor public service and media channels)
- FRS, GMRS (separate license required), or MURS handhelds
- Cellular telephone (even an unregistered phone can be used to call 911)
- Portable cassette tape recorder with VOX (for logging, recording important events)
- AM/FM radio (to monitor media reports)
- Portable television (to monitor media reports)
- Weather Alert radio with "SAME" feature (to provide specific alerts without having to monitor the channel continuously)
- Laptop computer with logging or EmComm-specific packet software

TESTING THE COMPLETE STATION

After making your equipment selection (or beforehand if possible), field test it under simulated disaster conditions. This is the fundamental purpose of the annual ARRL Field Day exercise in June, but any time will do. Operations such as Field Day, contests, and QSO parties can add the element of multiple, simultaneous operations on several bands and modes over an extended period. Try to test all elements of your system together, from power sources to antennas, and try as many variations as possible. For instance, use the generator, and then switch to batteries. Try charging batteries from the solar panels and the generator. Use the NVIS antenna while operating from batteries and then generator. This procedure will help reveal any interactions or interference between equipment and allow you to deal with them now — before proper operation becomes a matter of life and death.

PUBLIC SERVICE
Emergency Communication
READY ■ RESPONSIVE ■ RESILIENT

Handheld Transceiver Go-kit

Scott R. Gothard, W6SRG,
El Dorado County Amateur
Radio Club and El Dorado
County Amateur Radio
Emergency Service
srgothard@yahoo.com

As an Amateur Radio operator who's serious enough to be an active part of a club, most of us (maybe all of us) have a handheld transceiver. This may have been the first radio you started with when you got your license. This is a more-than-adequate communications device, especially when access to repeaters is available. But performance and usability can always be improved (see Figure 1).

One of the first places to start is with an aftermarket antenna. While the flexible antenna that comes with handheld transceivers is certainly adequate to the job when operating relatively close to repeaters or within a short distance of others operating on simplex, a "whip" antenna can extend the range of communications. As an example, Comet Antenna provides a good product, with models having both BNC and SMA connectors.[1] While somewhat longer than the stubby flexible antenna your radio came with, these whip antennas are still of a workable length while the radio hangs on your belt. A bit of wire clipped to the bare base of the antenna can complete the dipole, assuming your antenna allows for such an arrangement.

I also carry two other antennas in my kit. One is a telescoping Diamond RH789 with an SMA-to-BNC adaptor that gives my handheld an antenna equivalent to the mobile antenna on my vehicle.[2] This can be really handy if you're in an isolated area and need to access

Figure 1 — This is my entire handheld radio kit. Clockwise from top left: Mirage BD-35 dual-band amplifier, plastic storage box (1 of 2 to hold everything), rolled up "Slim Jim" J-pole antenna, parachute cord for mounting of J-pole antenna, extendable whip antenna, coax adaptors for radio, zip ties for mounting of J-pole antennas, belt pouch for handheld transceiver, pigtail coax adaptors, plug in speaker-microphone, cigarette lighter recharging chord for handheld radio and extension cable for amplifier. My handheld transceiver is shown in the center.

a distant repeater. The other antenna I carry, which gives me an antenna loosely equivalent to a base station antenna, is a dual band "Slim Jim" antenna, a custom made J-pole made out of common ladder line, which the manufacturer will outfit with the appropriate connector for you.[3] This unit rolls up into a conveniently small package and, when unfolded, can be suspended from a nail or by a section of parachute cord tossed over a branch.

A plug-in speaker-microphone is also very useful. This way, you get to keep your radio on your belt, rather than having to take it out each time you transmit. Make sure you don't "ground out" your antenna with your body.

In addition, if you mount your handheld transceiver on some kind of desktop cradle that allows it to sit up, you can use the speaker-microphone to turn it into a miniature "base station" radio, especially if you combine your handheld with an external amplifier.[4, 5]

In addition to the wall wart charging unit, I also have a vehicle recharger for my radio. You never know how long you'll be out in the field and away from a wall plug, so having the ability to recharge your radio in your vehicle is a virtual must.[6]

Another set of useful items to have along are a set of adaptors and a short "pigtail" extender that would allow you to attach your handheld transceiver to various types of transmission lines or antennas with connectors other than the style your radio uses. For instance, I use a handheld transceiver with an SMA connector that the antenna attaches to, so I have adaptors that include SMA-to-BNC and SMA-to-PL-259, as well as a "pigtail," a short section of RG-6 or RG-8 with a BNC connector on one end and a PL-259 connector on the other. This allows me to connect my handheld transceiver, by using an SMA-to-BNC adaptor and a "pigtail," to my Mirage BD-35 dual-band amplifier and use it as an alternate mobile or base station.

One last item to add to the list is some kind of small container to keep all of these items together in one place, as well as making it easy to pitch everything in your bug-out bag when you need to go somewhere quick.

[1]www.cometantenna.com
[2]www.hamcity.com/store/pc/viewPrd.asp?id category=306&idproduct=114
[3]www.n9tax.com/Slim%20Jim%20Info.html
[4]www.nifty-accessories.com puts out a nice little unit for about $35.
[5]A good example would be the Mirage BD-35 dual band amplifier that boosts the 5 W signal from your handheld to approximately that of a mobile radio, 45 W on 2 meters and 35 W on 70 cm.
[6]This assumes you don't have some kind of inverter mounted in your vehicle or one that can insert into a lighter plug.

Steve Ewald, WV1X Public Service Specialist sewald@arrl.org

Sidebar 12.1 – Reprinted from the March 2011 issue of *QST*.

The Amateur Radio Public Service Handbook

By Robert Hollister, N7INK

A Portable NVIS Antenna

As the Amateur Radio Emergency Service (ARES) District Emergency Coordinator (DEC) and Radio Amateur Civil Emergency Service (RACES) officer for Cochise County, Arizona, I am always looking for simple solutions to complex problems. The problem in this case was to provide easy-to-use antennas for our county Emergency Response Vehicle (ERV) (Figure 1). Our ARES/RACES unit operates the ERV communications unit for the county sheriff's department in support of the Emergency Services Office, Search and Rescue Team and other emergency missions. Located in the southeast corner of Arizona, Cochise County consists of high desert, higher mountains and deep canyons. This presents a difficult environment to maintain reliable radio communications. Although the Amateur Radio clubs and the sheriff's department have a good network of VHF/UHF repeaters, there are many areas where repeaters or cell phones just do not provide the coverage we need to maintain contact with the sheriff's dispatch center in Bisbee.

A Solution

I recently read a Web article published by W0IPL that described the use of mobile monoband whip antennas in a near vertical incidence systems (NVIS) configuration (www.w0ipl.com/ECom/NVIS/nvis.htm). I decided, therefore, to experiment with this NVIS technique and build a portable kit for use with our ICOM IC-706 installed in the ERV.

The first step was to develop a shopping list (Table 1) and purchase the necessary items I thought I would need to make it work. At a recent hamfest, I purchased two monoband mobile whip antennas for each of the two bands I wanted to operate, and a dipole adapter that I had seen at the Dayton Hamvention.[1] I then broke out my trusty MFJ antenna analyzer to help with the trimming and SWR measurements.

First, I set up antenna number 1 (75 meters) on my Outpost tripod to start the tuning process.[2] The target was a 5 MHz frequency that I wanted to use to connect with the Fort Huachuca Military Affiliate Radio Service (MARS) station digital bulletin board. Many of our ARES/RACE members are also active in the Army MARS program and we have a TNC in our ERV for both HF and VHF digital communications. I knew that the normal range of the mobile whip antenna covered the 75 meter portion of the band effectively, but that it would require a much shorter stinger than the 48 inch standard length to reach the target 5 MHz frequency.

Construction

Rather than trimming the stainless steel stinger that came with the antenna in steps, I started with a piece of wire coat hanger cut to 24 inches that fit into the collar on the top of the bottom section of the antenna. I scraped the varnish off the wire on the piece to be inserted in the collar for a good electrical connection (Figure 2). If you have ever tried to cut stainless stingers before, you will understand quickly why using the softer coat hanger wire is the preferred method. On the first test, the antenna was resonant at about 6.250 MHz. This told me that I was still much too long. I cut the wire stinger in half with my wire cutters to about a 12 inch stinger and tried again. I was now in the ballpark and only 500 Hz away. I continued cutting at half-inch intervals and tested the results.

I was soon at my desired resonant frequency. The stinger was now extended only 6 inches above the bottom section of the antenna. The antenna instruction sheet suggested that not more than 4 inches be inserted into the bottom section. The total length of the now resonant stinger was 10 inches. I used this trimmed piece of wire as a template and laid it next to the original stainless

Table 1
NVIS Antenna Parts

75 meter mobile monoband antenna	2 each = $30
Extra 48" stingers	2 each = $18
40 meter mobile monoband antenna	2 each = $30
Dipole adapter	1 each = $15
Stainless radiator clamps, 2"	2 each = $1
5 foot fiberglass pole, surplus	2 each = $15
RORO ("big foot") mount	1 each = $90
10" section, $^3/_8$" aluminum stock	1 each from junk box
10" section, 2^8 aluminum stock	1 each from junk box
	Total = $199

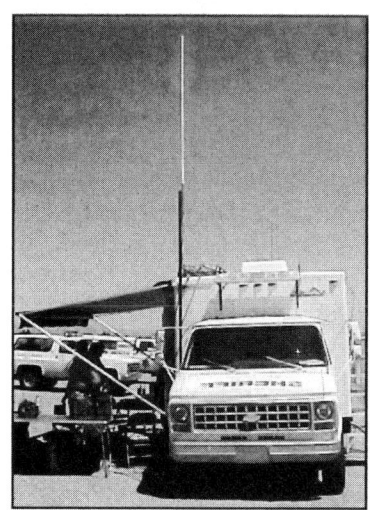

Figure 1—The Cochise County, Arizona emergency response vehicle (ERV).

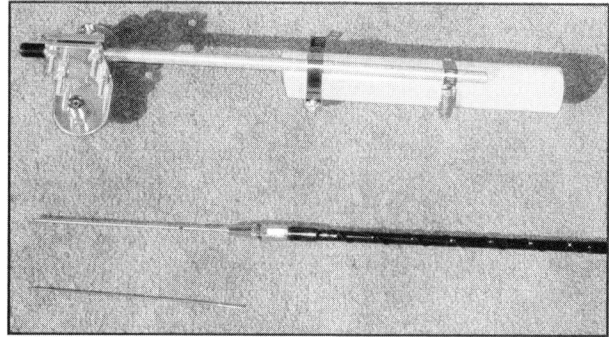

Figure 2—A coat hanger is used as a prototype "stinger" during testing and requires that the paint be scraped off its end. It is shown next to a new stainless steel tip. The dipole adapter is shown at the top.

Sidebar 12.2 – Reprinted from the January 2005 issue of *QST*.

stinger. After carefully marking the new length and allowing for adequate length to extend into the top of the lower section, I cut the stainless stinger to length with my hacksaw. After smoothing and rounding the cut end of the stainless steel whip with a diamond file, I rechecked the resonant frequency. It was right on. I took the new short stainless stinger and, using the same measurement as the first one, checked it with the second 75 meter antenna bottom section for resonance. It was close, but required a slight amount of adjustment to be brought to frequency. The second stinger was quickly cut and ready for testing (Figure 3).

NVIS Height

Since the dipole adapter I purchased did not come with any instruction sheet, I needed to figure out how to get the antenna to a suitable height above ground. I already had something we call our roll-on/roll-off (RORO) or "big foot" to use as an antenna support base plate (Figure 4).[3] The NVIS literature and field testing suggests that the optimum height above ground is not an exact science but that it should be approximately 10-12 feet high to reliably cover a range from 0 to 300 miles. Height above ground and the conductivity of the soil below the antenna will affect the optimum range and the resonant frequency. Minor adjustments to stinger lengths may be required at different locations, depending on ground conductivity. A height of 10 to 12 feet above ground also ensures that it is high enough so that most pedestrians and NBA basketball players are not likely to run into the ends.

As I owned a number of 5 foot military surplus aluminum and fiberglass antenna mast sections, I decided to use two of the fiberglass sections as the main antenna support.[4] These are the same type we use to hold up our spare dual band (144/440) antenna when we need additional range over the vehicle-mounted antenna. I then had to decide how to easily and quickly attach the adapter to these mast sections. A short piece of a scrapped aluminum mast section cut to 10 inches, two each 2 inch stainless hose clamps, and another 10 inch piece cut from some $1/2$ inch aluminum tubing fit the bill. This allows a quick slip-on fit over the top of the fiberglass mast section (see Figure 3).

The next step was to attach the two modified mobile antennas and coax to the dipole adapter. I slipped the adapter over the top of one of the fiberglass masts, inserted it into the "big foot" and reattached the coax to my analyzer. It was too much to hope that it would still be exactly on frequency and, unfortunately, I was correct. There was enough interaction with the added metal of the adapter to require some additional tuning. Fortunately, it was minor and only required shortening both of the elements approximately another $3/4$ inch to bring it back to my desired frequency. The next step: On-the-air testing.

Testing

With the two fiberglass mast sections inserted into the "big foot" base, the completed horizontal dipole was now approximately 11 feet in the air (Figure 5). Hooking up the radio, the computer and the TNC (terminal node controller), my first attempt to connect to the Fort Huachuca MARS BBS (bulletin board system) using PACTOR (a digital communications protocol) was a success. Solid S9 copy was obtained with approximately 40 W output. I quickly prepared a sample message to send, reconnected

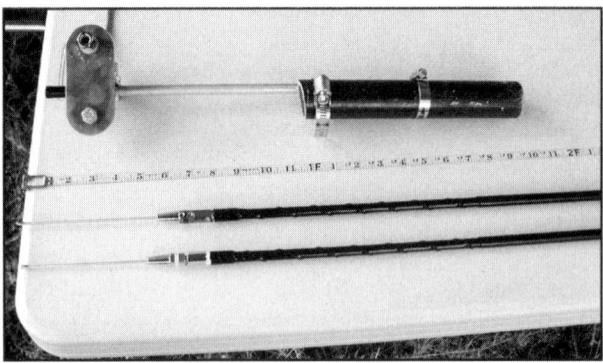

Figure 3—A top view of the assembled dipole adapter used to mount and secure both antennas, which are shown below it.

Figure 4—The RORO ("big foot") mount. This was custom-built for the author by a welding shop (see text).

Figure 5—The completed antenna in use in the Chiricahua mountains of Arizona.

Sidebar 12.2 – Reprinted from the January 2005 issue of *QST*.

to the BBS, uploaded the message and disconnected. Again success! Before disassembly to move to the next frequency of interest, each stinger and bottom antenna section were color coded with a band of red and white paint to ensure the correct stinger could easily be reassembled with the correct bottom section.

I had purchased a second set of stingers (of normal 48 inch length) for the 75 meter antennas and a pair of 40 meter antennas at the time I made the original purchase. Setting up for the other two primary frequencies I wanted to use was straightforward and required no trimming. Using two standard-size 48 inch whips and collets, these two stingers were adjusted for the Arizona RACES frequency of 3.990 MHz on the Outpost tripod. These are adjusted for our Arizona state RACES net on 3.990 MHz, and the third set is set for our alternate frequency of 7.248 MHz. There was adequate flexibility using the standard stingers to adjust to the state MARS nets just outside the amateur band with no additional trimming required.

The state RACES net has stations representing most state counties and reaches all corners of the state. I am located in the southeast corner with the furthest station about 250 miles away. Again, attaching the antennas to the dipole adapter required some minor shortening to achieve resonance (1.2:1). During a Sunday morning RACES net, I conducted on-the-air A/B testing. I compared the NVIS configuration to a Hustler vertical (previously used for portable operation), attached to the Outpost tripod. The NVIS configuration demonstrated an approximate 1-2 S unit improvement for most signals compared to the vertical.

Switching to 40 meters, and using the same procedure with two 40 meter antennas on the Sunday morning Arizona MARS net, showed a similar improvement. Most signals showed a 1-2 S unit improvement in signal strength using the NVIS dipole compared to the vertical. The biggest audible improvement was noticed in interference/noise levels. Some stations were totally lost in the noise on the vertical and came back to Q5 on the NVIS dipole. This net included stations in Arizona and southern California. Similar distances to those of the Arizona RACES net were covered. At that time of day (one hour later than the RACES net), the 40 meter coverage was generally better on all stations.

Conclusion

Two weeks later, I was able to run a few more tests from a remote location in the Chiricahua Mountains. Over the Labor Day weekend, the Cochise Amateur Radio Association conducts a special event operation from the ghost town of Paradise. This provided the most realistic test yet, as we deployed the ERV and operated portable in the mountains, much like a normal search mission. I tested all configurations on both voice and digital frequencies. The antenna was quick to set up and provided reliable communication to the stations I needed to contact. The complete antenna kit requires less than 5 minutes to assemble and erect. This antenna has proven to be a reliable addition to our emergency communications needs and it meets all of my expectations.

Notes

[1]The antennas are available from the Lakeview Company (**www.hamstick.com/9106.htm**) and WB0W (**wb0w.com**) as well as other sources. The dipole adapter is available from a number of sources, including: Atoc Technologies (Iron Horse) (**www.atoctechnologies.com/**) and Quicksilver Radio Products (**www.qsradio.com/products.htm**).
[2]**www.alphadeltacom.com/pg6.htm**.
[3]The RORO mount was built to specification by a local welder for $90. A drawing can be provided for others who are interested in having one fabricated. I have seen similar designs available at several hamfests.
[4]The mast sections came from The Mast Company (K4TMC), PO Box 1932, Raleigh, NC 27602; **www.tmastco.com/**. Henry offers a variety of military surplus mast sections in aluminum or fiberglass. I prefer the lighter fiberglass but I have used both with equal success.

Photos by the author.

Robert Hollister, N7INK, was first licensed as N2BCY in 1978, while serving with the US Army in Germany. He's also held the calls DA2HO and GM5ELQ. Bob retired from the Army in 1986 as a communications intercept technician. He has a BA from The State University of New York and an MS in Applied Management from Lesley College, Cambridge, Massachusetts. Bob also has an Amateur Extra class license and is a life member of the ARRL. He can be contacted at 5457 S San Juan Ave, Sierra Vista, AZ 85650, or at n7ink@cox.net.

Sidebar 12.2 – Reprinted from the January 2005 issue of *QST*.

Part 4: Nets and the National Traffic System

Chapter 13
The National Traffic System

The National Traffic System (NTS) is an organized effort to handle traffic in accordance with a plan that is easily understood. The NTS allows for rapid movement of traffic from origin to destination and trains Amateur Radio operators to handle written traffic and participate in directed nets. It employs modern methods of network traffic handling that are in general acceptance today. During disasters or other emergencies, radiograms are used to communicate information critical to saving lives or property, or to inquire about the health or welfare of a disaster victim. NTS operates daily, even continuously, with advanced digital links into the National Traffic System — Digital (NTS-D). NTS operators participate for one or two periods a week and some are active daily.

NTS is not intended as competition for the many independently organized traffic networks in existence. When necessitated by overload or lack of outlet for traffic, the facilities of such networks can function as alternate traffic routings where this is indicated in the best interest of efficient message relay and/or delivery.

One of the most important features of NTS is the system concept. No NTS net is an independent entity that can conduct its activities without concern for or consideration of other NTS nets. Each net performs its function and only its function in the overall organization. If nets fail to perform their functions or perform functions intended for other nets, the overall system may be adversely affected.

MEMBERSHIP IN NTS

Individual station participation in NTS is recognized by issuance of certificates, and by appointment to the field organization's traffic handling position, which is known as the Official Relay Station. Organizationally speaking, the "members" of NTS are the nets and digital nodes that participate therein. Most nets and many of the NTS-sanctioned nodes were created and organized for NTS purposes only and operate for specific purposes to be described later. Procedures are somewhat specialized, particularly at Region, Area, and Transcontinental Corps (TCC) levels.

Frequently, ARRL Headquarters is asked how a net or digital node (BBS) may become a part of NTS. This usually isn't easy, because NTS is not a "club for nets" that any existing net may join at will. In addition, making nets a part of NTS is less a matter of official action than a "state of mind" of the net itself. In this connection, the following points deserve mention:

- Voice or CW Nets or high frequency digital (HF) hubs or very high frequency (VHF) or ultra-high frequency (UHF) packet nodes (BBSs) operating within ARRL Section boundaries, or otherwise at local or Section level, may become a part of NTS by performing the functions of such.
- Nets whose coverage extends beyond Section boundaries but within region (roughly, call area) boundaries may become a part of NTS only by foregoing their general membership and setting up to operate as a session of the region net. Such nets would act as one of that region's net sessions and would be under the jurisdiction of the region Net Manager appointed by ARRL. All present NTS region nets were organized specifically at the outset for NTS region coverage.
- HF digital stations capable of storing and forwarding NTS messages in a system of such stations may be certified as NTS Digital Relay Stations by NTS Officials known as Area Digital Coordinators. They are responsible for handling volume or bulk NTS traffic to the same high standards as their counterparts in the traditional system. Also some NTS Digital Relay Stations are certified as Target Stations relaying traffic into the *Winlink 2000* system if stations cannot connect to a WL2K RMS station and send traffic into that system.

Since operation at the area level is so specialized, it is not possible for nets whose coverage extends beyond region boundaries to be a part of NTS at any level.

Any net or digital node which becomes a part of NTS is expected to observe the general principles of NTS procedures. Generally speaking, participation in NTS is best performed by individual-station participation in an already-existing NTS net, at any level.

Lack of recognition as an NTS net does not imply that such a net is without ARRL recognition or support. Many public service nets on which information is received are included in the ARRL's online Net Directory, and activities are often summarized in the appropriate part of the ARRL website. Although NTS is the ARRL-sponsored organization for systematic traffic handling, it is far from being the ARRL's only interest in public service communication.

MODE

The National Traffic System is not dedicated specifically to any mode or to any type of emission, nor to the exclusion of any of them, but to the use of the best mode for whatever purpose is involved. The aim is to handle formal written traffic systematically; by whatever mode best suits the purpose at hand. Whether voice, CW, RTTY, AMTOR, Pactor I, II, or III, packet, or other digital mode is used for any specific purpose is up to the Net Manager or managers concerned and the dictates of logic.

PRINCIPLES OF NTS OPERATION

The National Traffic System includes four different net levels that operate in an orderly time sequence to affect a definite flow pattern for traffic from origin to destination. The way a message flows through the National Traffic System can be compared to the way an airline passenger who starts out in a small residential town reaches a destination across the continent in another small town. One has to change carriers many times in the process, starting with a local ground conveyance to a feeder airline, to a transcontinental airline, to another feeder airline, then local transportation to deliver him to his destination. Similarly, the transcontinental message starts with the originating station in a local net, is carried to the Section net, the region net, the area net, via Transcontinental Corps (TCC) to a distant area net and then back down the line to delivery.

Of course the message, like the passenger, can "get on" or "get off" at any point if that's the origin or destination. Thus, a message from, say, New York to Detroit would never "get on" TCC, but rather would "get off" at area level. A message from San Francisco to Los Angeles would not go beyond region level, and one from Syracuse to Buffalo would remain inside the Section net.

Messages may also be passed through NTS-affiliated local and Section traffic nodes that employ digital modes with store-and-forward capabilities and bulletin board operations. Long hauls can be made by NTS Digital Relay Stations at HF that interface with Section traffic nodes, and the traditional nets of the system.

LOCAL NETS

Local nets are those which cover small areas such as a community, city, county or metropolitan area, not a complete ARRL Section. They usually operate by VHF (typically 2 meter FM) at times and on days most convenient to their members; some are designated as "emergency" (ARES) nets that do not specialize in traffic handling. The time slot designated for them is thus nominal and will vary considerably. Local nets are intended mainly for local delivery of traffic, inasmuch as such delivery could ordinarily be affected conveniently by non-toll telephone. Some NTS local nets operate on a daily basis, just as do other nets of the system, to provide outlets for locally originated traffic and to route the incoming traffic as closely as possible to its actual destination before delivery — a matter of practice in a procedure that might be required in an emergency.

Most local nets and even some Section nets in smaller Sections use repeaters to excellent effect. Average coverage on VHF can be extended tenfold or more using a strategically located repeater and this can achieve a local coverage area wide enough to encompass many of the smaller Sections. Since propagation conditions on the high frequencies are erratic, more use of VHF and repeaters is recommended at local levels.

A local net or node may also be conducted on a local packet BBS, where radiograms may be stored, forwarded and picked up by local operators for delivery. A Net (Node) Manager is appointed by the Section Traffic Manager to manage these functions, and assure that traffic is moved expeditiously in accordance with basic NTS principles, just like their counterpart nets on local repeaters.

SECTION NETS

Organizational and procedural lines begin to tighten at the Section net level. Coverage of the Section may be accomplished either by individual stations reporting in, by representatives of NTS local nets and nodes, or both. Ordinarily, all Section Amateurs are invited to take part; however, in a high-population Section with several metropolitan areas covered by local nets, representation may be by such liaison stations plus individual stations in cities or towns not covered by local nets.

The Section may have more than one net (a CW net, a VHF net, an SSB net, or even a Section packet BBS, for examples), or two or more Sections may combine to form a single net operating at Section level, if low population or activity seem to make this desirable. Section nets are administered through the office of the Section Manager, with authority for this function often delegated to an appointed Section Traffic Manager and / or designated Net Managers.

In the case of combined-Section nets, officials of the Sections concerned should collaborate on the designation of a qualified Amateur to manage the net. The purpose of the Section net is to handle intra-Section traffic, distribute traffic coming down from higher NTS echelons, and put inter-Section traffic in the hands of the Amateur designated to report into the next-higher NTS (region) echelon. Therefore, the maximum obtainable participation from Section Amateurs is desirable.

REGION NETS

Region nets cover a wider area, such as a call area. At this level the object is no longer mass coverage, but representation of each ARRL Section within the region.

Participants normally include:

- A net control station, designated by the region Net Manager.
- Representatives from each of the various Sections in the region, designated by their Section Net Managers.
- One or more stations designated by the region Net Manager to handle traffic going to points outside the region.
- One or more stations bringing traffic down from higher NTS echelons.
- Any other station with traffic.

There may be more than one representative from each Section in the region net, but more than two are usually superfluous and will only clutter the net; however, all Section representatives are required to represent the entire Section, not just their own net.

The purpose of the region net is to exchange traffic among the Sections in the region, put out-of-region traffic in the hands of stations designated to handle it, and distribute traffic coming to the region from outside among the Section representatives. Region nets are administered by managers who are elected by NTS Area Staff members.

AREA NETS

At the top level of NTS nets is the area net. In general, the area net is to the region net what the region net is to the Section net; that is, participation at area level includes:

- A net control station, designated by the area Net Manager.
- One or more representatives from each region net in the area, designated by the region Net Managers.
- TCC stations designated to handle traffic going to other areas.
- TCC stations designated to bring traffic from other areas.
- Any station with traffic.

The third and fourth points are functions of the Transcontinental Corps. There are three areas, designated Eastern, Central, and Pacific, the names roughly indicating their coverage of the U.S. and Canada, except that the Pacific Area includes the Mountain as well as the Pacific time zones. Area nets are administered by managers who are elected by NTS Area Staff members.

TRANSCONTINENTAL CORPS

The handling of inter-area traffic is accomplished through the facilities of the TCC. This is not a net, but a group of designated stations who have the responsibility for seeing that inter-area traffic reaches its destination area. TCC is administered by TCC directors — or as delegated to the Area Digital Coordinator — in each area who assign stations to report into area nets for the purpose of "clearing" inter-area traffic, and to keep out-of-net schedules with each other for the purpose of transferring traffic from one area to another.

DIGITAL STATIONS

The handling of traffic among Sections, regions, and areas can also be accomplished alternatively, on a supportive/cooperative basis, through liaison with the traditional aspects of the system, by the set of NTS Digital Relay Stations across the country. These stations, certified by their respective Area Digital Coordinators, handle traffic by digital modes at HF. The system structure is more loosely defined than is the traditional system. They serve to supplement the existing system, providing options and flexibility in moving traffic expeditiously across the country, especially in overload conditions.

CHECK-IN POLICY

National Traffic System nets at local and Section level are open to all Amateurs in the coverage area of the net. At region and area level, participation is normally restricted to representatives of Sections, and designated liaison stations. However, stations from outside the coverage area of the net concerned, or other not-regularly designated participants who report in with traffic will be cleared provided they can maintain the pace of the net as to procedure, speed, and general net "savvy." Such stations reporting in without traffic will immediately be excused by the NCS unless they can supply outlets unavailable through normal NTS channels. Visitors to NTS nets should bear in mind that NTS nets operate on a time schedule and that no offense is intended in observance of the above check-in policy.

FREQUENCIES

There is no specific NTS frequency plan. Each NTS net selects its own operating frequency in consideration of its requirements.

Because in an emergency it may be necessary to operate many NTS nets simultaneously that ordinarily operate at different times, it is desirable for nets within normal interference range of each other to use different center frequencies if possible. Within this consideration, it is also desirable to concentrate NTS operation on as few spot frequencies as possible to conserve frequency space and to make full use of those spot frequencies utilized in order to help establish occupancy. ARRL's online Net Directory records net frequencies and times and are useful to study in planning new nets.

MANAGER APPOINTMENTS

NTS net (packet node) managers at the local and Section level are appointed or designated by the STM. All other NTS managers, including Area Digital Coordinators, are elected by NTS Area Staff members. Net Managers and Area Digital Coordinators are appointed for no specific term of office. The Area Digital Coordinators are responsible for appointing the NTS Digital Relay Stations at HF.

For a Packet Node, the STM may look for an appointee who has the following qualifications:

1. Can dedicate a complete system (radio and Terminal Node controller) 24 hours a day, seven days a week.
2. Has a dependable radio that can operate on VHF or UHF frequency (Note: in some areas 440 MHz, 220 MHz, and 6 meters are used to exchange traffic between nodes)
3. When running a 1200 baud node is also capable of supporting a "backbone" packet system with another radio connecting to other nodes at 9600 baud on a different band (sometimes telnet connections are used or required and are used as a backbone between nodes instead of a 9600 baud connection.)

For Packet BBS on VHF the STM may look for an appointee who has the following qualifications:

1. Knows how to format NTS messages digitally.
2. Knows the how, what, and where of forwarding any type of traffic within and outside not only their local area but also your state, region, and area of the NTS and NTSD system.
3. Has a dependable radio that can operate on the VHF or UHF frequency. (Note: in some areas 440 MHz, 220 MHz, and 6 meters are used to exchange traffic between boards.)
4. Can dedicate a complete system 24 hours a day, seven days a week.

NTS CERTIFICATION

NTS certificates are available at local, Section, region, and area levels as well as for fulfilling TCC assignments. A participating station is eligible for an NTS net certificate when it has completed three months of performance (at least once per week), on an assigned basis, of one or more of three essential duties:

1. Regular participation as a net station. In the case of region and area nets, this means official representation of a Section or region within its respective region or area. No credit is given in region or area nets for random participation.
2. Liaison with other nets of the National Traffic System. This applies only to regular liaison in accordance with the NTS flow pattern as assigned by the appropriate Net Manager. In the case of Section nets, liaison with their proper region nets; in the case of region nets, liaison with their proper area nets; in the case of area nets, liaison with other area nets through regularly-assigned functions in the Transcontinental Corps.
3. Net Control Station. The NCS is the operator that presides over the net session.

Certification in the Transcontinental Corps is available through the TCC area director on completion of at least three months of regular performance of an assigned function.

Net Managers (or TCC directors) may use their discretion in excusing any station working for a certificate if that station is unable to perform its regular duty in any specific instance. Net Managers (or TCC directors) shall be the sole judges as to whether a duty, even though performed regularly, is performed adequately to merit certification.

Area Digital Coordinators issue certificates to NTS HF Digital Relay Stations.

NTS STANDARD NET PROCEDURES

The following procedures are recommended as NTS standards. Deviations from these procedures are made at the discretion of the Net Manager in cognizance of either necessity or desirability arising out of extraordinary circumstances, but always as a temporary expedient until standard procedure can be resumed.

The following procedures apply to all NTS nets:

- The net control station (NCS) transmits a net call-up promptly at the pre-established net meeting time.
- Stations reporting in indicate their function or the destination(s) for which they can take traffic, followed by the list of traffic on their hook, if any.
- Time-consuming pleasantries and other superfluous matters are not to be a part of the procedure while the net is in session.
- Net stations follow the direction of the NCS without question or comment if such directions are understood.
- Explanations of any kind are not transmitted unless they are absolutely essential to the net's conduct.
- Stations reporting into a net are held for 15 minutes, after which they are excused if there is no further traffic for them at that time. Stations in the net do not leave the net without being excused and do not ask to be excused unless absolutely necessary.
- All nets follow the general precepts of net operation outlined in the *ARRL Operating Manual*.

Section Nets

The random call-up method should be used in most cases. The clearing of traffic should commence as soon as stations reporting in the net have traffic for each other, rather than waiting until all stations have reported in. The use of side frequencies (QSY) should be used extensively. The QNA procedure (stations answering in prearranged order) should only be used in times of traffic overload, or for acknowledging the region net representatives at the beginning of the net.

The following additional procedures are used in Section Nets:

- Stations reporting in to the net with traffic, list the destination city first, then the number of messages for that city. Example: "W2RQ DE AA2Z QTC Paterson 1 AR." Traffic destined outside the Section is designated "through" (or "thru") followed by the number of "thru" messages. "Thru" traffic can also be listed for the appropriate region net.
- The region net representative is selected before-hand by the Section Net Manager, but nevertheless indicates his purpose in reporting in.
- Stations do not list their traffic until first recognized by the NCS.
- If a particular city for which there is traffic is not represented on the net, the NCS may inquire who will handle such traffic, direct that it be sent to the station who can take it to a local net or bulletin board for delivery, who is nearest to the destination or that it be mailed. In any case, there should be a minimum of discussion.

Region Nets

Stations reporting in indicate the Section they came from if they are officially reporting for the purpose of handling traffic to or from that Section. If their function is limited to sending or receiving, they should so indicate; otherwise the NCS will assume the station will do both.

Traffic for destinations within the region is reported by Section. If the destination is outside the region, the traffic should be designated "thru" or for the area net. For example, "DE W9QLW QIN QTC WIS 3 ILL 2 CAN 2 AR" tells the NCS that W9QLW represents the Indiana Section (QIN), and has traffic.

The area net liaison station (designated beforehand by the region Net Manager) receives all traffic designated for the area net.

Stations reporting in who are not authorized Section representatives or liaisons simply indicate the traffic they have to send. If they do not have any traffic, they are immediately excused by the NCS, unless they can provide an outlet not available on the net through regular NTS channels.

In the event that a particular Section is not represented in the region net, the NCS will use special liaison methods or any alternatives that are available for clearing traffic to that Section.

Area Nets

Stations reporting in indicate traffic by region if it is destined for a region within that area or by area if it is destined to a point outside that area. All stations reporting in with assigned receive functions indicate for which region in the area or for which other area they are authorized to receive traffic. (See NTS Policy Manual on the ARRL website for abbreviations of geographical areas.)

The TCC representative designated to take traffic for another area so indicates in his QNI (check-in). For example, "DE W2MTA PAN QRU AR" tells the NCS that W2MTA has been assigned to take any traffic destined to the Pacific area and that he himself has no traffic.

Send and Receive Stations

Many NTS nets have adopted the procedure of sending more than one representative to the next-higher NTS echelon — one to take the traffic up and report it in, another to receive traffic from the upper echelon and bring it back. More than one transmit or receive liaison may be provided if the traffic load is heavy and the personnel available is sufficient. It is perfectly permissible, and has many advantages in overload conditions — to the NCS and the net — for traffic both going and coming to be divided among two or more liaison stations.

Representatives who do not indicate which function they are performing will be assumed to be ready to perform both transmit and receive functions, at the discretion of the NCS. To indicate which you are performing, on CW send "RX" or "TX" after your QNI; on phone say "receive only," "transmit only," or "both."

ARRL PRECEDENCES AND HANDLING INSTRUCTIONS

All messages handled by Amateur Radio should contain precedences — that is, an evaluation of each message's importance, made by the originating station. Precedence is an "order of handling." There are four precedences in the ARRL message form: Emergency, Priority (P), Welfare (W), and Routine (R), in that order of handling. When and as they appear on a net or any other kind of circuit, messages will be handled in that order.

Emergency

An Emergency precedence message is a message of life-and-death urgency to any person or group of persons, which is transmitted by Amateur Radio in the absence of regular commercial facilities. This includes official messages of welfare agencies during emergencies requesting supplies, materials, or instructions vital to relief to stricken populace in emergency areas. During normal times, a message with Emergency precedence will be very rare. On CW, RTTY, AMTOR, and packet, this designation will always be spelled out. When in doubt, do not use this designation.

Priority

To indicate a Priority precedence message on CW, RTTY, AMTOR, and packet, use the abbreviation "P." This classification is for important messages having a specific time limit, official messages not covered in the emergency category, press dispatches, and emergency-related traffic not of the utmost urgency.

Welfare

This classification, abbreviated as "W" on CW, RTTY, AMTOR, and packet, refers to either an inquiry as to the health and welfare of an individual in the disaster area or an advisory from the disaster area that indicates all is well. Welfare traffic is handled only after all Emergency and Priority traffic is cleared. The Red Cross equivalent to an incoming Welfare message is DWI (Disaster Welfare Inquiry).

Routine

Most traffic in normal times will bear this designation. In disaster situations, traffic labeled Routine ("R" on CW, RTTY, AMTOR, and packet) should be handled last, or not at all when circuits are busy with higher-precedence traffic.

The precedence will follow, but is not a part of, the message number. For example, a message may begin with NR 207 R on CW, "Number Two Zero Seven, Routine" on phone.

Handling Instructions

Handling instructions (HX) are used less often but can be quite helpful in handling messages. They serve to convey any special instructions to handling and delivering operators. This "prosign," when used, is inserted in the message preamble between the precedence and the station of origin. Its use is optional with the originating stations, but once inserted is mandatory with all relaying stations. (See the NTS Policy Manual for more details about handling instructions).

NTS OPERATIONS DURING EMERGENCIES & DISASTERS

The National Traffic System is dedicated to communications during disasters on behalf of ARES, as well as the daily handling of third-party traffic. When a disaster situation arises, NTS is capable of expanding its cyclic operation into complete or partial disaster operation, depending entirely on the extent of the disaster situation and the extent of its effect. The normal cycles may be expanded as required by the situation, so that more traffic can be handled, and handled more rapidly. In the extreme case, the cycles can operate continuously, with required representation present in all nets continuously, with stations designed for this function replacing each other as others are dispatched to the higher or lower nets with which they make liaison.

In a situation like this, who alerts or activates NTS nets in a disaster and who determines which net or nets should be activated? ARRL Emergency Coordinators in disaster areas determine the communications needs and make decisions regarding the disposition of local communications facilities, in accordance with the need and in coordination with agencies to be served. The Section Emergency Coordinator, after conferring with the affected DECs and ECs, makes his recommendations to the Section Traffic Manager and/or NTS managers at section and/or region levels. The decision and resulting action to alert the NTS region management may be performed by any combination of these officials, depending upon the urgency of the situation.

While the EC is, in effect, the manager of ARES nets operating at local levels, and therefore makes decisions regarding their activation, managers of NTS nets at local, section, region, and area levels are directly responsible for activation of their nets in a disaster situation, at the behest and on the recommendation of ARES or NTS officials at lower levels.

Functions of the Section Traffic Manager and Section Net Manager

During a disaster situation, the SEC may contact the Section Traffic Manager and Section Net Manager to activate their section nets — whether NTS or not — either to provide section-wide contact or, in the case of NTS nets, to provide liaison with the "outside." For that reason, if you are in either of these positions, it's important to have some means of activating your net(s) at any time. Make it understood in your net that in the event of a disaster, net stations should monitor the net frequency. Some net stations, at locations badly needed, can be activated by telephone if phone lines are available.

Make contact with your NTS region Net Managers in the event that communications connected with the disaster transcend section boundaries, recommending extraordinary activation of the region net. You should have some prearranged method of contact for this purpose.

Designate net stations to conduct liaison with the NTS region net, either through another section net or direct. This is your responsibility, not that of the region Net Manager.

Functions of the Region Net Manager

If you are the Region Net Manager, one of the section officials in your region or another NTS region may contact you should a disaster situation develop. Try to predict such contact on the basis of circumstances and be available to receive their recommendation.

Make contact with your NTS Area Net Manager in the event that communications connected with the disaster transcend region boundaries, recommending extraordinary activation of the area NTS net. Have some prearranged method of contact for this purpose.

It is your responsibility to see that the region is represented in any extraordinary session of the area net, in addition, of course, to all regular sessions.

Functions of the Area Net Manager

There are only two Area Net Manager appointees for each area, but their function during and after disasters is of paramount importance.

If you are an Area Net Manager, it's important for you to maintain a high sensitivity to disasters in your area that extend or may extend beyond region boundaries. When one does, take the initiative to alert the region Net Manager involved to determine if extraordinary NTS operation is indicated.

In the event high precedence inter-area traffic is involved, contact the two TCC directors in your area to assist in making arrangements to clear the traffic to other areas.

Contact other NTS area Net Managers to confer on possibilities of their having extra net sessions if deemed required to handle the traffic reaching them through NTS inter-area handling. Under some circumstances, direct representation or "hotlines" may be indicated.

Maintain close contact with all region Net Managers in your area and make decisions regarding overall NTS operation in consultation with them.

Functions of the Transcontinental Corps Director

These NTS officials will be involved only where traffic of a precedence higher than routine is to be handled between NTS areas, or when extreme overloads are anticipated.

Be ready to alert your TCC crew and set up special out-of-net schedules as required.

You may be called upon by the area Net Manager to set up "hotline" circuits between key cities involved in heavy traffic flow. Bear in mind which of your TCC stations is located in or near enough to large cities to man such circuits.

Functions of the Area Staff Chair

The three Area Staff Chairs administratively oversee the NTS Officials and their operations, and will advise their TCC Directors, Area and Region Net Managers when appropriate. Their advice may be based on information forwarded by ARRL Headquarters.

It is wise for the Area Staff Chair to maintain a high sensitivity to disasters and other emergencies that may develop. The Area Staff Chair should contact the other Area Staff Chairs via the International Assistance and Traffic Net and other prearranged schedules.

GENERAL POLICY

NTS operators should be self-alerting to disaster conditions that might require their services, and should report into an appropriately assigned net or other function without being specifically called upon. That is, the assignment should have been worked out with your Net Manager in advance. Each NTS operator should ask himself or herself: "What is my disaster assignment? If I hear of a disaster condition, what should I do?" If he or she cannot answer the question, he or she should seek an answer through the appropriate Net Manager. The answer may be as simple as reporting into a certain net on a certain frequency. If the operator concerned is highly specialized, the answer might be to report to the TCC director in a certain net on a certain frequency for a special assignment. Such an assignment might be an extra TCC function, or it might be as a functionary in a "hotline" point-to-point circuit needing special abilities or equipment. Flexibility is needed, but a definite assignment pertaining to disaster operation is something that all NTS operators should have. If you don't have a specific assignment, push the matter with your Net Manager. NTS should be the front line of available Amateur Radio disaster communication.

HEALTH AND WELFARE TRAFFIC

One of the biggest problems in a disaster is the handling of so-called "health and welfare" traffic, or "disaster welfare inquiries." The ARRL-recommended precedence for this type of traffic is "W," or "Welfare," which refers to an inquiry as to the health and welfare of an individual in the disaster area or an advisory from the disaster area that indicates all is well. The influx of "W" traffic into the disaster area may be large, and NTS may be called upon to assist with this overload.

The NTS policy with respect to the handling of "W" traffic is to handle as much of it as possible, but to adhere to its precedence. Higher-precedence traffic must be handled first, and "W" traffic only when the circuit is free. Routine ("R") traffic is not normally handled by an NTS net operating under disaster conditions, because they usually more than have their hands full with traffic of higher precedence, but should a disaster circuit be temporarily available, there is no reason why "R" traffic cannot be handled until the circuit again becomes occupied with higher-precedence traffic.

In a widespread disaster situation, it is seldom possible to handle all the Welfare traffic with efficiency and dispatch. Sometimes, in fact, such traffic piles up alarmingly, to the extent that much of it is never delivered. There are a number of ways in which this can be controlled, but few of them are consistent with public relations objectives.

The best way to handle such situations is to maintain close contact with the Red Cross or the Salvation Army as appropriate, since most inquiries are handled through these organizations. Civil preparedness organizations also can often set up procedures for handling such traffic. In the past, special digital circuits have been established with great success. Unless means for handling such traffic are established, it is usually wisest not to accept it from the general public, or to do so only with an explicit understanding that chances of delivery are not guaranteed.

OPERATION OF THE DIGITAL SYSTEM: HF DIGITAL NTS OPERATIONS

Radiogram-formatted NTS traffic on HF is being handled by digital means on the so-called "National Traffic System Digital" system. This system is a group of mailbox, store-and-forward (MBO) station operators spanning the country. Many of these stations have the capability of receiving and sending traffic via several digital modes including HF PACTOR I, II, III, by HF and VHF packet, all interchangeably. For example, a message received on HF Clover or PACTOR can be forwarded via VHF packet without modification. This flexibility is one of the system's assets, as it allows for forwarding along the best path at the time, resulting in the highest efficiency and reliability.

The chief concerns of the digital system, of course, are responsibility and accountability. Most MBO system operators are not concerned about NTS traffic that passes through their systems. Radiograms arriving at their stations may be in some cases passed out the VHF port into the packet forwarding system or forwarded onto the *Winlink 2000* system or forwarded onto a telnet server that will get to their destination. It is possible in some rare cases that these messages end up in the infamous "black hole" or "bit bucket." Naturally, this violates the most basic principle of NTS: getting the message through all the way, from originator to addressee.

What's the answer? The solution is to introduce responsibility and accountability into the mix, just as is done with traditional NTS nets and operators. Interested MBO system operators are now certified as "NTS Digital Relay Stations" or Target Stations as such, they accept responsibility for relaying traffic only to other NTS Digital Relay Stations or NTS-approved nets or nodes or the WL2K system. They are appointed by, and are accountable to, their Area Digital Coordinator, who is elected by the Area Staffs and who serves as an NTS Official and official member of the Area Staff. The ADCs maintain and publish a roster of these stations and report their activity to HQ. This way, the traffic stays "in the family," giving it the best chance for proper delivery.

The network of NTS-approved HF Digital Relay Stations is a mighty tool that, when used properly, can provide support to the traditional NTS nets, at any level. The digital network can pick up and move traffic when normal NTS nets cannot, due, for example, to an overloading situation, or lack of normal liaison operators.

VHF PACKET RADIO BULLETIN BOARDS

NTS can take advantage of local packet radio bulletin boards for their ability in buttressing Local and Section NTS Nets and getting channels closer to traffic origination and destination points. Again, however, the concern is that traffic might end up in a "black hole." Fortunately, many Section Traffic Managers have recognized this potential problem, and are working to ensure that packet BBSs (PBBS) are being cleared of radiogram NTS traffic each day.

Some STMs have found it extremely effective to appoint "Net" Managers to manage the NTS element of these PBBSs. "Net" members, including Official Relay Stations specializing in packet traffic handling, ensure that traffic is forwarded properly, or remove traffic from the boards and either deliver it or bring it to Section- and Local-level NTS nets for handling.

Some STMs have gone so far as to affiliate a major PBBS as an NTS Local or Section "Net." If every Section were as vigilant, packet would be a more reliable resource for moving traffic expeditiously, and with accountability.

ON GETTING MORE TRAFFIC

A concern has been expressed that the addition of digital stations will "rob" traditional NTS nets of their lifeblood, traffic. This is not true, for the most part. Yes, digital stations handle lots of NTS traffic, but the majority of the traffic handled by digital stations is filtered down to the Section nets and below. A look at the current net statistics will show that some, but not all region nets are handling only a few messages per session. In some cases it's not the fact that the digital stations are handling the traffic, it's the fact that some sections do not want to handle certain types of general bulk messages or that they limit these message types. There is no question that without traffic, NTS nets will starve and die. Each and every one of us, therefore, must do our part to support the origination of more traffic. Here are some ways we can ensure that this happens:

- Each Section or Local NTS net in the Section should sponsor a message fair at a public place or event once or more each year. Nets should coordinate their scheduling of events throughout the year. This may also be a good project for a Region Net, which could sponsor a message booth at a large, regional exposition, for example. Not only would this program of public contact generate more traffic, but it would also provide a great social event for Net members themselves, resulting in closer bonding and higher morale.
- Each Official Relay Station appointee should be encouraged to originate a minimum of 10 messages per month.
- Each traffic handler in the Section should be encouraged to bring at least one message per Net check-in. Checking in QRU is to be frowned upon, akin to going to a pot-luck supper without your contribution of a dish! "Each one, bring one," should be every net's motto.
- Section Traffic Managers should install Local NTS Nets on repeaters to gain access to new Technicians. New traffic handlers mean more traffic. A side benefit is that there will be new candidates for

upper-level liaison and NCS functions following service of their apprenticeships at the Section level and below. Far too many Region Nets, for example, rely on one or two Net members to carry the workload over the majority of sessions per week.
- PBBS software should be incorporated on NTS-cooperative boards to prompt/teach users how to originate/send a message in radiogram format. A separate, stand-alone program should also be developed for use "off-line" to accomplish the same function.
- Basic educational/motivational articles should appear more regularly in *QST* and other ARRL publications. Send your contributions to ARRL HQ for use in *QST*.
- The traffic handling community is encouraged to take advantage of the NTS awards program (PSHR, BPL, etc.) sponsored by ARRL.
- NTS Officials are to be encouraged to put on traffic handling seminars for new hams at conventions, hamfests, and club meetings to generate interest in the activity.

Chapter 14
Nets and Message Handling

During an emergency or public service event, most of our time is spent communicating on behalf of someone or some agency — in other words, passing messages. Even in our day-to-day Amateur Radio activities, we are passing messages and traffic; contests have exchanges, DXing has quick signal reports, VHF/UHF and satellite operating have grid squares, even a routine QSO generally starts with RST, name, and QTH.

Amateur Radio has a long history of making sure critical messages get from point A to point B as quickly as possible. This is such a large part of who we are that it is even represented in the name of our national association, the American Radio *Relay* League. Nets and message or traffic handling do not exist independent of each other. Relaying messages and passing traffic requires a well-organized network. Individual nets make passing messages and traffic possible.

In this chapter we will explore the world of Amateur Radio message handling with a look at nets, formatting messages, and more.

NET OPERATING GUIDELINES

Every organization needs an executive-level manager to oversee the entire operation and ensure that everything runs smoothly. In the case of nets, that executive-level person is the Net Manager. Depending on the type of net, the Net Manager is responsible for recruiting and training Net Control Station (NCS) operators, liaison stations, and other net members (see **Figure 14.1**). The Net Manager sets up the net's schedule and makes sure that one or more qualified NCS operators will be available for each session of the net. Even the most loyal of net members will find it difficult to be present on the air on a consistent and regular schedule if the net does not happen as scheduled. In a long-term emergency net, the Net Manager may also arrange for relief operators and support services. Some Net Managers may be responsible for more than one net.

Any net that is established with the primary objective of serving an emergency need is generally going to have routine on-air events and exercises that don't deal directly with emergency situations at all. "Emergency Nets" require regular practice, perhaps weekly, during which the net is conducted by a NCS under the direction of the Net Manager. A true Emergency Net may only occur every few years in many areas, the nature of the weekly net is to practice all the steps involved in a net, so that when an actual emergency does occur, everyone involved will know exactly what to do.

Some people have asked, "So what happens during the net besides giving your call sign and being acknowledged by someone?" Answer: "Nothing." The purpose of the net is fulfilled when each station is heard and knows that their signal is audible at a desired location.

HF nets most often require operators to be licensed above that of Technician and require the correct radio, antenna system, possibly a tuner, and sometimes an amplifier. However, one must use the best equipment that they have available at the time. Most emergency situations in local areas start with 2 meter nets and may involve 70 cm. In practice it is found that almost all emergency communications on ham radio are on 2 meters, other than those requiring HF ranges. When repeaters are operational, they are extremely useful. However, it is required that NCS operators be familiar enough with their radios

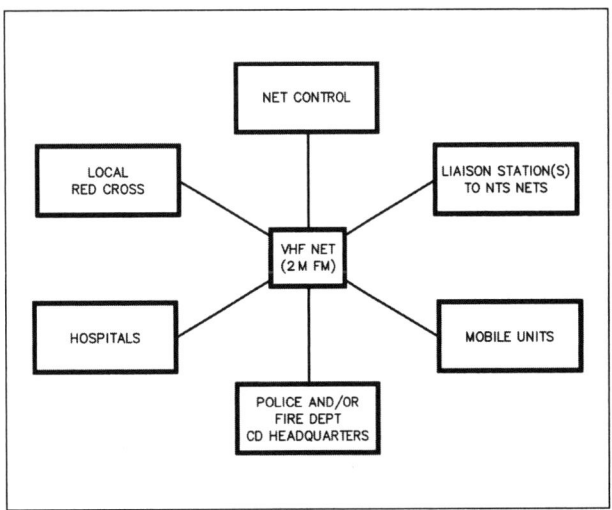

Figure 14.1 — Typical station deployment for local ARES net coverage in an emergency.

and equipment to be able to utilize simplex procedures to maximally communicate when situations are at their worst. NCSs may need to use 2 meter simplex, local repeaters, as well as HF communications. Commercial power must be assumed to be unavailable.

Rosters are commonly used to keep track of net members. Generally, the Net Manager will prepare the roster and distribute it to the NCSs as well as to others who have an interest in the roster. The order of roster listings is determined by the Net Manager according to the purposes of the net. Possible methods include: (a) alphabetically by last name or by first name; (b) alphabetically by the suffix of the call sign (the first letter after the number); (c) by geographic area or region of the country; or (d) by some other method determined by the Net Manager. Net rosters can be very simple, including only call sign, first name, and dates involved with the net, or the roster can include other information such as address, phone number, e-mail address, etc.

When nets are called by region, there are likely to be small pileups with more than one station responding at a time. On HF nets, the NCS will often hear parts of the call signs of two or more stations at nearly the same time, yet be able to sort out part of the call sign to use to reply to the calling stations.

When there are several stations responding at once, it can be helpful for the NCS to ask for stations to respond with the first letter of their suffix. This gives a bit of separation in the calling stations for the NCS.

In any net, when there are two stations that need to communicate with one another during the net, it is reasonable to ask them to tune to an available non-net frequency to communicate, often to a pre-identified frequency that can be confirmed as available before they leave the net frequency. Upon returning to the net frequency, they would be expected to check in with the NCS.

THE NCS

Think of the NCS operator as a "ringmaster" or "traffic cop" who decides what happens in the net, and when. If the EOC has a Priority message for Red Cross Shelter 1, and Medical Station 4 has an Emergency message for Mercy Hospital, it is the NCS's job to make sure that the Emergency message is sent first. He or she decides when stations will check in, with or without traffic, and whether messages will be passed on the net's frequency or a different one. The NCS needs to be aware of everything going on and handle the needs of the net, its members, and served agency as quickly and efficiently as possible. In a busy net, this can be a daunting task.

The NCS can be located anywhere, but should be in a position to hear most, if not all, stations in the net. This helps avoid time-consuming "relays." Some groups place their NCS at the EOC or command post, while others like to keep them away from the noise and confusion.

The NCS is in charge of one specific net, but should not be responsible for the entire EmComm operation. That is the job of the EC or similar EmComm manager. It is not possible to be in command of all aspects of an emergency response and still run a net effectively, since both jobs require 100% of a person's attention.

The Backup NCS

A backup NCS needs to be readily available should there be an equipment failure at the primary NCS location, or if the primary NCS operator needs to take a break. All members of the net should be made aware of the backup NCS assignment early in the net's operation. There are two types of backup NCS. Either the Net Manager or the primary NCS operator, depending on the situation, appoints both.

The first type of backup NCS is at the same location as the primary NCS operator. The second type of NCS is a station at a different location that maintains a duplicate log of everything happening during the net. Whenever possible, an offsite backup NCS should be maintained, even if an onsite backup is present. This is especially important during an emergency in which antennas can be damaged or power lost. Equipment can fail even during less demanding operations.

Acting as a "Fill-In" NCS

Even before you have had a chance to be trained to act as a NCS operator, an opportunity might arise in which you must handle the job temporarily. During an emergency, anyone and everyone can be asked to take on new and unfamiliar tasks in order to deal with a rapidly changing situation. Fortunately, basic NCS skills are not difficult to teach or learn. Here are some basic dos and don'ts:

- Remember that although you are in control of the net, you are not "God." Treat members with respect and accept suggestions from other experienced members.
- If you are taking over an existing net, try to run it in the same manner as the previous NCS.
- Always follow a script, if one is provided. Write your own if necessary, but keep it short and to the point.
- Handle messages in order of precedence: Emergency — Priority — Welfare.
- Speak clearly and in a normal tone of voice. Use good mic technique.
- Make all instructions clear and concise, using as few words as possible.
- Keep notes as you go along. Do not let your log fall behind.
- Write down which operators are at which locations. When one leaves or is replaced, update your notes.
- Ask stations to pass messages off the main net frequency whenever possible.
- All the reading and study in the world will not replace actual experience. You should look for opportunities to practice being the NCS operator well before an emergency occurs.

NET SCRIPTS

Many groups open and close their nets with a standard script. The text of the script lets listeners know the purpose and format of the net. Using a standard script also ensures that the net is run in a similar format each time it operates, regardless of who is acting as the NCS. The usual parts of a script often include:

- The name of the net
- A call to member stations, those who want to join and to those listening
- The usual, or "nominal," frequency the net will use
- The purpose of the net
- Usual date and time of the net (consider giving times in UTC/Zulu to avoid confusion with a variety of time zones and DST changes)
- Identification of the name, call sign, and location (city/state) of the NCS for that date and time
- Directed or non-directed net
- Indicate when visitors will be recognized
- Describe the steps for joining the net
- Include a call for emergency traffic and early check-ins before the roster is called
- Include suggested closing comments for the NCS to use after the roster is completed and the net is being closed

A typical net script might look like this:

Opening: This is [call sign], net control station for the New Hampshire ARES/RACES Emergency Net. This is a directed emergency net for liaison stations from all New Hampshire ARES/RACES regions. Please transmit only when requested to, unless you have emergency traffic.

Any station with emergency traffic, please call now. *(Stations call in and emergency traffic is passed.)*

Any station with priority traffic, please call now. *(Stations call in and priority traffic is passed.)*

All other stations with or without traffic, please call now. *(Stations call in and any traffic is passed.)*

Closing: I would like to thank all stations that checked in. This is [call sign] securing the New Hampshire ARES/RACES Emergency Net at [date and time] returning the [repeater or frequency] to regular use.

There are more examples of net scripts in the appendix of this book.

TACTICAL TRAFFIC

Tactical traffic is first-response communications in an emergency situation, involving a few operators in a small area. It may be urgent instructions or inquiries, such as "Send an ambulance," or "Who has the medical supplies?" Tactical traffic is generally unformatted and seldom written down, but it is particularly important in localized communications when working with government and law enforcement agencies. Logs should be kept by any hams passing tactical traffic, as logs may be relevant later for law enforcement or other legal actions, and can even serve to protect the Amateur Radio operator in some situations.

The 146.52 MHz FM calling frequency — or VHF and UHF repeaters and net frequencies — are typically used for tactical communications. This is a natural choice because FM mobile, portable, and fixed-station equipment is so plentiful and popular. However, the 222 and 440 MHz UHF bands provide the best communications from within steel or concrete structures, have less interference, and are more secure for sensitive transmissions.

One way to make tactical net operations clear is to use tactical call signs — words that describe a function, location, or agency. Their use prevents confusing listeners or agencies that are monitoring. When operators change shifts or locations, the tactical calls remain the same; that is, the tactical call remains with the position even if the operators change position. Amateurs may use tactical call signs like, "Parade Headquarters," "Finish Line," "Red Cross," "Net Control," or "Weather Center" to promote efficiency and coordination in public service communications activities. However, amateurs must identify with their FCC-assigned call sign at the end of a transmission or series of transmissions and at intervals not to exceed 10 minutes.

It's also a good idea to use the 12-hour local-time system for time and date when working with relief agencies, unless they understand the 24-hour or UTC systems.

Taking part in a tactical net as an ARES team member requires some discipline and following a few rules:

- Report to the Net Control Station (NCS) as soon as you arrive at your assigned position.
- Ask the NCS for permission before you use the frequency.
- Use the frequency for traffic, not chit-chat.
- Answer promptly when called by the NCS.
- Use tactical call signs.
- Follow the net protocol established by the NCS.
- Always inform the NCS when you leave service, even for a short time.

In some relief activities, tactical nets become resource or command nets. A resource net is used for an event that goes beyond the boundaries of a single jurisdiction and when mutual aid is needed. A command net is used for communications between EOCs and ARES leaders. Yet with all the variety of nets, sometimes the act of simply putting the parties on the radio (instead of trying to interpret their words) is the best approach.

NET MEMBERS

Operators at various sites are responsible for messages going to and from their location. They must listen to everything that happens on the net, and maintain contact with the served agency's people at the site. They assist the served agency with the creation of messages, put them into the appropriate format, and contact the NCS when they are ready to be sent.

Whenever possible, two operators should be at each site. When the station is busy, one can handle logging and message origination, and work with the served agency's staff while the other monitors the net, sends messages, and copies incoming traffic. During slower periods, one member can be "off duty" for rest, meals, or personal needs.

BULLETIN STATIONS

In some nets, the NCS does not send out bulletins and other incident-related information. That is the role of the "bulletin station," which relays ARRL bulletins or those authorized by the served agency to all stations in the net. They may also be transmitted on a preset schedule, such as at the top and bottom of each hour. The bulletin station must be located at the served agency or have a reliable communication link to them.

LIAISON STATIONS

Liaison stations pass messages between two different nets. The NCS or Net Manager, depending on the type of organization, usually assigns these stations. Messages may be passed as needed, or on a pre-set schedule. In some cases, a liaison station will monitor one net full time. When a message must be passed to another net, they leave the net temporarily to pass it, and then return. The other net has a liaison station that does exactly the same thing, but in reverse. In other situations, a single liaison station may need to handle messages going both ways between two nets. There are two ways to do this. You can use two radios to monitor both nets at the same time — a difficult task if either or both nets are busy. The radios antennas must be separated sufficiently to prevent interference between radios when one is used to transmit. In the second method, one radio is used, and the liaison station switches between the two nets on a regular schedule.

RELAY STATIONS

While not a regular net position, a relay station is one that passes messages between two stations in the net that cannot hear each other. Relay stations are generally designated by the NCS on an as-needed basis. If you can hear a station or stations that the NCS cannot, it is okay to volunteer to act as a relay station.

WORKLOAD AND SHIFT CHANGES

Although it happens frequently, no operator should try to work excessively long hours. When you become tired, your efficiency and effectiveness decline, and your served agency is not getting the best possible service. Net managers and NCS operators should work with the EC or other EmComm manager to ensure that all net members get some rest on a regular basis.

It is a good practice for any replacement NCS, liaison, or net member to monitor the net for at least 15 minutes and review the logs with the present operator before taking over. This assures continuity in the net's operation.

INTERFERENCE PROBLEMS

No one has exclusive use of a frequency unless such has been stated by an official of the FCC by emergency declaration. The principle of "first come, first served" does apply. Many nets have experienced the kindness shown by some operators when a net is scheduled to start on a frequency where people are talking informally, and the ragchewers agree to move so the net can start on its usual, or "nominal," frequency. However, if the station(s) on the frequency first choose(s) not to move, the net should move up or down 2 or 3 kHz to a frequency that will not interfere with the "first" station. If the net moves too far, it may be difficult for member stations to find the net. The first station will sometimes agree to re-direct net members to the net's "new" frequency when the net members are heard. Some nets arrange several alternate frequencies in advance, which they may use if their usual frequency is unavailable. In this way, net members have an idea where to look for a relocated net.

If your net experiences interference, the NCS has several options. If the interference is coming from adjacent or co-channel stations that may be unaware of the emergency net, the NCS should politely inform them of the net and ask for their cooperation. Alternatively, the NCS might ask an HF net to move over a few kHz. If the problem cannot be resolved in this manner, each net should have one or more alternative frequencies to which it can move, as required. If possible, the frequencies themselves should not be published or mentioned on the air.

Never discuss, acknowledge, or try to speak with an intentionally interfering station. Many years of experience has proven that this only encourages the offender. If the interference is making communication difficult, simply announce to the net that everyone should move to the alternate frequency and sign off. Better yet, put a plan in place ahead of time so that when interference occurs, all net members know to move to the alternate frequency without being told to do so on the air.

If intentional interference persists, the Net Manager or NCS can contact an elected League official or an Official Observer Station, and ask that the FCC be notified of the interference. In some cases, they may be able to track down and contact the station responsible for the interference.

NON-VOICE MODES

Packet modes include FM packet, HF packet, and PACTOR. Because packet modes can provide an automatic connection between two stations, it is not really proper to speak of a "packet net." Although messages can be transmitted between two stations "keyboard to keyboard" as with RTTY or PSK31, it is usually better to transmit them as "traffic," using the bulletin board or mailbox facility of the terminal node controller (TNC). Packet messages are automatically routed and stored without any action by the receiving station's operator or an NCS. This makes packet a powerful tool for traffic handling, especially for messages with detailed or lengthy text, or messages that need to be more securely transmitted than voice modes allow. An operator can prepare and edit messages offline as text files and send them error free via packet in just seconds.

Non-packet digital modes are not automatic, and may require a NCS operator to manage the net in much the same way as a phone or CW net. These include RTTY, PSK31, AMTOR, and GTOR.

CW Procedures

Clean and accurate code sent at 10 words per minute is better than sloppy code sent at 30 words per minute. Sending speed is not a true measure of effectiveness, but accuracy is. When propagation or interference make communication difficult, or when the receiving operator cannot keep up, it is time to reduce the sending speed. Always send at a speed that the receiving station can copy comfortably.

There are variations used when passing traffic via CW, especially when both stations are operating in "full break-in" mode (both stations are capable of receiving signals between each Morse character sent). The receiving station can "break" (stop) the sending station at any point for needed fills, instead of waiting for the entire message to be sent. There are additional special prosigns used, and interested amateurs should be familiar with ARRL Publication FSD-218, available at **www.arrl.org/fsd-218-amateur-message-form**. This publication, sometimes referred to as the "pink card," contains CW net procedures as well as a description of the Amateur Message Form, message precedences, and handling instruction abbreviations.

When formatting an ARRL Radiogram message, use abbreviations and prosigns consistently and appropriately. For instance, do not send "R," meaning you have received everything correctly, and then ask for repeats like "AA" (all after) or "AB" (all before).

FORMAL vs INFORMAL MESSAGES

Both formal (written in a specific format, i.e. ARRL) and informal (verbal or written but not in a specific format) messages have their place in emergency communication. In general, informal messages are best used for non-critical traffic that is delivered directly from the author to the recipient. Formal messages are more appropriate when two or more people will handle them before reaching the recipient, or where the contents are critical or contain important details. The most common formal message format is that used by ARRL's NTS, discussed below.

Informal Verbal Messages

Some emergency messages are best sent informally in the interest of saving precious seconds. If you need an ambulance for a severely bleeding victim, you do not have time to compose and send a formal message. The resulting delay could cause the patient's death.

Other messages do not require a formal written message because they have little value beyond the moment. Letting the net control station know where you are or when you will arrive need not be formal. The message is going directly to its recipient, is simple and clear, and has little detail. Many of the messages handled on a tactical net fit this description.

However, this does not mean that accuracy is any less important. If someone gives you a short message to relay to

---

PROCEDURE SIGNALS (PROSIGNS) FOR MORSE CODE

BK	Invite receiving station to transmit (break)
CL	Going off the air (clear)
CQ	Calling any station (literally, Come Quick)
K	Go, invite any station to transmit
R	All received OK
AA	Separation between parts of address or signature
AR	Over, end of message
AS	Please stand by
BT	Separation (break) between address and text; between text and signature
KN "X"	Go only, invite a specific station "X" to transmit
SK	End of contact (send before sending your call)

Abbreviations

Fill	Term used to describe missing items (words, characters, numbers etc.) when handling messages in the National Traffic System
AA	All after (use to get fills)
AB	All before (use to get fills)
ADEE	Addressee (name of the person to whom the message is addressed)
ARL	(Used with "check" — indicates use of ARL numbered message in text)
BN	Between
SIG	Signed; signature (last part of message)
WA	Word after
WB	Word before

someone else, you should repeat it as closely to the original as possible. Messages that will be relayed more than once should always be sent in ARRL format to prevent multiple modifications. If you are not careful to maintain the precise meaning of the original message, a message such as, "The shelter manager says she needs fifty cots and blankets at Hartley Hill School by tonight," could, after being passed through several people, easily turn into, "He says they need a bunch more cots and blankets at that school on the hill."

Formal Written Message Formats

A standard written message format is used so everyone knows what to expect. This increases the speed and accuracy with which you can handle messages.

The ARRL message form, or "Radiogram," is a standard format used for passing messages on various nets, and is required for all messages sent through the National Traffic System. While this format may not be perfect for all applications, it serves as a baseline that can be readily adapted for use within a specific served agency.

There can be requirements for formats that are unique to an individual agency or type of emergency. A good example is the popular Incident Command System (ICS) form ICS-213 used by most government agencies. This message form is intended to be used for emergency messages between two parties not in the same location. The content of the form may be sent electronically, including via Amateur Radio.

Though not a separate format in and of itself, there is a specific type of message of which you should be aware: A "service message" is one that lets the originating station know the status of a message they have sent. A service message may be requested by a handling instruction (HX), or may be sent by any operator who has a problem delivering an important message. During emergencies, service messages should only be sent for Priority and Emergency messages.

Regular practice with creating and sending messages in any standard format is recommended.

COMPONENTS OF A STANDARD ARRL RADIOGRAM

The standard Radiogram format is familiar to most hams from the pads of yellow-green forms available from ARRL Headquarters. The form has places for the following information:

The *Preamble*, sometimes referred to as the "header," consists of administrative data such as the message number, originating station, message precedence (importance), and date and time of origination. The combination of the message number and the originating station serves as a unique message identifier, which can be traced if necessary.

The *Address* includes the name, street address or post office box, city, state, and zip code of the recipient. The address should also include the telephone number with area code, since many long-distance Radiograms are ultimately delivered with a local phone call.

The *Text* of the message should be brief and to the point, limited to 25 words or less when possible. The text should be written in lines of five words (10 if using a keyboard) to make it easier and faster to count them for the "check." Once complete, the receiving operator compares the word count with the check. If okay, the message is "rogered" — if not, the message is repeated at a faster reading speed to locate the missing or extra words.

Care should be taken to avoid word contractions, as the apostrophe is not used in CW. If a word is sent without an apostrophe, its meaning could be lost or changed (consider "I'll" vs "Ill!") Contractions are also more difficult to understand when sent by phone, especially in poor conditions. Commas and other punctuation are also not used in formal messages. Where needed, a "period" can be sent as an "X" in CW and digital modes, and spoken as "X-RAY." The "X" may be used to separate phrases or sentences, but never at the end of the text. Question marks can be used as needed, and are usually spoken as "question mark," and sometimes as "query." Both the X and question mark should be used only when the meaning of the message would not be clear without them.

The *Signature* can be a single name, a name and call sign, a full name and a title, "Mom and Dad," and occasionally a return address and phone number — whatever is needed to ensure that the recipient can identify the sender and that a reply message can be sent if necessary.

Details of the Preamble

The preamble or "header" is the section of the ARRL message form where all the administrative details of the message are recorded. There are eight sections, or "blocks," in the preamble. Two of them, "time filed" and "handling instructions," are optional for most messages.

Block #1 — Message Number

This is any number assigned by the station that first puts the message into ARRL format. While any alphanumeric combination is acceptable, a common practice is to use a numeric sequence starting with the number "1" at the beginning of the emergency operation. Stations who are involved in day-to-day message handling may start numbering at the beginning of each year or each month.

Block #2 — Precedence

The precedence tells everyone the relative urgency of a message. In all but one case, a single letter abbreviation is sent with CW or digital modes. On phone, the entire word is always spoken. Within the ARRL format, there are four levels of precedence:

Routine — Abbreviated with the letter "R." Most day-to-day amateur traffic is handled using this precedence. It is for all traffic that does not meet the requirements for a higher precedence. In a disaster situation, routine messages are seldom sent.

Welfare — Abbreviated as "W." Used for an inquiry as

to the health and welfare of an individual in a disaster area, or a message from a disaster victim to friends or family.

Priority — Abbreviated as "P." For important messages with a time limit; any official or emergency-related messages not covered by the EMERGENCY precedence. This precedence is usually only associated with official traffic to, from, or related to a disaster area.

EMERGENCY — There is no abbreviation. The word EMERGENCY is always spelled out. Use this precedence for any message having life-or-death urgency. This includes official messages from agencies requesting critical supplies or assistance during emergencies, or other official instructions to provide aid or relief in a disaster area. The use of this precedence should generally be limited to traffic originated and signed by authorized agency officials. Due to the lack of privacy on radio, EMERGENCY messages should only be sent via Amateur Radio when regular communication facilities are unavailable.

Block #3 — Handling Instructions

This is an optional field used at the discretion of the originating station. The seven standard handling instructions (HX) prosigns are:

HXA – (Followed by number.) "Collect" telephone delivery authorized by addressee within (X) miles. If no number is sent, authorization is unlimited.

HXB – (Followed by number.) Cancel message if not delivered within (X) hours of filing time; service (notify) originating station.

HXC – Report date and "time of delivery" (TOD) to originating station.

HXD – Report to originating station the identity of the station who delivered the message, plus date, time and method of delivery. Also, each station to report identity of station to which relayed, plus date and time.

HXE – Delivering station to get and send reply from addressee.

HXF – (Followed by date in numbers.) Hold delivery until (specify date).

HXG – Delivery by mail or telephone — toll call not required. If toll or other expense involved, cancel message, and send service message to originating station.

If more than one HX prosign is used, they can be combined, as in HXAC. However, if numbers are used, such as with HXF, the HX must be repeated each time. On voice, use phonetics for the letter or letters following the HX to ensure accuracy, as in "HX Alpha."

Block #4 — Station of Origin

This is the FCC call sign of the first station that put the message into NTS format. It is not the original author of the message. For instance, you are the radio operator for a Red Cross shelter. The fire station down the street sends a runner with a message to be passed, and you format and send the message. You are the "Station of Origin," and the fire station is the "Place of Origin," which will be listed in Block 6.

Block #5 — The Check

The check is the number of words in the text section only. Include any "periods" (written as "X," spoken as "X-Ray"). The preamble, address, and signature are not included. After receiving a message, traffic handlers count the words in the message and compare the word count to the check number in the preamble. If the two numbers do not agree, the message should be re-read by the sending station to verify that all words were copied correctly. If the message was copied correctly and an error in the check number exists, do not replace the old count with the new count. Instead, update the count by adding a "slash" followed by the new count. For example, if the old count was five, and the correct count was six, change the check to "5/6."

Block #6 — Place of Origin

This is the name of the community, building, or agency where the originator of the message is located, regardless of whether the originator is a ham. This is not the location of the station that first handled the message, which is listed in Block 4, "Station of Origin."

Block #7 — Time Filed

This is an optional field, unless handling instruction "Bravo" (HXB) is used. HXB means "cancel if not delivered within X hours of filing time." Unless the message is time sensitive, this field may be left blank for routine messages, but completing the time field is generally recommended for Welfare, Priority, and Emergency messages. Many hams use Universal Coordinated Time (UTC) for messages and logging. During emergencies, it is better to use local time and indicators such as PST or EDT to eliminate confusion by served agency personnel.

Block #8 — Date

This is the date the message was first placed into the traffic system. Be sure to use the same date as the time zone indicated in Block 7.

Header Example

This is how a complete header might look for a CW or digital message:

NR207 P HXE W1FN 10 LEBANON NH 1200 EST JAN 4

This is how the same header would be spoken:

"Number two zero seven Priority HX Echo Whiskey One Foxtrot November One Zero Lebanon NH One Two Zero Zero EST January four."

A brief pause is made between each block to help the receiving station separate the information. Note that the title of each block is not spoken, with the exception of the word "number" at the beginning, which tells the receiving station that you are beginning the actual message.

PROWORDS AND PROSIGNS

When sending formal traffic, standard "prowords" or "prosigns" (CW) are used to begin or end parts of the mes-

MESSAGE HANDLING PROWORDS, PROSIGNS, AND ABBREVIATIONS

Proword	Prosign (CW)	Meaning or Example
BREAK	BT *	Separates address from text and text from Signature.
CORRECTION	HH *	"I am going to correct an error."
END	AR *	End of message.
MORE	B	Additional messages to follow.
NO MORE	N	No additional messages. In CW can also mean "negative" or "no."
FIGURES	Not needed	Used before a word group consisting of all numerals.
INITIAL	Not needed	Used to indicate a single letter will follow.
I SAY AGAIN	IMI *	Used to indicate a repeat of a word or phrase will follow.
I SPELL	Not needed	"I am going to spell a word phonetically."
LETTER GROUP	Not needed	Several letters together in a group will follow. Example: ARES, SCTN.
MIXED GROUP	Not needed	Letters and numbers combined in a group will follow. Example: 12BA6.
X-RAY	X	Used to indicate end of sentence, as with a "period."
BREAK	BK *	Break; break-in; interrupt current transmission on CW.
CORRECT	C	Correct; yes.
CONFIRM	CFM	Confirm ("Please check me on this.")
THIS IS	DE	Used preceding identification of your station
HX	HX	Handling instructions, single letter to follow — optional part of preamble.
GO AHEAD	K	Invitation for specific station to transmit.
ROGER	R	Message understood. In CW, may be used for decimal point in context.
WORD AFTER	WA	"Say again word after..."
WORD BEFORE	WB	"Say again word before..."
BETWEEN	—	"Say again between...and ...."
ALL AFTER	AA *	"Say again all after..."
ALL BEFORE	AB	"Say again all before..."

* The two or three letters are sent as one character.

sage, and to ask for portions of the message to be repeated. In addition to adding clarity, the use of standard prowords and prosigns saves considerable time.

Some prowords and prosigns tell the receiving station what to expect next in the address, text, and signature portions of the message. They are not used while reading the header, since the header follows a pre-determined format. Examples of commonly used prowords are "figures" (sent before a group consisting of all numerals), "initial" (indicates that a single letter will follow), and "break" (signals the transition between the address and the text, and the text and the signature).

SENDING A MESSAGE WITH VOICE

When the receiving station is ready to copy, read the message at a pace that will allow the receiving station to write it down. Once you are done, if the receiving station has missed any portion of the message, they will say, "say again all after____," "say all before," or "say again all between ____ and ____." In some nets, the practice is to say "break," and then unkey between sections of the message so a station can ask for missing words to be repeated before going on (these repeated words are also known as "fills"). In many nets, the entire message is read first before any fills are requested, to save time. All numbers in groups are spoken individually, as in "three two one five," not "thirty-two fifteen," or "three thousand two hundred and five."

Here is the entire message in the Radiogram in **Figure 14.2** as it would be spoken:

Number two zero seven Priority HX Echo Whiskey One Foxtrot November One Zero Lebanon NH one two zero zero EST January four
Mark Doe
American Red Cross Disaster Office
Figures one two three Main Street
Rutland VT figures zero five seven zero one
Figures eight zero two five five one two one two
Break
Need more cots and sanitation kits at all five shelters
Break
Joan Smith Shelter Manager
End No more

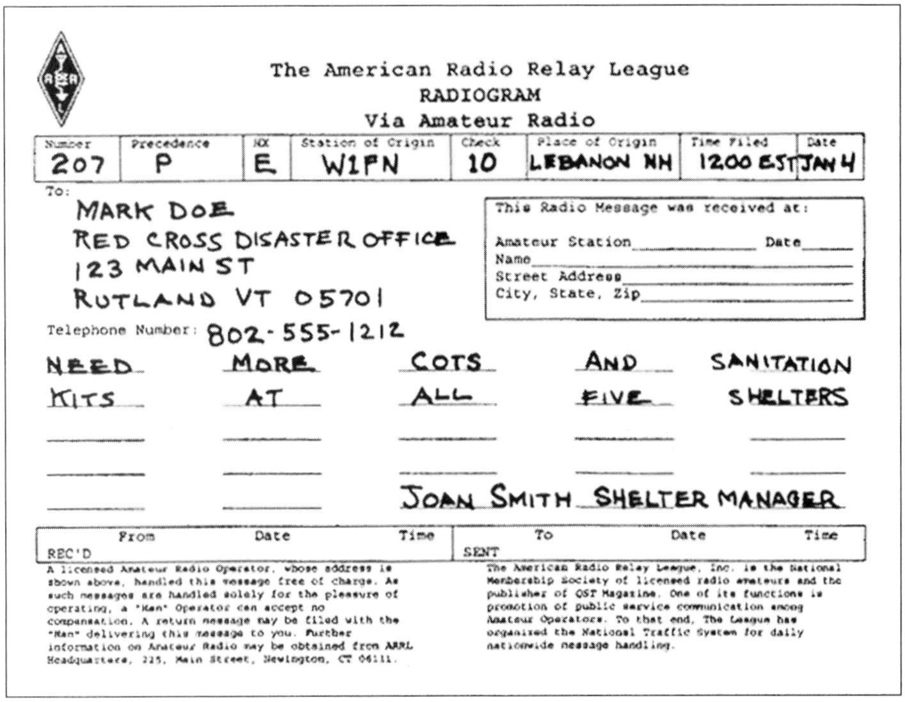
The American Radio Relay League
RADIOGRAM
Via Amateur Radio

Number	Precedence	HX	Station of Origin	Check	Place of Origin	Time Filed	Date
207	P	E	W1FN	10	LEBANON NH	1200 EST	JAN 4

To:
MARK DOE
RED CROSS DISASTER OFFICE
123 MAIN ST
RUTLAND VT 05701
Telephone Number: 802-555-1212

This Radio Message was received at:
Amateur Station ________ Date ____
Name ________
Street Address ________
City, State, Zip ________

NEED MORE COTS AND SANITATION
KITS AT ALL FIVE SHELTERS

JOAN SMITH SHELTER MANAGER

	From	Date	Time		To	Date	Time
REC'D				SENT			

A licensed Amateur Radio Operator, whose address is shown above, handled this message free of charge. As such messages are handled solely for the pleasure of operating, a "Ham" Operator can accept no compensation. A return message may be filed with the "Ham" delivering this message to you. Further information on Amateur Radio may be obtained from ARRL Headquarters, 225, Main Street, Newington, CT 06111.

The American Radio Relay League, Inc. is the National Membership Society of licensed radio amateurs and the publisher of QST Magazine. One of its functions is promotion of public service communication among Amateur Operators. To that end, The League has organized the National Traffic System for daily nationwide message handling.

Figure 14.2 — An example of a typical ARRL Radiogram.

When passing formal traffic, do not add unnecessary words. Since the parts of the header are always sent in the same order, there is no need to identify each of them. The only exception is the word "number" at the beginning of the header.

Here is an example of how *not* to read the header of a message on the air:

"Number, two zero seven precedence, Priority handling instructions, HX Echo station of origin, W1FN check, one zero place of origin, Lebanon NH time, one two zero zero EST date, January 4."

This example adds many unneeded words to the message, including "station of origin," "check," "time," and other block titles. If there is something about the message that deviates from the standard format, or if an inexperienced operator is copying the message without a pre-printed form, then some additional description may be necessary, but in most cases, it just wastes time. (Note: "Number" is always spoken before the message number.)

MESSAGE HANDLING RULES

Do not speculate on anything relating to an emergency! There may be hundreds of people listening to what you say (other amateurs, as well as the media and general public using scanners) and any incorrect information could cause serious problems for the served agency or others. You do not want to be the source of any rumor.

If your served agency requests an estimate, you can provide that information as long as you make it very clear at the time you send it, that you are only making an estimate. For example, saying, "The estimated number of homes damaged is twelve," would be acceptable.

Pass messages exactly as written or spoken: In addition to passing messages quickly, your job as a communicator is to deliver each message as accurately as possible. Therefore, you must not change any message as you handle it. If the message is longer than you would like, you must send it anyway. Misspelled words or confusing text must be sent exactly as received. Only the original author may make changes. If you note an inaccurate word count in a NTS format message, you must maintain the original count and follow it with the actual count received at your station, i.e.: "12/11."

Should you return a message to the author before first sending it, if it seems incorrect or is confusing? This is a judgment call. If the apparent error will affect the meaning of the message and the author is easily contacted, it is probably a good idea to do so. Whenever possible, it is a good practice to read each message carefully in the presence of the author before accepting it. This way, potential errors or misunderstandings can be corrected before the message is sent. Much of the tactical information being passed during a major emergency will not be in ARRL format. It may have much of the same information, but will be in a non-standard format or no format at all. These messages should also be passed exactly as received. If necessary, use the ARRL format and place the entire non-standard message in the "text" section.

THE IMPORTANCE OF THE SIGNATURE

During an emergency, the messages you handle can easily contain requests for expensive supplies that have a very limited "shelf life" (such as blood for a field hospital), or for agencies that will only respond to properly authorized requests (i.e. for medevac helicopters). For this reason, it is critical that you include the signature and title of the sender in every message.

LOGGING AND RECORD KEEPING

An accurate record of formal messages handled and various aspects of your station's operation can be very useful, and is required by law in some cases. Lost or misdirected messages can be tracked down later on, and a critique of the operation afterward can be more accurate. All logs should include enough detail to be meaningful later on, especially the date and an accurate time. With some agencies, your log becomes a legal document and may be needed at some later time should an investigation occur. In this case, logs should be completed and turned in to the appropriate person for safekeeping and review.

What to Log

Log all incoming and outgoing messages. Record the name of the sender, addressee, the station that passed the message to you, the station to whom the message was sent, the message number, and the times in and out. Keep the written copy of each message in numerical order for future reference.

Also, log which operators are on duty for any given period, and record any significant events at your station. These might include changes in conditions, power failures, meals, new arrivals and departures, equipment failures, and so on. In addition to the log, copies of all messages should be kept and catalogued for easy retrieval if needed later for clarification or message tracking. Many operators make notes about when the message was received and sent, and to and from whom, directly on the message form itself. This helps speed up tracking later on. Never rely on your memory.

Should informal messages be logged? This is usually up to the stations involved, and depends on the circumstances. Even informal messages can contain important details that may need to be recalled later. Emergency or Priority messages of any kind, even unwritten messages, should always be logged. Some net control operators like to log every message or exchange, no matter how inconsequential. Others like to log only those with potentially important details.

Log Formats

At a station with little traffic, all information can be included in one chronological log. However, if a large number of messages are being handled and you have a second person to handle logging, separate logs can make it faster and easier to locate information if it is needed later. You might keep one log for incoming messages, one for outgoing messages, and a third for station activities. The NCS will also need to keep a log of which operators are assigned to each station, and the times they go on and off duty.

Who Should Log

At the net level, logging can be handled in several ways. If activity is low, the net control operator can handle logging. In busy nets, a second person can keep the log as the net's "secretary" and act as a "second set of ears" for the NCS. The logger can be at the NCS, or they might be listening from a different location.

If an "alternate NCS" station has been appointed, they should keep a duplicate log. If they need to "take over" the net at any point, all the information will be at hand, preserving the continuity of the net.

In addition to logs kept at the net level, each individual operator should keep his or her own log. This will allow faster message tracking and provides duplicate information should one station's logs become lost or damaged.

In a fast-moving tactical net, keeping a log while on the move may be impossible for individual stations. In this case, the net control station may decide to keep one log detailing the various informal messages passed on the network.

Logging is a good position for a trainee with limited experience, or an unlicensed volunteer. Two experienced and licensed operators can also alternate between on-air and logging duties to help combat fatigue.

Writing Techniques for Message Copying and Logging

Your logs should be clear and legible to be of any use. If only you can read your handwriting, the log will be of little value to the operator who takes the next shift or to the served agency as a legal document. Print in neat block letters on lined paper or a pre-printed log form. Pre-printed forms generally allow for five words per line, which assists when counting words for the check. For this reason, it's a good idea to maintain five-word lines even when writing on regular lined paper. A firm writing surface with support for your forearm will reduce fatigue and improve legibility.

Keep both pens and pencils on hand since each works better under different conditions. Logs that will become legal documents should always be written in permanent ink. Some operators prefer special "diver's" pens that will write on wet surfaces at any angle.

Logs should be kept in notebooks to prevent pages from becoming lost. In the case of pre-printed log sheets, use a three-ring binder. If more than one log is kept, each should be in its own notebook to prevent confusion and accidental entries. Logs that will become legal documents should be kept in hardbound books with pre-numbered pages so that missing pages will be obvious.

In fast-moving situations, it can be difficult or impossible to keep a log of any kind. If a message, exchange, or event should be logged, try to do it as soon as possible afterwards, or ask the NCS to add it as a notation in his log. If there are enough operators to do so, one may be assigned the sole task of logging the net's operations, thus freeing up other net participants to handle messages more quickly.

MESSAGE AUTHORING — THEM OR US?

One of the oldest arguments in emergency communications is the question of whether or not EmComm personnel should author (create) agency-related official messages. If your job is strictly communication, and the message is not about the communication function you are providing, the best answer is "no." "Pure" communicators are not generally in a position to create messages on behalf of the served agency. They have no direct authority and usually lack necessary knowledge.

However, you should always work with a message's author to create text that is clear, to the point, and uses the minimum number of words necessary. Once you do this with most agency personnel, they will be happy to send you appropriate messages, since it saves them time, too. If the author tells you to "just take care of the wording for me," it is still a good idea to get their final approval and signature before sending the message.

If you have additional training for an agency-specific job that involves message origination, this is quite different from the situation of a "pure" communicator. In this case, you may be able to generate an official message if you have been given the authority to do so.

Other messages that can and should be generated by all EmComm operators are those that deal solely with communication. Examples would be messages about net operations and frequencies, and requests for relief operators, radio equipment, supplies, and food and water for EmComm personnel.

MESSAGE SECURITY AND PRIVACY

Information transmitted over Amateur Radio can never be totally secure, since FCC rules strictly prohibit us from using any code designed to obscure the meaning of a message. Anyone listening in with a scanner can hear all that is said on voice nets. The federal Communications Privacy Act does not protect Amateur Radio communications, and anything overheard may be legally revealed or discussed. Reporters in disaster-prone areas have been known to purchase digital-mode decoding software for laptops in order to intercept ham radio communications during disasters.

However, this does not mean that you can discuss any message you send with others. Messages sent via Amateur Radio should be treated as privileged information, and revealed only to those directly involved with sending, handling, or receiving the message. This must be done to offer at least a minimum level of message security. You cannot prevent anyone from listening on a scanner, but you can be sure they do not get the information directly from you.

Your served agency should be made of aware of this issue, and must decide which types of messages can be sent via Amateur Radio, and using which modes. The American Red Cross has strict rules already in place. In general, any message with personally identifiable information about clients of the served agency should be avoided — this is a good policy to follow with any agency if you are in doubt. Messages relating to the death of any specific person should never be sent via Amateur Radio. Sensitive messages should be sent using telephone, landline fax, courier, or a secure served-agency radio or data circuit.

While we can never guarantee that a message will not be overheard, there are ways to reduce the likelihood of casual listeners picking up your transmissions. Here are some ideas:

- Use a digital mode: packet, PSK31, fax, RTTY, PACTOR, digital phone, etc.
- Pick an uncommon frequency — stay off regular packet nodes or simplex channels.
- Do not discuss frequencies or modes to be used openly on voice channels.
- Avoid publishing certain ARES or RACES net frequencies on websites or in any public document.

Some agencies use a system of "fill in the blank" data gathering forms with numbered lines. To save time on the radio, all that is sent is the line number and its contents. A casual listener might hear, "Line 1, 23; line 5, 20%; line 7, zero." The receiving station is just filling in the numbered lines on an identical form. Without the form, a casual listener will not have any real information. As long as encryption is not the primary intent, this practice should not violate FCC rules.

Chapter 15
Network Theory and the Design of Emergency Communication Systems

The study of information transfer between multiple points is known as "network theory." Network theory can be thought of as the process of matching a particular message to the best communication pathway, with the best pathway being the one that can transfer the information with the most efficiency, while tying up communication resources for the least amount of time, and getting the information transferred most accurately and dependably. By incorporating some fundamental concepts about network theory into the planning of emergency communication systems, hams can take steps ahead of time to be sure that efficient and appropriate communication modes are available when an emergency strikes, thus providing a more valuable service to the public.

CHARACTERISTICS OF MESSAGES

During an emergency, messages vary greatly in terms of length, content, complexity, and other characteristics. Similarly, the available communication pathways vary in how well they handle messages having different characteristics. Here are some of the characteristics a message may exhibit, along with suggestions for the appropriate communication pathways.

Single Versus Multiple Destinations

There are major differences between broadcasting and one-to-one (exclusive) communication channels. Some messages are for one addressee, while others need to be received by multiple locations simultaneously. And some messages addressed to one destination can be useful to "incidental" listeners, such as the National Weather Service. A specific instruction to a particular shelter manager is a completely different kind of communication than an announcement to all shelters, yet it is common to hear such messages on the same communications channel.

High Precision Versus Low Precision

Precision is not the same as accuracy. All messages must be received accurately. But sending a list of names or numbers requires precision at the "character" level, while a report like "the lost hiker has been found" does not. Both may be important messages and must be transferred accurately, but one involves a need for more precision.

Over low-precision communications channels (such as voice modes), even letters of the alphabet can be misinterpreted unless a phonetic system, feedback, or error-correcting mechanism is used. Conversely, typing out a low-precision message like "the delivery van containing the coffee has arrived at this location" on a high-precision packet link can be more time-consuming (and inefficient) than a simple voice report.

Complexity

A doctor at a hospital may use a radio to instruct an untrained field volunteer how to splint a fractured leg. A shelter manager may report that he is out of water. The level of complexity varies greatly between these two messages.

Some messages are so long and complicated that the recipient cannot remember or comprehend the entire message upon its arrival. Detailed maps, long lists, complicated directions, and diagrams are best put in hard copy or electronic format for storage and later reference. This avoids the need to repeat and ask for "fills," activities that tie up the communication channel. Some modes (such as fax and packet radio) by their very nature generate such reference copy; others (such as voice modes) do not, and require a time-consuming conversion step.

Timeliness

Some messages are extremely time-critical, while others can tolerate delays between origination and delivery without adverse effect. Relief workers and their communicators can be very busy people, and requiring a relief worker to handle a non-time-critical message may prevent him or her from handling a more pressing emergency. Also, a message might need to be passed at a time when the receiving station is tied up with other business, and by the time the receiving station is free the sending station is then occupied. In these cases, provision can be made for "time shifting;" that is, the message can be left at a drop point for pickup when the receiving station becomes free. Conversely, highly time-critical messages must get through without delay.

Timeliness also relates to the establishment of a communications link. Some modes, such as telephones, require dialing and ringing to establish a connection. An operator of a base station radio may need to track down a key official at the site to deliver a message. What matters is the total elapsed time from the time the message originates to the time it is delivered to its final party.

Priority

The concept of priority as used within network theory is better known to hams as QSK — the ability to "break in" on a communication in progress. For example, say a communication pathway is in use with a lengthy, but low-priority, message when a need suddenly arises for a high-priority message. Can the high-priority message take precedence and interrupt the low-priority one to gain access to the channel? Some communications modes allow for this; others do not.

CHARACTERISTICS OF COMMUNICATION CHANNELS

Now that we have looked at the different message characteristics, let's consider the communication channels that might be used in an emergency. In addition to the concepts of destination, precision, complexity, timeliness, and priority, communication channels also can be evaluated in terms of their reliability and ease of use.

Landline Telephones

The pathway most familiar to non-hams is the telephone. This voice-based mode is surprisingly reliable and can be operated without the need for specialized communication volunteers. It is often fully operational with plenty of unused capacity during localized and small-scale emergencies, but it can quickly become overloaded during large-scale disasters.

The telephone system is very good for transferring simple information requiring low precision. Since this mode utilizes the human voice, transferring a large amount of high-precision data (such as spelling a long list of names or numbers) can become tedious and time consuming.

The telephone system is a one-to-one communication pathway, meaning it cannot be used for broadcasting. But the one-to-one relationship between sender and receiver makes it ideal for messages containing sensitive or confidential information, such as casualty lists.

The exclusive nature of most telephone circuits makes it difficult or impossible to break in on a conversation to deliver a higher-priority message. The need for break-in usually precludes leaving the channel open continuously between two points, resulting in the need to dial and answer each time a message needs to be sent.

The major drawback to using telephones during emergencies is that the sending and receiving stations are not self-contained. The system requires wires and cables that can be damaged or destroyed during severe weather. When the central switching center goes down or becomes overloaded, all communications on this mode come to a halt, regardless of priority or criticality.

Cellular Phones

Cellular phones offer advantages that make them attractive: they are simple to operate and do not require a separate, licensed communication volunteer. They are lightweight and can be carried in a pocket, eliminating the need for tracking individuals as they move around.

Like landlines (and unlike devices used in Amateur Radio), cellular phones are ideally suited to one-to-one communications, avoiding distraction to stations not involved in the message exchange. They are unsuitable for multiple-recipient messages that are better handled on a broadcast-capable communications mode.

Like the landline telephone system, cellular phones are not self-contained communications units. They are reliant on a complex central switching and control system that is subject to failure or overloading. If the central base station goes down, or if its links with the other components of the phone system fail, cellular phone communication comes to a halt. There is no "go to simplex" contingency option with cellular phones.

Fax

Fax machines overcome the limitations of voice communications when it comes to dealing with high-precision, lengthy, and complex information. A four-page list of first-aid supplies, for example, can be faxed much faster than it can be read over a voice channel and transcribed. Fax machines can transfer drawings, pictures, diagrams, and maps — information that is practically impossible to transfer over voice channels.

Today, fax machines are widely available. Most organizations use them as a routine part of their business communications. It is likely that a fax machine will be found at the school, church, hospital, government center, or other institution involved in emergency or disaster-relief efforts.

Another advantage of fax machines is their production of a permanent record of the message as part of the transfer process. They also facilitate time-shifting. Their disadvantages are that they rely on the phone system and add one more piece of technology and opportunity for failure. Except for laptop modems, they generally require 120 V ac current, which is not always available during emergencies unless plans have been made for it.

Two-Way Voice Radio

Whether on the public service bands or ham frequencies, whether SSB or FM, via repeater or simplex, voice radio is simple and easy to operate. Most units can operate on multiple frequencies, making it a simple matter to increase the number of available communication circuits as the need arises. Most important, the units are generally self-contained, enhancing portability and increasing reliability of the system in adverse environmental conditions.

Radios are ideal for broadcasting. On the flip side, though, while a message is being transferred between two stations, the entire channel is occupied, preventing other stations from communicating. Using radio for one-to-one communication can be very distracting to stations not involved in the exchange. (The most common example of inefficient use of communication resources is a lengthy exchange between two stations on a channel being shared by a large number of users.) Also, radios suffer from the low precision inherent in voice modes of communication.

Trunked Radio Systems

These systems are becoming highly popular with public service agencies. They are similar to the standard voice radio systems described above, but with two exceptions. Unfortunately, both exceptions have a direct (and adverse) impact on the use of trunked systems in emergency and disaster situations.

The first difference has to do with the fundamental purpose behind trunking. Trunked systems came into being to allow increased message density on fewer circuits. In other words, more stations could share fewer frequencies, with each frequency being utilized at a higher rate. Under normal circumstances, this results in more efficient spectrum use, but when an emergency strikes and communication needs skyrocket, the channels quickly become saturated. A priority queue results, and messages are delayed. Medium- and low-priority messages, and even some high-priority messages, might not get through unless important stations are assigned a higher priority in the system's programming.

The second difference deals with the way in which frequencies are shared. Trunked systems rely on a complex central signaling system to dynamically handle the mobile frequency assignments. When the central control unit goes down for any reason, the entire system — base and mobile units — must revert to a pre-determined simplex or repeater-based arrangement. This fallback strategy is risky in emergencies because of the small number of frequencies available to the system.

Packet Radio

As already mentioned, voice modes are ideal for low-precision messages. Digital data modes, on the other hand, facilitate high-precision message transfer. Modes such as packet radio ensure near-perfect accuracy in transmission and reception. And, like fax machines, packet has the ability to provide a relatively permanent record of the message for later reference.

The packet mode has another advantage when dealing with information that is in electronic form: there is no need for a conversion step before transmission. This is especially valuable when the information being sent is generated by machines, such as automated weather sensors, GPS receivers, or shelter management computers.

Packet stations are generally self-contained, and if located within line of sight, do not need a central switching system.

Unlike fax machines, packet radio systems are perfect for the distribution of high-precision information to a large number of destinations simultaneously. Furthermore, the automated retry feature means that several connections can share a single frequency simultaneously, effectively increasing the capacity of the channel.

Among the mode's disadvantages is that real-time packet messages require the operator to use a keyboard. This makes the mode unacceptable for low-precision but lengthy messages, such as describing an injury or giving a status report, especially when the operator is not a fast typist. Also, its need for perfect transmission accuracy means that it may not be reliable along marginal RF paths. And again, unlike fax machines, most of today's common packet protocols are inefficient when transferring precision graphics, drawings, and all but the most rudimentary maps.

Store-and-Forward Systems

Sometimes considered a subset of packet radio, store-and-forward systems (bulletin boards, messaging gateways, electronic mailboxes, etc.) can handle non-time-critical messages and reference material, enabling communication in situations where sender and receiver cannot be available simultaneously. These systems also increase the effective capacity of a communication channel by serving as a buffer. When a destination is overloaded with incoming messages, the store-and-forward unit can hold the messages until the receiver is free.

Note that store-and-forward systems are not limited to digital modes. Voice-answering machines and even a National Traffic System (NTS)-like arrangement of liaison stations can function as voice-based store-and-forward systems.

Other Modes

Slow-scan television, fast-scan television, satellite communications, human couriers, the Internet, e-mail, and other modes of communication all have their own characteristics. Space limitations prohibit more discussion here, but by now you get the idea of how communications channels relate to different types of messages.

PLANNING AND PREPARATION: THE KEYS TO SUCCESS

Serious communication planners should give advance thought to the kinds of information that might need to be passed during each kind of emergency they wish to consider. Will maps have to be transferred? What about long lists of names, addresses, supplies, or other detailed information? Will the communications consist mostly of short status reports? Will the situation likely require transfer of detailed instructions, directions, or descriptions? Will they originally be in verbal, written, or electronic form?

Planners should next consider the origins and destinations of the messages. Will one station be disseminating information to multiple remote sites? Will there be many one-to-one messages? Will one station be overloaded while others sit idle? Will a store-and-forward system, even via voice, be useful or necessary?

The content of the messages should also be considered. Will a lot of confidential or sensitive information be passed? Will there be a need for break-in or interruption for pressing traffic, or can one station utilize (tie up) the communications link for a while with no adverse consequences?

Along with the message analysis described above, the frequency of occurrence (count of messages) of each type should also be estimated.

Then, in the most important step, the characteristics of the high-volume messages should be matched to one or more appropriate communication pathways.

Once you have identified the ideal pathways for the most common messages, the next step is to take action to increase the likelihood that the needed modes will be available during the emergency. Hams take pride in their "jump kit" emergency packs containing their 2-meter radios, extra batteries, and roll-up antennas. How about doing the same thing for some additional communication modes, too? Put a list of critical phone numbers (including fax numbers, pager numbers, and cellular numbers) in your kit. Make sure your local packet digipeater has battery backup. If you are likely to be assigned to a school, church, or office building, see if you can get a copy of the instructions for using the fax machine to keep in your kit. If the phones are out, know how to interface the fax machine to your radio.

Advance scouting may be needed. It is a good idea to see if fax machines are in place and whether they will be accessible in an emergency. Is there a supply of paper available? Are the packet digipeaters within range of every likely communication post? Can computers be made available, or will hams have to provide their own? How will backup power be provided to the computers? Can a frequency list be developed, along with guidelines of when and how to use each frequency?

Contingency planning is also of critical importance. How many times has a repeater gone down and the communicators found themselves wishing they had agreed in advance on an alternate simplex frequency? What will you do if you need to send a map and the fax machine power fails? Suppose you are relying on cellular phones and the cellular network fails? Remember, if you plan for problems, they cease to be problems and become merely a part of the plan.

The final step is training. Your manning roster, assignment lists, and contingency plans need to be tied in to the training and proficiency of your volunteers. Questions you might want to ask include: Who knows how to use a cellular phone? Who knows how to use fax software? Who knows how to upload or download a file from a packet BBS? Who knows how to touch-type?

By matching your needs with your personnel, you can identify areas where training is needed. Club meeting programs and field trips provide excellent opportunities for training as well as for building enthusiasm and sharing knowledge of plans. You will be surprised how far a little advance planning and effort can go in turning a volunteer mobilization into a versatile, effective, professional-level communication system.

ALTERNATIVE COMMUNICATION METHODS

Amateur Radio may not always be the only or best radio service for the job. Sometimes it is better to hand an official a radio he can use to stay in contact with the ARES team on site, and not saddle him or her with a ham radio "shadow." This is particularly true for officials who must regularly deal with sensitive issues.

The radio services discussed in this chapter are in general use and access to them is commonly available at low cost. Other voluntary agencies may use these services in their own operations, and many emergency communications volunteers may already own the required equipment. Amateur emergency communication groups should be equipped to communicate with them.

Legal Considerations

Some radio services require licenses and others do not. General Mobile Radio Service (GMRS), for instance, requires a license that is relatively easy to obtain, although not free. However, in a true emergency *as defined by the FCC*, this may not be a problem. FCC rules give everyone special permission to use "any means necessary" to communicate in order to protect life and property — *but only when no other normal means of communication is possible*. Please do not assume that this means you can just modify your radio and call for help on the local police frequency the next time you see a car crash on the highway. Law enforcement agencies are not bound by the FCC's rules. Hams who have called for help on police frequencies have been convicted of "interfering with a police agency" under state and local laws, even though the FCC had taken no enforcement action. In one case, the judge ruled that by modifying his radio in advance, the amateur had committed "pre-meditated" interference, a serious charge. If you are in a position to save someone's life or property, be sure you are ready to defend your actions in court if necessary — and possibly lose — before pressing the mic button.

If your group is planning to use licensed radios, obtain your license well before any emergency and keep it current. If you own a radio, but no license, a judge could claim premeditation if you use it and disturb licensed users.

Using Modified Ham Radios

While it is easy to modify many VHF and UHF amateur radios for operation in nearby public service and business bands, it is not legal to do so for regular emergency use. Radios used in those bands must be "Type Accepted" by the FCC for the purpose, and amateur radios are not. If you plan to use non-ham radio frequencies discussed in this chapter, it is better to purchase the proper radio. However, if the need arises and your ham radio is all you have, the FCC will probably not prosecute you for using it — if the use falls within their strict rules for emergencies, as outlined above.

PERMISSIBLE MODES ON THE OTHER RADIO SERVICES

Note: In most of the radio services listed below, only voice communication is permitted. Packet and other forms of data or image transmission are illegal.

Citizens Band (CB) Radio

As a widespread system of casual communication, CB radio is still quite popular among the public in general and truckers in particular. Since the 1950s, CB has been available to anyone for the purpose of short-range business and personal/family communication. No licensing is required, and tactical or self-assigned identifiers are acceptable. A recommended method promoted by the FCC is the letter "K," followed by the user's first and last initials, followed by his or her zip code. If you had a valid Class D License before the mid 1980s, you may continue to use your old CB call sign. *Do not use your amateur call!*

CB radios operate in the 11 meter band, on 40 designated channels from 26.965 to 27.405 MHz, with a maximum output power of 4 W. Most use amplitude modulation (AM), but a few also offer single sideband (SSB). The effective range between two CB mobile stations averages two to eight miles. Depending on antennas, terrain, and propagation, base to mobile communication is possible up to 25 miles. The use of SSB can significantly increase range, but SSB use is not widespread due to the extra cost. FCC rules permit communication to a maximum of 75 miles.

In many remote areas with little or no telephone service, families rely on CB radios for basic day-to-day communications. Many rural police and sheriff's organizations still monitor CB traffic. In a number of states, highway patrol officers install CB units in their patrol cars with the blessing of their agencies. However, many departments that used to monitor channel 9 have given up the practice. Radio Emergency Associated Communications Teams (REACT) groups in a given area may still be monitoring.

In disaster situations, great emphasis is placed on the timely movement and distribution of supplies by truck. By far, the largest group of CB users is the trucking community. Channel 19 has been the unofficial trucker channel since the late 1960s, and in some areas is as good as channel 9 when calling for assistance.

Channel 9 is reserved for emergency and motorist assistance traffic only. There are other organizations in many parts of the world that monitor channel 9 and other designated distress channels. In some countries, Citizens Radio Emergency Service Teams (CREST) teams serve the same functions as REACT.

Multi-Use Radio Service (MURS)

With little fanfare, the FCC added a new, unlicensed citizen's radio service in 2000. Both personal and business operation is permitted, with a maximum power of 2 W. The MURS frequencies are 151.820, 151.880, 151.940, 154.570, and 154.600. While base operation is not specifically prohibited at this time, the service is primarily intended for mobile and portable operation.

For about 20 years, certain businesses have been able to obtain licenses for operation on what the FCC calls "itinerant" frequencies. These channels became commonly referred to as the "color dot" channels (a color dot label on the packaging identifies the frequency of the walkie-talkie).

One of the former itinerant channels, 154.570 MHz (blue dot), is now a MURS channel. This means that a number of these low-cost 1 or 2 W output "itinerant" radios — which are usually user programmable for itinerant channels only — could be utilized for MURS. This allows you to equip unlicensed volunteers with a VHF portable having much the same simplex capability as a 2 meter handheld.

Family Radio Service (FRS)

Almost anywhere, in most every situation, you can find FRS radios in use. These portables are useful, effective, and inexpensive. Like CB, the FRS is designed for short-range personal communications. Campers, hikers, vacationers, and families on weekend outings use FRS units to keep in touch. There are 14 available UHF channels, and 38 different CTCSS codes to limit background chatter and noise. Output power is from 100 to 500 mW, depending on the model.

In an effort to standardize the ability to call for help using FRS, REACT recommends the use of FRS channel 1 (462.5625 MHz) with no CTCSS tone as an emergency calling channel. REACT is also lobbying the manufacturers of FRS equipment to suggest this plan in the user's information packed with new radios. A petition to the FCC requesting that this be made official was denied in late 2001. Monitoring the channel is recommended to all persons in outdoor areas whenever possible.

The first seven FRS channels are shared with the General Mobile Radio Service (GMRS). Although the original rules seem to prohibit it, a later FCC Report and Order explicitly permits communication between the two services. The chance of a distress call being heard on either service is greatly increased on these seven common channels.

Most FRS radios are available with 2 or 14 channels, although single channel radios can be found. It is important to note that the channel numbers on each radio are not always interchangeable between these units. Single channel radios are usually on channel 1, which corresponds to channel 1 in the 14-channel units.

General Mobile Radio Service (GMRS)

The GMRS consists of 15 UHF frequencies between 462.5625 and 462.7250 MHz. Eight of these are paired with matching repeater inputs 5 MHz higher, as with amateur and commercial systems. Seven "interstitial" channels are shared with FRS, and operation there is restricted to simplex with a maximum of 5 W; power on the other channels is limited to 50 W. GMRS stations have the option of working only simplex modes if desired, even on paired channels. There is no frequency coordination, and users must cooperate locally to effectively use channels. CTCSS codes are the same as for FRS, and the first seven channels are common to both services. FM voice operation is permitted, but digital modes and phone patches are not.

Operating a GMRS station will require a low-cost system license from the FCC. You can apply using FCC Form 574, or apply online. FCC online licensing information can be obtained at **www.fcc.gov**. System licenses are currently granted only to individuals. A system comprises any and all radios operated by family members, and may include fixed, mobile, and repeater equipment. Use under the license is restricted to members of the licensee's immediate family. Licenses to entities other than individuals are no longer issued, but non-individual entities licensed before July 31, 1987, may continue to renew their licenses, but may not increase or modify their use.

The frequency of 462.675 MHz is recognized for emergency and travel information use, and is monitored by many REACT teams nationwide. Many teams operate repeaters on this and other frequencies.

Current uses for GMRS involve mostly personal and family communications. Hiking, camping, and convoy travel are all common GMRS applications. GMRS use for emergency services is limited by the licensing requirements, but it could be pressed into service in a disaster situation. One or more members of a volunteer group might wish to become licensed if use of GMRS is likely, especially for liaison with locally active REACT teams.

Public Safety Radio

There are instances where the use of police and fire radio frequencies is possible. The agency itself might allow and train you for such use, or an individual officer may ask you to use his radio to call for help when he cannot. Keep your transmissions short and to the point. Do not tie up the channel with long explanations, and cease transmitting if instructed to.

Cellular and Personal Communications Service (PCS) Phones

In a widespread disaster situation, these phone systems can quickly become overloaded. In smaller emergencies, they may still be usable. If a message is too sensitive to send via any two-way radio, try your cell phone. Cellular and PCS phone transmissions, especially digital, are considerably more secure. In addition, it is possible to send low-speed data or fax transmissions over the cellular network.

Aviation Radio

AM radios operating in the 108–136 MHz band are used in aircraft and in certain limited vehicles and ground stations. FCC licenses are required for all stations. Emergency Locator Transmitters (ELTs) are automatic devices that transmit a distress signal on 121.5 and 243.0 MHz (121.5 is the civilian distress channel and 243.0 is its military counterpart). These frequencies are also used for marine Emergency Position Indicating Radio Beacons (EPIRBs) and the new land-based Personal Radio Beacons (PRBs). While it is unlikely that you will ever need to use an aircraft band radio except where it is provided by the served agency, it is good to be familiar with the radio service. Monitoring 121.5 for ELT, EPIRB, and PRB signals and distress calls is always a good idea.

Marine Radio

FM marine radios operate on internationally allocated channels in the 160 MHz band. HF SSB radios operate on a variety of International Telecommunication Union (ITU) channels between 2 and 30 MHz. Operation of FM stations for vessels in U.S. waters does not require a license, but operation on the HF channels does. The most common marine radio mode is VHF-FM (156 to 162 MHz), with an effec-

FREQUENCIES (MHz) FOR KEY MARINE VHF CHANNELS		
FM 9	156.45	Calling
FM 16	156.8	Calling/Distress
FM 17	156.85	State/local govt. shore station
FM 18	156.9	Commercial Intership
FM 21	157.05	Coast Guard
FM 22	157.1	Coast Guard — NOTAMS
FM 23	157.15	Coast Guard
FM 68	156.425	Intership
FM 69	156.475	Intership
FM 83	157.175	Coast Guard Auxiliary

tive range from ship to ship of 10 to 15 miles, and ship to shore of 20 to 30 miles. Vessels that routinely travel outside this distance generally have MW/HF-SSB, satellite communications or both. CW communication on MW/HF is no longer used.

No license is currently required for pleasure boats operating on the FM channels in U.S. territorial waters. The FCC limits VHF-FM marine radios to a maximum of 25 W. Radios are also required to be capable of 1 W operation for short range and in-harbor use. For more regulatory information, visit **http://wireless.fcc.gov/services/index.htm?job=service_home&id=maritime**.

When working in coastal areas, along major rivers or the Great Lakes, it may be a good idea to have an FM marine radio in your group's inventory. During major storms you can monitor channel 16, the distress channel. If you hear a vessel in distress whose calls are going unanswered by the Coast Guard, you may legally answer them from an unlicensed land-based station under the FCC's emergency communications rules. If the Coast Guard is in communication with the vessel, do not transmit. Most other land-based operation is illegal, except where authorized by an FCC coast station license. For information on how to handle distress calls, see **www.navcen.uscg.gov/?pageName=mtBoater**.

Part 5: The Response

Chapter 16
Managing the Response

For those who have never activated for an emergency, there are numerous questions to be answered. What exactly does one do "when the call comes in?" First of all, to be called at all you must be registered with a local EmComm group in advance of an emergency in order to be on their notification list. Spontaneous, unrequested volunteers are extremely difficult to integrate into an already confusing emergency response. Join a group well in advance of any emergency and take advantage of any training opportunities they offer, in order to be ready when the call comes.

INITIAL NOTIFICATION BY THE SERVED AGENCY

In most cases, three or more members serve as "activation liaisons" to the served agency, meaning they are called first when EmComm volunteers are needed. Never rely on a single point of contact. The served agency should have all possible telephone numbers for an individual, including mobile, as well as e-mail addresses, with all information listed in the order in which it is to be used. Contact information should be updated and verified regularly.

GROUP ALERTING SYSTEMS

Once the served agency has notified the activation liaison, the liaison must alert members of the EmComm group. A number of group alerting methods may be used. The most common ones are described below. No one method should be relied upon, since emergency conditions may render it useless. Commercial paging systems and ham repeaters might be off the air, phone lines down, and Internet service disrupted. To safeguard against every eventuality, a written plan and checklist should be developed well in advance, and updated periodically.

Telephone Tree

In this system, the liaison calls two members, who each call two other members, and so on until the entire group has been notified. If any one person cannot be reached, the person calling must then call the members the unreachable person would have called, had they been reached. This method insures that the "tree" is not broken. Messages should always be left on all answering machines and voice mailboxes.

Cellular Telephones

Increasingly, individuals and agencies are depending on cellular telephones. There is no doubt that as cellular communications have advanced over the last decade they have become essential communications tools. They are also useful in notifying your ARES group of an activation, but their limitations must also be considered.

Cellular phones offer several methods for notifying members of an ARES activation. Traditional voice communications are enhanced by extreme portability. Text messaging capability allows for mass notification and can be combined with photo and video attachments. Smartphone applications can be leveraged to provide ARES members with additional situational awareness. Even e-mail and Internet can be supported. All of this comes with one big caveat: these features are available as long as the network is functioning.

Cellular networks can become overloaded or go down completely during large and small disasters. A localized bridge collapse in one city proved that a network can be rendered useless just by too many people trying to call 911 at the same time. During a major hurricane, cellular networks can be down for days. However, most of the time, the network is functioning and can be used as part of a comprehensive ARES notification system.

E-Mail

While e-mail might not immediately reach members wherever they happen to be, it is a good backup method as long as it continues to function. Many people have full-time high-speed Internet connections at home and the office, and most people check their e-mail frequently. Someone who has otherwise been unreachable may check their e-mail even several hours later, just as they might check an answering machine or voice mailbox.

Self-Activation

If you become aware of an incident or situation that might require the activation of your EmComm group, you should take immediate steps to make yourself available. Depending on your group's activation plan, this might mean monitoring the assigned net or served agency frequencies, or making contact with one or more appropriate persons in the EmComm group or served agency. SKYWARN members might also monitor National Weather Radio. Remember, if you are not specifically authorized to directly contact served agency personnel, do not do it. Know your plan and follow it.

FAMILY FIRST

The first thing you must do before you report to an assignment is secure your home and family. There are times when circumstances at home may take precedence over your commitment to your EmComm group. Of course, family must come first, and only you and your loved ones can evaluate whether you can be excused from whatever situation is transpiring at home during an emergency. It's a good idea to discuss potential scenarios with your loved ones before an emergency situation occurs, and come up with a general plan of which everyone is aware. Whenever you do activate for an emergency, family members, and perhaps even friends or neighbors should know where you are going, when you plan to return, and how to contact you in case of an emergency at home. More information on ensuring the safety of your property and your family while you are on assignment can be found in the chapter on Personal Safety, Survival, and Health.

KNOW YOUR RESOURCES BEFORE DEPLOYMENT

Unlike commercial and public safety radio users, amateurs have a vast amount of radio spectrum at their disposal when serving during an emergency. Most local and regional EmComm communications take place on 2 meter or 70 centimeter FM, or on 40, 60, or 80 meter SSB/CW. The choice made is based on the locations to be covered, the availability of repeaters, as well as the distance, terrain, and band conditions.

VHF and UHF FM are preferred for most local operations because the equipment is common, portable, has a clear voice quality, and the coverage is extended by repeater stations. VHF and UHF communication range is determined by terrain, antenna height, and the availability of repeaters.

For larger areas or in areas without repeaters, HF SSB may be needed. Most local EmComm operation is on the 40 or 80 meter bands using Near Vertical Incidence Skywave (NVIS) propagation. For long-haul communication needs and international operations, 15 or 20 meter nets may be the best option.

Many EmComm groups will have pre-selected a number of frequencies for specific purposes. The complete list of these frequencies should be in your jump kit, and pre-programmed into your radios.

Become familiar with the coverage and features of each permanent repeater and digital message system in your area, and pre-program your radios with the frequencies, offsets, and CTCSS tones. Ask your EC or AEC which repeaters are used for emergency communications in your area. Will they be available for exclusive EmComm use, or must they be shared with other users? Information to find out includes:

- How does the repeater identify itself?
- Are there any "dead spots" in critical areas? How much power is required to reach the repeater with a clear, quiet signal from key locations?
- Does the repeater have a courtesy tone and, if so, what does it sound like? Do the tones change depending on the repeater's mode?
- How long is the "time-out timer?"
- Is the repeater part of a linked system of repeaters? What features does it have, and which touch-tone commands or CTCSS tones activate them?

For net frequencies that support digital communication systems, such as packet radio bulletin board messaging systems, PACTOR, PSK31, and RTTY:

- Which software do they use? *ARESPACK, Fnpack, FNpsk*?
- Do the digital systems have mailboxes or digipeater functions?
- Which other nodes can they connect to? Can traffic be passed over an Internet link automatically or manually?
- How many connections can they support at once?

EN ROUTE

There are several things you may need to do while you are headed home to pick up your jump kit or other gear, or while you are on your way to your assigned location. Fill your vehicle with fuel and pick up any supplies you may need, including alkaline batteries for radios and lights; food and water; and other supplies on your checklist. Check into and continue to monitor the activation net for further information or instructions. Contact your spouse, children, or other family members to let them know what is happening and where you will be. Give them any instructions they will need in order to be safe. Tell them when you will next try to contact them, and how to contact you if necessary. Knowing that your loved ones are all right can let you do your job without needless worry, and, of course, the same is true for them.

FIRST STEPS

Your group's activation plan should tell each member what steps to take immediately after learning of emergency communications activation. In most cases, the first step should be to check in on a specific frequency or repeater. If a repeater is used as the primary gathering point for members, a backup simplex frequency (the repeater's output frequency works well) should be specified in the event that the repeater is no longer operating. One of the liaison stations should be available on the net to provide additional information from the served agency and directions to members as they check in. If a member is pre-assigned to act as NCS for the "activation" net, that person should take over the task as soon as possible, to free up the liaison to work with the served agency or take other action. Some groups simply have the first person signing on act as a temporary NCS until an assigned NCS checks in. Again, it is important to have more than one person assigned to take on the NCS duties, in the event that someone else is unavailable.

In other cases, some members may have specific assignments. These might include making contact with the served agency, going directly to a specific location such as an EOC, or making certain preparations. These members should quickly check into the "activation" net to let EmComm managers know that they have been reached and are responding. Members without a standing assignment should check into an activation net and make themselves available for an assignment. It might be a "resource" logistics net if one is active, or the general "tactical" command activation net. Local procedures vary widely, so familiarize yourself with your group's specific plans and procedures well in advance of a potential activation.

ARRIVING AT THE SITE

In some cases, once you arrive on site, you may be asked to proceed to a "staging" or "volunteer intake" area to wait for an assignment. This could take some time, especially if the situation is very confused. Often, the development of the response to the emergency is unclear and it will take some time to develop a cohesive and uniform response plan for that incident. You should expect the situation to be fluid as each incident is unique and to respond accordingly. Be prepared to wait patiently for a determination to be made and an assignment to be given.

If you are assigned to a facility operated by the served agency, such as a shelter, introduce yourself in a brief and to-the-point manner to the person in charge. Identify yourself and explain that you have been assigned to set up a communications station for that location, and tell them by whom. Inform them that you would like to set up your equipment and get on the air. Ask if another communicator has already arrived. Ask if they have a preference for the station's location, and explain your needs.

If you are the first communicator to arrive, be prepared to suggest an appropriate location — one that can serve as both an operating and message desk, has feed line access to a suitable antenna location, access to power and telephone, and is just isolated enough from the command center to avoid disturbing each other.

Ask if there are any hazards or considerations in the immediate area that you should be aware of, or that may cause you to relocate later.

If no building or other suitable shelter is available, you may need to set up your own tent, or work from your car. Choose a location that provides shelter from wind, precipitation, and other hazards, and is close enough to the served agency's operations to be convenient, but not in the way.

WHO IS IN CHARGE?

At each station on site, the EC or other EmComm manager should appoint one member of the group to take a leadership role as "station manager," with full responsibility for all operations at that site. This person serves as a point of contact, handling information and decisions for the team with the incident commander and with other groups aiding in the response. This helps avoid confusion and arguments. This position should not be confused with the role of control operator.

When you accept a position as an emergency communications volunteer, you do so knowing that you will often need to follow the directions of another person. Cooperation and good teamwork are key elements that result in an efficient and effective emergency communications operation. As the situation arises, you may have to step into a role of a leader to keep the operation moving forward.

Expect to work with others. Expect that there are times when you are the follower. Expect that at other times, you may be the leader.

BEING A GOOD GUEST

In many cases, you will be occupying a space that is normally used by someone else for another purpose. Respect and protect their belongings and equipment in every way possible. For instance, if you are in a school and will be using a teacher's desk, find a way to relocate all the items from its surface to a safe place for the duration of operations. A cardboard box, sealed and placed under the desk usually works well. Do not use their office supplies or equipment, or enter desk drawers or other storage areas without specific permission from a representative of the building's owners. Some served agencies will seal all filing cabinets, drawers, and doors to certain rooms with tamper-evident tape upon arrival to protect the host's property and records.

When installing antennas, equipment, and cables, take care not to damage anything. For instance, avoid using duct tape to fasten cables to walls or ceilings, since its removal will usually damage the surface. If damage is caused for any reason, make note of it in your log and report it to the appropriate person as soon as possible.

INITIAL SETUP AND GATHERING OF INFORMATION

In most cases, your first priority will be to set up a basic station to establish contact with the net. Pack that equipment in your vehicle last, so you can get to it first. If you arrive as a team of two or more, station setup can begin while others carry in the remaining equipment.

Set up and test the antenna for proper SWR, and then check into the net. Test to find the lowest power setting that produces reliable communication, especially if you are operating with battery or generator power, to conserve power for extended operations. To prevent interference with other radio systems, telephones, and electronic equipment, high power should also be avoided whenever lower power will work just as well.

Once your basic station is on the air, you can begin to work on other needs, such as:

- Check for working telephones, faxes, Internet, and other means of communications.
- Learn about the served agency's operations and immediate needs at that site.
- Install additional stations or support equipment.
- Make a list of stations within simplex range.
- Identify possible alternative message paths.
- Find sanitary facilities.
- Determine water and food sources, and eating arrangements.
- Review overall conditions at the site, and how they will affect your operations.
- Find a place to get some occasional rest.

As soon as possible, ask a member of the served agency's staff to spend a few moments to discuss the agency's operational needs. What are the most critical needs? With whom do they need to communicate, and what sort of information will need to be transmitted? Will messages be short and tactical in nature, or consist of long lists? Will any messages be too confidential for radio? Are phones and faxes still working? What will traffic needs be at different times of day? How long is the site anticipated to be open? Will there be periodic changes in key agency staff?

You may also need to provide agency staff with some basic information on how to create a message. Show them how to use message forms and instruct them on basic procedures to follow. Be sure to let them know that their communications will not be private and "secure" if sent by Amateur Radio, and discuss possible alternatives.

NETWORK COVERAGE CONCERNS

Most EmComm managers rely on simplex operation when planning their VHF or UHF FM nets for one reason: repeaters often do not survive disasters or are overwhelmed by the amount of traffic. Repeaters that do survive and are usable are considered a bonus. Since simplex range is limited by terrain, output power, and antenna gain and height, operation over a wide area can be a challenge. Almost any structure or hill can block signals to some degree.

To avoid last-minute surprises, your group should pre-test all known fixed locations in your area for coverage. For instance, if you are serving the Red Cross, test simplex coverage from each official shelter to the Red Cross office and the city's EOC or other key locations, and mobile coverage in the same areas. If needed, there are several ways to improve simplex range:

- Use an antenna with greater gain.
- Move the antenna away from obstructions.
- Use a directional antenna.
- Increase antenna height.
- Increase transmitter output power as a last resort.

In a fast-moving situation with poor simplex coverage and no repeater, it can be helpful to place a mobile station on a hilltop or office building where it can communicate with, and relay for, any station in the net. A mobile relay station can also allow communications to follow a moving event, such a wildfire or flash flood. That station becomes, in effect, a "human repeater." Although an expedient "work-around," this slow and cumbersome process can reduce net efficiency by more than half. A modern aid to this kind of operation is the "simplex repeater." This device automatically records a transmission, and immediately re-transmits it on the same frequency. Remember that FCC rules do not allow unattended operation of simplex repeaters, and that you must manually identify it.

A better solution is a portable duplex repeater that can be quickly deployed at a high point in the desired coverage area. The coverage of this repeater does not have to be as good as a permanent repeater — it just has to reach and hear the stations in your net. Portable repeaters have been used successfully from the back seat of a car parked on a ridge or even the top floor of a parking garage, using a mobile antenna. Portable masts and trailer-mounted towers have also been used successfully.

If all stations in the net have dual-band radios or scanners, a strategically located mobile radio may be operated in "cross-band repeater" mode. If you use your dual-band mobile in this manner for an extended period, use the low or medium power setting to avoid overheating and damaging your radio. Consider using a fan to further reduce the likelihood that your radio will be damaged from overheating.

For a permanent repeater to be useful in a disaster, it must have emergency power, and be in a location and of such construction that it can survive the disaster. Agreements with repeater owners should be in place to allow emergency operations to the exclusion of regular users.

FREQUENCY AND NET RESOURCE MANAGEMENT

While we may have a large amount of frequency resources, in actual practice our choices are limited to the available operators and their equipment. Net managers may occasionally need to "shift" resources to meet changing needs. In the early stages of an emergency, the tactical nets may require more operators, but in later stages, the health and welfare traffic might increase.

In addition to the main net frequency, each net should have several alternate frequencies available. These should include one or more "backup" frequencies for use in the event of interference, and one or two frequencies to be used to pass traffic "off net."

MESSAGE RELAYS

When one station cannot hear another, a third station may have to relay the messages. Although this is a slow and cumbersome process, it is often the only way to reach certain stations. If relays must be used, move the stations involved off the main net frequency to avoid tying up the channel for an extended period.

RADIO ROOM SECURITY

To protect your equipment and the messages you handle, as well as to prevent unnecessary distractions, it is best to allow only the operators who are on duty to be in the room. Avoid leaving the radio room with equipment unattended and accessible. It is never a good idea to allow members of the press to be in the room without specific permission from the served agency.

RECORD-KEEPING

Most served agencies will expect you to keep records of your operations. These records will certainly include original copies of any messages sent, station logs, memos, and official correspondence. Some may even require you to keep "scratch" notes and informal logs. Depending on agency policy, you may be required to keep these records in your own possession for a time, or to turn some or all records over to the agency at the end of operations. In some agencies, your station records are permanent and important legal documents, and must be treated as such. It is important to know your served agency's policy on record-keeping in advance, so that you can comply from the very beginning of operations.

Your station operating logs should probably contain the following information:

- Your arrival and departure times.
- Times you check in and out of specific nets.
- Each message, by number, sender, addressee, and other handling stations.
- Critical events — damage, power loss, injuries, earth tremors, other emergencies.
- Staff changes — both EmComm and site management, if known.
- Equipment problems and issues.

Every individual message or note should be labeled with a time and date. In the case of scratch notes, place dates and times next to each note on a sheet, so that information can be used later to determine a course of events.

If you expect to operate from the location for more than a day or two, establish a message filing system so that you can retrieve the messages as needed. A "portable office" type of file box, expanding file, or any other suitable container can be used to organize and file the messages. This is also an efficient way to allow another operator to pick up where you left off quickly, even if he or she arrives after you leave.

DEALING WITH STRESS AND EGOS

Any unusual situation can create personal stress — disasters create incredible amounts of it. Most people are not used to working under extreme stress for long periods, and do not know how to handle it. They can become disoriented, confused, unable to make good decisions (or any decisions at all), lose their tempers, and generally behave in ways they never would at any other time. Nervous breakdowns are common among those who get overwhelmed and have not learned to manage stress and stress-causing situations.

Especially in the early hours of a disaster, the tendency is to regard every situation or need as an "emergency" that requires an immediate response. You might get a barrage of requests for action. You might not have the extra seconds it requires to fully consider the options, and to prioritize your actions. The result is an overload of responsibility, which can lead to unmanageable levels of stress.

While you cannot eliminate disaster-related stress, you can certainly take steps to reduce or control it. Here are some tips to help you manage excessive stress and stressful situations:

- Delegate some of your responsibilities to others. Take on those tasks only you can handle.
- Prioritize your actions — the most important and time-sensitive ones come first.
- Do not take comments personally. Mentally translate personal attacks into constructive criticism — and a signal that there may be an important need that is being overlooked.
- Take a few deep breaths and relax. Do this often, especially if you feel stress increasing. Gather your thoughts, and move on.
- Watch out for your own needs — food, rest, water, medical attention.
- Do not insist on working longer than your assigned shift if others can take over. Get rest when you can, so you will be ready to handle your job more effectively later on.
- Take a moment to think before responding to a stress-inducing challenge. If needed, tell the person issuing the call that you will respond in a few minutes.
- If you are losing control of a situation, bring someone else in to assist or notify a superior. Do not let a problem get out of hand before asking for help.
- Keep an eye on other team members, and help them reduce stress when possible.

STRESS MANAGEMENT

Emergency responders should understand and practice stress management. A little stress helps you to perform your job with more enthusiasm and focus, but too much stress can drive you to exhaustion or even death. Watch for these physiological symptoms:

- Increased pulse, respiration, or blood pressure
- Trouble breathing, increase in allergies, skin conditions, or asthma
- Nausea, upset stomach, or diarrhea
- Muffled hearing
- Headaches
- Increased perspiration, chills, cold hands or feet, or clammy skin
- Feeling weakness, numbness, or tingling in part of the body
- Feeling uncoordinated
- Lump in throat
- Chest pains

Cognitive reactions may next occur in acute stress situations; many of these signs are difficult to self-diagnose:

- Short-term memory loss
- Disorientation or mental confusion
- Difficulty naming objects or calculating
- Poor judgment or difficulty making decisions
- Lack of concentration and attention span
- Loss of logic or objectivity to solve problems

Perhaps the best thing to do as you start a shift is to find someone you trust and ask him or her to let you know if you are acting somewhat "off." If, at some time, they tell you they've noticed you're having difficulty, then perhaps it's time to ask for some relief. Another idea is to have some sort of stress management training for your group before a disaster occurs.

Some people within the emergency response community have big egos, and still others feel a need to be in full control at all times. Both personality types can be problematic at any time, but they can seem worse under stress. Take time ahead of time to consider how you will respond to the challenges these personalities present. If your automatic response to certain behaviors is anger, make a conscious decision to come up with a different and more positive response strategy. Depending on the official position of the problem person, you might:

- Do your job as best as you can, and deal with that person after the emergency is over.
- Politely decline to serve with that person, and state your reasons.
- Refer the issue to a superior.
- Choose in advance to volunteer in another capacity and avoid that person altogether.

LONG-TERM OPERATIONS

As soon as it becomes clear that the situation is not going to return to normal for a while, you and your group should make plans for extended EmComm operations. Hopefully, your EmComm group and served agency have prepared contingency plans for this, and all you will have to do is put them into action. If not, here are some potential needs to consider:

- Additional operators to allow for regular shift changes, and to replace those who go home.
- Replacement equipment (as operators leave with their own gear or if equipment fails).
- Food and water.
- A suitable place to sleep or rest.
- Generator fuel.
- Fresh batteries.
- Sanitation facilities.
- Shelter.
- Message handling supplies and forms.
- Alternate NCS operators, backups.
- Additional net resources to handle message traffic.

BATTERY MANAGEMENT

If you are operating on battery power, you will eventually need to recharge your batteries. Some batteries need more time to recharge than others, and this needs to be taken into account in your planning. Deep cycle marine batteries, for instance, can require a full day or longer to fully recharge. Sealed lead-acid (SLA) batteries, also known as "gel-cells," require up to 18 hours to recharge, depending on the size of the battery. NiCd, Li-ion, and similar batteries can be recharged quite quickly, although repeated rapid charge cycles can reduce overall battery life.

If you are using slow-charging batteries, you may need to have enough on hand to last the entire length of the operation. If your batteries can be charged quickly, some means must be provided for doing so. Some chargers can be powered from a vehicle's 12 V system, and are a good choice for EmComm. If no local means of charging is available, your logistics team may need to shuttle batteries back and forth between your position and a location with power and chargers.

GENERATOR AND POWER SAFETY

Take some care in the placement of generators so that they will not be a problem for others. Engine noise can make it difficult for shelter residents and volunteers to get much-needed rest, and for anyone trying to do their job. Exhaust fumes should not be allowed to enter the building or nearby tents or vehicles. A position downwind of any occupied location is best. Even when vehicles are not included in the statistic, internal combustion engines are still the number one cause of carbon monoxide poisoning in the United States. Propane powered engines produce as much (or more) CO as gasoline or diesel engines.

Earth grounding of portable or vehicle-mounted AC generators is not required as long as only plug- and cord-connected equipment is used, and the generator meets National Electrical Code (NEC) standards listed in Article 250-6. The main exception is for generators that will be connected, even temporarily, to a building's permanent electrical system. For further details on grounding ac electrical systems, please refer to Article 250 of the NEC.

Ground Fault Interrupters (GFIs) add a further degree of safety when working with generators and portable power systems. GFIs detect any difference between the currents flowing on the hot and neutral conductors, and open the circuit. Also, be sure to test any GFI device to be used with or near HF radios to be sure that the GFI will function properly while the radio is transmitting.

AC extension cords used to connect to generators or other power sources should be rated for the actual load. Consider radios, lights, chargers, and other accessories when calculating the total load. Most extension cords are rated only for their actual length, and cannot be strung together to make a longer cord without "de-rating" the cord's capacity. For example, a typical 16-gauge, 50-foot orange "hardware store" cord is rated for 10 amps. When two are used to run 100 feet, the rating drops to only 7 amps. Choose a single length of cord rated for the load and the entire distance you must run it. If this is not possible, you can also run two or more parallel cords to the generator in order to reduce the load on any single cord. For more information on portable power cord requirements, consult Article 400 of the NEC.

While some groups have used Romex-type wire for long extension cords, this is actually a violation of the National Electrical Code, and a dangerous practice. Repeated bending, rolling, and abrasion can cause the solid copper conductors and insulation to break, resulting in a fire and electrocution hazard. Use only flexible insulated extension cords that are UL rated for temporary, portable use.

SHOULD YOU LEAVE EQUIPMENT BEHIND?

You are exhausted, and ready to head for home, but the EmComm operation is far from over. You brought along a complete station and, when you leave, the next operator is not nearly as well equipped. Should you leave your equipment behind for the next operator?

You have several options here. No one can, or should, tell you to leave your equipment behind. If you feel comfortable that someone you know and trust will look after your gear, you may choose to leave some or all of it behind. If you do, be sure every piece is marked with at least your name and call sign. Do not leave behind anything that the next operator does not truly need. Also, remember that even if you leave the equipment in the possession of someone you know, you still have the ultimate responsibility for its operation and safety. Emergency stations are difficult places to control and monitor. If your equipment is stolen, lost, or damaged, you should not hold anyone responsible but yourself. Conversely, if someone leaves their equipment in your care, treat and protect it better than you would your own, and be sure it is returned safely to its owner.

ACCEPTING SPECIALIZED ASSIGNMENTS

In the world of modern EmComm, you may be asked to handle other assignments for the served agency that may or may not include communicating. At one time, most EmComm groups had strict policies against doing other tasks, and this is still true of some. In the days when radios were difficult to operate under field conditions and required constant attention, this was important. The other common reason given is that hams have volunteered to serve as communicators, not "bed pan changers." It is true that some agencies' staff will abuse the situation when they are short of help, but if both agency staff and the EmComm group are clear about any limits beforehand, the problem should not arise.

Today, most EmComm groups will permit their members to be cross-trained for, and perform, a variety of served-agency skills that also include communicating. Examples are SKYWARN weather spotting, Red Cross damage assessment, and many logistics jobs. If your group still has a "communication only" policy, are you really meeting your agency's needs? Is it necessary to have a damage assessment person and a communicator to do that job? What would happen to your agency if each driver also had to bring along a dedicated radio operator? Can one person do both jobs?

WHAT TO EXPECT IN LARGE-SCALE DISASTERS

What happens to critical communications assets during the onset of disaster conditions? First, there is a huge increase in the volume of traffic on public-safety radio channels, accompanied by prolonged waiting periods to gain access. As the disaster widens, equipment outages occur at key locations. Messages are not handled in order of priority, and urgent messages are often lost.

As agencies respond, the need arises for agencies to communicate with one another. Meeting that need is an uphill battle as these agencies have incompatible radio systems, and use unfamiliar or unattainable frequencies, names, terms, and procedures. Exacerbating the situation is the fact that most agencies are reluctant to use another agency's system, or to allow theirs to be used by others.

In a large-scale situation, a need arises to contact locations at distances beyond the range of a given radio or system (50 to 350 miles or more). Message reply delays are experienced, leading to deferred decisions on crucial matters, message duplication, and confusion. A need arises to generate and decipher handwritten messages sent through relaying stations.

Different modes of communication are required, in addition to voice:

- Volume data in printed form — data modes, high-speed packet and facsimile.
- Morse code or PSK31 under difficult reception conditions.
- Encoded data for extreme privacy.
- Television — mobile, portable, aeronautical and marine.
- Telephone interconnections from/to radio systems.

Simultaneously with a high volume of message traffic, stations must cope with messages having widely differing priorities. Also, priority and precedence designations differ among agencies, if any are used at all.

Operational problems arise, such as:

- High-volume traffic circuits with no supply of message forms.
- Using the only printed forms available, which may have been designed for a different, unrelated agency or function.
- Attempting to decipher scribbling from untrained message writers; using scribes who cannot understand radio parlance or read through QRM.
- Becoming inundated with traffic volume so heavy it results in confusion over which messages are to be sent, which were sent, which have been received for delivery, and which have been received to be filed for ready reference.

What Happens in the First 72 Hours?

In the early hours of an emergency turning into a major disaster, it takes precious time to overcome the obstacles to placing fully activated mutual aid resources into operation. Communication is one of those vital resources.

The greatest concentration of relief efforts is generally found in the incorporated cities served by agencies with paid professionals — assuming their equipment, facilities, and personnel remain operable. While urban areas experience more concentrated damage, suburbs and isolated areas of a county suffer from remoteness from fire departments, public works, law enforcement, and the services of all other agencies. All organizations scramble to respond to an unprecedented demand for service within their authorized jurisdiction.

In these circumstances the public is often isolated, unable to call for help or determine the nature and extent of the disaster so that they can make plans to wait it out, prepare to evacuate, actually evacuate (with some possessions) to a safe place, obtain physical aid, or offer aid to others.

Lack of information results in further attempted use of the telephone when the system is already saturated, if it is still operating at all. Calls can often be received from out of town but not made across town. The opportunity to call for help is often unavailable to most citizens during the first 72 hours. Occasionally, a passing public safety vehicle or one equipped with an operational commercial, utility, amateur, or CB radio can be flagged down to make a call — assuming it can contact a person who can help.

Too little information is gathered about the public's immediate needs, and ways to meet them. Distorted public perceptions develop through misinformation. At the same time, essential damage-assessment report data is needed by state and federal agencies to initiate relief aid from outside the disaster area.

Broadcast stations (those still on the air) initially disseminate rumors in the absence of factual information. Those few people who possess an operating battery-powered broadcast band radio can tune until they find a local station that can provide helpful information. Others receive such information second hand, if at all.

Everywhere, people walk aimlessly seeking a route to family and friends. Many, fearful of looting, remain in hazardous buildings, or return, as do shopkeepers, to salvage valuables. As darkness falls, rumors of looting are generated – some true.

Word circulates about shelter locations. Some displaced persons stay at homes of friends, relatives, or strangers. Others are housed at public shelters into the fourth day, still searching for family members elsewhere, and without communication. The opportunity to notify concerned distant relatives is not afforded except via Amateur Radio and the American Red Cross.

Later, often too late, information trickles in about problem areas or cases that have been overlooked due to the lack of communication. Some potential evacuees are overlooked. Once the immediate threat to life has passed, survival instincts prevail, printed "What to Do" instructions are located and followed, and people operate essentially on their own for an indefinite period while public agencies respond to the most urgent problems of which their communications make them aware.

Aftershocks, fire flare-ups, weakening or breaking of dams and rising new flood crests, buildup of winds, etc., result in some relief work being undone and the posing of new threats. Inter-agency communication is poor to non-existent. At the end of 72 hours, the disaster area remains in virtual isolation except for helicopter service for known critical cases and official use.

Little centralized information is available. Amateur Radio operators from neighboring counties and states offer to help but are often unable to cross the roadblocks established to limit access by sightseers and potential looters. Disorganized local volunteers often lack essential skills and orientation. Costly mistakes are made and systems bog down.

The dead pose a serious health problem. Stress rises among the citizenry. Little overall assessment emerges in the first 72 hours about available emergency resources and relief supplies. Shortages are apparent and growing.

Travel continues to be difficult and slow. Relief supplies trickle in to uncertain storage locations. Some supplies are useless.

Restaurants remaining open are unable to cook without gas or to serve the masses that flood them. Food and water shortages have become critical. Normal water sources may have been cut off or contaminated.

Eventually, essential functional communication networks evolve as priorities are asserted and clusters of traffic emerge. Relief efforts are mounted when someone takes charge, makes a decision and directs the efforts of others. The command and control process of directing requires communication - the ingredient in short supply in all disasters.

At critiques following a disaster, as always, the cry is heard: "Next time we must be better prepared!"

ENDING OPERATIONS

Operations may end all at once, or be phased out over time. Several factors may affect which operations end and when, including damaged communication systems being restored and returned to service, traffic loads being reduced to a point at which they can be handled with normal systems, and the closing of shelters and other locations.

How you are notified to end operations will depend on the policies of your EmComm group and served agency, as well as the specific situation. For instance, even though a shelter manager has been told to shut down by the served agency, your orders may normally come from a different person who may not be immediately aware of the shelter's closing. In this case, you might need to check with the appropriate EmComm manager before closing your station. Once the decision to close your station has been received and verified, be sure that the person in charge of the location is aware that you are doing so and, if necessary, why.

File and package all messages, logs, and other paperwork for travel. Return any borrowed equipment or materials. Carefully remove all antennas and equipment, taking care to package and store it correctly and safely. Avoid the temptation to toss everything into a box with the intention to "sort it out later," unless you are under pressure to leave in a hurry. In the event you are re-deployed quickly, this will save time in the end.

DEPARTURE

Several actions may be necessary when leaving. First, be sure to leave the space in which you worked in as good a condition as possible. Clean up any messes, remove trash, and put any furniture or equipment back where it was when you arrived. If you sealed desktop items in a box for safekeeping, simply place the box on the cleaned desk. Do not unpack the items and attempt to replace them on the desk.

This will provide proof to the desk's owner that you took steps to protect their belongings, and helps keep them secure until the owner takes possession again. Do not remove tamper-evident tape or similar seals placed by others unless told to do so by the appropriate person, or in accordance with the agency's policy.

Thank all those who worked with you. Even a simple verbal "thanks" goes a long way, compared to hearing not a single word. Do not forget the building's owners or staff, the served agency staff or others you worked with, and any other EmComm personnel. This is also the time for any apologies. If things did not always go well, or if any damage was caused, do your best to repair the relationship before departing. These simple efforts can go a long way toward protecting relationships between all groups and individuals involved.

THE DEBRIEFING

After each operation, your EmComm group, and perhaps even the served agency, will probably want to hold a meeting to review the effectiveness of the operation. There may have been issues that occurred during operations that you will want to discuss at this meeting. Events may have occurred within the served agency that involved communications you handled. If you try to rely entirely on your memory or logbooks, you will probably forget key details or even forget certain events altogether.

To prevent this from happening, keep a separate "debriefing" diary, specifically for use during this meeting. Some entries might only refer briefly to specific times and dates in the station operating log, or they may contain details of an issue that are not appropriate in the station log. If you will be required to turn over your station logs immediately at the end of operations, your debriefing diary will need to contain full details of all events and issues for discussion.

Such information might include:

- What was accomplished?
- Is anything still pending? Note unfinished items for follow-up.
- What worked well? Keep track of things that worked in your favor.
- What needed improvement?
- Ideas to solve known problems in the future.
- Key events.
- Conflicts and resolutions.

During the debriefing, organize the session into (a) what worked well, and (b) what could be improved for the next operation. Frame criticisms and judgment statements as constructively as possible by saying, "This method might have worked better if…" rather than, "This method was stupid." Also, avoid personal attacks and finger-pointing. In most cases, interpersonal issues are dealt with most effectively away from the group meeting.

Chapter 17
ARES PIO: The Right Stuff

More and more ARRL Sections are appointing ARES-specific Public Information Officers (PIOs) from within the ranks of their ARRL-appointed PIOs. These ARES PIOs are specialists in covering public relations when ARES units are deployed in an emergency or community service operation. While generalist PIOs may also do this work, the entire emergency field is becoming more complex and special training is not only advisable, but strongly recommended.

The window of time during which Amateur Radio provides the only means of communications in emergency situations is shrinking as other communications options are starting to "heal faster." In 2011, the fires in Texas and the tornadoes in Alabama, Mississippi, and Missouri showed that hams were most needed in the hour immediately before the events and for about two days afterward, at which point other systems started coming back online. It was during this initial time frame that the news, both local and national, was interested in the activities and special skills of hams, but there was no one to tell our story. "After-action" reports came in many days, and even weeks, later. Even pleas for timely information from ARRL HQ went unanswered. By then, the news had moved on to other things and hams wondered why they did not get recognition for their efforts. We must have people whose first — and only — job is to get and share information in a deployment, or accept the fact that we will be ignored.

The PIO must provide the correct information to the right people at the right time, so that those people can make the appropriate decisions. For hams to do justice to the goal of providing public information for Amateur Radio, they must wear only one hat. They cannot handle emergency traffic, direct operations, be effective ECs, and be PIOs at the same time.

DO YOU HAVE THE RIGHT STUFF TO BE AN ARES PIO?

To act effectively as a PIO with ARES in emergency settings, you need to:

1. Understand your job; you are representing a critical program of the ARRL. Even if you are not an official ARRL-appointed PIO, you need to know the expectations you are taking on. See the sidebar, "Job Description for an ARRL Public Information Officer (PIO)."
2. Take the ARRL's PR-101 course; see **www.arrl.org/pr-courses**
3. Communicate with your section Public Information Coordinator (PIC) and know what information you are looking for. You'll need to ask:
   - What are the hams doing?
   - How many are involved?
   - Why are they active?
   - Where are they located?
4. Get information from the Section Emergency Coordinator (SEC), District Emergency Coordinator (DEC), or Emergency Coordinators (ECs) involved. They have a job to do, but you have yours too. Your job is to get information from them by:
   - Monitoring the nets
   - Calling them on the phone
   - Finding them — make an in-person visit if possible
   - Setting up a way to keep updated on events — without current information you cannot fulfill your responsibilities
5. Have prepared releases with blanks to be filled in (see **Figure 17.1** for an example). Much of the basic information can be prepared ahead of time.
6. Have a list of media people, news outlets, and a way of accessing supporting media for your outreach. Remember that media is not just print anymore. If there are no reporters at hand or readily available, don't just sit there. Go get ham-specific pictures, videos, and audio recordings. Media folks will want these as "B-roll," or supplemental materials.
7. Maintain up-to-date lists of other important contacts, such as Amateur Radio leadership.

Your Appearance

Let's begin with one facet that should be obvious, but is too often overlooked. As a PIO, you are the face of Amateur Radio. You are the one who will be seen on TV, by politi-

Media Information Form

Date _____

The _____ Amateur Radio Club has been asked to assist with primary/auxiliary communications for this event. The group is coordinated by _____ (name and callsign of leader).

_____ is working with the _____ and following agency(ies):

_____
_____
_____

The group is providing communications links between:
_____
_____

Amateur Radio operators are stationed at the following locations to provide communications assistance:

_____
_____
_____

_____(#) of Amateur Radio operators are at the sites

_____(#) of additional Amateur Radio operators are on standby for additional communications needs

(Insert boilerplate paragraph about your local group)

For more information contact
_____ (name of acting PIO)
_____ (e-mail)
_____ (phone and cell numbers)

Figure 17.1 — A sample of an easy-to-use release form for the media. Such a template simplifies the PIO's task of including the most important information for public distribution.

JOB DESCRIPTION FOR AN ARRL PUBLIC INFORMATION OFFICER (PIO)

ARRL Public Information Officers (PIOs) are appointed by and report to the ARRL Section Public Information Coordinator (PIC) with the approval of the Section Manager (SM). ARRL PIOs are usually chosen from club publicity chairpersons, must be full ARRL members, and are recommended to have completed the ARRL's PR-101 course or its successor. Additional training for PIOs should be provided regularly on a sectional or regional basis by the PIC and/or other qualified people.

Good "grass roots" public relations activities involve regular and frequent publicizing of amateur activities through local news media plus community activities, school programs, presentations to service clubs and community organizations, exhibits and demonstrations, and other efforts which create a positive public image for Amateur Radio.

Recruitment of new hams and League members is an integral part of the job of every League appointee. Appointees should take advantage of every opportunity to recruit a new ham or member to foster growth of Field Organization programs, and our ability to serve the public.

Specific Duties of the Public Information Officer

1. Establishes and maintains a list of media contacts in the local area; strives to establish and maintain personal contacts with appropriate representatives of those media (e.g., editors, news directors, science reporters, etc.). Understands how stories should be submitted to media outlets and knows the rules for successful media submissions by media type.
2. Acts as a contact for the local media and assures that editors/reporters who need information about Amateur Radio know where to find it.
3. Keeps informed of activities by local hams and identifies and publicizes those that are newsworthy or carry human interest appeal. (This is usually done through news releases or suggestions for interviews or feature stories).
4. Attempts to deal with and minimize any negative publicity about Amateur Radio and to correct any negative stories incorrectly ascribed to Amateur Radio operators.
5. Generates advance publicity through the local media of scheduled activities of interest to the general public, including licensing classes, hamfests, club meetings, and Field Day operations.
6. Helps individual hams and radio clubs to develop and promote good ideas for community projects and special events to display Amateur Radio to the public in a positive light.
7. Keeps current and familiar with available materials including ARRL Public Service Announcements (PSAs), knows appropriate websites and social media to post viral media, brochures, and audiovisual materials; contacts local radio and TV stations to arrange airing of Amateur Radio PSAs; secures appropriate brochures and audiovisual materials for use in conjunction with planned activities.
8. Seeks to constantly improve their skills by attending regional training sessions, PR Forums and other training activities endorsed by section PICs.
9. Submits articles and photographs to ARRL for *QST* and online applications.

Working with Others

1. Works with Local Government Liaisons to establish personal contacts with local government officials where possible and explain to them, briefly and non-technically, about Amateur Radio and how it can help their communities.
2. Works with the section PIC to identify and publicize League-related stories of local news interest, including election and appointment of local hams to leadership positions, *QST* articles by local authors, or local achievements by amateurs.
3. Keeps the section PIC fully informed on their activities and places PIC on news release mailing list.
4. Assists the section PIC in recruiting hams for public speaking engagements and promotes interest among community and service organizations in finding out more about Amateur Radio.
5. Quickly informs the ARRL Media and PR Manager of any issue or significant event noted which may either enhance or damage the reputation of the ARRL.
6. Networks with other PIOs to facilitate coordination.

Work in Emergencies

Note: It is highly recommended that the PIO should *not* also be the Emergency Coordinator.

1. Is familiar with local emergency protocols and establishes relations with the lead governmental response agency in their home area and their requirements for the PIO's participation in a Joint Information Center in an emergency.
2. Helps local clubs and/or Section-designated ARES PIOs (PIOs who are specifically trained and tasked by the SM and/or PIC to work with ARES units) prepare emergency response media plans.
3. Provides aid to Section-appointed ARES PIOs or assume that role if none is available.
4. Works with the PIC, aids clubs and others involved in emergency response operations in the preparation and contents of PR kits containing vetted information about Amateur Radio in their local area that may be distributed in advance to local Emergency Coordinators for use in dealing with the media during emergencies. During emergencies, these kits should be made available to reporters at the scene, the JIC, or at a command post.
5. Maintains contact with the local Emergency Coordinator and/or District Emergency Coordinator and League officials (local stories often can quickly turn national).
6. Summarizes Amateur Radio activity in an ongoing situation, and follows up any significant emergency communications activities with accurate and prompt reporting to the media of the extent and nature of Amateur Radio involvement. The PIO should not speak or give information for our served agencies.
7. Takes free online FEMA courses in the National Incident Management Structure/ICS/Disaster PIO or seeks similar training from the state emergency management office.

cians, by media, and by representatives of our served agencies. You must dress the part! You may be an "amateur," but in this respect you are a pro, so look like a pro. Wear neat clothing that you would wear to an office. T-shirts with slogans, old jeans, and hunting vests are not appropriate. We've all seen too many hams on TV who looked disheveled or worse — don't be one of them! Pass this message along to the hams doing EmComm with served agencies, also: Appearance matters, even in the EOC or out in the field. For a helpful visual on how to dress and act, see **www.youtube.com/watch?v=Abman9_kzvE&feature=youtu.be** The hams in it look good, speak well, and get their main points across in a short time.

Your Most Important Network

News is now — not a day later, or even hours later. If we want the public to know what we do, gain their understanding and support, minimize antenna restrictions, protect our frequencies and promote Amateur Radio, we need to be in the news. To accomplish this, you will serve as the media's source for current information, and you can only get that from the SEC, DEC, or ECs involved. Sitting back and waiting for them to come to you will not work; you must be proactive in this. You are not "bothering" them with trivia, because this is important information. Getting the right information out saves lives and property, prevents rumors, and helps in recovery, just as much as our other work does. Hopefully, you already have a good working relationship with ARES leaders in your area and may even have participated with them in drills. Regardless, one way or another, you need current information from them or you will be completely ineffective as a PIO. Set up a way to get regular updates so you stay current on events.

Your Role Within the ICS and JIC

In addition to the regular duties and tasks of establishing media relationships, informing the public, and attracting new members, the ARES PIO may have the opportunity to become an integral part of the Incident Command System (ICS). For those unfamiliar with ICS protocol, following is a quick overview.

As a disaster or emergency begins to develop, typically the first responder to the event becomes the Incident Commander; that is, the "boss" of the situation. As the scope of the situation unfolds and more responders and agencies become involved, a Unified Command (UC) is activated. The UC comprises groups of trained and qualified individuals who work together to lead and discharge the effort. One component of the UC is the PIOs who, representing the various responders, agencies, and disciplines, together form a Joint Information Center (JIC). It is the duty of the JIC to establish a unified message and serve as the voice of the event, providing consistent information, dispelling rumors, and providing a central location for media to receive updates and ask questions.

As an ARES PIO you are likely to be invited to represent Amateur Radio within the JIC. As such, it will be your responsibility to obtain, manage, and disseminate information. Your job is to be the "expert" on Amateur Radio's involvement in the effort. Again, that means having the current information on details such as the number of ARES personnel involved and the locations of ARES stations. You may be assigned multiple other tasks within the JIC to assist the Lead PIO, which you would be expected to carry out simultaneously with your ARES PIO duties. Should the media inquire about ARES or Amateur Radio involvement, the Lead PIO will call on you to provide the relevant facts and figures, and you'll be expected to answer any related questions.

Prepared releases should relate directly to the reporting of events. A joint news release statement might include information like number of shelters designated or set up, the number of individuals being sheltered, services provided, such as beds, meals, medical care, and the number of amateurs supporting the shelters. The activities of other agencies, such the Salvation Army, the American Red Cross, and Baptist Disaster Relief and may use ham radio and should also be covered.

Keep in mind that the PIO supports the local activity and works as part of the local ARES leadership team. If there is disagreement, the PIO's opinion is trumped by the EC in charge.

There are also subjects that the PIO must never address. Under no circumstances should you ever discuss with the media the gravity of the situation, speculate as to its outcome, or provide any information about victims, including names and medical. This is the job of others, not the ARES PIO. Speak to the media only about Amateur Radio's role, the number of ARES personnel involved, and the kinds of communications supported by ARES.

Let's also dispel a dangerous myth: During an emergency situation, with risk to life and property, there is no such thing as "off the record." Anything you say directly or within earshot of the media, even in jest, can lead to disastrous results that could jeopardize the entire operation, cause your dismissal, and possibly risk exclusion of Amateur Radio from future incidents.

HOW TO BE THE BEST PIO YOU CAN BE

Remember that it is the job of the ARES radio operator to gather and provide the right information to the right people at the right time so they can make the right decisions. Your responsibility as ARES PIO is to inform the media that a group of trained and dedicated ARES operators are on the job and are involved with providing emergency communications.

Work in conjunction with the PIOs of your served agencies. Our Memoranda of Understanding (MoUs) are only as good as we make them. As ARES volunteers we pledge to work together and respect each other's organization and skill areas. We breathe life into that pledge when we support each other. Besides, a joint story line or release is much more effective in getting attention. Following are some other ideas for increasing your chances of publicizing your efforts.

Tips for the Effective PIO

1. Take images of hams at work, showing their call signs and appropriate logos.
   - Photograph them at their communications equipment
   - Photograph them operating communications equipment
   - Photograph them in the act of communicating
   - Photograph them working with served agencies
   - Have a shot list of pictures or videos that you currently have available, as well as what you want to acquire
   - Don't waste time taking general pictures of rubble or similar visuals — there will be enough of those taken by others, and they are not all that helpful in our cause.

2. Document Amateur Radio activities.
   - Get the who, what, when, where, and the why
   - As Amateur Radio operators pass information, get on their distribution list
   - Get audio recordings
   - Look for ways to highlight hams working with served agencies
   - Think new media for stills, video, and audio

3. Practice, Practice, Practice.

This last point is critical. As someone wise once said, "A failure to plan is a plan to fail," and on-the-job-training as an ARES PIO during a crisis is a rough way to learn. Establish your relationships and action plans and determine what works best in your area before your services are needed. Then test it all in drills and community service activities by:
   - Participating in planned exercises as a PIO
   - Using the exercises to refine your function as PIO
   - Updating contact lists, call signs, e-mail addresses, cell, home, and work phone numbers.
   - Using predesigned releases to publicize exercises
   - Using the exercise to educate the ham community of your duties as a PIO

Your Reward

In addition to the most important motivation for volunteering your skills — helping those affected by an emergency — there is another good reason to do so: You will also be helping the Amateur Radio Service itself. By volunteering as a PIO, you are rewarded by the knowledge that you are promoting public recognition of Amateur Radio activities, informing local/extended communities of beneficial Amateur Radio activities, supporting future recruitment, and advancing the protection of operating privileges.

LEARN MORE

There is a wealth of information available to you regarding the ARES PIO, including background documents, press releases, and statistics at **www.arrl.org/pr-tools-for-pics-and-pios**.

There are Basic and Advanced PIO classes offered both online and in the classroom by many state emergency management agencies, as well as by the FEMA Emergency Management Institute in Emmetsburg, Maryland.

Basic PIO

The Basic Public Information Officers course is aimed at the new or less-experienced PIO, including those individuals who have functioned in a role of secondary responsibility. Course topics include an overview of the job of the PIO, understanding the media, interview techniques, writing a news release, and conducting public awareness campaigns. This course is conducted by the emergency management agency each state. Contact your state's agency to find out when and where the course will be offered. See **www.fema.gov/about/contact/statedr.shtm** for a list of the training offices of state emergency management agencies.

Advanced PIO

The Advanced Public Information Officers course (E-388) covers the application of public information skills to a major emergency or disaster situation. This is accomplished with a series of lecture presentations and exercises over the four days of the course. It is recommended that students complete FEMA's Basic Public Information Officers course (G290) at the State level before they attend the Advanced Public Information Officer course, which is conducted at the Emergency Management Institute, Emmitsburg. See **www.training.fema.gov/emicourses/crsdetail.asp?cid=E388&ctype=R**.

ADDITIONAL RESOURCES

Also visit the following websites for further relevant information:

Recommended Additional Training
PR-101 from ARRL
   www.arrl.org/pr-courses
NIMS 700–800
   http://training.fema.gov/EMIWeb/is/is700a.asp
   http://training.fema.gov/EMIWeb/Is/is800b.asp
ICS 100, 200
   http://training.fema.gov/emiweb/is/is100b.asp
   http://training.fema.gov/emiweb/is/is200b.asp

Part 6: Public Service

Chapter 18
Organizing Amateur Radio Communications for a Public Service Event

Amateur Radio has a rich history of providing emergency communications during public service events as well as during natural disasters. Public service events such as bicycle races, road races, marathons, parades, college football games, the Special Olympics, and even dog sled races require communications volunteers and make a superb training exercise for handling communications in an actual disaster scenario.

There are many similarities between a disaster scene and a public service event. In both cases, many people show up at a given time and location needing food, water, shelter, and instructions. The main difference between a disaster and a public service event is that one event is planned and the other is not. Public service events have been characterized as "planned catastrophes" in that they both involve large numbers of people and a great deal of chaos.

The organization responsible for responding to an emergency situation or hosting a large public event will have a great many logistics to communicate and people to keep safe. The hosting organization, also referred to as the served agency, also usually requires a large number of volunteers who are trained in facilitating communications in an ad hoc, confused environment.

How do you know if a public service event is appropriate for Amateur Radio? Since there are no age restrictions on becoming an Amateur Radio operator, it is important to make sure that events are family friendly. Additionally, Part 97 of the FCC Rules and Regulations states that "Communications in which the station licensee or control operator has a pecuniary interest" are prohibited, so it is advisable to limit participation to non-profit events. Most importantly, make sure the event is safe for everyone involved.

PLANNING TO SERVE AN EVENT

The key to success is planning. This cannot be overemphasized! It is a great policy to draft a written understanding of what will be required from each participating organization, such as a community group, town government, or sponsoring group. A Memorandum of Understanding (MoU) should be written and signed between each organization. An MoU can be a very simple yet detailed document. It should serve as a clearly worded, well-defined blueprint of the entire event and should state the relationship and responsibilities of each party. This will prevent misunderstanding, confusion, and finger-pointing during and after the event.

The goal here is to establish clearly defined areas of responsibility. This should prevent anyone from assuming that someone else is responsible for a particular task. Pre-event planning should be a combined effort between agencies to ensure a smooth working relationship before, during, and after the event. Specific individuals should be tasked with the responsibility of serving as liaison between your organization and other agencies. Define formal communication channels at all levels between agencies so there is a clearly defined communication path during planning and execution, as well as post-event activities.

As the saying goes, the devil is in the details. Be sure to think through the requirements necessary for a particular task. For instance, if you are tasked with staging the participants in a parade so they are all in a published sequence, think about what the task might require. Before the people show up, it would be wise to have the staging parking lot organized. Signs numbering the positions will need to be posted, so people can put themselves in order. Of course, more tractors and scouts than you expect will inevitably show up, so flexibility, patience, communication, and cooperation are essential ingredients to your plan.

Often success comes down to finesse. The ability to identify a problem, devise a solution, and communicate the solution is more of an art than a skill. When the plan fails, then you resort to thinking on your feet. Patience, understanding, and empathy along with clear communication will take care of most situations. Of course, a great plan is much easier than solving a problem in the field; do not skimp on planning and preparations.

RESOURCE REQUIREMENTS

Once you know what your served agency expects from your organization, you can answer the following questions:
- How many stations are necessary for the event?
- Where should they be located?
- What capabilities are needed?

When you have a feel for the number, location, and capabilities of the stations necessary, then you can decide what type of station is necessary for each task. Types of stations include but are not limited to fixed, automobile, motorcycle, recreational vehicle, aeronautical, bicycle, pedestrian, and equestrian. Modes can include HF, FM, digital, and data with capabilities such as voice, data, CW, position reporting, and photographs; all should be considered.

When deciding on the type of station and mode to use, common sense is the rule. Keep it as simple as possible, and think the situation through. First decide if the station should be fixed or mobile. Then decide how much air time and power the station will need. Will battery power work, or do you need another power source? Consider all the factors at your disposal, and make choices from there. Your decisions will be tested during the site survey and testing phases of preparation. The event will serve as the final exam.

At each event, there is typically a location where the top level event management is located for both the served agency and the amateur volunteer leadership. Leadership from other agencies participating in the event will also likely be located here. Every event is different; this location may be called Command Post at one event, Incident Command at another, and Operations Center at another (see **Figure 18.1**). However, the principle is the same, regardless of what the location is called. For purposes of this chapter, this central location where event leaders are stationed is referred to as the Operations Center (OC). Depending on the size and capacity of the OC, it may or may not be appropriate to have Net Control Station (NCS) located in the same area. If there are a lot of folks stationed in the OC, then it might be better to have an NCS liaison to act as a runner between the OC and NCS.

For events that have a large operations area and a number of mobile vehicles such as bicycles, it's a nice touch to provide a map on a big screen with Automatic Position Reporting System (APRS) data of key vehicles continuously updated (**Figure 18.2**).

Site surveys should be conducted well in advance of the event. This is the time to determine what you will need to provide at each site. Do you need to provide tables, chairs, food, water, shelter, power? What type of antenna is necessary? Do you have cell phone coverage and Internet service?

Be sure that all operators, at all sites, have headphones that are noise resistant. These will enable operators to hear the radio in noisy environments. Additionally, with the use of headphones, people around the operator will not be both-

Figure 18.1 — Operations Center communications room for the 2010 MS Bike ride from Atlanta, Georgia to Athens, Georgia, and back to Atlanta. Amateur Radio Emergency Service members from around the region provided communications over the 100-mile course for this two-day event. ARES members also staffed 10 sag vans (equipped with D-STAR radios) and aid stations. Dr. Stan Edwards, WA4DYD, former Deputy ARES SEC, Georgia Section, pictured here, served as amateur communications coordinator for the event. Here, WA4DYD salutes to the success of the event.
[SCOTT HARTLAGE, KF4PWI]

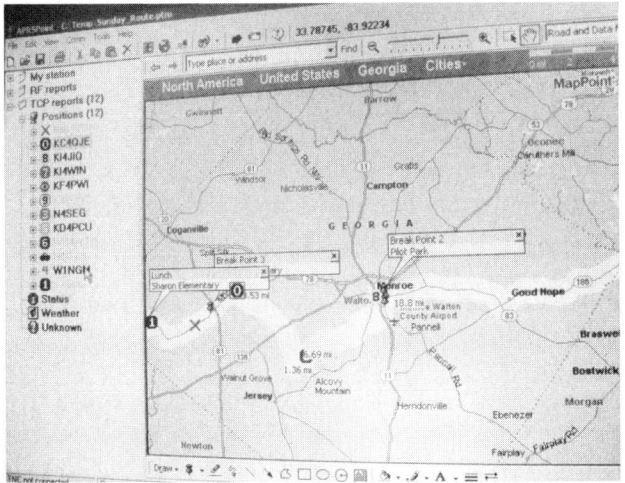

Figure 18.2 — Real-time event map for the 2010 MS Bike ride in Atlanta, Georgia, updated with APRS data. For this event, ARES deployed D-STAR in all locations, with the GPS position data being transmitted to this map in real time every time a station called in to the Operations Center. This system was highly effective, to the point where net control was even able to give a lost driver turn-by-turn, real-time directions back to the race course. [SCOTT HARTLAGE, KF4PWI]

ered by the sound of the radio traffic if they are not interested. Headphones also help promote the confidentiality of radio traffic.

Propagation testing should be conducted at each site. Test the path to other sites that you will need to communicate with during the event. Test each mode of operation you are planning to use with each site you will need a link with. You don't want to show up on the day of the event to find out that you are unable to communicate from point A to B, when everyone is depending on that link to work. Do not make assumptions; test diligently and find out in advance!

Determine what equipment is necessary to perform the required task at each site. During the site survey and propagation testing, it will become apparent if a handheld will do or if a mobile radio is necessary. The size, height, and placement of antennas will also need to be considered and tested. Does the function of a site require the transmission of a large volume of lists, names, and numbers that will require a digital mode? Is a printer or copier necessary?

Before the site survey, determine which of the equipment requirements are critical to the mission of that site. Then arrange for redundant pieces of hardware, which will perform the same functions, if necessary, and ensure that these backups are tested when the primary rigs are tested. Handheld transceivers and other battery-operated stations should have a backup battery available.

If you are planning on using repeaters for any of your nets during the event, by all means, secure permission to use that or any resource ahead of time. Again, an MoU should be in place with each organization that is contributing a resource, including the organization or owner allowing repeater use.

Even though there is an MoU covering the use of a repeater, for example, contention may still occur for the resource. The first step to avoid conflict is to publicize that the repeater will be used in this fashion during the event. Make announcements on the organization's nets leading up to the event to politely remind folks and take the opportunity to thank the organization for the use of the resource.

During the event, make periodic announcements that the repeater is in a net condition. Even though these precautions are used, there may still be folks who will pop up on the frequency and make a call like there is no net in progress. Please remember that you are a guest on a resource that they use on a regular basis. Have the net control station politely return their call and cordially inform them that there is a net in progress, and once again thank the sponsoring organization for the generous use of the resource. If the station is meeting someone on frequency, try to accommodate them and let them pass their traffic so they may move to another resource. The Golden Rule should be applied in these situations, as this is a perfect situation in which to demonstrate how to be a courteous operator.

Cell phones make a great contingency mode of communication. Text messaging is another effective method of communication when there is an issue to work around with radio frequency communications. Develop phone directories of all the landline and cell phone numbers available for use during the event by all levels in the chain of command, covering all agencies, to be used as a backup mode of communication should your primary methods falter. These directories should be available to event leadership, net control stations, and to volunteers, with instructions to use them if normal methods fail. This list should include both the Amateur Radio volunteers as well as the served agency personnel. While the master list should contain all the available numbers, sub-lists may be necessary for different participants, based upon their need to know.

Folks are depending on amateurs to provide communications streams as necessary. Make sure there is at least one level of redundancy for each communication segment. Two levels of redundancy are desirable for the critical segments in the design. Assume that there will be failures, and design the system to self-heal when failures are experienced.

EVENT LOGISTICS

It's a good idea to establish a band plan that lists all the frequencies that will be used during the event, as well as the purpose of each frequency. This should include, but not be limited to, HF, digital, repeaters, and simplex. Be sure to include contingency and backup frequencies and modes in case your primary plan runs into trouble.

Establish a simplex frequency for stations to use around local sites and for en route communications between vehicles. This is very efficient for coordination among members of a caravan and, upon arrival at a local site, for team members to coordinate parking and facility entry. This frequency can also be used for local site management communications, providing a channel for local resource and operational traffic without tying up the event-wide operations and resource nets.

You'll need to set up an Operations Net where calls will be made during the event to convey the information that the served agency requires as stated in the MoU. Then, set up a Resource Net to coordinate volunteers and other resources that are separate from the Operations Net. The Resource Net is a good place for talk-in and for volunteers to check in and out of the event. Coordination of logistics without tying up the Operations Net will allow the Operations Net to work efficiently.

How many nets are necessary? It is probable that more than one net is necessary. A parade, for instance, may need a net for the marshaling area that is separate from the net along the parade route. In this example, it may be that the marshaling area net can be on a simplex frequency and not require a repeater because the coverage area is small.

The Net Control Station (NCS) needs to be fully informed on all aspects of the event. They are the key to a smooth operation and they need to be staffed by highly competent operators who can remain cool under pressure and are able to think on their feet. This operator gets the big picture and sets the tone of the event by staying calm and lighthearted, no matter what kind of storm is raging in the field.

An Alternate Net Control Station (ANCS) is an operator that should constantly follow every step on the net in real time and be ready to take over from net control at any time. It is essential that the ANCS is using a completely independent equipment configuration from the NCS, and a separate site location is also desirable. This way, if anything happens at the NCS site or with their equipment, the ANCS will be completely independent and able to assume NCS duties without interruption to the operation. The ANCS will also be available to allow the NCS to take breaks.

The first duty of an ANCS, once promoted to NCS, is to establish their own ANCS in case anything happens to the newly promoted NCS in their capability to conduct the net. Part of the ANCS planning should be to have a station willing and capable of being their alternate lined up in a contingency fashion.

Use of tactical call signs reduces confusion. Call signs like "Operations Center" or "Aid Station One" will help everyone on the net keep up with the traffic. This way, people do not have the burden of trying to remember where individual call signs are stationed. This concept does not come naturally to most operators and should be consistently demonstrated and reinforced before the event.

Be sure to remind participants that FCC Part 97 rules for station identification still have to be observed while using tactical call signs. Part 97.119 states that each station "must transmit its assigned call sign on its transmitting channel at the end of each communication, and at least every 10 minutes during a communication." Since most traffic will be shorter than 10 minutes long, it is a good practice to use tactical call signs as necessary until you feel as though you are making your last transmission in your current exchange, at which point you should use your own call sign along with the tactical call sign to clear; for example: "Aid Station One, (your call sign), clear."

VOLUNTEERS

Volunteers make the event happen. The key to success is to start recruiting volunteers early. This will allow you to staff strategic positions with your best people. It will allow those vital people time to recruit their staff, get to know their volunteers, and train for specific duties.

An added benefit to recruiting early is that you gain commitments from volunteers. If the event is on their calendar, it will reduce the likelihood of them scheduling something else on the same date. Too often, people who are contacted late will say: "I'm sorry. I wish I had known; I would love to help, but I already promised someone that I would do something else." Be the first on your volunteer's calendar!

Always remember that volunteers are special people who donate their time and energy to worthwhile causes. Be sure to treat them with respect and appreciation at all times. When, how, and what you communicate will dictate whether they show up at your event or if "something more important" comes up. Unfortunately, there are usually a certain percent-

Figure 18.3 — Valuable volunteers come in all shapes and sizes. Here, 13-year-old Andrea Hartlage, KG4IUM, tells a military tanker where to park at the 2002 Dacula, Georgia Memorial Day Parade. [SCOTT HARTLAGE, KF4PWI]

age of volunteers that do not show up to the event, for whatever reason. To avoid coverage gaps during the event, recruit and train backup staff for key positions. If "extra" volunteers are available during the event, they can provide relief during breaks and can augment the more challenging assignments.

Local nets are great place to start promoting the event and recruiting volunteers. Start checking into every net you can before you need to start recruiting. Then, when it is time to start recruiting, you will be a regular on the net, will understand its ebb and flow, and will know when and how to start making regular announcements requesting volunteers. Be sure to write out your announcement ahead of time; be concise and make sure you cover all the points that encourage folks to step up and volunteer.

Join the various e-mail reflectors in your local amateur community in advance of an event. Help folks in posts when you are able to do so. "Build stock" on the reflector before you need to ask for volunteers.

Local Amateur Radio clubs are great sources of volunteers (see **Figure 18.3**). Join the local clubs in your area. Attend the meetings and build relationships with the members. Pitch in and help whenever you can. People help folks they know and respect, so become a member in good standing. When it is time for you to stand up in front of a club to invite participation in your upcoming event, you and your cause will be well received.

CREATING AN INFORMATION PACKAGE

Create a written information handout package with everything a volunteer needs to know well in advance of the event. Allow plenty of time to get the packet to volunteers in advance, allowing time for them to become familiar with the material. Be sure to include everything from an overview of the entire event to directions to specific sites. Include staff rosters of volunteers as well as pertinent members of the served agency. The more folks know, the better decisions they can make when the event is in operation.

If your event is a parade, road race, or bicycle race you will most likely need several different maps. Most of these are usually provided by your served agency for the promotion of the event and the orientation of the participants. Be sure to augment their maps or create your own with Amateur Radio-specific information like tactical call signs, frequencies, and staffing information for the various sites.

The handout package of information should also be available in a form that can be attached to an e-mail. This will serve as a convenient way to get the information to volunteers that you do not see on a regular basis. While a website is not a requirement for a well-planned event, it can be a great tool for recruiting and communicating with volunteers. It allows you to display all the information concerning your event, along with pictures from past events, which is a great way to showcase the benefits of being involved. If you do choose to create a website, keep in mind that the Internet is a public forum that can be viewed by anyone and everyone; be sure to use discretion when deciding what information to publish on the web. For example, cell phone numbers of event organizers might be better left to hard copies that are distributed to your volunteers on the morning of the event.

SAFETY

Safety is job one. Everyone should be trained to consider what impact each choice has upon safety. Every decision made in planning by the event organizers and each decision made in the field should first begin with a consideration of how that decision will impact the safety of the volunteers, participants, and spectators. During emergencies and public service events, people tend to get tunnel vision with regard to accomplishing the mission. Focusing on the mission is all well and good, but it can lead to decisions that overlook normal safety guidelines. It is tempting to take shortcuts to make something necessary happen. Avoid mistakes like running a coax, wire, or rope across the ground, creating a trip hazard; or erecting an antenna near a power line, endangering participants with electric shock. Make sure your volunteers know it is better to not do something than to attempt to provide a solution in an unsafe manner.

Ensure that NCS, ANCS, and management have access to the National Weather Service and preferably weather radar so they can stay apprised of current weather conditions. Having accurate data will allow them to make informed decisions when severe weather is threatening.

Establish communications methods and modes for acquiring emergency services, when necessary, during the event. Know the locations, contact information, and capabilities of each local aid station. ANCS and NCS should know the locations of and procedures for dispatching local ambulance, fire, rescue, and law enforcement resources, whether stationed at the event or external to the event; they should know which hospitals are nearby and the procedure for contacting them.

Your volunteers should arrive with a go-kit that provides the necessities needed for each participant to be self-reliant. Our mission as communication volunteers is to provide assistance to the served agency in conducting a safe and successful event. Often times we are relaying messages that connect participants who need assistance with a part of the organization that can provide the necessary support. One of the last things we want to do is to become a participant who needs help.

Each volunteer's go-kit should contain water and food with high calorie content in compact packaging. If the event organization provides tasty food for the volunteers — great! But if no food is provided or a volunteer's assignment does not allow convenient access to event food, then he or she should be self-sufficient and not add an extra burden to the event organizers.

Each volunteer should be aware of his or her individual health needs. They should arrive with the kind of medicines and over-the-counter health aids such as headache remedies, pain relievers, allergy symptom relievers, etc. A good go-kit will also contain the essentials for a first aid kit. A small kit that allows a volunteer to take care of his or her own potential first aid needs will suffice. The kit will serve a dual purpose of allowing a volunteer communicator to provide aid to a participant if the need arises. Do so only to your ability; your first responsibility is to call for help.

Most public service events are outdoors. They are often run in any kind of weather. It may be freezing in the morning and stifling hot in the afternoon. Sun, rain, snow, wind, or insects may be encountered; plan solutions for all of these challenges. A smart choice of seasonal clothing will consist of multiple layers that can be adjusted as necessary.

Often times at a public service event, the communication volunteer will need to interact with the participants and the public at large. Sometimes it is necessary to get to the front of a group of people in a line to take care of an issue. If you look like everyone else, then the public may not yield to your effort to get to the front of the mob to solve a problem. This is where a safety vest and uniform components like hats and ID badges come in handy. The public is conditioned to follow instructions of public servants like police and fire, who are identified by a uniform. Communication volunteers at a public service event would be well-served by adopting a unifying look that makes them easily distinguishable from the rest of the crowd.

Being more visible will also help volunteers keep safe from motorists and other hazards throughout the day. Another benefit of a unified look is that event participants will be able to identify the communicator when they need help. Consult the event coordinator on appropriate dress. Some events require all volunteers to be dressed the same.

DELEGATION

One of the hardest tasks for a leader to master is delegation. Leaders are often people who are used to taking action when something needs to be accomplished. To be an effective leader you have to inspire others to do what needs to be done. As a leader, it is your responsibility to focus on the big picture and to anticipate what is good for the entire organization and the event at hand. If you are focused on accomplishing a task that could have been delegated, then you are not focused on the big picture and will miss something that could have been prevented. Keep your eye on the big picture!

The National Incident Management System (NIMS) offers a great model for managing a public service event. Visit **www.fema.gov/emergency/nims/** for more information about this highly regarded management system. Not only is it great practice for emergency communications, but it is a very practical model to use in managing a successful Amateur Radio public service event.

Early in the planning stages, focus should be concentrated on installing leaders in key positions in your organization. These leaders will be able to contribute to planning, organization, and the recruiting and training of all of the volunteers necessary for the event.

Empower leaders to plan their area of participation. The leaders should be encouraged to help in recruiting the volunteers who will work on their team. Provide general guidelines for the organization as a whole, and let the team leader work from the general guidelines. This will empower your team leaders to contribute the bulk of the work and to worry about the specifics of planning and recruitment; it will also help prevent micro-management.

PROCEDURES FOR MANAGING EVENT COMMUNICATIONS

Make sure all net control stations and their assistants know they need to regularly announce that there is a controlled net being run in support of the event. Anyone requiring use of the frequency should call net control; emergency traffic of any nature will always be given first priority.

Radio discipline is the key to effective net operations; the frequency should only be used for passing traffic necessary to the event. Sometimes folks think that they need to key up on the frequency to fill the dead air and ragchew; they may need help in resisting this temptation through instruction and example. Likewise, it is important for net control to establish clear guidelines regarding what does and does not need to be reported to the net. Otherwise, well-intentioned volunteers may fall victim to over-reporting, calling net control about their every action, such as when they take a restroom break. In some cases, the net may need to know this, but in others, these exchanges only tie up the frequency unnecessarily. Establishing clear guidelines will ensure that everyone is on the same page. A quiet frequency is ready and available to pass traffic when needed by the served agency. A great net will be quiet the majority of the time; this indicates that the bandwidth is not overloaded and is an available resource.

Avoid words that require spelling; this will keep the traffic brief. When you do have to spell words for clarity, use the worldwide standard phonetic alphabet published by the International Telecommunication Union. A copy of the ITU alphabet can be found online at **www.arrl.org/fsd-220-handy-operating-aid**. Using this internationally recognized standard alphabet will help your message be received more accurately.

Volunteers should be trained in composing short, efficient messages that convey exactly what is necessary. Compose thoughtfully constructed messages that promote clear, concise communication with little misunderstanding and minimum use of net time.

Clearly communicate the chain of command; all of your volunteers should know to whom they report so everyone knows where to direct questions. Following the chain of command will keep the leaders at the top from being overwhelmed.

No matter how much planning is done beforehand, questions are bound to arise. Sometimes these questions will rear their heads in the form of conflict. Everyone needs to be trained to avoid conflict and to find ways to overcome differences and find solutions. If a conflict should arise, volunteers should be trained to pass the issue up the chain of command before there is a chance for ill will to begin forming. Guide conflicts off the net; they should be calmed down via a secondary mode or frequency.

The key word to remember here is *cooperation*. There is a fine line between taking initiative and providing the necessary assistance. Emphasize that volunteers should not try to dictate how to run the show. Volunteers may make suggestions, but the volunteer should adopt the attitude of accomplishing what is asked of them. A volunteer should never insist on his or her personally conceived notion of how to help. A great volunteer will do whatever is necessary from the leadership's point of view. The basic example of this mentality is, "How can I help?" Remember, at the end of the day, a communicator is there to serve a very specific purpose — to communicate. Going above and beyond that role is acceptable, but only when invited and only after the primary goal of communication has been accomplished.

CONCLUSION

Once the event starts, it is up to the leadership of the organization to station themselves where they can monitor as much of the activity as possible, keeping up with what is going on while staying out of the way. Constant supervision should be employed in recognizing emerging challenges and taking strategic steps to mitigate them before they become problems.

Try to circulate among the various sites and volunteers while remaining available to the organization. Keep an eye out for potential safety issues. Make sure volunteers are taking care of themselves; ensure that they stay hydrated, and watch them for signs of fatigue or stress. It's easy for a volunteer to get caught up in his or her duties and to neglect his or her own health and welfare. By staying reserved and accessible, the volunteer leadership is in the perfect position to observe the status of the volunteers, recognize negative trends, and implement corrective actions before troubles begin.

If you are very diligent, and even more fortunate, the leadership will be able to sit back and watch the event unfold without having to do a single thing. The well-prepared team will take care of everything!

That, however, is the "perfect world" scenario; in the real world, chaos will rule and everything unforeseen will become painfully apparent. This is a great reason to keep a notebook of lessons learned. You can use these lessons to make the next event go even more smoothly.

Follow up with the served agency and volunteer organizations that worked with you to see how well the amateur participation worked from their perspective. Also follow up with your ham volunteers to get their take on things. What lessons can be learned? What went well? What was missing? What can be done better next time?

Be sure that, as you participate in a public service event, to keep it light-hearted. Find humor in the difficult situations and help others keep a positive outlook. Find enjoyment in helping yourself and others in honing your skills in handling communications in a chaotic environment. If everyone walks away from an enjoyable and fulfilling experience, they will be willing to volunteer at the next event.

Chapter 19
Planning and Organizing Ham Communications for the Boston Marathon®

Public service events are some of the best training grounds for Amateur Radio operators who are likely to be operating during emergencies. The skills that are developed doing public service events are directly applicable to emergency communications. For this reason, public service events are sometimes referred to as "preplanned disasters." Though the events go smoothly — and Amateur Radio operators are often integral to that success — the skills needed to make an event successful are some of the same skills that are valuable in a disaster situation.

The Boston Marathon® is the oldest annual marathon, with the 116th running being completed in 2012. The field is limited to about 27,000 participants, with a requirement of having qualified in a previous marathon. The Boston Marathon has always run the same course. It is unique among marathons because it is run in a straight line, a factor that affects the communications required for the event.

WHAT COMMUNICATIONS WILL BE NEEDED?

One of the most important planning activities for a public service event is deciding exactly who your customers are and what kind of communications they will need during the event. Many people who organize public service events are unaware of the skills and capabilities of Amateur Radio operators. So, to some extent, participating in a public service event is a sales activity as well as a planning exercise. It's interesting that sometimes the amateurs who are helping during a public service event have had more experience with them than the people who are actually running them.

Using the Boston Marathon as an example, one would think that the Boston Athletic Association (BAA), which organizes the marathon, would be the primary customer for communications. But for the course part of the event, primary communications is really in support of the American Red Cross, which runs the first aid stations for the Boston Athletic Association. The key communications customer also will vary over time during the event. Most of the communications before the event starts and after the event has finished are of a logistical nature, making certain that equipment, personnel, and other elements crucial to the event are where they are supposed to be. During the event, the character of the communications can change significantly, becoming more centered on medical emergencies, managing transportation, or controlling the flow of pedestrian and vehicle traffic.

Part of preparing for a public service event is making sure that all the people who will be using the communications are aware of what Amateur Radio operators can do. We also need to understand what event staff and other volunteers will be doing, and how we can help with communications. We need to understand and plan how the communications will be conducted. Frequently some level of security may be needed, and a method other than Amateur Radio may be more appropriate. Cell phones often are used in this role, but there is the possibility that the cell phone network will become overloaded. After the September 11, 2001 attacks in New York City, it was found that Amateur Radio was probably the most reliable form of communications, but there were times when it did not work as well as a cell phone. So it's important for us as Amateur Radio operators to understand that the best means of communication may not be Amateur Radio, though normally it is.

RECRUITING

One of the best ways to recruit Amateur Radio operators to work for the event is to get a radio club to actively support and operate the event. It does not mean that only club members are allowed to participate; rather it means the club would supply the leadership and organization in the planning for the event. The club would be the primary communications support contact for the organization that is putting on the event. The other advantage of the club is they do have a list of members that you can try to recruit. There are other ways in which you can contact Amateur Radio operators who might be interested in participating. One of these is to attend club meetings and do a presentation on the event to further encourage members to participate. The use of Section-wide mailing lists or other mailing lists that allow you to use e-mail to recruit is very effective and takes less work.

Once volunteers are in place, communicate known information to them often. Giving information over time helps to keep the volunteers engaged and increases the likelihood they will read the information you supply.

It is important to communicate to potential volunteers about the necessary time commitment and required radio equipment for the event. These kinds of activities seem to require more time for setup and tear-down than one might expect, so make sure your time commitment estimate includes getting to the final locations, setting up equipment as needed, tearing down the equipment, and returning. When making assignments, it's important to consider the physical mobility of the operators. An operator who has mobility problems might not be the best choice for a "shadow" — a stationary assignment might suit him or her better. If possible, have a brief pre-meeting during which everyone can meet each other face-to-face and have their questions answered. Supplying coffee and doughnuts at these meetings encourages people to show up on time and gives everyone a chance to talk informally. The Boston Marathon hosts an after-event pasta dinner, which is very informal but gives the best opportunity to do after-action reports. It is also a good opportunity for the organizers to thank the amateurs for their support. Including these pre- and post-activities in your recruiting information may encourage more people to participate.

NUMBER OF HAMS NEEDED

You need to make sure that you know how many Amateur Radio operators you need. This can be a difficult parameter to come up with initially. If this is the first time the event has ever taken place, discussions with the event planners should give you some insight into what is needed as far as communications. For events like the Marathon, most of the time-critical communications are medical or logistical in nature. Some questions to ask are: *What kind of medical support staff will be there? What are the potential forms of medical emergencies? What simple medical activities would be involved?* You will also want to understand any logistics regarding the delivery or installation of equipment. Sometimes there are unusual requirements for communications or uses of communications the event organizers might not have considered. Beware of your client inadvertently turning hams into general volunteers with no real communications duties.

For the Boston Marathon, there are three Amateur Radio operators at each first aid station on the course. The rationale for this is the eventuality that a medical team may have to be sent up or down the course to check on a downed runner. Each one of these teams would require communications capability, so an Amateur Radio operator would go with them. And of course the first aid station itself still needs communications. Another reason for this setup is, it permits more experienced operators to train newer hams in becoming familiar with public service activities. One interesting phenomenon that has occurred with the dropping of the Morse code requirement, is that many Red Cross people have obtained Amateur Radio licenses, allowing them to be the third amateur at a first aid station. In this case their primary role is first aid, but if two teams are sent out on the course, the first aid station will still have communications if the Red Cross person is a licensed Amateur Radio operator.

FREQUENCIES AND REPEATER NEEDS

How many frequencies will you need? This question depends upon two important factors; the first is coverage area. The number of repeaters or simplex frequencies needed will depend upon the physical size of the event and whether the Amateur Radio operators will primarily be using HTs or mobile radios. The course of the Boston Marathon uses five separate repeaters because of the distances involved. Unlike most marathons, Boston is run in a straight line. This means that HT communications are needed over a 26-mile distance. Using the repeaters also allows the stations to communicate on alternate frequencies if the primary channel is tied up with priority traffic. If there are no repeaters in the area, simplex can be used. A prearranged relay station that can relay from one location to the other along the course is very important. Some pretesting of repeater coverage is very worthwhile and could prove to be useful in a real emergency as well. Make sure you get permission from the repeater owners to use them for the public service event. The last thing you want is to find out that two groups are using the same repeater for public service events on the same day. It's also important to tell participants that they will need to bring equipment that's better than a rubber duck antenna. Spare batteries are always required.

The second factor you will need to consider in order to determine how many frequencies you will need is the amount of information to be communicated. Go through the process of determining how much communications you will be doing on a given frequency to make sure you can handle the workload. Make sure you always have one or two standby frequencies to use if you suddenly realize that too much traffic must occur on a given frequency — and keep in mind that it's not always obvious when high traffic will occur. Sometimes the simple mental exercise of determining how long it will take you to communicate one piece of the communications can give you some real insight in how long it might take. Plan on not occupying more than 25% of the time doing communications.

TACTICAL CALL SIGNS

Another part of the planning is the assignment of tactical call signs. This is a very powerful concept that is used to aid in understanding where the communications are occurring. We are all very familiar with the tactical call of "Net Control," because that is normally the center of the activity. It's equally important to use descriptive tactical calls that identify the other points of communications as a first aid station, water station, or food station. Assign them in a sequential order so you can have a relative idea what their position is on the course of the event. Because people can change roles during the event, operating one kind of station at one point, and then moving to another, tactical call signs can help eliminate the confusion of hearing a familiar voice or amateur call coming from a new station at the event. It's also worthwhile to identify stations that are mobile, versus stations that are fixed. Often the organization that you're supporting already has some sort of designator they use for various areas. Use those as a basis for tactical call signs.

A tactical call is used to identify a location, not necessarily a single Amateur Radio operator. Again, "Net Control" is a good example. This also applies to other locations as well. For example the first aid stations on the Boston Marathon have three Amateur Radio operators present, but all of them use the same tactical call sign when at the first aid station. A different tactical call sign is assigned when they become part of an away team, but that call sign includes the call sign of the first aid station, with a suffix applied.

Predefine the appropriate identification procedure for the use of a tactical call sign. For example, the first aid stations on the Boston Marathon use the letter F as the type designator, followed by a two-digit number. So the appropriate identification F10 is predefined as "Fox one zero." It is important that people are consistent in using this method of identification and not to use "F ten." Because people on both ends of the communications are expecting two numbers preceded by a letter, there will not be confusion with numbers like nine versus nineteen. Water stations (or, more correctly today, fluid stations) use the letter W as the designator, followed by a two-digit number. The same approach for identification is used.

Do not forget that the tactical call does not eliminate the need to identify with your FCC-issued call sign. This is normally done at the end of a sequence of transmissions. If the communication will take longer than 10 minutes, giving the operator's FCC call will be required before 10 minutes have passed. It is only necessarily to give the FCC call of the transmitting station and not the station you are communicating with.

Tactical identifiers are also used for all frequencies, which makes it easy to ask a station to move to a different frequency. This also solves the problem of repeaters that are linked together. All of those repeaters have the same tactical call sign. The amateur then can decide which one is closest or which one he or she can actually access without having to communicate all that information.

PRIVACY CONCERNS AT EVENTS

One issue to consider at these events is participant privacy. The general public and news organizations are listening to our communications. Some reports, such as those regarding lost children and medical problems, cannot go out in the clear. If you work with medical personnel, the basics of the Health Insurance Portability and Accountability Act (HIPAA) will become familiar to you. Discretion must be used regarding what information does go out. There is certain publicly available data that can be used, such as an event registration number or the bib number of a runner. On the Boston Marathon, we use the bib number as an identifier for all the runners. Remember, when communicating via Amateur Radio, there can be no reasonable expectation of privacy.

ORGANIZATION AND DELEGATION

Managing a large event requires more organization than managing a smaller event. For an event like the Boston Marathon, one individual cannot manage all the activities that

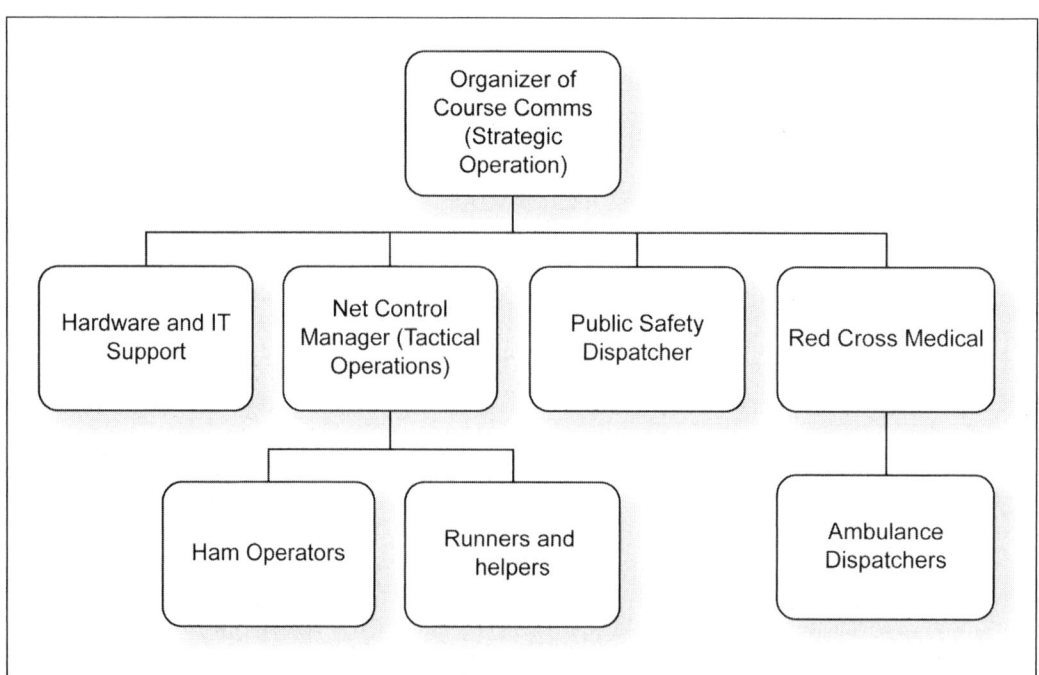

Figure 19.1 — The structure of net control for the Boston Marathon course.

need to be performed when as many as 250 Amateur Radio operators are involved. In a case like this, the work is divided into three major groups: the start of the race, the finish of the race, and the course. One person is responsible for organizing each of these parts of the event. These people have a group working under them, and each has their own net control. Each group further divides the Amateur Radio operators into various groups that have unique responsibilities. During the event, people may move from one of these groups to another depending upon needs and the phase of the operations.

The organization of the Boston Marathon course net control is a good example of applying delegation (see **Figure 19.1**). It is staffed with 20 Amateur Radio operators and representatives of several of the organizations for whom communications are supplied. One person is responsible for the tactical management of net control. This includes assigning operators to operating positions, as well as making sure they take breaks and are fed. Another person is responsible for managing the strategic operation of the net control. This includes communicating with the other net control positions at the start and finish, addressing issues involving requests from the course stations and communicating that information to the various agencies represented at that control. Another ham is responsible for maintaining and repairing any technical issues with the communications equipment. It is important to clearly articulate the roles and responsibilities of the people who are managing each one of these areas and to have clear lines of communications that include how to escalate issues appropriately so they can be handled in a timely manner.

NET CONTROL AND OPERATIONS CENTERS

Net control is usually the operations center or the center of primary communications for the event. During the event, one of its primary functions is managing ambulance requests for the first aid stations. Having the ambulance company dispatchers at net control helps to speed up the process and minimize duplicate calls for the same runner.

The concept of managing the frequency is a concept that most other organizations do not have. We often think that a dispatcher for the police and fire is like a net control operator, but the reality is, they are much more of an information supplier than a frequency manager. Frequencies are managed by the military, police, and fire more by rank than they are by the dispatchers.

For the course net control of the Boston Marathon, net control operators are arranged in a U shape with each of them each facing out (see **Figure 19.2**). This allows runners

Figure 19.2 — Net control for the Boston Marathon. Note the bulletin board at left. The cards are color-coded to represent different stations: blue cards are water stations, red are Red Cross stations, and yellow are command vehicles. [ROBERT PHINNEY, K5TEC]

to "run" on the outside of the U, passing papers or information between the various net control operators. There is also a bulletin board that all the net control operators can see, and to the left of each row is the designator for their frequency. Beside them are 3 x 5 cards with the tactical call of the stations that are on that frequency. This way everybody can determine which control operator to talk to in order to communicate with that location.

EQUIPMENT NEEDS

It's important to communicate with all of the amateur volunteers as to what equipment they will need for the event. Getting people the frequency information in advance allows them to pre-program their radios. This is especially important when dealing with repeaters, so hams can make sure they have programmed the appropriate offset and tone information. Two areas that often need emphasis are the kinds of antennas to be used, and the use of earphones. For example, if the ham is going to be in a bus, can a mag mount antenna be used outside the bus? Or are the buses fiberglass bodies, making a mag mount ineffective? In such a case, gaffer's tape can be used to hold down a magnetic plate that also acts as a ground plane for quarter-wave antennas. The stock antenna that comes with most HTs is not adequate for communicating any significant distance. A full-length whip antenna on the HT makes a big difference in its ability to communicate and, more importantly, the effect of the surrounding area on how well the antenna works with the radio. The use of earphones also helps reduce noise so the amateur can hear better and keeps all the people around them from having to listen in. Many speaker mics have an option of plugging an earphone into them, rather than broadcasting communications through the speaker.

Batteries can also be an issue. It is important that all amateur operators have an extra battery pack or set of batteries. One procedural thing used to compensate for the problem of batteries running down is to do a periodic roll call of all the stations on frequency, perhaps once an hour. One of the interesting phenomena regarding battery life is that a receiver will continue to work on lower batteries than a transmitter will. Here's a typical scenario: five hours into the event, the roll call will include several stations that do not respond right away. What has happened is, those stations' batteries ran down and they heard the call, but when they tried to transmit, the transmitter would no longer work. This is especially true in situations where the radios have to operate at full power all the time.

The use of a remote base mobile radio can have a significant impact in the use of the batteries on an HT as well as the ability to communicate. Many of the dual-band or multiband mobile radios have this capability. This allows a very low-power HT to communicate with the mobile unit, which retransmits its signal on a different band at a higher power and possibly in a better location as well. This is very important in a situation where simplex is being used over a significant distance because of the lack of availability of repeaters.

One interesting use of this technology was at the Boston Marathon downtown, two floors below ground level in an area where runners could get a massage. An Amateur Radio operator was stationed at that location, and there was also a vehicle with a remote base mobile in it on the street right near that building. The ham could communicate with very low power to the vehicle and the vehicle then retransmitted the signal, using the "cross band repeating" feature of many dual/multiband mobile radios. The professionals present are always surprised that hams are able to communicate when their radios won't work in that location.

It's also important to remind participating Amateur Radio operators of some of the personal items they will want to bring: a folding chair, sunscreen, hat, name tag, snacks, and water. In some cases, some of these things are supplied, but it's always a good idea for volunteers to bring their own. It is often necessary to have a government-issued ID, such as a driver's license, and hams should also bring a copy of their amateur license as well. Some of the more mundane things are often forgotten: pad of paper, pencils, pens, felt-tip marker, and tape. Of course there are some Amateur Radio-specific things like a spare radio, extra batteries as mentioned above, electrical tape, and diagonal cutters (the portable restrooms have plastic ties holding the door shut during transport, that need to be cut upon delivery to the site).

Dress appropriately. You have to earn respect quickly. Many other volunteers, professionals, and the public are observing you and your actions. Red Cross volunteers follow a dress code of long dark blue or black pants (no jeans; shorts only if it is very hot out) and a white shirt (without logos of any kind), making it very easy to look professional and responsible.

At net control it is desirable to have full-duplex communications with repeaters, meaning the ability of the net control operator to hear the repeater output when they are transmitting. This allows them to know when they are actually able to communicate with the repeater. This becomes very important when there is a stuck mic, which seems to happen at most events — at one point or another, someone accidentally sits on the push-to-talk button on his or her microphone. If the net control station can hear the output from the repeater, they can determine if they can override that signal and tell all the other stations on frequency to move to one of the backup frequencies so communications can continue. Some repeaters have a small delay between the received signal and the transmitted signal that can make communicating very difficult for the net control operator if they can hear the delayed signal in their ear at the same time. If this is the case, it is often possible for the feature to be defeated during the event. To support this full-duplex operation normally requires separate receive and transmit antennas that have some separation, or a hybrid coupler that has significant isolation between the two ports. Of course it does require a separate transmitter and receiver for each net control station. The receiver needs to have very good intermodulation characteristics. It is often as simple as having an HT on your belt with your body between the transmit antenna and the HT.

PRINTED FORMS

An advantage of a public service event over an emergency event is we can anticipate some of the major types of communications that will occur and make it easier for Amateur Radio operators to collect the required information by building a custom form. By giving the radio operators in the field these forms, hams can accurately collect all the information before beginning the communications. The form can be used to track the activity as it progresses.

For example, an ambulance request form is used at the Boston Marathon. The request is given an ID by the first aid station before communications has begun. This includes the first aid station's number and a sequential request number. A duplicate of that request is generated by net control, which is where the ambulance dispatchers are. The amateur at the first aid station fills in the initial part of the form and then calls net control to communicate that information. The net control operators fill out the initial part on their duplicate copy of the request, and request a runner to deliver it to the dispatchers. The ambulance company dispatchers then use a telephone or their own radio system to dispatch an ambulance to the first aid station. The amateur at the first aid station records the time of arrival of the ambulance and communicates that to net control. When the ambulance leaves, the amateur also communicates the final destination of the ambulance to the Boston Athletic Association, who will be able to inform family members of the medical status of their loved one. Having a form helps the information flow at net control, because it allows each person to have an actual piece of paper containing the information they need to communicate. After the event is over, the forms are handy for analyzing the activity load and time frames of activity during the event.

Other examples of useful forms are: transportation request, supply request, and National Traffic System (NTS) message forms. The NTS message form is a convenient way of recording almost all forms of communications involving third parties. Using NTS forms does not require you to use formal messages. The forms are just a convenient way of recording all the information that is normally needed: time of transmission, time of reception, who the message is to, and who it's from. That way there is a paper trail that allows the organization to whom the request was sent to have a record of it. It really helps to have people on both sides of the communications trained in passing formal traffic, even if formal messages are not going to be used. This is because the use of formal traffic pro-words like "I spell" is helpful even when sending tactical messages. Using the standard phonetic alphabet for spelling out words should also be required.

LOGGING

It is important for net control to keep a log of activity on a given frequency. This does not need to be a verbatim log, but it should include times and basic communicated information. Significant events such as ambulance requests and station openings and closings should definitely be recorded. This allows you to re-create the timeline of the event in fairly good detail. More importantly, during the event it allows you to see if all requests have been serviced, determine if it is taking too long to service requests, or monitor the activity level on a frequency. Sometimes it is advantageous to assign a person whose only task is to maintain the log for the net control operator. This is especially true if there is a great deal of activity on the frequency. This helps the net control operator keep focused on managing the frequency instead of worrying about the status of activities. All formal traffic should be logged both on a message form and in the control operator log.

Computer programs can be used for logging. Make sure you use one that is intended for public service events and not contesting. Contest log programs have a very different goal from public service event logging programs. Logging software for public service events includes the important features of being able to go back in time to check on events. They often have the ability to share logs between operators, and they help keep the list of stations on a particular frequency. One of the nicest things about them is they automatically record the time of log entry. One example of a good program is *Netlog*, developed by Eric Horwitz, KA1NCF. It is available at **http://netlog.ka1ncf.org**.

AFTER THE EVENT

It is very important to thank all of the Amateur Radio operators who participated in the event and to solicit both negative and positive feedback on how the event was executed. Remember, to err is human, so freely admit your mistakes and solicit help on matters of improving the planning and execution of the event. If a particular operator had difficulties, constructively communicate with him or her in private on how to improve. Often you do not have the full information about what went on in the field during the event. The same is true in reverse. Activities at net control can be very busy sometimes, which may result in a less-than-desirable response time to requests from the field. The only way for both groups — net control and the field — to fully appreciate that, is to communicate what was going on.

When supplying feedback, always try to say positive things in addition to giving constructive criticism. People will listen more attentively if there is positive information as well. It is also important to encourage bidirectional communications on the negative issues, because it is not always clear exactly what happened.

CONCLUSION

Organizing and running a public service event can be both difficult and fun. You get to meet some very interesting people who are sometimes extremely dedicated to the causes the public service event is supporting.

Remember to plan for the worst and hope for the best. Have a Plan A and a Plan B, but plan on executing some combination of the two.

Part 7: Digital Modes

Chapter 20
NBEMS Best Practices

Narrow Band Emergency Messaging System (NBEMS) is an integrated suite of programs that can be used for both Emergency Communications and recreational operations. The software is designed to run on nearly any modern computer and can easily interface with almost any radio. In fact, it's possible to operate NBEMS at times without any radio-to-computer interface at all.

NBEMS consists of the following components:

Fldigi (Fast Light Digital modem application) generates and decodes digital signals using either the sound card of your computer or an external sound card connected to your computer. *Fldigi* also can be used for rig control and logging. In addition to modes designed for emergency communications use, such as MT63, *Fldigi* can be used for popular recreational modes like PSK31 and RTTY.

Flwrap embeds a checksum in files that are transmitted using *Fldigi*. This allows you to verify with nearly 100% certainty that file was received without error.

Flmsg is used to compose messages in a variety of popular formats, such as ICS-213 and ARRL Radiogram. *Flmsg* can also be used to compose and send simple text messages. Using *Flmsg* greatly simplifies the workflow for sending and receiving messages.

Flarq (Fast Light Automatic Repeat reQuest) allows you to send a file to another station using ARQ handshaking for file transfer verification.

Flrig is a rig-control application that works with a variety of recent transceivers. The rig control in *Flrig* is much more powerful that what is provided in *Fldigi*.

NBEMS is in active development, with regular releases of new versions containing additional features and bug fixes. Consider what follows a high-level overview of NBEMS. For the latest information, please visit the official NBEMS website at **www.w1hkj.com**. Training presentations and help sheets may be found at the website of Western Pennsylvania ARES at **http://wpaares.org/html/nbems.html**.

ADVANTAGES OF NBEMS

NBEMS has been rapidly growing in popularity for a variety of reasons. Training operators in use of NBEMS is relatively easy. Much of the workflow is either drag-and-drop or requires just a few mouse clicks. For example, it's possible to send a message and an embedded checksum with just one mouse click. The user interface is much simpler than other digital modem and rig-control programs. NBEMS can be configured to automatically open and display incoming messages either in *Flmsg* or your web browser without operator intervention.

NBEMS excels on the existing analog FM repeater network. This means we don't need a dedicated digital network; we can instead leverage existing repeaters and turn them into a digital network as needed. This allows us to easily combine voice and digital operations on bands where permitted by FCC regulations. NBEMS also works very well on HF. Through the use of macro keys, it's easy to customize NBEMS to simplify local net operation practices, for example.

The most popular mode on VHF/UHF FM for emergency communications, MT63, works well without requiring a hardwired interface between radio and computer when using a technique known as acoustical coupling, which we will discuss later in this chapter. The most difficult aspect of NBEMS to configure is the rig-control feature, in which you use your computer to control many of the functions of your radio's front panel, such as changing frequency and mode. Rig control is not necessary for emergency communications, and use of acoustical coupling completely eliminates this problem. It also means you don't need to navigate a complex series of wires and cables to operate the radio or risk forgetting a cable.

All components of NBEMS are released under the Gnu Public License (GPL). It is unencumbered by licenses or patents, so you can easily and freely redistribute it. It is possible to carry around a copy of NBEMS on a USB thumbdrive so that it can be installed on computers as needed during an incident. Hams are able to modify or extend NBEMS because it is open source, with the code available to all.

NBEMS runs on *Windows*, Linux, and Mac OS X operating systems, as well as nearly any radio on either VHF/UHF or HF. You also do not need expensive proprietary external hardware. NBEMS is technology that is inclusive of the amateur community.

Because NBEMS contains recreational modes like PSK31 and RTTY and features like contest logging and callbook lookup, you can use NBEMS between disasters, deployments, and drills. When we are called upon to serve the public, we know we will be ready because we will be using the same tools we use every day for our recreational ham activities, and we will know our equipment is in working order.

NBEMS CONCEPTS AND FUNCTIONAL OVERVIEW

Fldigi uses the either the sound card of your computer or an external sound card connected to your computer's USB port to generate digital audio signals that are output from your computer's speakers or headphone jack and transmitter. Incoming signals are routed to your computer's sound card by a microphone connected to the computer (either built in or connected to the mic input) or via the line-in input and then decoded and displayed by the program. In the case of using an external sound card, data is input and output through a USB port.

Some newer radios like the ICOM IC-7200TM and Kenwood TS-590STM contain a built-in USB port that can be connected directly to your computer. *Fldigi* can use this port for input and output of both audio signals and rig-control commands without the need of an additional interface.

If you're using acoustical coupling to interface your radio to your computer, you do not need an external interface between the computer and radio. You also don't need an external Terminal Node Controller (TNC) as you do with the various forms of packet radio.

The *Flmsg* program is tightly integrated with *Fldigi* and the rest of NBEMS. It uses a common set of folders to interface with *Fldigi* to ease a user's workflow. We'll be discussing *Flmsg* later on.

OBTAINING AND INSTALLING NBEMS

NBEMS can be downloaded from **www.w1hkj.com**. Each component (*Fldigi*, *Flwrap*, etc.) is available as a separate download. You can also download the C++ source code, help files, and files that define radio-to-computer interfaces for the RigCAT™ rig-control library. Detailed and up-to-date instructions for installing NBEMS may be found at **www.w1hkj.com** and **http://wpaares.org/html/nbems.html**.

The first time you execute *Fldigi* on any platform, you will enter a configuration wizard. You will be prompted to enter information such as your name and call sign, asked to configure the interface to your sound card, and, finally, configure whatever rig control (if any) you may be using.

It is very important that you select the correct devices on the wizard page for configuring your audio devices. If you're using your internal sound card, just select your default sound card device. You may also need to go into your computer's control panel for audio devices to select whether to send audio output to headphones or speakers, or to input audio from your computer's internal mic or line-in jack.

If you're using a USB sound card device or a rig with a USB sound card interface, you will want to select "USB audio CODEC" as your sound device. As we will see, this is the simplest way to interface your computer to your rig.

THE *FLDIGI* SCREEN

Figure 20.1 is a screenshot of the *Fldigi* screen, which consists of the following parts (described from the top down):

The top part of the screen contains menus, rig and frequency controls, and fields for logging contacts.

The large empty box just below the menus is the RECEIVE pane, where received text appears. On your computer, this pane will appear yellow.

The large empty box below that is the TRANSMIT pane, where you can type data to be transmitted. On your computer, this pane will appear blue.

Below the TRANSMIT pane are the macro buttons. Macros are shortcuts used to minimize the typing of common activities such as changing modes, and starting and stopping transmitting. Macro keys are easily customized for contest exchanges, sending your station info, or easing the workflow of a net control station.

Below the macro keys is possibly the most interesting part of the display — the waterfall. The waterfall shows signals within your audio passband by displaying signal strength, frequency, and time. Portions of your passband with no signal are displayed in black. Weak signals are blue, medium strength signals are yellow, and strong signals are red. You'll want to adjust the audio level coming from your rig to your computer so that most signals are blue with just a bit of yellow. *Fldigi* works better with lower audio levels; too much yellow and any red is an indication of too high an input audio level. The portion of the waterfall that contains the signal you want to copy is shown by the red cursor you see in the screenshot. Just use your mouse to move this red cursor over the signal you wish to decode.

At the bottom of the screen are additional controls and status information. For details on this area, please read the official *Fldigi* documentation at **www.w1hkj.com**. One important item is the audio frequency of the signal you are trying to decode. In our screenshot, this is 1000 Hz. The

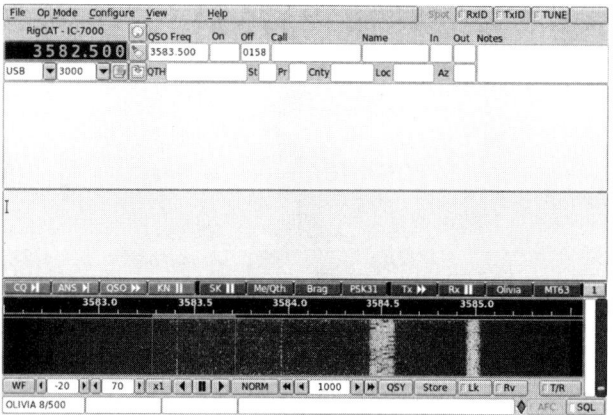

Figure 20.1 — The basic *Fldigi* screen.

arrows to either side of this may be used to fine-tune the selection of the signal to receive. The diamond at the bottom is an indication of your overall audio level. Black indicates no signal, green indicates correct audio level, yellow indicates a bit too strong, and red means excessive level. The vertical bar on the right is a signal confidence bar. To the right of this bar is a squelch slider control. The level of green in the confidence bar will increase as *Fldigi* is better able to decode the incoming signal. The squelch allows you to control at what level of confidence you will begin decoding text. To start, move this squelch as low as possible without receiving "garbage" characters.

To transmit, first select a mode using the mode menu. PSK31 is a good mode to start out with because it is the most popular of the sound card modes. Next, select an empty place in the band pass to transmit; just click on an area that does not contain any signals. Place your cursor inside the TRANSMIT pane and type your message. You're now ready to transmit. Press the TX macro key. To quit transmitting, either press the RX macro key or hit the ESCAPE key on your keyboard.

When transmitting, *Fldigi* transmits up to the position of your cursor in the TRANSMIT pane and no further. A common beginner error is to click somewhere inside the TRANSMIT pane while transmitting. *Fldigi* will transmit text up to your cursor and not finish the rest of your transmission. Always be sure to have your cursor at the end of the text you want to transmit.

ACOUSTICAL COUPLING

If you follow any of the NBEMS mailing lists, it seems like one of the most common problems is interfacing a radio to a computer. Even experienced operators have problems with hard-wired interfaces from time to time. Then, there are all the darn cables you've got to connect with some interfaces. Acoustical coupling between your radio and computer completely eliminates this problem. To use acoustical coupling, all you need to do is the following:

To transmit, hold your radio's mic up to your computer's speakers, and hold down your radio's PTT button as you transmit your NBEMS audio.

To receive, place your radio's speaker either up to your computer's internal mic or to a mic connected to the computer's line-in plug.

That's it! Many beginning NBEMS operators scoff at acoustical coupling until they give it a try. In a world where we equate complexity with technical sophistication, it seems too simple to work. Once people realize it works and is reliable, they become proponents of the method. It is actually a preferred interface method for some operators when in the field.

The M63-2000 mode on VHF/UHF FM, which will be discussed in detail later, works extremely well with acoustical coupling. MT63 has the advantage of not requiring very precise setting of audio level, and the large amount of redundancy built into the MT63 signal covers any loss of data. Background noise does not seem to be an issue.

Despite the advantages of acoustical coupling, a hard-wired interface is recommended under the following circumstances:

- On HF, where signal-to-noise ratios are much lower than VHF/UHF FM.
- Whenever you're using a mode other than MT63.
- In an EOC or other environment like a permanently installed station where noise from a speaker would be unwelcome.
- If using *Flarq* for one-to-one communications. This is because *Flarq*'s handshaking requires precise timing between the two stations that are connected.

HARD-WIRED INTERFACING TO RADIO

There are three common ways to interface your computer to a radio. Each method has its own strengths and weaknesses.

Direct Connect Interfacing

Figure 20.2 shows how to connect a radio using a RIGblaster Nomic™ interface. The Nomic is very inexpensive, with a street price around $60. It connects to your radio using the radio's mic connector for PTT and transmit audio

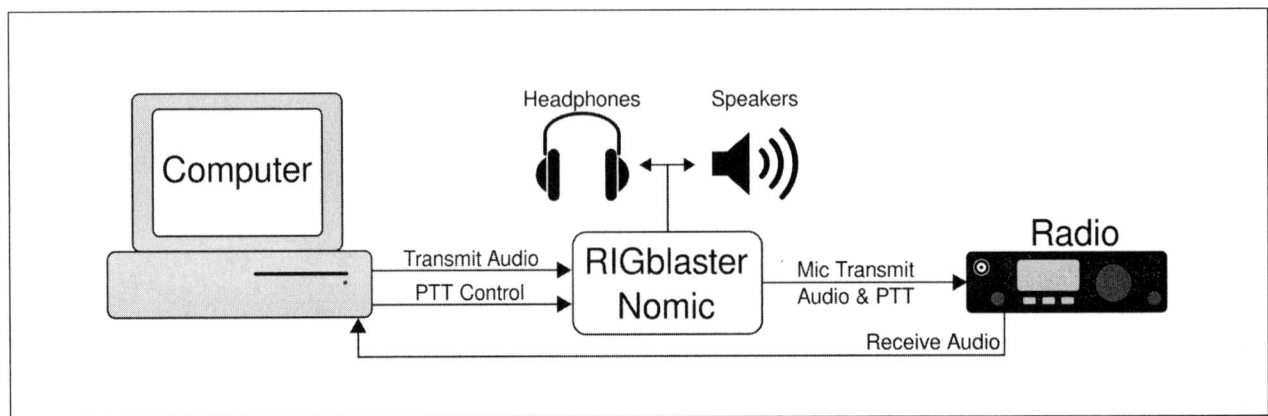

Figure 20.2 — Connecting a radio to a computer using a RIGblaster Nomic™ interface.

and uses the radio's headphone or speaker jack for input audio, although an aux connector on many radios may be used for input audio. The Nomic comes with a set of internal jumper wires along with instructions for configuring the jumpers for nearly any amateur transceiver.

The Nomic is powered by the RS-232 cable that goes between your computer and the Nomic, so no "wall wart" is required for power. Many newer computers are dropping the RS-232 interface for USB ports. The Nomic comes with a USB-to-RS-232 adapter cable for these newer computers. However, *Windows* does not automatically recognize the USB device driver. The driver must be downloaded and installed from the Internet. This is a bit of a pain, and if you plan on using the Nomic in the field and installing it on a different computer, you must either download the driver, or better yet, carry the driver on a USB thumbdrive along with the rest of NBEMS.

Many of these direct interfaces use your radio's speaker jack and mic connector to interface with your computer. This does work, but unless you connect an external speaker to the interface or add a device to allow you to switch mics, you cannot hear incoming signals and you cannot use your mic. Use the aux port on the back of your radio for computer interfacing, if your radio has one. However, for older radios, using the speaker jack and mic connector may be the only way to interface with your computer.

One RIGblaster™ that has been used with success is the RIGblaster Plug and Play. It plugs into your radio's aux and CAT ports so you can easily hear incoming signals, and it's powered by your computer's USB port. However, at around $100, it's more expensive than the Nomic, you must download the USB device driver, and, as with any device involving full rig control, configuring rig control for use with *Fldigi* can be tricky and frustrating.

If you're comfortable with a soldering iron and love homebrewing, Howard "Skip" Teller, KH6TY (a co-author of NBEMS), designed an excellent interface that was described in the March 2011 issue of *QST* (see the article at the end of this chapter).

USB Sound Card Interface

The next method for interfacing your radio to a computer is through an external sound card device such as the SignaLink USB™. Use of the sound card in the SignaLink frees up your computer's sound card for other applications. It's even possible to use SignaLink to control transceivers while using your computer's built-in sound cards for voice communications during webinars. The SignaLink also controls your radio's PTT through a VOX circuit built into the device. Whenever the SignaLink detects an audio signal from your computer, its VOX will trigger automatically and key your transmitter. This means that you do not need to configure *Fldigi* for any rig or PTT control. Only two cables are required. One is a generic USB cable. The other is a cable that goes from the SignaLink USB to your radio. These cables may be bought commercially for around $10–15, or may be easily constructed. In addition to being easy to interface to your computer, the SignaLink contains potentiometers on the front panel that allow you to control incoming and outgoing audio levels. Yes, you can control levels using the audio controls on your computer, but with some operating systems, particularly *Windows*, these controls are hidden away in the control panel and hard to find, especially when you're trying to send priority traffic. To configure *Fldigi* to use the SignaLink USB for audio, just change the CAPTURE and PLAYBACK drop-downs to use the USB Audio device. Unlike other methods, you do not need to configure any other rig control.

A common setup issue with the SignaLink is not having the VOX trigger to control your radio's PTT. The problem is usually insufficient audio level from your computer to trigger the VOX in the SignaLink. The solution is to open your computer's audio controls for the USB Audio device and increase the level of the output until the VOX triggers. Note that this does not affect the level of audio going to your radio; this is controlled by the output potentiometer on the front of the SignaLink.

Use of a USB sound card interface is a good method for interfacing your computer to a radio. It's easy to configure, only two cables are required, and it's convenient to have the volume controls on the front of the device.

USB Connection to Radio with USB Port

The final method of connecting your computer to a radio is to use a USB cable with a transceiver like an ICOM IC-7200 or Kenwood TS-590S that contains a USB port. This is illustrated in **Figure 20.3**. This has the obvious advantage of requiring only one cable — a generic USB cable. You may also configure full rig control through the USB cable.

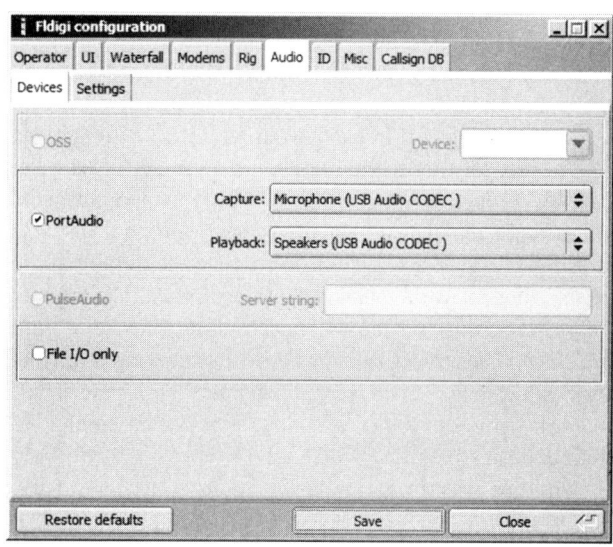

Figure 20.3 — Connecting a radio using a USB cable with a transceiver that has a USB port. The ICOM IC-7200 and Kenwood TS-590S are two such transceivers.

Although the cabling requirements are minimal, you still need to configure *Fldigi*'s rig control and change appropriate radio settings for rig control to take effect. You must dig through your radio's menus to tell it to transmit the audio that is provided by the USB device and not from the radio's mic. This can be frustrating and require some trial and error. Audio interfacing is the same as with the SignaLink USB; the radio appears to the computer as a USB audio device. However, the IC-7200 does not have the ability to control the level of output audio going to your computer, and there doesn't seem to be a way to control this level in *Windows* XP and OS X. Other operating systems do not have this issue.

In general, rig control for emergency communications purposes is difficult to configure and troubleshoot. You do not want to be trying to debug rig control during a disaster, so forgo it in those cases. However, you may find that you enjoy recreational operations using rig control.

Many laptops designed for business use come with software to "improve" the quality of audio from your computer's mic when using VoIP programs like *Skype*. This software is optimized for the human voice, not the audio signals generated by *Fldigi*. It is very important that you disable and/or remove these applications before using NBEMS, as they can lead to garbled text.

IMPORTANCE OF SOUND CARD CALIBRATION

Fldigi requires somewhat precise timing in your sound card to encode and decode data. Many sound cards, particularly ones in SignaLink devices, are dead-right-on, but some require adjustments. Price is no guarantee of having an accurate sound card. There are $200 netbooks that are highly accurate, and $1500 business-class machines that will not decode properly without calibration.

There are three common symptoms of out-of-calibration sound cards:
1. Garbled text when using MT63
2. Upper-case characters when expecting lower case in MT63
3. "Walking" off frequency during PSK-31 contacts

The easiest way to calibrate your computer is with the *CheckSR* program, which is available for download from the Internet. To use this utility, first enter *Fldigi* and use the HELP/AUDIO DEVICE INFO menu to determine the default audio sample rate for your sound card. In nearly every computer we've seen, this will be 44,100 Hz. Plug this value into *CheckSR*. Then run *CheckSR* for a few minutes until the error values begin to converge. Just enter these values into *Fldigi* by going to the CONFIGURE/SOUNDCARD menu and selecting the SETTINGS tab.

Another way to calibrate your sound card is to use *Fldigi*'s WWV modem. This is very accurate, but requires you to connect your computer to an HF radio, which may not be available if you are installing *Fldigi* in the field. Instructions for this method of calibration may be found at www.w1hkj.com.

COMMONLY USED EMCOMM MODES

Fldigi allows you to operate with a bewildering array of modes, but it's important for emergency communications groups to settle on a small number of modes to improve interoperability and simplify training. For example, Western Pennsylvania ARES uses the following:

- MT63-2000 on VHF/UHF FM
- Olivia 8/500 or 16/500 for keyboard chat and check-ins on HF nets
- MT63-1000 or MT63-500 for HF bulletins
- PSK-500R or PSK-250R for transferring very large messages

Each of these modes employs a method known as Forward Error Correction (FEC) to improve the odds of receiving a message without error. FEC involves sending redundant data so that the station receiving the message can attempt to reconstruct any parts that may have been lost or garbled. This is one reason not to use the popular PSK31 for emergency communications. As you may have noticed, the slightest fade or static crash causes loss of data with PSK31. The modes listed above have the ability to recover from this.

MT63

Western Pennsylvania ARES uses three versions of MT63: MT63-2000, which is 2000 Hz wide; MT63-1000, which is 1000 Hz wide; and MT63-500, which is 500 Hz wide. MT63 consists of 64 tones within the designated bandwidth. Many of these tones contain redundant data in both frequency and in time. This makes MT63 very robust. You can lose up to 25% of the signal and still have 100% copy. There have been instances where MT63-2000 was transmitted through a repeater that had a notch filter in the audio for control signals, and the MT63-2000 data was copied 100%. There have also been instances where voice signals were too noisy to copy but data sent using MT63-2000 was copied 100%. **Figure 20.4** shows an MT63-2000 signal in the waterfall, and **Figure 20.5** shows Western Pennsylvania ARES's preferred *Fldigi* configuration for MT63.

The group has found that MT63-1000 and MT63-500 are ideal for transmitting bulletins on HF and are resistant to static crashes. When using MT63 through a repeater, it is very important to obtain the permission of the repeater trustee ahead of time and to hold some training nets so that repeater users know how to recognize an MT63 signal. To those unfamiliar with it, an MT63-2000 signal sounds like a repeater malfunction. You also want to be sure ahead of time that there are no issues with the repeater's audio path.

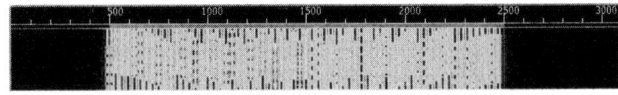

Figure 20.4 — An MT63-2000 signal in the waterfall.

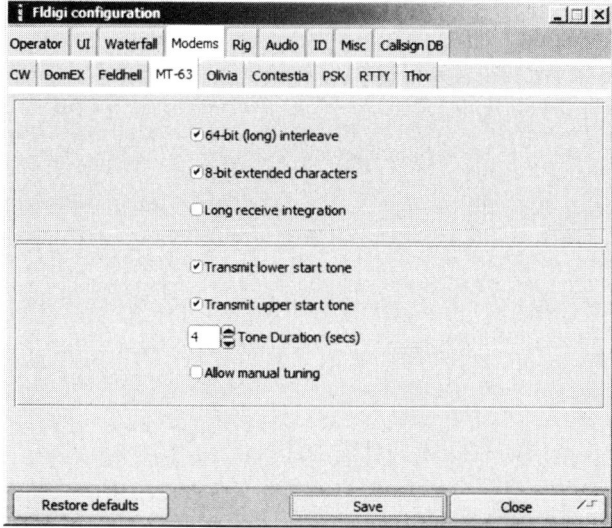

Figure 20.5 — Western Pennsylvania ARES's preferred *Fldigi* configuration for MT63. The setting for 64-bit long interleave is checked because that setting increases the amount of error correction. They leave the ALLOW MANUAL TUNING setting unchecked for VHF/UHF operations. By default, the lower edge of an MT63 signal will be placed at 500 Hz in your waterfall. By unchecking ALLOW MANUAL TUNING, there is little likelihood of not reading the entire MT63 signal using a radio with fixed channels as with a typical VHF/UHF FM transceiver.

Olivia

Olivia is Western Pennsylvania ARES's preferred mode for keyboard-to-keyboard chat and net operations on HF. Olivia signals are identified by two parameters; the first is the number of tones in the signal, and second is the bandwidth of the signal. For example, an 8/500 Olivia signal consists of eight signals in a 500 Hz wide bandwidth. **Figure 20.6** shows an 8/500 signal in the waterfall. They will typically start HF operations on 8/500 Olivia and then change to either 16/500 or even 8/250 if conditions deteriorate.

Olivia is very noise-resistant and has no problem dealing with the static crashes often heard on 80 meters during the summer. It also has excellent weak signal characteristics (it's possible to have 100% copy of an Olivia signal that cannot be seen in the waterfall), and it's easy to tune. You do not have to be exactly on frequency to receive a signal with 100% accuracy.

The main drawback to Olivia is its speed. It is a very slow mode, and it's twice as slow if you double the number of tones within the same bandwidth, and twice as slow again if you cut the bandwidth in half. But Olivia works better than just about any other mode when band conditions are poor. If you can't get through on Olivia, consider the path to the other station to be dead!

More info about Olivia may be found in "Ghost QSOs — Olivia Returns From the Noise," a December 2008 *QST* article by Gary L. Robinson, WB8ROL.

USE OF *FLWRAP* FOR CHECKSUM VERIFICATION

We know from experience that digital modes that use FEC are very reliable, and we can usually determine if there was 100% copy by inspecting the text in the *Fldigi* RECEIVE pane. However, there are times when we must be 100% certain that a message was received without error. This might be data that contains lots of numbers, lists of required medications, spreadsheets, or any transmission where our served agencies demand 100% accuracy. There are times when we need to be able to look our EMA director in the eye and say, "Yes sir, this message was received with 100% accuracy." This problem of guaranteeing 100% accuracy is solved by the *Flwrap* program.

Configure *Fldigi* and Install *Flwrap*

1. Configure *Fldigi* to capture incoming files that have been processed by *Flwrap* (see **Figure 20.7**).
   CONFIGURE/MISC/NBEMS
   Check the top checkbox under NBEMS DATA FILE INTERFACE. You may optionally check the OPEN MESSAGE FOLDER box. This will cause the folder that contains the incoming messages to open automatically. This is recommended for beginning users.
2. Install *Flwrap*:
   Download *Flwrap* from **www.w1hkj.com**.
   Double click on the installer and go through the install process.
   When finished, you will have an *Flwrap* icon on your desktop.

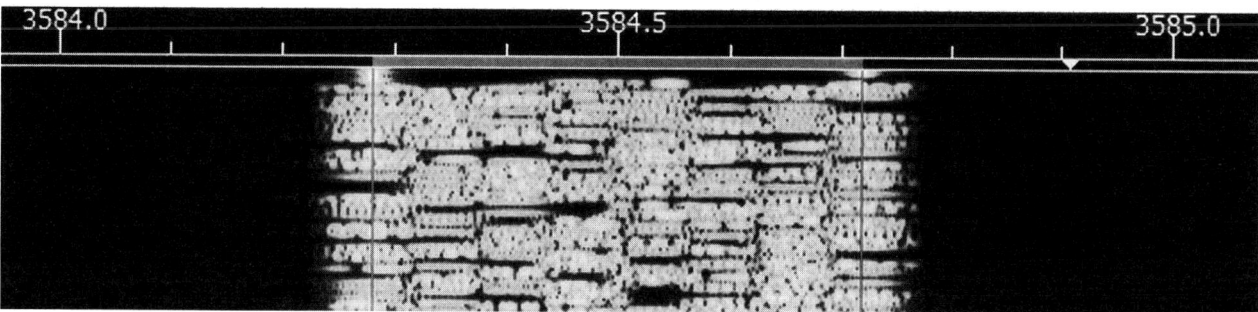

Figure 20.6 — An eight-tone, 500 Hz Olivia signal.

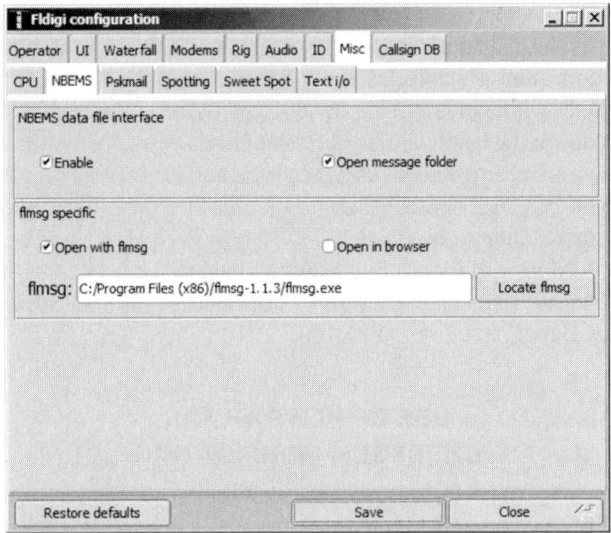

Figure 20.7 — Configuring *Fldigi* to use *Flwrap*.

Sending a File Using *Fldigi* and *Flwrap*

Figure 20.8 illustrates the steps for sending and receiving a CSV file using *Fldigi* and *Flwrap*.

1. Create your file using either a text editor or by exporting a spreadsheet in CSV format.
2. Drag and release this file over the *Flwrap* icon on your desktop. A file with a *Flwrap* checksum embedded in it will be created in the same directory as your original file. You will see a pop-up window indicating that the file was wrapped.
3. Drag this "wrapped" file into your *Fldigi* transmit window pane. By default, this is the lower large pane, which is blue.
4. Hit the TX macro key, followed by the RX key.

Receiving a File Using *Fldigi* and *Flwrap*

1. Watch the incoming data for the [WRAP:end] designator. You'll also see an indication at the bottom of your *Fldigi* screen that a file is being extracted.
2. Go to the folder that contains incoming wrap files: In *Fldigi*:
   FILE/FOLDERS/NBEMS FILES...
   Double-click on the WRAP folder.
   Double-click on the RECV folder.
3. *Fldigi* saves extracted wrap files using the naming convention "extract-yyyymmdd-hhmmss."
   Find the most recent extract file.
   Drag and drop it over the *Flwrap* icon you created previously.

You will see a message indicating success or failure of checksum verification (see **Figure 20.9**). If the checksum was verified, the original file that was transmitted will appear in the RECV folder. You will also be notified if there is a checksum failure.

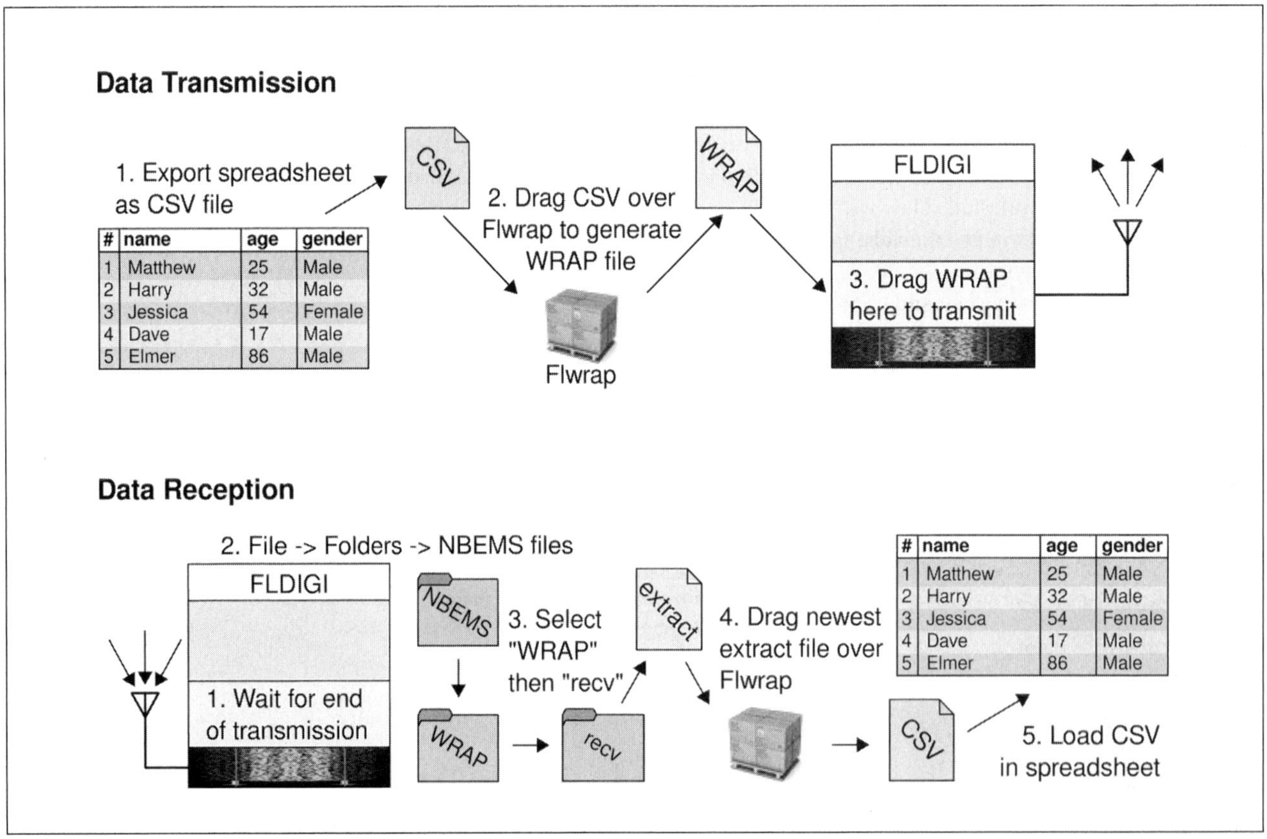

Figure 20.8 — Sending and receiving a CSV file using *Fldigi* and *Flwrap*.

Figure 20.9 — File verification. *Flwrap* has successfully sent this file.

How it Works

Flwrap embeds the name of the original file and a checksum into a new file that is transmitted. The receiving station runs *Flwrap* to extract the original file and the embedded checksum. A checksum calculation is then run against the incoming data. If this checksum is identical to the embedded checksum, we can be practically 100% certain that the file was received without error.

This is a CSV file of a shelter roster that was exported from an *Excel* spreadsheet:

```
Wrist Band ID, Last Name, # of family @ shelter, # of family unaccounted for
02-004,Duncan,1,1
01-002,Duncan,1,1
01-001,Boston,3,0
02-003,Clarkin,3,0
06-001,Charnock,1,0
```

This is the same shelter roster CSV file after being processed by *Flwrap*. Note that we have inserted a checksum of 4162 and "wrapped" the data in special text that is used by *Fldigi* to identify that this is wrapped data. The [wrap:beg] string indicates the beginning of the checksummed data. The checksum and [wrap:end] indicate the end of the checksummed data. The checksum itself is found in the [wrap: checksum 4162] string.

```
[WRAP:beg][WRAP:crlf][WRAP:fn Shelter Data.csv]Wrist Band ID,Last Name,# of family @ shelter,# of family unaccounted for
02-004,Duncan,1,1
01-002,Duncan,1,1
01-001,Boston,3,0
02-003,Clarkin,3,0
06-001,Charnock,1,0
[WRAP:chksum 4162][WRAP:end]
```

Guidelines and Tips

Send text only. You won't have the bandwidth to send large binary files. Remember, a 2 kb file takes two minutes to transmit using MT63-2000. This is a typical repeater timeout period.

Work with your served agency ahead of time to set their expectations for our capabilities. You do not want them to be surprised during a drill or an actual disaster.

If handed a *Word* file, try to export to plain text, or copy and paste the contents of the document into *Notepad*. You can also use *Flmsg*, as we will describe later.

Export *Excel* spreadsheets as "CSV (comma delimited)" before wrapping.

To make life a little simpler, place a shortcut to *Flwrap* in your WRAP\recv folder.

To more easily find the most recent incoming wrap file, change the view of your WRAP\recv folder to Detailed with VIEW/DETAILS.

NBEMS BENCHMARKS — HOW LONG WILL IT TAKE TO SEND THAT MESSAGE?

Rather than publish theoretical values for throughput of various NBEMS modes, tests were made by transmitting two benchmark messages with several modes, and timed the transmissions with a stopwatch. One message was a 2 kb file consisting of plain text. The other was a combination of plain text and spreadsheet data. The effects of *Flwrap* compression were also noted. **Table 20.1** is a summary of these benchmarks.

One question that this data can help answer is the usefulness of *Flwrap*'s ability to compress data. Take a look at the time to transmit the 2 kb bulletin using PSK-125R with and without compression. Compression increases the transmission time from 190 to 215 seconds! Why is this happening? This is occurring because the PSK modes use a form of encoding known as Varicode, which is optimized for plain English text. Letters that are more commonly used are shorter in Varicode than letters that are used less often. For example, "e" is represented in Varicode by "11," and "q" is represented by "110111111." As "e" is much more common in

Mode/Method	2kb Bulletin	6kb Bulletin
MT63-2000	115	320
MT63-2000 compressed data	95	215
8/500 Olivia	715	N/A
16/500 Olivia	1070	N/A
PSK 125R	190	615
PSK 125R compressed data	215	520
PSK 250R	95	310
PSK 250R compressed data	110	265
PSK 500R	45	155
PSK 500R compressed data	55	130
PSK 500	25	85
PSK 500 compressed data	30	80
PSK 125R *Flarq* 256 block	235	710
PSK 125R *Flarq* 64 block	315	980
PSK 125R *Flarq* 16 block	652	1970
PSK 500 *Flarq* 256 block	40	115
PSK 500 *Flarq* 64 block	60	175
PSK 500 *Flarq* 16 block	145	440

Table 20.1 — Measurement of transmission time of various **NBEMS** modes.

plain text than "q," this means that on average, text containing a lot of e's will take less time to transmit than text containing a lot of q's. Varicode may be thought as a form of data compression.

When we compress a file with *Flwrap*, the resulting text no longer has an average distribution of letters. The odds of "e" appearing are now the same as "q." So, it is entirely possible that there may be more 1s and 0s to transmit in the resulting compressed file, especially if it is a short file.

In short, data compression is not worth the trouble. As you can see, it does not work well with modes that use Varicode. It also means that should there be a transmission error in even one bit of data, it is not possible to recover any data by viewing the received "extract" wrap file. Finally, for compression to be effective, a sizable amount of data must be compressed. Most of the files that we will be sending will be too small to be effectively compressed.

TACTICAL vs LOGISTICAL TRAFFIC

During deployments and incidents, Amateur Radio operators send two different kinds of traffic: tactical and logistical.

Tactical messages are short and to the point, and are usually best sent by voice. Some example of tactical messages are:

"We need help now!"
"Busses have arrived at my location."
"Water's rising, we're evacuating."

Logistical messages are more detailed and can be troublesome when sent by voice. Examples of logistical messages are a roster of evacuees, a list of required medications, directions to a staging area, or phone numbers and e-mail addresses of critical personnel. Sending logistical messages by voice is troublesome for a variety of reasons:

- Reading a long message over the air is time-consuming and prone to error. Imagine reading a shelter population list over the air, along with addresses and phone numbers.
- Operators record messages by hand on paper. There's only one original copy of the message, and bad handwriting can lead to errors. Messages must be retyped to be entered into an agency's automated messaging system or to be e-mailed. If Amateur Radio does not fit into an agency's workflow, we are not likely to be used.
- There is no 100% certain way to be sure that a message was received without error.
- There is no reliable way to go back and review received traffic during a "hotwash" or after-action review. It's easy in the heat of an incident for a stressed-out operator to forget to record the time a message was received, or to misplace a sheet of paper.
- Should the operator become distracted or forced to temporarily leave his or her post, any incoming traffic will be missed.

Logistical messages are best sent digitally, using whatever standard methods an ARES group has agreed upon.

Flmsg

Flmsg is a component of NBEMS that greatly simplifies the workflow to send formal messages like ARRL Radiograms and ICS-213 forms. All you do is select a form, fill it out, tell *Flmsg* to transmit the message, and NBEMS does all the rest, including automatically embedding a checksum for file transfer verification and keying your transmitter! Receiving and verifying 100% copy of an incoming message can even be done automatically. You have a variety of options of displaying and saving the message, including a pretty-print format that will impress your EMA director. And, you can automatically extract and save files, should you need to step away from your station. When you return you will find your incoming messages displayed in either *Flmsg* or your default web browser.

Flmsg Overview

Flmsg works by presenting you with a form to fill out, which you then save in a file. Once the form has been saved, you may then transmit the message. The easiest way is to tell *Flmsg* to AutoSend the file. AutoSend embeds an *Flwrap* checksum in the file, passes the file to *Fldigi*, and instructs *Fldigi* to automatically key your transmitter and send the message. It takes just one mouse click and selecting a couple of pop-up menus to initiate an AutoSend.

Receiving a file is just as simple. Once a message has been received, you tell *Flmsg* to import a file from where *Fldigi* stores incoming files. *Flmsg* opens the folder, you select the newest file, and *Flmsg* reads the file, selects the correct form, and automatically populates the form. Should you need to step away from your station, or if you become preoccupied with another activity, messages are still saved by *Fldigi* and you can easily view them later at your convenience.

Figure 20.10 is a screenshot of a blank *Flmsg* display.

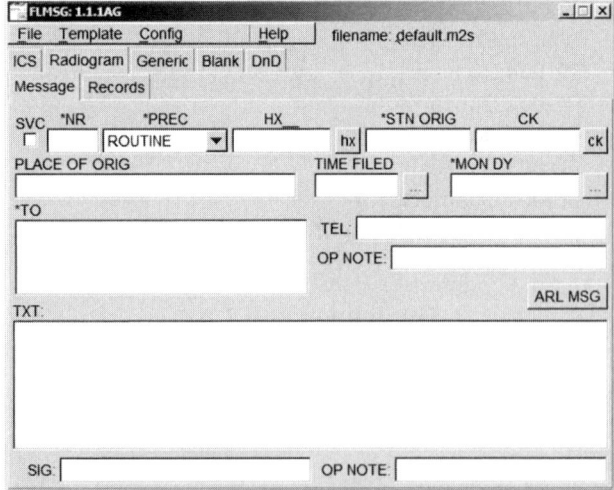

Figure 20.10 — Initial *Flmsg* screen showing two rows of tabs.

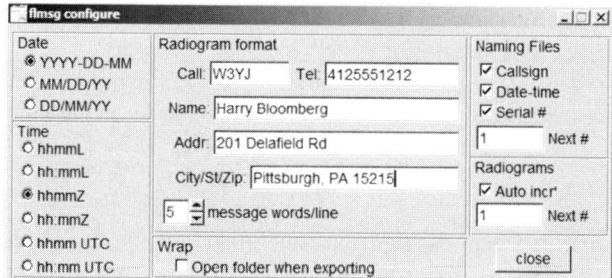

Figure 20.11 — *Flmsg* configuration menu showing the sign, date-time stamp, and serial number boxes checked, to ensure that each of your files will have a unique file name.

There are two rows of tabs. Clicking on the top tab selects the message type. Clicking on the bottom tab selects a message subtype. For example, if you want to transmit an ICS-205 form, you will first select ICS from the top row of tabs and then 205 from the bottom row of tabs. By default, *Flmsg* opens in an ICS-213 format, which is the most common ICS message form.

Before sending any forms, you should first configure *Flmsg* (see **Figure 20.11**). This is done from the *Flmsg* CONFIG menu. You will be presented with parameters that you may customize. Among the most important are the formats for date and time, your personal station information, and how to name files. It's a good idea to name files using your call sign, date-time stamp, and a serial number so your file names will always be unique, for example, W3YJ-20101225-195609Z-1.213. This is important because you can imagine the confusion in a large deployment, should everybody send a file with a fairly common name such as "message.txt." The default file name format allows receiving stations to easily identify who sent the message as well as the date and time of the message. You can also modify the file name to contain additional descriptive text such as a tactical call sign (W3YJ-PARKVIEW-SHELTER-20101225-195609Z-1.213).

Sending ICS Forms

Flmsg is currently capable of sending the following NIMS-compliant ICS forms, with more forms always being added. Among the more popular forms available are:

ICS-203 Organization Assignment List
ICS-205 Incident Radio Communications Plan
ICS-206 Medical Plan
ICS-213 General Message
ICS-214 Unit Log
ICS-216 Unit Log

Figure 20.12 shows a filled-out ICS-213 form ready for transmission. Note that for ICS-213 there are tabs for Originator and Responder. When sending a reply to this message, just click on the RESPONDER tab, fill out the responder form, and transmit the entire form. This is in line with how ICS-213 is supposed to work.

Figure 20.13 shows use of AutoSend to send the message. This will cause an *Flwrap* checksum to be added to the file, and *Fldigi* will automatically transmit the file. That's it! You no longer need to edit the file in a text editor, manually run *Flwrap* against the file to add a checksum, then drag it into the *Fldigi* transmit pane, and manually initiate a transmission as under the previous workflow.

ICS Message Templates

Let's suppose you know ahead of time that your team will transmit information about the population at evacuation shelters. To ease your work and allow for a standard format for the shelter reports, you can create a form ahead of time and save it as a template. You can then distribute the template to the operators at the shelters, either ahead of time via e-mail or USB jump-drive, or over the air.

Creating a template is simple. Just fill out an ICS-213 form

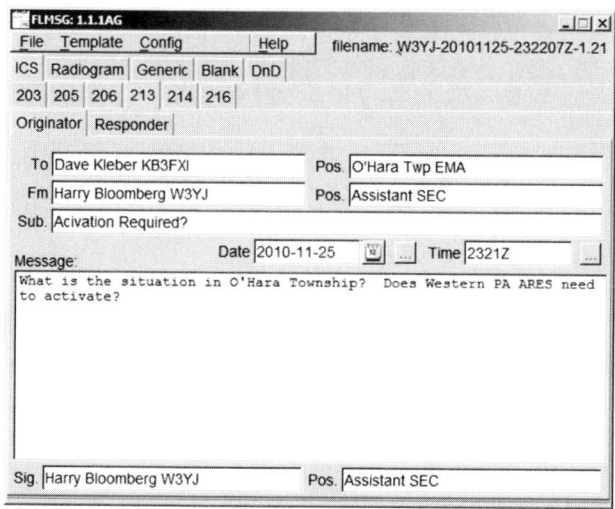

Figure 20.12 — A filled-out ICS-213 form ready for transmission. To respond, click the RESPONDER tab and type into the form.

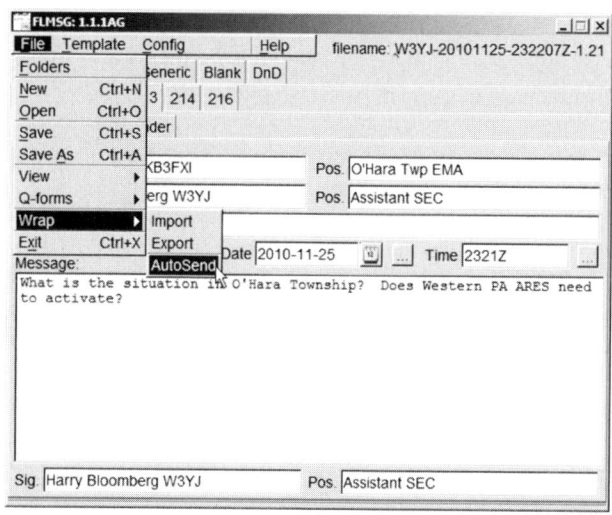

Figure 20.13 — Sending a message in *Flmsg* using AutoSend.

and leave blanks for places where operators are to provide data. Once the message form is complete, click on the TEMPLATE menu and save the template file. You can then distribute the template to other operators by simply transmitting the template over the air and instructing them to save the template with a specified template file name.

Opening up a template is easy also. Just go to the TEMPLATE menu and select LOAD.

Manually Receiving ICS Forms

Once you see from your *Fldigi* screen that you have received an incoming message, it's easy to import it into *Flmsg*. Just go to the WRAP menu and select IMPORT. You will be presented with a listing of wrap files; just select the newest one, which will be at the bottom. The selected file will have its checksum automatically verified with *Flwrap* and the message will be loaded into *Flmsg*.

On rare occasions, you may have a checksum error on receive. If at all possible, request a retransmission, as you won't want to take chances with corrupted data. But there may be occasions when this may not be possible, due to band conditions or a time-critical situation. In this case, *Flmsg* will tell you that you have a checksum error and will ask if you want *Flmsg* to try to load the message anyway.

Once you have received a message, you may decide to display it in HTML format using the FILE/VIEW menu. This is recommended if your workflow requires hard-copy output. This is much easier to read and more professional looking than chicken scratches on a paper tablet!

Automatically Opening and Displaying Incoming Messages

A new feature in *Fldigi* and *Flmsg* is the ability to have incoming messages automatically opened in either *Flmsg* or your default web browser. The ability to automatically open messages in a web browser is especially useful for a station in a critical location like an EOC. Messages are easier to read in a browser, and it's very easy to print messages from a browser.

Auto open is enabled in *Fldigi* by going to the CONGFIGURE/MISC/NBEMS menu and checking the following boxes as described below:

ENABLE — Checking this box tells *Fldigi* to identify and save incoming files that have been formatted with the *Flwrap* program. It is very important that this box be checked, otherwise incoming files will not be saved.

OPEN MESSAGE FOLDER — This checkbox will cause the folder that holds incoming messages to be automatically opened. Some will find this useful, some will find this a nuisance.

In the "*Flmsg* specific" section are two checkboxes and an input box.

OPEN WITH FLMSG — When this is checked, incoming messages will be automatically opened and displayed using *Flmsg*.

OPEN IN BROWSER — Causes the message to be automatically displayed in your system's default browser.

The FLMSG: input box is the path to the *Flmsg* program on your computer. The easiest way to fill this box is to push the LOCATE FLMSG button, navigate to the folder where *Flmsg* is installed, and then click on the *Flmsg* program. In *Windows* machines, the folder is named C:/PROGRAM FILES and the *Flmsg* program is named FLMSG.EXE. Important: you must click on the FLMSG.EXE file; it is not sufficient to just select the PROGRAM FILES folder. If you're not sure where *Flmsg* is installed on your computer, pay attention to the installation wizard when you install *Flmsg*. The wizard will inform you where the FLMSG.EXE file is being installed.

Flmsg and ARRL Radiograms

If you click on the *Flmsg* RADIOGRAM tab, you will be taken to a blank NTS message form, as you can see from **Figure 20.14**. This form was developed in consultation with local NTS experts and is believed to be in conformance with NTS standards.

Radiograms can be confusing to newcomers. It often seems like you need a manual in hand to properly format messages. For example, there's a huge variety of ARRL messages to choose from, and you may have questions as to which handling instruction is the most appropriate, and how does one compute the CK (check) field. *Flmsg* attempts to automate and ease the use of Radiograms.

Figure 20.15 shows how to select the handling instruction from a drop-down menu. To see a list of available handling instructions, just click the HX button. Note that the pop-up menu includes a description of each handling instruction. There's a built-in tool for ARL messages. Just click the ARL MSG button to see a list of all available ARL messages. *Flmsg* will even help you fill in blanks in ARL messages that contain blank fields.

As you fill out the form and add ARL messages, the important CK (check) field is updated. But if you go back and

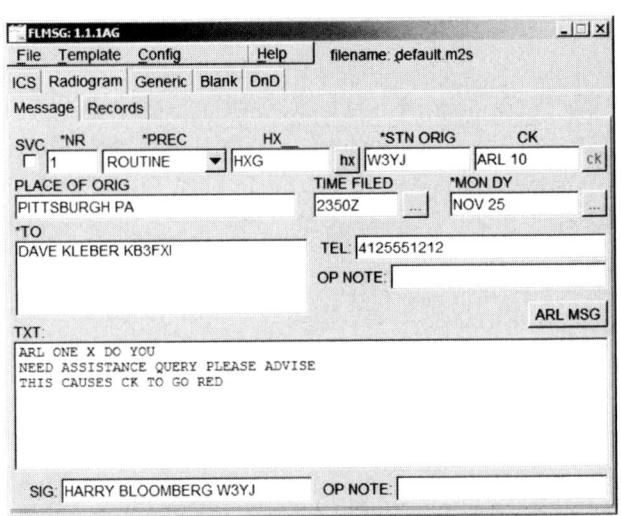

Figure 20.14 — An NTS message form.

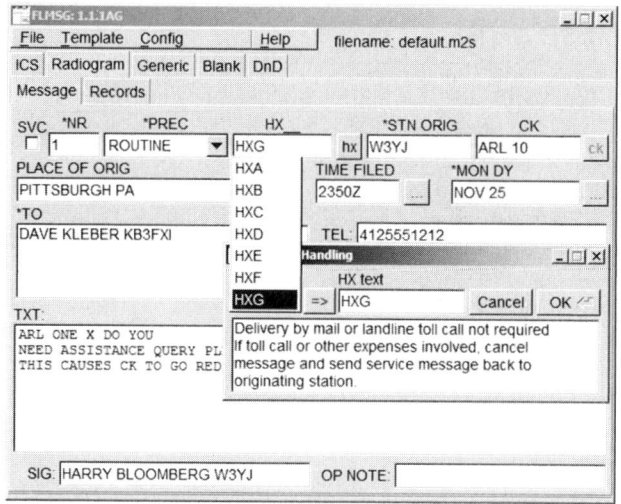

Figure 20.15 — Click the HX button to select handling instructions.

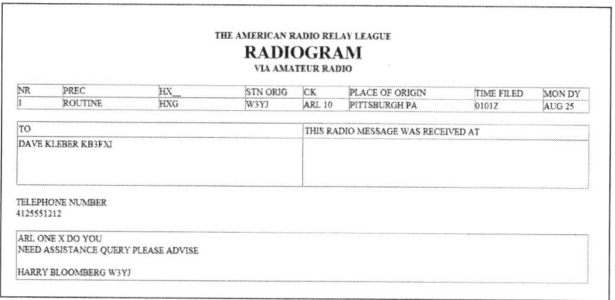

Figure 20.16 — ARRL Radiogram in html format.

change the contents of the message, this will cause the CK to no longer be accurate. When this happens, the CK button changes color to red to let you know that the CK is invalid. You then simply press the CK button to recompute the CK.

As you move around the *Flmsg* Radiogram screen, *Flmsg* automates much of the work necessary to format the message for NTS. For example, note that text is automatically capitalized and that Xs are inserted in the message text. Once you receive a message, it may be displayed for output in HTML, just as with ICS messages. See **Figure 20.16** for an example.

There are times when a message must be relayed from a digital net to a more traditional voice or CW net. *Flmsg* contains a message view that simplifies this. If you go to the FILE/PLAIN TEXT menu, you will see your message formatted in plain text that can be easily read over the air or sent via CW.

PLAIN TEXT MESSAGES WITH *Flmsg*

Flmsg can be used to send messages in plain text. Why would one want to do this when one can type directly into the *Fldigi* TRANSMIT pane? Quite a few hams do not know how to use text editors like *Notepad* on their computers, and the workflow sending and receiving checksum-verified files

using *Flwrap* is confusing to some. The biggest hurdle is in navigating a computer's file system to find incoming files that have been wrapped.

Flmsg contains two different types of plain text messages. The Generic form includes some simple message header info, such as the name of the sender and recipient. The Blank form is even simpler, and just contains a large blank box. The operator types the message in the blank box, then uses AutoSend to insert a checksum and transmit the message. As with the other message forms, one can read an incoming file using the FILE/WRAP/IMPORT menu.

USE OF NBEMS IN UNATTENDED STATIONS

The NBEMS software is designed for use between manned stations, utilizing what we like to call "human error correction." The power in NBEMS is its simplicity and flexibility in allowing manned station operators to use their radio skills and experience to dynamically adapt to changing conditions.

But, an often overlooked aspect of NBEMS is its ability to work in a monitor-only state, even at an unmanned station. This type of operation should not be confused with automated unmanned operations where the remote/unmanned station automatically transmits and responds to incoming calls.

The NBEMS unmanned monitoring station, through some simple setup and configuration, can be set to automatically receive and save both the plain text of keyboard-to-keyboard transmissions as well as extracting and saving messages and bulletins on a predetermined net or emergency communications frequency. This ability to record and save received messages without operator intervention is of great value in emergency communications.

Those who have been involved in emergency communications exercises and deployments should have no trouble recalling situations where communications are repeated — often several times — to stations and responders just coming onboard the nets. Some common examples are responders or net participants that are delayed in joining a voice communications net and have to ask for previously transmitted information such as ingress routes, staging locations, and general situation reports.

In the example of a rapidly expanding localized emergency situation, an NBEMS designated channel on a VHF or UHF frequency with NBEMS monitoring stations can help solve some of these problems. For instance, a governmental Emergency Operations Center (EOC), early in an emergency situation, could simply turn on a receiver (such as a police scanner) and a computer equipped with NBEMS prior to the arrival of an ARES volunteer at the station.

Upon arrival at the station, an ARES volunteer can visually scan all the text transmitted on the designated NBEMS frequency. More importantly, they may also view and confirm 100% receipt all of the higher-priority and emergency traffic sent with *Flmsg*, which eliminates their need to interrupt a busy net seeking previously transmitted information.

> ### TIPS FOR NBEMS USE ON HF
>
> Touchpads (pointing devices found on the keyboards of many modern laptop and notebook computers) do not work in strong RF fields. Instead, use either an external mouse or a USB mouse. There is nothing more frustrating than losing control of your computer during an HF transmission.
>
> When transmitting on HF, disable your speech compressor, noise blanker, noise reduction, and all other audio processing. All these are optimized for voice communications, not digital signals.
>
> By convention, all transmissions are done using upper sideband (USB), even on bands like 40 meters and 80 meters, where voice communications are done with lower sideband (LSB). There is no good technical reason for this, other than "everybody else does it."
>
> Be very careful to not overdrive your audio. Adjust your audio level to your transceiver in one of two ways (both are more-or-less equivalent):
>
> Adjust ALC so that it just moves as you transmit.
>
> Start with zero audio level into the transceiver and monitor the output power. Gradually increase the audio level going to the transceiver until the output power no longer increases.
>
> Because of the digital signal processing done by *Fldigi* to read weak signals, you do not need to run a lot of power. Under most circumstances, 50 watts is plenty. This will also help you conserve power if operating under Field Day-like conditions.

What's even more powerful with NBEMS is that tens or even hundreds of monitoring stations all have the same ability to receive and confirm 100% receipt of critical information, from a single transmission, through the use of NBEMS/*Fldigi* and the embedded checksum feature of *Flmsg* and *Flwrap*.

TRANSMISSION OF BINARY FILES OR LARGE FILES

Transmission of binary files such as *Word* documents or *Excel* spreadsheets is to be discouraged because of the large size of these files and our limited bandwidth. Instead, we should convert *Word* files to text format and *Excel* spreadsheets to Comma Separated Values (CSV) text format.

USING RSID TO AUTOMATICALLY CHANGE MODES

Reed Solomon id (RSID) is an eight-tone identifier that, when enabled, produces tones prior to the beginning of data transmissions to help identify the mode in use to receiving stations. Each digital mode is assigned a unique RSID code, which is standardized among all the digital sound card mode software packages, including NBEMS/*Fldigi*. *Fldigi* can be configured to automatically change to the proper mode upon receiving RSID. *Fldigi* provides the ability to select RSID auto detect over the entire waterfall passband or to only a specific spot on the passband (+/- 200 Hz). When "detector searches entire passband" is enabled, when an RSID signal of a strong enough strength is received, *Fldigi* detects the RSID code, switches to the proper mode and shifts the receive spot to where the RSID signal and the following data mode is transmitted on the waterfall. In addition to automatically changing modes and shifting to the spot on the waterfall, *Fldigi* also gives the option to insert a detector marker in the received text window which timestamps and details the mode change. An additional setting allows the receiving station to be configured to only detect certain modes which can be selected/deselected in the "Received Modes" option under the RSID settings in *Fldigi*.

RSID is very useful in unattended station monitoring as no operator intervention is required in order to follow mode changes and shifts on the waterfall. *Fldigi* can be set with "detector searches entire passband" disabled (to only detect the RSID code within the marker spot set on the waterfall) to automatically detect scheduled bulletins and *Flmsg* message on a fixed call frequency without the need for an operator being present at the station. **Figure 20.17** illustrates how to configure *Fldigi* to automatically change modes based upon reception of RSID tones.

Even a basic receiver such as a scanner can be set to monitor a call frequency with *Fldigi* to capture all the traffic on the given frequency. For a practical example, during a scheduled net, operators who aren't available at the time of the net can leave their stations on to monitor call frequencies while they're away from their station. Upon returning to the station, they have a record of all received text with the RSID timestamp and mode change markers, as well as the extracted *Flmsg*/*Flwrap* files sent during the net.

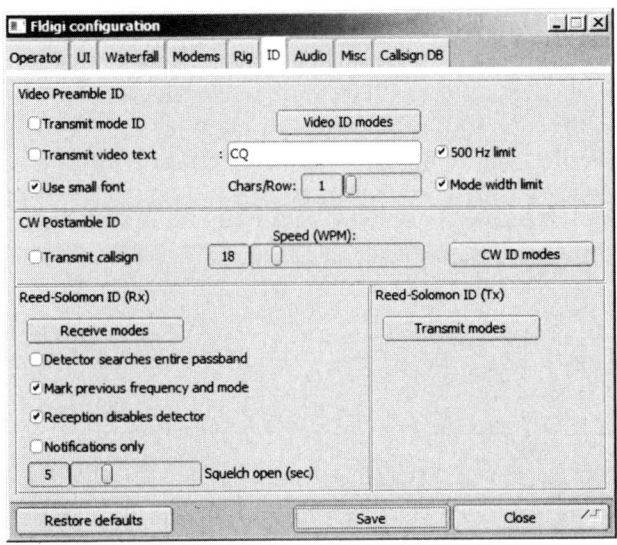

Figure 20.17 — Configuring *Fldigi* to automatically change modes based upon reception of RSID tones.

The Amateur Radio Public Service Handbook

Digital VOX Sound Card Interface

Here's an interface that will get you on the air with PSK31 and other digital modes regardless of the type of computer you are using — no serial or USB cable required!

Howard "Skip" Teller, KH6TY

Most sound card interfaces are powered either by a voltage from the computer serial port, by a voltage taken from the computer accessory port or microphone port, or by a "wall-wart" ac adapter. If the computer has no serial port, and most computers these days have USB ports instead of serial ports, it is also necessary to use a USB-serial adapter to generate a *virtual* serial port that the communications software can use for push-to-talk operation.

It would be more convenient if no dc voltage were needed to power the interface, and also if no serial port or USB-serial adapter were needed, so I wanted to find a way to eliminate both the need for a serial port or USB-serial adapter and a dc voltage. I then realized that computer sound cards have evolved from having both a high level speaker output jack and a line-level audio output jack, to usually having just an earphone/headphone jack and a microphone jack. Measuring the maximum audio output level of this jack on several computers, and also on an external sound card, such as "USB sound adapters" commonly sold to provide microphone and earphone jacks via a USB connector, I found that it was generally around 2.5 V peak-to-peak — not enough to power a switching transistor in an interface.

By connecting the earphone/headphone output to the center tap of a 600:600 Ω isolation transformer, however, that voltage is doubled across the full secondary winding of the isolation transformer to 5 V peak-to-peak — enough to rectify and power a transistor switch for push-to-talk switching. Since all this occurs at the secondary winding of the transformer, the transformer still isolates the computer earphone output and ground from the transceiver itself, thereby preventing any hum or ground loops from disturbing the transmit or receive audio.

By using another isolation transformer for the receive audio, the computer is totally dc isolated from the transceiver, both on the audio input lines for transmit, the receive audio output line for receive, and the push-to-talk switching line for transmit/receive switching. The schematic diagram in Figure 1 shows how this isolation is provided by the transformers.

How It Works

See Figure 1. Sound-card-based digital communications software such as *DigiPan*, generates a WAV audio signal when placed in the transmit mode. This WAV audio contains the modulation that you use to communicate with on digital modes, such as PSK31 and others, and is used to modulate the transceiver in the same way that you modulate the transceiver audio with the microphone when operating SSB. Since this WAV audio comes out of the earphone or speaker jack of the computer, that jack is connected to isolation transformer T1, but only between one side of the primary winding and the center tap of the primary winding. When this audio is coupled across the transformer to the full secondary winding, it appears at twice the value that is present between the primary center tap and either end of the primary winding, because the turns ratio of the transformer in that case is no longer 1:1, but 1:2.

The WAV audio is then used to modulate the transceiver through the data, accessory or microphone jacks for digital operating. The audio level is adjusted by means of potentiometer VR1.

In order to switch the transceiver from receive to transmit, this same double-value ac voltage is applied to capacitor C1 and resistor R1, which isolate the transformer audio from the switching action of the following voltage doubler circuit, D1 and D2. These diodes form a classic dc voltage doubler rectifier circuit, which conducts on both positive and negative cycles of the WAV audio. C2 then charges up to the peak value of the ac voltage and holds that charge long enough to drive the base of switching transistor Q1, through current limiting resistor, R2, causing the collector of Q1 to saturate and pull the push-to-talk pin of the transceiver to ground. Q1 gets its operating collector voltage from the push-to-talk circuit in the transceiver, which must be designed, as almost all transceivers are these days, for an "open collector" switch for transmit/receive switching.

Transformer T2 is used to isolate the receive audio to the computer from the transceiver, and the primary is connected to the computer microphone input. The receive audio output voltage is fed from the transceiver through an L-pad consisting of R4 and R3, attenuating the high audio output of the transceiver data jack, or earphone jack, to a suitable level for the microphone input of the computer. It is this input to the microphone of the computer that creates the "waterfall" display of the typical digital communications program that also decodes the WAV audio being transmitted by the other station into characters on the screen. Resistors R4 and R3 can be exchanged in position if the computer audio input requires a higher audio level from the transceiver.

Assembling the Interface

The following steps assume you are building the kit that is available on my Web site at **https://sites.google.com/site/kh6tyinterface/**, although much of this information is also helpful if you are building the interface from scratch.

If you are using the circuit board

Reprinted from the March 2011 issue of *QST*.

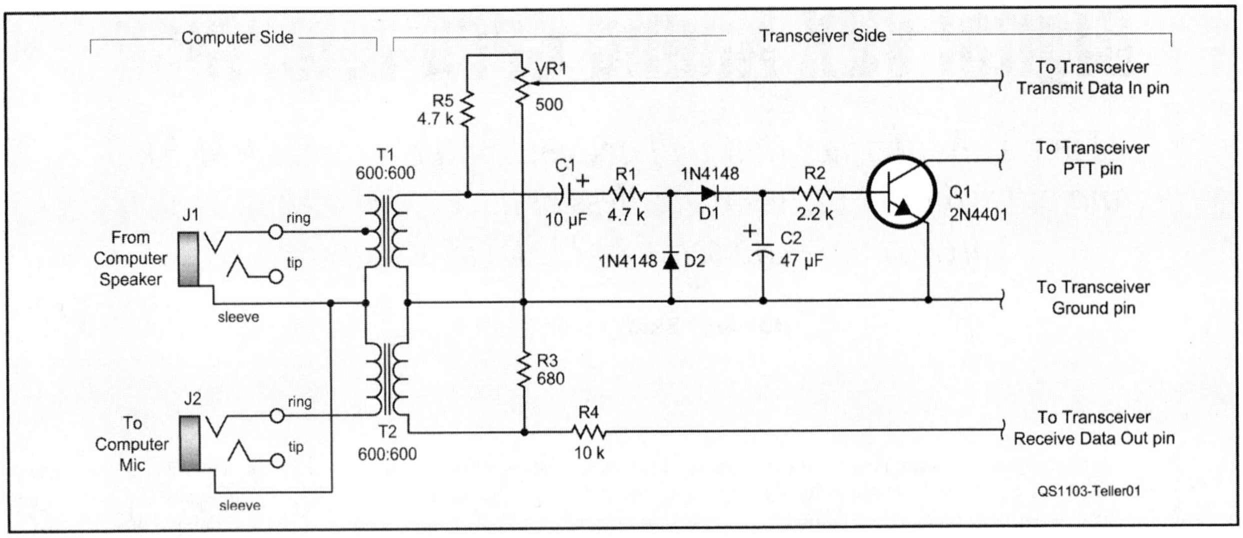

Figure 1 — Schematic diagram of the Digital VOX Sound Card Interface. Vendor part numbers are shown in parenthesis. Components can be obtained from Mouser Electronics, tel 800-346-6873; www.mouser.com. A double-sided, plated-through hole circuit board with legend, or a complete kit with all cables, is available for a limited time at https://sites.google.com/site/kh6tyinterface/.

C1 — 10 µF, 50 V electrolytic capacitor (Mouser 647-UVR1H100MDD1TA).
C2 — 47 µF, 50 V electrolytic capacitor (Mouser 647-UVR1H470MED1TD).
D1, D2 — 1N4148 diode (Mouser 512-1N4148).
J1, J2 — ⅛ inch stereo jack (Mouser 161-3507-E).
Q1 — 2N4401 or other switching transistor (Mouser 512-2N4401BU).
R1, R5 — 4.7 kΩ, ¼ W resistor (Mouser 291-4.7K-RC).
R2 — 2.2 kΩ, ¼ W resistor (Mouser 291-2.2K).
R3 — 680 Ω resistor (Mouser 291-680-RC).
R4 — 10 kΩ, ¼ W resistor (Mouser 291-10K-RC).
T1, T2 — 600CT:600CT isolation transformer (Mouser 42XL016-RC).
VR1 — 10 kΩ potentiometer (Mouser 652-3362-1-501LF).

from the kit, just mount and solder all parts, following the legend printed on the board. Use good soldering techniques and make sure there are no solder bridges between adjacent holes. In the kit the transformers have been partially mounted in place for protection during shipping, so be sure to complete the soldering of all six pins of each transformer. If any hole is accidentally filled with solder, heat the area and thrust a sharp-pointed round wooden toothpick into the hole. On the opposite side of the circuit board, melt away any solder around the toothpick and the hole should be cleared for resoldering.

When installing the resistors, save the cutoff leads, some of which will be used for wire tie points, which are inserted after all other parts have been soldered. It is not absolutely necessary to use the tie points; the wires can just be inserted into either of the two holes for a tie point. However, using the tie points makes it easier to solder the transceiver cable wires to the interface. To create a tie point, bend a cutoff lead in half, making a hairpin shape, place both ends into the two holes on the circuit board, push down until about ⅛ inch over the circuit board, spread the leads sticking out the bottom of the circuit board to hold the wires in place, and solder both holes.

The kit comes with a transceiver cable. Decide how long you want the cable to be and cut off the end with the female connector, leaving about 6 inches on the female connector end in case you are using a different accessory connector, or a microphone connector, rather than the 6 pin MiniDIN connector supplied. In this case, plug the male end of the cable into the female end and strip and tin the wires for connecting to the connector you are going to use. On the cut end of the cable strip off about ⅜ inches of insulation and then strip each wire to expose ³⁄₁₆ inches of conductor. Tin the wires.

Making Connections

If you are using a microphone plug or an accessory plug that has a dc voltage on it, refer to the transceiver manual for the correct pin connections. If your transceiver uses a MiniDIN jack that is in the same configuration as the kit cable, the easiest way to identify where each wire is connected to the interface is to first plug the cable into the transceiver and then touch each of the wires, in turn, to the electrical ground of the transceiver. The wire that makes the transceiver go into transmit is the push-to-talk wire, which will be soldered to the tie point marked PTT on the circuit board, but before doing that, touch each of the remaining wires to the now-identified push-to-talk wire. The one that makes the transceiver go into transmit is the wire that goes to the GROUND on the circuit board. Now solder both the ground wire and the PTT wire to the correct tie point on the circuit board.

Connect a stereo audio cable from J1 to the computer speaker or earphone/headphone jack and another stereo audio cable from J2 to the computer microphone jack.

Run digital mode software such as *DigiPan* (**www.digipan.net**) and adjust the audio input level to maximum. If the software you've chosen doesn't provide a means to do this, you'll need to go into *Windows* Control Panel and access the record-audio levels there. Touch each remaining cable wire to the point marked "Data Out Receive Audio" on the kit board. Once you locate the wire that causes a continuous noise signal to appear on the waterfall display, solder it to the "Data Out Receive Audio" connection.

Place the transceiver in SSB mode and the software in transmit mode. There should be no power output from the transceiver because there is no WAV audio coming from the interface. Quickly touch each of the remaining wires to the point marked "Data In Transmit Audio" on the kit board. Solder the wire that causes the transceiver to have RF output to the "Data In Transmit Audio," and cut off the remaining wires.

Now lay the cable down between the two unused holes next to the "Data In" and "Data Out" markings on the circuit board and place one end of a cutoff lead into each hole. Solder one end, pull the other end tight around the wire and solder it also. This will provide a strain relief for the cable. This completes the assembly and wiring of the circuit board.

Reprinted from the March 2011 issue of *QST*.

Modifying the Enclosure

Place the wired circuit board into one half of the enclosure. It will not settle down fully because the jacks and cable holes need to be drilled. Take a pencil and make a mark where each jack and the cable touch the edge of the enclosure. Remove the circuit board, snap on the other half of the enclosure, and drill a small pilot hole where the edges of the enclosure top and bottom meet at the place where you made the pencil marks. Leaving the enclosure snapped together, enlarge the pilot holes to ¼ inch diameter. Open the case by inserting a small screwdriver in the indentations where the enclosure halves meet and applying a gentle twisting motion. Do this on both indentations of one side and the enclosure will pop open. Clear the holes of extra plastic if necessary.

Adjusting the Transmit Level

Before connecting the audio cables, you should be able to hear the WAV audio in the computer speakers when the software is in the transmit mode. Plug in both cables between the computer and transceiver. With the software waterfall cursor around 1500 Hz and the software in transmit, raise the audio output level until the transceiver switches into transmit. Once again, if the software you are using doesn't offer the means to make this adjustment you'll need to access the audio levels within *Windows* Control Panel.

When the transceiver switches into transmit, increase the audio output level about one more notch to insure there is enough audio for push-to-talk to consistently activate. Adjust the transceiver RF power control for maximum, or if using the microphone jack on the transceiver, to the normal position for SSB phone operation. Now adjust VR1 until the RF output of the transceiver is about 30% of rated maximum power. It should not be necessary to make this adjustment again. Just raise the audio output level control to obtain a little more output power if desired, but at 30% power, you should automatically have a clean PSK31 signal.

After setting the RF power level, place the wired circuit board into the enclosure and snap the two halves together. This completes the assembly and adjustment of the interface. If you find it necessary to adjust the interface very often, it might be convenient to drill a small hole in line with VR1 so a screwdriver can be inserted into the enclosure to adjust VR1. In most cases, however, it should be possible to just set VR1 and leave it alone, doing all the fine power setting adjustments with the software audio level controls. Near the extreme edges of the IF passband, the audio output of the transceiver may decrease because of the shape of the IF filter, so it may be necessary to increase the audio output level to maintain push-to-talk action. Remember to recheck the power output when retuning to the center of the passband.

The digital VOX interface works well with all digital modes such as PSK31 and AFSK RTTY, but does not work correctly with sound-card CW or Hellschreiber. If you intend to operate those modes, the Classic Sound Card Interface described on page 37 of the July 2010 *QST* is a better choice because its serial port switching keeps the transceiver in transmit until the software returns it to receive. It is not dependent upon the audio tones being present to keep it in transmit.

Howard "Skip" Teller, KH6TY, is an ARRL member and was first licensed in 1954. He received his commercial First Class Radiotelephone license in 1959 and worked his way through college as chief engineer of several radio stations. He holds a BS degree in electrical engineering from the University of South Carolina and is retired from running a factory in Taiwan, where he manufactured the weather-alert radio he originated in 1974 and is still sold by RadioShack and many other companies now. Skip enjoys developing digital software, such as DigiPan and NBEMS, and designing VHF/UHF antennas. He is currently studying the potential of working 432 MHz DX using the Contestia digital mode. You can contact Skip at 335 Plantation View Ln, Mt Pleasant, SC 29464; skipteller@gmail.com.

An interior view of the Digital VOX Sound Card Interface.

Reprinted from the March 2011 issue of *QST*.

NBEMS RESOURCES AND SUPPORT

- Official W1HKJ NBEMS site for downloads and documentation: **www.w1hkj.com**
- The NBEMSham Yahoo group is an excellent mailing list for help with difficulties
- "Ghost QSOs — Olivia Returns from the Noise," December 2008 *QST* by Gary L. Robinson, WB8ROL
- "NBEMS — a Digital EmComm Tool," August 2009 *QST* by Dave Kleber, KB3FXI and Harry Bloomberg, W3YJ
- "A Digital Simulated Emergency Test," June 2010 *QST* by Dave Kleber, KB3FXI and Harry Bloomberg, W3YJ
- "Digital VOX Sound Card Interface," March 2011 *QST* by Howard "Skip" Teller, KH6TY

Chapter 21
Ham Radio on the Internet:
Hints and Tips for *EchoLink* Users

EchoLink software for *Windows*, Android, and iPhone was created by Jonathan Taylor, chief engineer of the Synergenics Corporation, known as K1RFD to his ham friends. Jonathan gave *EchoLink* as a gift, free of charge, to the ham radio community. He also created *EchoStation*, a repeater-control program for *Windows*, for which he charges a small, reasonable fee. In this chapter, we'll discuss how to get the most out of *EchoLink*.

GOOD OPERATING PRACTICES

Although most hams are eager and willing to help those new to *EchoLink*, it might save some time and frustration if you first connect to the *EchoLink* "Test" Server (click the STATION menu, then select CONNECT TO TEST SERVER). See **Figure 21.1**.

The reason? Here you can transmit, then listen to what you just sent (called your "echo"). This allows you to hear exactly what others will hear and tweak your audio levels, if required. This ensures readability before you attempt to communicate with others. Some hams monitor for new connections and immediately connect to those with high node numbers, to offer assistance, but it is best not to depend on that. Although you can reasonably assume that anyone shown as "active" wants to chat, it's best to think of *EchoLink* as if it were a typical ham band. How often have you heard a station calling CQ but failed to get a QSO going? If you are scanning the active user list for a QSO and you succeed in getting a CONNECT, please don't wait for the other ham to start the QSO after you make the connection. It is considered proper form to talk first with an invitation to engage in a QSO. Would you knock on a door and then, after it opens, just stand there silently?

If you're running *EchoLink*, but temporarily leave your terminal unmonitored, put *EchoLink* in Busy mode by clicking the white hand in the toolbar. There also seems to be a great deal of impatience when users get a CONNECT, but get no reply within 10 seconds or so. Wait one full minute before blasting off. The ham on the other end may simply be away from his or her keyboard (AFK) for a short time that doesn't warrant going into Busy mode.

If your intention is to connect to another user, link, or repeater station with the desire to only read the mail, that's fine and even encouraged. Many new radio amateurs starting out with *EchoLink* do not have microphones connected or operating properly. No matter what your reason for connecting, it is still considered courteous to at least announce your presence and intention, if not by voice, then by a text message. This way, no one will be left guessing as to the purpose of your visit.

Regarding station ID requirements, peer-to-peer or direct connects on *EchoLink* do not require signing with calls; connects to Link Stations and repeaters do. However, as a matter of habit and courtesy, most hams on *EchoLink* sign; if not every time, then every so often.

If you are a busy person, you probably want to do other things on your PC while chatting on *EchoLink*. Here's an operating tip that will allow you to confirm *EchoLink*'s transmit or receive status in a flash. This can be useful when you think you toggled back to receive, yet you're still in Transmit mode. Reposition the *EchoLink* window so that its status bar is just above the *Windows* taskbar at the bottom of your desktop. Now, reposition your browser, e-mail clients, etc., so their status bars rest just above *EchoLink*'s status bar. Now *EchoLink*'s status bar will always be visible on your desktop, above your task bar, indicating transmit activity by the red [TX] Transmit flag on the right side of *EchoLink*'s status bar or green [RX] when in receive mode.

Figure 21.1 — Station menu showing the location of the "Connect to Test Server" option.

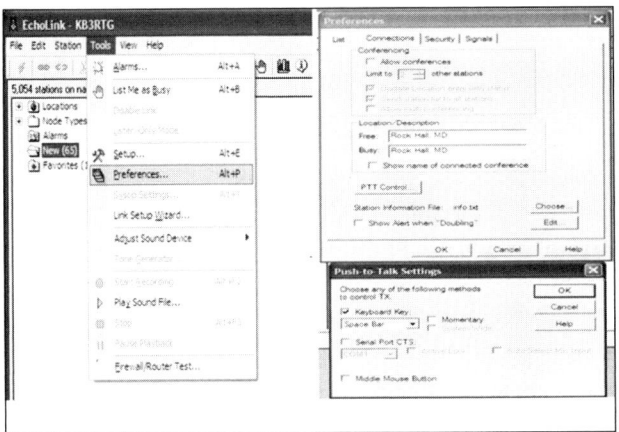

Figure 21.2 — The PREFERENCES submenu, with PTT preferences visible. The screenshot shows the space bar being used for PTT.

In addition, you can avoid having to return "focus" (a Microsoft term meaning the window that is active, or "on top of" all others showing on your desktop) to *EchoLink* in order to regain control of the transmit toggle. Just configure *EchoLink* to use the "Enter" key (or another key of your choice) on the numeric keypad to provide "system-wide" PTT functionality. Open the TOOLS menu, then the PREFERENCES submenu, then click the CONNECTION tab at the top, followed by clicking the PTT button (see **Figure 21.2**). When making this change, remember it is important to check the SYSTEM-WIDE box in the setup menu.

CONNECTION PROBLEMS

A frequent error message when attempting to connect to another user is "Cannot Connect to [IP address] — No Route Available." This means the user you tried to connect with has blocked ports. Typically these are new users (check their node number — if it's over 500,000 the user is relatively new). If you are interested in being a "good Samaritan," you can go to **www.QRZ.com**, enter their call, check to see if they included an e-mail contact address and then write to them, offering assistance.

The solutions to curing port blocks vary a great deal with your particular setup (that is, type of modem/router and software firewalls used). Because of the large number of possible combinations, a good generic solution is offered here. Many newer DSL modems (the Westells™, in particular) have a rudimentary internal firewall that blocks all inbound traffic except on the major ports used by the web and e-mail clients. You should DMZ (DMZ stands for "demilitarized zone," and in the world of computers refers to setting a buffer zone that separates the Internet from your equipment) or set the modem to "Port Follow-through" mode. Next, DMZ the router (block nothing) on the LAN IP of the PC *EchoLink* that is running on. Finally, stop blocking (if using XP SP2+ internal firewall) when the "Blocking Security" alert pops up. Caveat: If you DMZ, it is important that you run a software firewall.

Enabling the *Windows* XP SP2+ built-in firewall is your best bet. If anyone is intent on using their router's NAT as a hardware firewall and going through the process of port forwarding, there is an excellent website with many setup menus for almost all manufacturers' hardware at **www.portforward.com**.

The setup menus to change modem or router configurations are reached using your web browser and the IP address of the device's internal web server. 192.168.1.1 and 192.168.1.254 are commonly used addresses; check your hardware manual if those don't work.

EchoLink requires that your modem, router, or firewall allow (DMZ) or forward inbound and outbound UDP traffic to ports 5198 and 5199, and outbound TCP traffic to port 5200. If you do not DMZ, you must configure your router to forward UDP ports 5198 and 5199 to the node or PC on which *EchoLink* is running. Typically few firewalls block outbound traffic, but it doesn't hurt to check yours.

ICON REPRESENTATION

A "human face" means *EchoLink* is running in either standard (peer-to-peer) or conference mode. A pair of "chain links" represents a Link Station — someone (usually at home) with a transceiver on a simplex frequency, who is connected to their home computer running *EchoLink*. A "set of gears" represents a repeater that is connected to a PC running *EchoLink*. A "PC with two faces" is a reflector; that is, a PC connected to the Internet on a high-bandwidth connection that is primarily intended to connect Link Stations and repeaters as well as many single users. A big advantage to using a conference is that it won't dump your friends who are connected to you when you close your connection. A popular software package for running a reflector or conference is *theBridge*, which can be found at **http://cqinet.sourceforge.net/thebridge.shtml**.

NODE NUMBERS

Your node number is assigned to you when you become an authorized *EchoLink* user. These numbers are assigned in ascending order and, as of this writing, are topping 500,000. Obviously then, users with the highest nodes are the newbies. However, just as the FCC will issue vanity calls for a fee, the *EchoLink* author will sell you a low node number. Some users take great pride in advertising how long they have been using *EchoLink*. Unfortunately, with low node numbers for sale, having a low node number has lost its importance, unless keying in four numbers is more efficient for you than six.

There are two ways you can connect to another *EchoLink* user. If you are on a PC running *EchoLink*, you may use the other station's call or node number. If you are mobile and trying to connect to an *EchoLink* user via an *EchoLink* repeater or Link Station, you click the node numbers of the other *EchoLink* user into your DTMF pad, and if they have *EchoLink* running — that is, if they are either in conference mode or not Busy — you will be connected.

THE SPACE BAR

You already know that, by default, the space bar acts as an RX/TX toggle and that you should not hold it continuously unless you reset it for "push to talk — release to listen" under TOOLS/CONNECTIONS/PTT. But are you aware that the space bar's operational function depends on not losing "focus" (that is, changing the active or "on top" program on your desktop)? You shift focus any time you place the cursor and click in any other dialog box or window. Restoring focus to *EchoLink* only requires that you left-click the mouse inside the tan window where the horizontal audio level indicator resides.

YOUR PROFILE

Adding a profile is a great way to introduce yourself as well as keep a QSO moving along. Why *EchoLink* effectively hides the profile edit button is a mystery, but go to the TOOLS/PREFERENCES/CONNECTIONS tab (see **Figure 21.3**) and *voila*! There it sits at the bottom right of that menu tab. If you

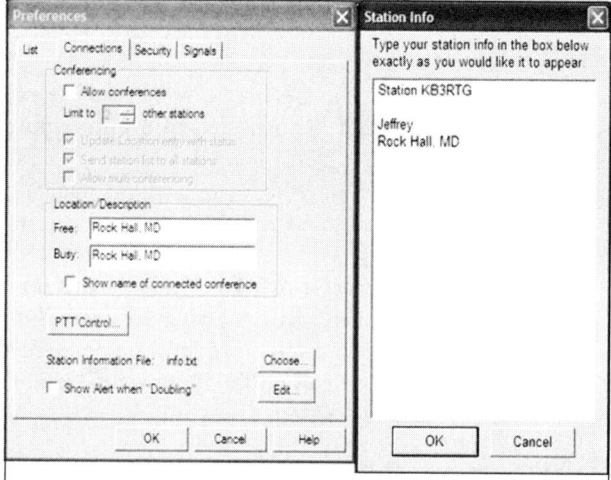

Figure 21.3 — The button for editing your *EchoLink* profile can be elusive. You'll find it in the PREFERENCES submenu, at the lower right of the CONNECTIONS tab.

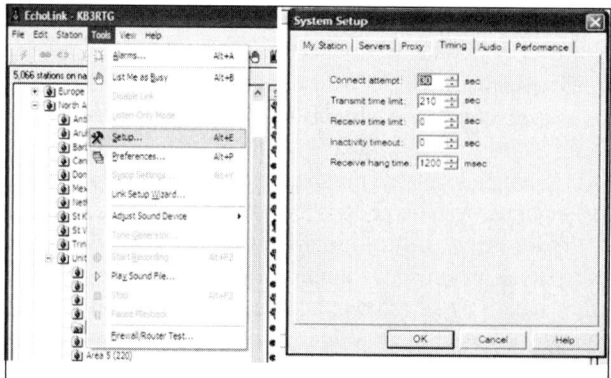

Figure 21.4 — Placing a zero in the timeout boxes disables the default two-minute timeouts that can be troublesome in peer-to-peer communications.

want to see how your profile looks to others, just connect to yourself (yes, you can do that!).

TIMEOUTS

Are you tired of the default two-minute xmit or recv timeouts? Not a problem. Although important in repeater and Link Station operations to avoid locking up a system if you get distracted, it is totally not necessary for peer-to-peer communications. Go to TOOLS/SETUP/TIMING (see **Figure 21.4**) and place a zero in the appropriate boxes (zero means NO timeouts, or disable). Some people use the timeout bell to warn themselves when they have talked too long.

CONNECTION ATTEMPT TIMEOUT

Generally, if you fail to get a connection to your intended contact within six or seven seconds, you will not be able to make that connection. Or, if you eventually do connect, the latency or delays on the net won't allow for 59 communications anyway. Try setting the delay to time out of the attempt at seven seconds rather than the default of 30.

BUSY MODE (THE HAND)

Busy mode allows you to be seen on the various active *EchoLink* users' listings, but prevents any user from having the ability to contact you directly. If you are in Busy mode and you are also connected to a reflector (a special multi-chat server), this reflector connection can prevent sharing of your bandwidth by blocking other users from connecting directly to you. When you set Busy mode on, your text color turns blue rather than the normal black. The Busy status affords you a certain amount of privacy, but does not stop you from connecting to others who are not either in Busy mode or otherwise blocking.

CONFERENCING

By default, conferencing is off. The effect is that once you connect to another user, no one else can get through to you (unless your contact has conferencing enabled on their *EchoLink* and someone connects through their side). Call this a privacy mode if you will. However, a great tradition in ham radio is the roundtable, or ragchew. Unfortunately, if you are on a dialup to your ISP, you won't have the bandwidth for solid roundtables or conferencing. But if you have a DSL or digital cable connection, conferencing can add a lot more fun to the entire *EchoLink* experience! See **Figure 21.5**.

Side Note: Say you are station A and you connect to station B, who is operating in conference mode, and a third station, C, connects through station B and joins your QSO. If station B elects to leave (or disconnect), you will lose both stations B and C. This can sometimes be frustrating. The only solution when running roundtable ragchews is for the so-called "control station," in this case, station B, to remain connected until all parties disconnect. Another option is to

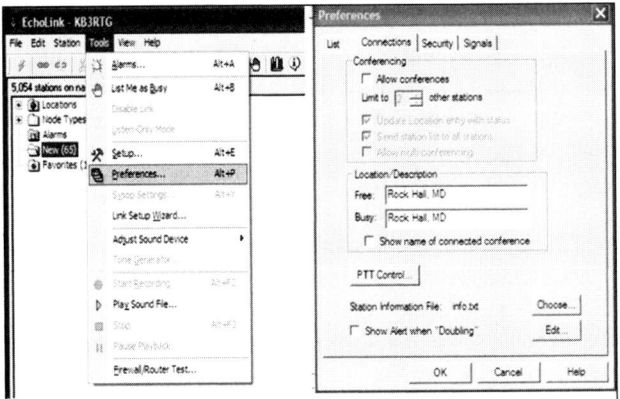

Figure 21.5 — By default, conferencing is turned off. You can enable it via the PREFERENCES submenu, in the CONNECTIONS tab.

use one of the many reflectors (called "Conferences") available on *EchoLink*.

TEXT CHAT

A frequently overlooked feature, the text chat window, can be used any time after a connection is established. Its real usefulness is akin to a break-in, when you wish to comment before having the mic passed back to you. An audible alarm (on by default) will inform you when a text message has arrived. You many even assign your own homebrew WAV file for this purpose. By general agreement, typing three plus symbols (+++) is a request for an immediate break-in.

CALL SIGN LOG

By default, *EchoLink* records (in a standard text file) every contact you make, as well as any connection attempts that you are not around to answer (see **Figure 21.6**). If you are very active on *EchoLink*, these logs and their Search function make recalling contacts a snap. You'll find the CALLSIGN LOG submenu under the VIEW menu.

TEST SERVER

Many users find *EchoLink* runs perfectly right out of the box. However, if you ever decide to make changes to your audio levels, etc., and want to hear how you sound to others, don't forget the Test (or echo) Server under the STATION menu. This test server will echo back anything you transmit, allowing you to check and tweak your volume and microphone levels (see **Figure 21.1**).

THE ALARM

When a user is in your Alarms list and they join or leave *EchoLink*, you will get a notifying pop-up window and a system sound. *Tip*: From either the Index or Folder view, right-click to pop up a menu for easily adding a user to either your Alarms or Favorites.

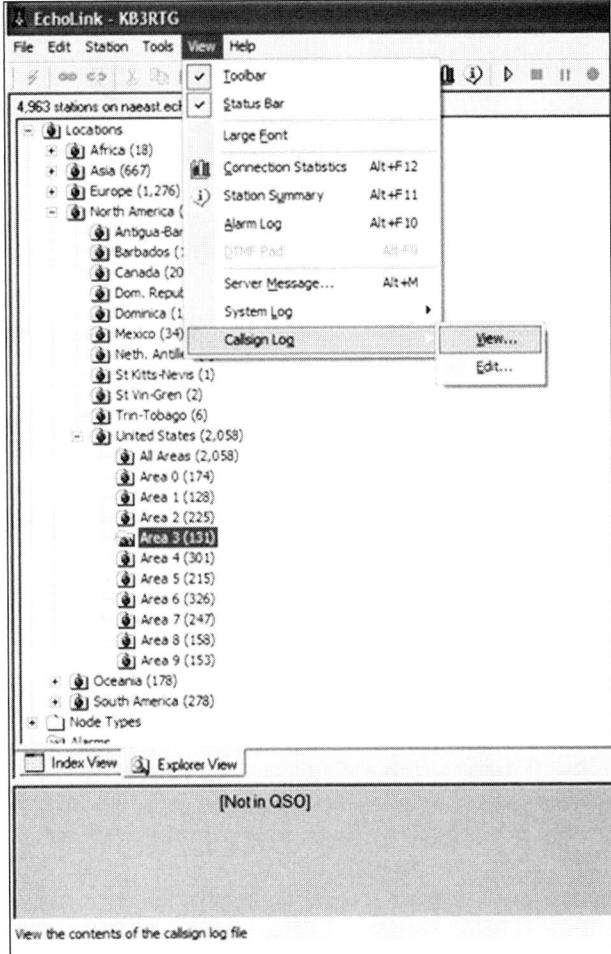

Figure 21.6 — *EchoLink* keeps a log (in a standard text file) of every contact you make, as well as connection attempts that you didn't answer. You'll find the CALLSIGN LOG submenu under the VIEW menu.

SECURITY

The software's Security feature, found under the TOOLS/PREFERENCES/SECURITY tab, allows any user to block another user from making contact (see **Figure 21.7**). In addition, you can block, say, countries with a primary language other than the one you speak.

PORT BLOCKS AND PROXY SERVER CURES

Firewall issues are discussed elsewhere in this chapter; however, a common problem users are experiencing involves ISPs who block incoming connections to your modem/router. This is especially troublesome with WiFi or wireless connections in hotels, coffee shops, and airport lobbies. When experiencing this *EchoLink* problem, you can connect to, say, Joe on *EchoLink*, but Joe can never connect directly to you. If you don't mind having unknown parties seeing all of your traffic, there is a solution for your connection problem. The solution is to use an *EchoLink* proxy server.

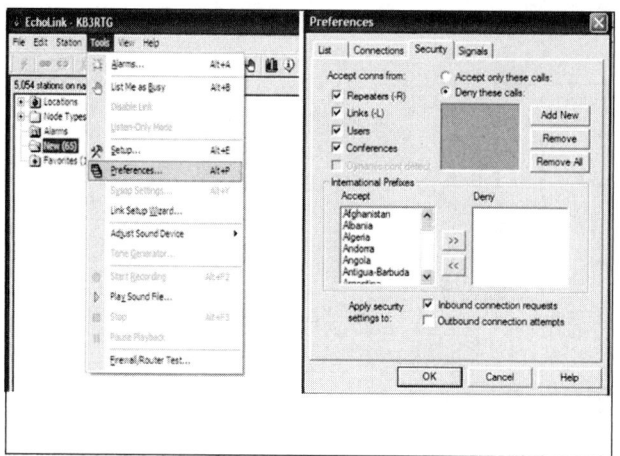

Figure 21.7 — You can set your Security preferences to block or allow contacts from certain people, and even certain countries.

The problem: you can connect to a peer out on the net but they cannot make direct connect to you. The solution: you connect to a third-party proxy server. The third-party proxy server in turn receives connections for you and relays them to your connection. In many ways the proxy server operates like a typical ham radio repeater.

The two most negative aspects of this solution are proxy availability when you need it and security. When using public Internet access, such as WiFi, any information on your computer that is set for sharing, as well as any Internet traffic, will have the potential to be visible to anyone using that access point. Using an *EchoLink* proxy server is no more secure than using an Internet access point that is not your own. *EchoLink* (version 1.8 and above) uses public-key cryptography to encrypt your login information, and *EchoLink* Proxy uses a challenge-response, digest-based authentication mechanism. Passwords are never sent "in the clear." These features help ensure security over a shared public network.

Enjoy your journey with *EchoLink*!

Chapter 22
The *Winlink 2000* System:
Its Use in Emergency Communications

Winlink 2000 is a free-of-charge worldwide radio e-mail network system that interfaces the de facto Internet e-mail system with HF, VHF, and UHF radio. The *Winlink 2000* system has become widely used in EmComm, and has become a reliable means for providing contingency communications for all levels of government and non-government organizations involved with Emergency Communications. This chapter will first describe *Winlink* at a high level, followed by real-world examples of first-class implementations for different communities with different constraints and needs.

Winlink is a system of programs that route e-mail between the Internet and Amateur Radio for both local and distant locations. At a local level, *Winlink* software is used to communicate with a local *Winlink* server station using packet radio. For communications beyond a local area, *Winlink* connects with a distant station over an HF link running PACTOR. The system also supports a sound card mode known as WINMOR for HF. WINMOR is an ARQ OFDM (Automatic Repeat Request; Orthogonal Frequency Division Multiplexing) protocol similar to PACTOR 3, but without the advantages of robust hardware. Unlike Forward Error Correction (FEC) systems that do not communicate directly with the recipient, *Winlink 2000* uses true data modes that maintain error correction between the two connected stations. In addition, *Winlink 2000* uses a compressed binary format that significantly reduces transmission times.

Winlink 2000 is real e-mail following Internet standards, and supports multiple recipients and attachments. In the post-9/11 world, *Winlink* usage has skyrocketed as a means to send attached ICS forms for served agencies. **Figure 22.1** lists the collection of programs available in the *Winlink* system. The *Airmail*, *Paclink*, and *RMS Express* programs are "client" programs used to retrieve mail from a *Winlink* server. *RMS Packet*, *RMS PACTOR*, *RMS WINMOR*, and *Linux RMS Gateway* are "server" programs that deliver mail to user

WINLINK PROGRAM	FUNCTION
Airmail	Client for Telnet, Packet and PACTOR. Limited TNCs supported. Self-contained email. Point-to-point supported.
Paclink	Client for Telnet, Packet and PACTOR. Many TNCs supported. Uses separate email program. RMS Relay supported.
RMS Express	Client for Telnet, Packet, SCS Robust Packet, PACTOR and WINMOR. Many TNCs. Self-contained email. Point-to-point supported. RMS Relay supported.
RMS Packet	Server for Packet. Many TNCs supported. RMS Relay supported
RMS PACTOR	Server for HF PACTOR 1, 2 and 3. Requires SCS TNC. Many HF radios supported for frequency scanning.
RMS WINMOR	Server for HF WINMOR. WINMOR is a sound card-based transfer protocol approaching PACTOR 2 speeds.
Linux RMS Gateway	Independently developed Winlink Packet server for the Linux operating system.
RMS Relay	Internet gateway for Winlink servers that can use a local mail database during internet outages. Supports PACTOR 3 HF mail forwarding.

Figure 22.1 — The *Winlink* system of software components.

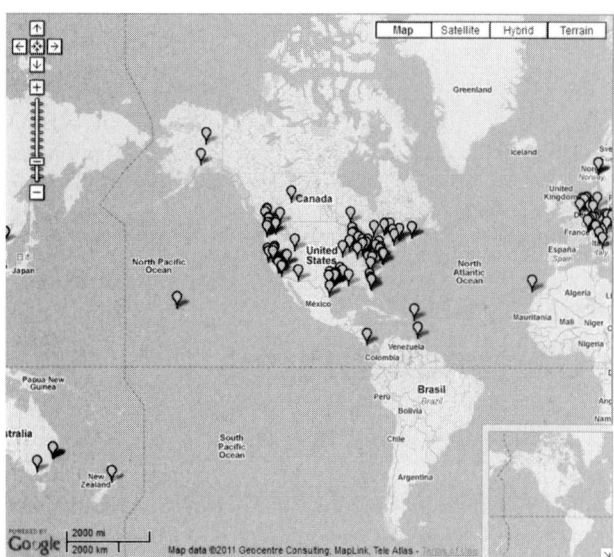

Figure 22.2 — Worldwide active *RMS PACTOR* and WINMOR HF *Winlink* servers.

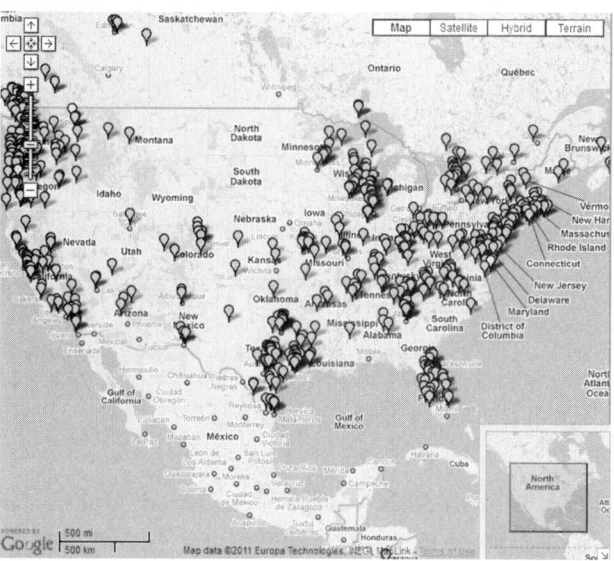

Figure 22.3 — *RMS Packet* VHF/UHF *Winlink* servers in North America. Positions of RMS Packet gateway stations are calculated by the WL2K servers and plotted randomly within the six-character grid square configured in the RMS station's packet channel. This map is intended as a tool for finding on-air stations, not as a way to pinpoint exact locations of gateway stations.

"clients." *RMS Relay* is an optional program that can allow for continued operation of a *Winlink* server during an Internet outage. *RMS Relay* considerations will be discussed later.

At a high level, *Winlink* supports amateur operators through both public servers intended for any amateur and dedicated stations intended for emergency communications with unpublished HF frequencies. *Winlink* is used for maritime emergencies almost weekly and in major disasters such as the Indonesian and Japanese tsunamis. *Winlink* also supports several government agencies, including the International Red Cross and the Canadian Forces Affiliate Radio System (CFARS), as well as the Military Auxiliary Radio Service (MARS) civil agencies at all levels. *Winlink* also works with a separate and extensive set of HF servers operating on DoD MARS frequencies (see "Army MARS Implementing *Winlink 2000* with Airmail Network," ARRLWeb, March 2, 2006). The EmComm, public, and DoD MARS systems interoperate seamlessly to provide complete interoperability. Since amateur HF bandwidth is a precious commodity, the local VHF Amateur and MARS HF *Winlink* assets are an important part of the system.

Figure 22.2 is a screenshot of the RMS HF stations worldwide, including public and EmComm PACTOR and WINMOR servers. As you can see, it is a very active system. **Figure 22.3** shows active RMS Packet stations in North America. The number of these stations has increased dramatically over the last decade.

GETTING STARTED WITH *WINLINK*

Winlink has "client" and "server" software. You connect to a *Winlink* e-mail server to send and receive e-mail over the system. Other *Winlink* servers are RMS PACTOR and *RMS WINMOR* for HF e-mail using PACTOR or the sound card-based WINMOR protocol.

Paclink, which may use multiple originating users, can connect to the *Winlink* system using Telnet, VHF/UHF Packet, or HF PACTOR stations. Another program, *RMS Express*, can do what *Paclink* does but also supports sound card-based WINMOR. Unlike *Paclink*, which uses external e-mail software, *RMS Express* has a built-in e-mail interface. At the time of this writing, WINMOR is not recommended for Emergency Communications. To use WINMOR successfully under stressful situations, WINMOR sound card operators need specialized skills and practice in making multiple adjustments to the software, computer, and radio simultaneously. Amid the stress of emergency situations, this is often difficult for those not familiar with this program. Another *Winlink* "client" is the venerable *Airmail* program authored by Jim Corenman, KE6RK. Links to all this software, including server versions, are available at **www.winlink.org**. All of these programs have extensive and well-written help files that make installation, configuration, and use easy.

EXTENDING *WINLINK* CAPABILITY: *RMS RELAY*

Winlink servers use the Internet to route *Winlink* traffic to and from the redundant *Winlink* Central Message Servers (CMS) around the world. But what if the Internet is lost to the Radio Message Server (RMS)? There's a simple answer: *Winlink* servers without the Internet refuse traffic. Local areas often choose to set up multiple *RMS Packet* servers using a diverse selection of Internet providers to reduce the chances of this occurring, but if there is a complete Internet outage in the area, you get stuck for local connections. The *RMS Relay* program resolves this. *RMS Packet* stations can be set up to work through RMS Relay to allow continued operation during an Internet outage. When the Internet is lost, *RMS Relay* uses a local database to store traffic until the Internet is restored. Since a large percentage of emergency communications traffic is local, local client stations can still exchange most traffic through this Internet-stranded *RMS Packet* station using *RMS Relay*. *RMS Relay* also supports HF forwarding via PACTOR 3, allowing this locally stored traffic to also be routed outside the area as well as routing in mail for any recent users of the system. As with any HF connection, it is very important to listen to the channel before connecting, to avoid interference. *RMS Relay* requires this HF connection to be done manually by the sysop. *RMS Relay* is a wonderful addition to the *Winlink* system of tools, but it should be used thoughtfully. A local *RMS Packet* station running *RMS Relay* in an outage when other Internet-connected *RMS Packet* stations are available can result in stranded mail, as can running multiple sites with *RMS Relay*. Many locations run *RMS Relay*, but don't check the ENABLE RMS RELAY box in the *RMS Packet* site properties setup page unless the sysop decides its use is warranted. This "standby" configuration keeps *RMS Relay* updated with any new versions.

DESIGNING A *WINLINK* EMCOMM SYSTEM FOR YOUR AREA

The best way to use *Winlink* will vary with your local situation. Let's examine some different scenarios for implementing an effective *Winlink* system. Here we will draw on some examples from Texas, one of many parts of the country making good use of the *Winlink 2000* system.

Case 1: McCulloch County, Texas

McCulloch County, Texas has only 8200 people, with most concentrated in the town of Brady, population 5500. Other towns are spread out. The terrain is very hilly and there are no adjacent counties with ARES activities. This is a very isolated place with a county EOC that would be the communications customer to back up. This is an ideal location for an HF *Winlink* setup. There are a number of *RMS PACTOR* EmComm servers in the section, which makes a horizontal NVIS antenna the best bet to minimize interference with other *Winlink* traffic. *Airmail*, *Paclink*, or *RMS Express* could be used for this application.

Case 2: Orange County, Texas

Orange County, Texas has about 10 times the population of our first example (around 85,000 people) and is situated on the coast with terrain much more friendly to local propagation. They have elected to run *RMS Packet* in conjunction with two additional *RMS Packet* servers in separate parts of neighboring Jefferson County. A high digipeater in nearby Beaumont also provides good connectivity to an *RMS Packet* server in Sabine County, 75 miles away. This wide geographic diversity greatly increases the chance of maintaining communications after a hurricane. In addition, they run *RMS Relay* on standby and they are adding an SCS PACTOR 3 TNC to enable HF e-mail forwarding via *RMS Relay*. In previous hurricanes, they have relied on a separate PACTOR 3 client setup, but the ability to route all the mail in and out with *RMS Relay* HF forwarding will be a big step up for them. They are in a good position to add additional *Winlink* VHF client locations and to maintain portable packet-equipped stations that can be deployed to shelters or other needed locations.

Case 3: Central Texas

Central Texas includes the much more populated counties of Travis (population over 1,000,000) and Williamson (population over 400,000), plus surrounding counties with dramatically lower population. The area has a number of EOCs with *Winlink* capability as well as a system of 25 *Winlink*-equipped hospitals. With this number of users scattered over a wide area, a good deal of planning and experimenting went into creating a frequency and server plan to make all of this work well together. **Figure 22.4** shows all the *RMS Packet* servers in Central Texas as the plan has evolved. The different shades indicate stations on different frequencies to increase the bandwidth.

Originally all the 25 hospitals used simple (non-*Winlink*) packet to do simple messaging back to the Combined Transportation Emergency Communications Center (CTECC) in Austin with a single frequency and a number of long range digipeaters. Communications was all point-to-point and even a low level of traffic was painfully slow. Today, this primitive system has been replaced by *Winlink*. A requirement set by the hospitals was for the system to function without the Internet. To accomplish this, CTECC runs *RMS Packet* with *RMS Relay*. To increase the bandwidth, CTECC added a second *RMS Packet* server on a separate frequency. The next step in the evolution of this system was adding more servers. Each box in **Figure 22.4** indicates an *RMS Packet* server

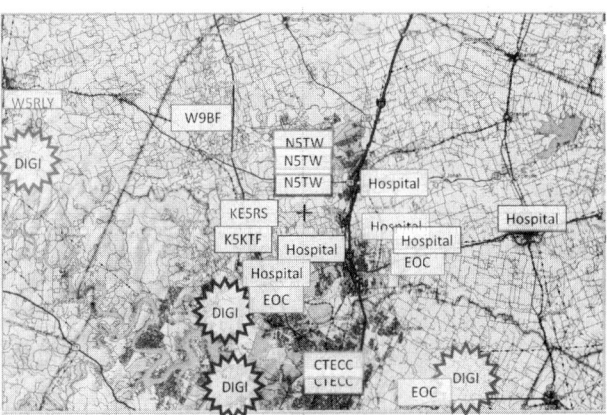

Figure 22.4 — *Winlink* implementation in central Texas. This map is intended as a tool for finding on-air stations, not as a way to pinpoint exact locations of gateway stations.

in operation today. All hospitals in Williamson County have Internet connections and run servers on additional frequencies. As long as a hospital continues to have a working Internet connection, its server can support other hospitals that might have lost their Internet connection and user mail can use the much faster Telnet option to send traffic. They do the same thing in a couple of EOCs, and a number of individual amateur stations also run servers.

Amateur Radio station N5TW runs *RMS Relay* with HF forwarding and three instances of *RMS Packet* on three frequencies. Since this is the only station in the area with HF forwarding, it was beneficial to run three stations to maximize bandwidth. The station uses four-bay antennas on tall towers and all area hospitals and EOCs have a good path to it. Using a duplexer allowed two of these stations to share one of the two antennas. The station has both battery and generator backup power. The CTECC facility also has plans to add HF forwarding to their installation and this will eliminate any single point of failure in this system.

This central Texas system has incredible redundancy and a very wide geographic footprint that makes an overall Internet failure unlikely. *RMS Relay* with HF forwarding covers even that eventuality. Multi-hour exercises have run with all facilities occupied and simulated an area-wide Internet outage with excellent results. Even with all facilities using the three N5TW *RMS Packet* frequencies, significant levels of traffic were handled efficiently and the HF forwarding was able to keep up. Central Texas also has portable stations called the "packet cavalry," which can be dispatched to shelters or other temporary stations. All servers are also set up as digipeaters and this means the "packet cavalry" can deploy portable servers to a location with the Internet and be reached by one of these digipeaters. What could make the system even better? Some areas are providing high-speed links using mesh nodes or other techniques to allow for Telnet *Winlink* speeds even during outages. Packet then backs up these links, should something cause a failure.

OVERALL SOUTH TEXAS EXAMPLE

Figure 22.5 shows the Texas *RMS Packet* stations. South Texas has been emphasizing *Winlink* since SEC Jerry Reimer, KK5CA, began promoting it in 2003. During that time, he wrote a two-part series published in *QST* entitled "*Winlink* for ARES" (*QST*, August 2004, September 2004).

Since then, the number of stations has soared from a cluster in the Houston area to stations in most populated parts of the state. Each of these servers has local clients in EOCs and hospitals depending on them.

Why not just use HF PACTOR instead of relying on local Internet connectivity? The amount of HF spectrum in which PACTOR 3 is allowed for unattended operation is tiny and would soon saturate with a small number of users. Even using attended servers in emergency conditions has limits on what can be supported reasonably. Another reason is the expense of VHF installations is much lower and these setups can be operated by all license class operators. Finally, most

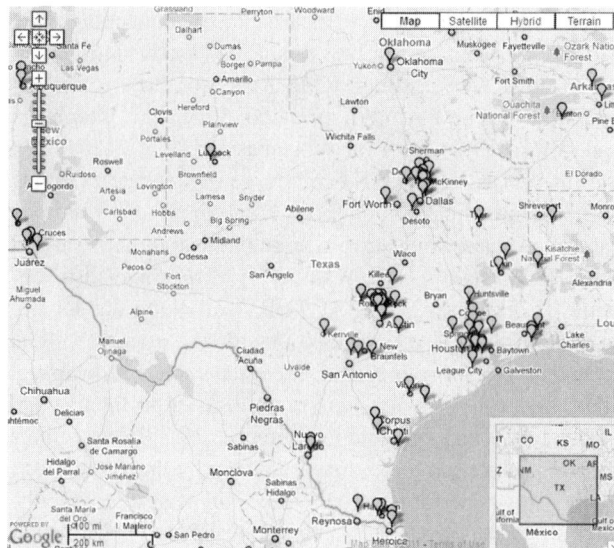

Figure 22.5 — Texas *RMS Packet* Servers.

emergency traffic is going to be local. A local installation with one station running *RMS Relay* with HF forwarding means the local traffic can be handled with no waiting and traffic for outside areas can still be accommodated. Save the HF bandwidth for small, isolated stations and for *RMS Relay* HF forwarding.

After a major hurricane, large areas can be without power and Internet coverage, and this means HF is necessary and the bandwidth needed can be substantial. While there are many RMS PACTOR servers across the county, we made a decision in South Texas to add a number of relatively close-in RMS PACTOR servers to support this higher bandwidth. More local servers mean that lower frequencies can be used with NVIS antennas where interference with longer-distance, higher-frequency stations can be minimized. We have had excellent experience using 80 meter frequencies in the middle of the day to carry PACTOR traffic at very reasonable rates using frequencies not being otherwise used. South Texas has five EmComm RMS PACTOR servers and an increasing number of WINMOR servers.

THE TEXAS RAPID RESPONSE TASK FORCE — FORWARD-DEPLOYED *WINLINK* SERVERS

Like many progressive states, the state of Texas created a collection of resources to deploy assets to an affected area immediately after a hurricane. ARES and Texas Army MARS both participate in providing backup communications for these "Rapid Response Task Forces." North and South Texas sections work together to staff these teams which stage out of Dallas, Waco, Austin, and San Antonio and then deploy when winds drop below 39 MPH. We were fortunate to have the state of Texas provide eight SCS PACTOR 3 TNCs to be placed in eight HF go-kits. Both Texas Army MARS and ARES team members bring both voice and HF *Winlink* to these task forces. This, of course is key to the concept of

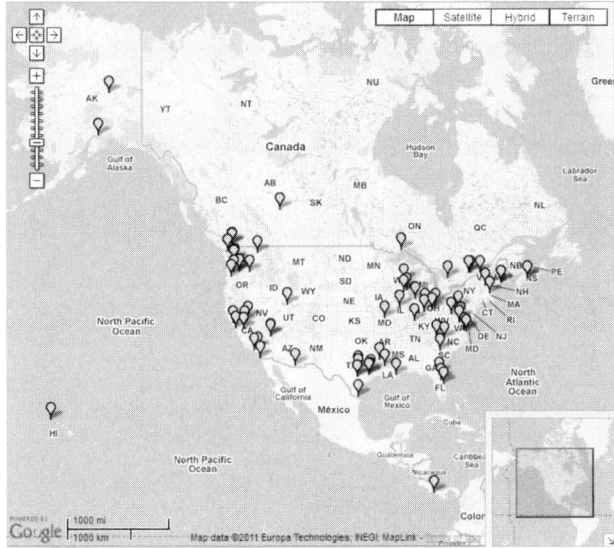

Figure 22.6 — MARS *RMS PACTOR* Stations.

agreed to provide satellite Internet when available, which will make the HF forwarding only needed as a backup.

MARS AND THE *WINLINK* SYSTEM

The Military Auxiliary Radio System (MARS), with the help of its Federal, State, UASI, District, and DHS tiered non-government Agencies has built an extensive *RMS PACTOR* HF network (see **Figure 22.6**). Each of these servers scans five channelized frequencies. The *Winlink* HF bandwidth is simply incredible compared to what can be done in the crowded, small slivers of frequency available to regular HF operations. An ever-increasing number of agencies are taking advantage of this excellent resource. The government and the regular amateur *Winlink* system work seamlessly with each other to provide complete interoperability for an effective radio e-mail bridge into the Internet e-mail system.

interoperability. We decided to take our ARES *Winlink* efforts a step further to create forward-deployed RMS Packet servers with *RMS Relay* plus HF forwarding. Many of our coastal *RMS Packet* stations do not have HF forwarding and this provides a way to connect these stranded *Winlink* assets to the outside world. We have proven this to be a very useful capability in large exercises. ARES task force members carry two completely redundant setups and this adds additional capability if the teams split up. *Winlink* packet-equipped Texas Army MARS members can also use these ARES *RMS Packet* servers. The Texas military forces have

IN CONCLUSION

The *Winlink* system provides an outstanding backup e-mail system with the flexibility for both local and distant coverage. In the post-9/11 world, sending ICS forms, disaster assessment photographs and other attached data such as spreadsheets has become a necessary part of the National Incident Management System process. As the examples have shown, the *Winlink* system can scale to match the needs of communities small or large and can be implemented to deal with widespread Internet outages.

The Yahoo group WL2KEmComm is a good resource for overall questions and discussion about the *Winlink 2000* system.

Chapter 23
D-STAR for Emergency Communications

D-STAR, or Digital Smart Technology for Amateur Radio, is an open standard for digital voice and data transmission developed by the Japan Amateur Radio League (JARL). D-STAR offers some unique capabilities for assisting in emergency communications. Since D-STAR is based on an open standard, new features will continue to be developed.

The D-STAR signal transmitted is a 4800 bps digital signal with 50 percent allocated for voice, 25 percent allocated for Forward Error Correction of the voice, and 25 percent for data. The D-STAR signal is more spectrally efficient than FM, occupying only 6.25 kHz of bandwidth, or roughly half of a traditional FM signal. Commercially available D-STAR radios are capable of both D-STAR and FM modes and include a serial port for data transmission without an external TNC. As with Automatic Packet Reporting System (APRS), the data stream can be used for sending text or GPS data for location information. A station's call sign is always sent with every transmission.

D-STAR on the 2 meter, 70 cm, and 23 cm bands is used in ways similar to the traditional FM mode. Communications between two or more stations can be made on simplex frequencies with a range similar to FM. D-STAR repeaters increase the range of stations, again, as in FM. An added feature of many D-STAR repeaters is the ability to connect to the Internet for linking two or more repeaters together.

In D-STAR mode, the operator quickly notices the clear and consistent audio quality whether operating simplex or working a station across the world via a linked repeater. When first hearing voice on D-STAR, the audio quality may sound slightly different from the FM mode, but the ear quickly adjusts to the difference. In FM, as the received signal strength from a station decreases, the static level increases until it is difficult to understand the voice; in D-STAR, the voice quality remains constant, with no static as the signal decreases. At the point that the signal is very weak, the voice becomes garbled and then mutes.

In the 23 cm (1.2 GHz) band, D-STAR offers a higher-speed digital data (DD) transmission mode with data speeds of up to 128 kbps. This mode may be used between two DD mode-capable radios as a wireless Ethernet bridge. Repeaters equipped with a 1.2 GHz DD module can provide Internet

Figure 23.1 — The WD4STR D-STAR repeater is located in Lawrenceville, Georgia, in the northeast suburbs of Atlanta. It has been in service since 2008 and is used by Gwinnett County ARES as part of its voice and data communications system. [JOHN DAVIS, WB4QDX]

connectivity from a remote radio. While the faster data speeds are available in the 23 cm band, the propagation characteristics in this band are considerably different and are more sensitive to obstructions and distance limitations. D-STAR implements a wider bandwidth modulation in the 23 cm band, which provides a 128 kbps data rate. It is data only and is referred to as the digital data (DD) mode.

USING D-STAR IN EMERGENCY COMMUNICATIONS

D-STAR offers some unique advantages for emergency communications, but it should be considered just one tool in the toolbox of Amateur Radio equipment and modes. It will not be appropriate for all situations, just as HF can be more suitable than 2 meter operation in a particular application. Some may believe that D-STAR may not be appropriate for emergency communications since its linking capabilities use the Internet, as does EchoLink or Internet Radio Linking Project (IRLP). However, with the loss of the Internet, the local repeater (**Figure 23.1**) would still be available with the same voice and data capabilities, minus the repeater linking feature. In the event of repeater loss, simplex operation would still be possible with the same features.

In an EmComm environment, D-STAR can have several advantages. First, the ability to easily link repeaters to other areas enables rapid establishment of communications between areas. Linking to reflectors can set up repeater networks over multiple or large areas. Reflectors are like conference bridges for D-STAR and are server-based devices in locations with access to higher-bandwidth connections, such as data centers or large networks. When linked, anything transmitted on one repeater is heard on all linked repeaters with no degradation in voice quality. Linking a local repeater to another repeater or reflector can be initiated by a station on the local repeater with simple commands (**Figure 23.2**). This allows ad hoc networks to be quickly configured without special infrastructure as in most linked repeater networks.

D-STAR DATA

One attractive quality of D-STAR is its built-in data communications. All D-STAR radios are equipped with a serial data port and can easily connect to a laptop or other computer with just a cable. No additional equipment, such as a TNC, is necessary. Sending data between two or more stations on the same repeater or through linked repeaters is easy, and simple keyboard-to-keyboard messaging can be sent with only simple terminal software.

More advanced data communications among D-STAR stations is possible with readily available software programs. One example is *D-RATS*, a free program written by Dan Smith, KK7DS, that is available for *Windows*, Mac, and Linux com-

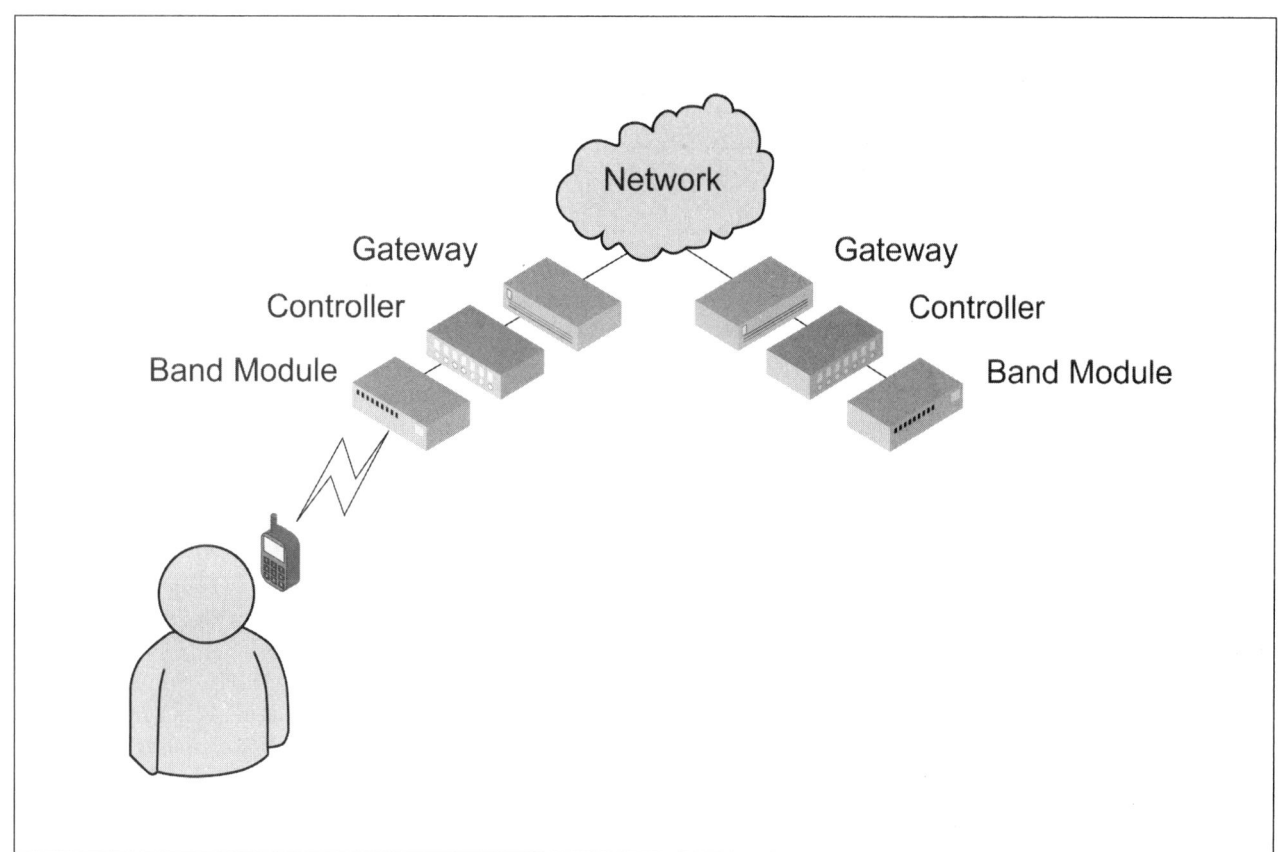

Figure 23.2 — D-STAR linking between two repeaters via the Internet. [DSTARINFO.COM]

D-RATS: A COMMUNICATIONS TOOL FOR USE WITH D-STAR

The *D-RATS* Communications Tool is a multi-purpose computer program for D-STAR (Digital Smart Technologies for Amateur Radio) that does not require a single piece of equipment other than a computer. The name of the software is a play on words. D-STAR spelled backwards is RATS-D. By moving the trailing D to the front of the acronym, you get *D-RATS*.

This free, open source program was written, maintained, and copyrighted 2010 by Dan Smith, KK7DS. Updates and beta versions adding new features are released periodically. The software can be downloaded at **www.d-rats.com**. This is also the place to find documentation, news, bug tracking, and recent updates, and join the mailing list for all things related to *D-RATS*.

This project has been in active development since at least early 2009. There is a link on the website if you would like to contribute to its development. The program will run on modern versions of Linux/Unix, *Windows*, and the Mac OSX operating systems.

D-STAR is often thought of as a new Amateur Radio voice mode. The other side of the technology allows for efficient data transmission, and this is the focus of *D-RATS*. The program enables users to transport data in a myriad of forms via radio frequency (RF), local area network (LAN), Internet, and terminal node controller (TNC), and also serves as a *Winlink 2000* gateway.

Simple data can be exchanged in a chat window. E-mail and messages can be formatted in a variety of formats, including National Traffic System (NTS) radiograms and Incident Command System (ICS) forms. Computer files can be exchanged among users. The application also takes advantage of the Global Positioning System (GPS) capabilities of the D-STAR mode and will track the position of users, calculate the distance/direction of users, and support static beaconing, all on an integrated map viewer with an offline caching capability.

While this software was designed around utilizing a variety of Amateur Radio gear, you do not have to have any equipment in order to use it. All you need to do to initially employ and test the functionality of this program is to download it and install it. You can use it as-is over your LAN or the Internet. Once you download *D-RATS*, you are only minutes away from being able to utilize this robust communication suite.

---

puters. It adds easy-to-use features for messaging between stations (including e-mail to SMTP addresses), file transfer, and mapping using GPS. For messaging (e-mail and chat), a familiar-looking e-mail interface allows the user to start using *D-RATS* quickly. A feature called Repeater Proxy allows bridging a D-STAR radio to the Internet to allow users without D-STAR radios to utilize all the program's functions via an Internet connection and bridge to other radio users.

In EmComm applications, *D-RATS* allows the easy sending of messages, forms, and files between stations. A typical use would be in communicating from a shelter location to an EOC. Messages composed in either a memo or e-mail format can be addressed to a station's call sign or a tactical call sign entered in the PREFERENCES screen. Small attachments can also be included with a message as in e-mail.

Another *D-RATS* feature that is particularly useful in EmComm is the ability to transmit forms. Several pre-defined forms, such as FEMA's Incident Command System ICS-213 and the ARRL's National Traffic System NTS Radiogram, are included, but custom forms can be designed by the user. For example, a form to report shelter occupancy could be quickly developed, allowing the originating station to fill in the form, address it to a station, and send the data. The receiving station will open the form like an e-mail message, view the report in the same format, and print the message for delivery to the served agency.

D-RATS also offers unattended file transfer. In the program setup, each station can designate a folder for the sharing of files. As in FTP programs, users can connect to another station, view the files in the shared folder, and transfer a file from their computers to the remote station or download a file from the remote station to their local computer. File transfers should be limited to smaller file sizes since the transmission speed is 1200 baud. For more information on *D-RATS* and to download the program, visit **www.D-RATS.com**.

For applications requiring higher data speeds, recall that D-STAR's DD mode allows speeds of about 128 kbps in the 1.2 GHz band, though the propagation characteristics of this band can limit transmissions to shorter distances and is affected by obstructions.

Served agencies, especially emergency management organizations can benefit from D-STAR's high-speed data mode in sending photos from an aircraft while airborne for near-real-time transmission. An aircraft equipped with a D-STAR 1.2 GHz radio (such as ICOM's ID-1), a laptop, and a camera can fly over an affected site, take a photo, and send it to an Emergency Operations Center (EOC) via a repeater site providing the Internet connectivity (**Figure 23.3**). The radio acts as a wireless bridge to the repeater and the Internet from the laptop using its standard Ethernet port connected to the radio. Transmissions from an aircraft allow sending over longer distances — typically 50 to 75 miles, depending on altitude - by increasing the line-of-sight path. Uploading photos to a website allows easy viewing by a web browser, or downloading for printing.

GPS ON D-STAR

GPS information can be transmitted over the data port of a D-STAR radio. The radio's serial port is compatible with NMEA-formatted GPS data available from many GPS receivers. The GPS information can travel along with the voice on every transmission for position reporting or beaconing at user-defined intervals. Beaconing is not usually recommended when using a repeater, since the radio sends a transmission over the air and through the repeater, and that can interfere with ongoing voice transmissions. Configuring a radio to send GPS data with every voice transmission allows a station's position to be displayed using one of the many APRS mapping programs or Web sites. Using the GPS-A configuration in a D-STAR radio formats the GPS strings to an APRS-compatible format. Most D-STAR repeaters will gate the GPS information to APRS-IS (Internet service) systems for use with Web-based mapping and display programs, just like APRS digipeaters with I-GATE functions. This allows a D-STAR station's position to be displayed along with stations transmitting APRS. Sending position information from a D-STAR radio does not require an additional radio and TNC to handle both voice and position data.

Figure 23.3 — Scott Hartlage, KF4PWI, is shown preparing for a flight using D-STAR's high-speed data transmission mode as a part of the 2010 Simulated Emergency Test by Gwinnett County Georgia ARES. The flight demonstrated how airborne photos could be taken and immediately transmitted via D-STAR to the Gwinnett County Emergency Operations Center. The photos were also immediately available to the Georgia Emergency Management Agency State Operations Center. [SCOTT HARTLAGE, KF4PWI]

Part 8: Other Relevant Organizations

Chapter 24
Hams and Military Working Together: How MARS Serves National Security

Editor's Note: Amateur Radio operators who are also volunteer MARS members created this entry as a public service. This entry does not represent official positions of the Department of Defense, the Department of the Army, the Department of the Navy, or the Department of the Air Force.

Mars is known as the Roman god of war. MARS is also the acronym for the Military Auxiliary Radio System and, appropriately so, as the organization's role expands from a military-only operation, to one of serving national security and emergency management efforts. Mars was all-powerful, according to ancient lore, but he did serve his subjects well, just as MARS' expanded role allows the organization to serve our country better.

It was Army Captain Robert L. Gabardy, K4TJ (SK), who renamed what had been the Army-Amateur Radio System (A-ARS) before World War II to the Military Amateur Radio System (MARS).

A BIT OF HISTORY

In 1902, when radio was a "new technology," it was the Amateur Radio operator who proved it worked. Marconi's Trans-Atlantic transmission from Wellfleet, MA to the United Kingdom was such proof. Further key developments were also the result of amateur experimenters. For example, the triode tube, developed by Lee de Forest, made radio more reliable.

By 1914, Amateur Radio was thriving, though it had still not been formally recognized by federal authorities. Amateur Radio was formally recognized as a function of the Commerce Department just as World War I began. Radio communication was to have a profound effect on the conduct of the war.

American observers in Europe had seen the value of radio to the military as Europe's troops battled in the trenches of France and Belgium. The interest shown was, in no little part, the result of the visionary work by the Army Signal Corps, which had worked with privately owned amateur stations in their experimental work.

The military value of radio communications made collaboration between the military and Amateur Radio pioneers inevitable. Within 10 days of the United States' entry into World War I on April 7, 1917, approximately 500 Amateur Radio operators — or nearly 10 percent of the nearly 6,000 US amateur operators licensed — had enlisted in the Navy. With the signing of the Armistice on November 11, 1918, amateurs began clamoring to have wartime restrictions on their activities lifted. The Commerce Department worked to remove the wartime restrictions so that by 1921, amateurs were fully back on the air. Commerce had used a gradualist approach to reinstating the Amateur Radio Service as frequency allocations had changed with the addition of commercial interests to broadcasting.

Incorporation of Amateur Radio operators into specialized units dates to 1925 when both the Naval Communications Reserve and Army-Amateur Radio System (A-ARS) were established. The first offered actual rank and uniforms to eligible amateur licensees. The Army opted for a less formal alliance with civilian amateur licensees, and this arrangement eventually prevailed across all three MARS Services.

On August 7, 1925, Major General Charles McKinley Saltzman, Chief Signal Officer of the US Army, wrote to the American Radio Relay League's (ARRL) founder and president, Hiram Percy Maxim, formally proposing a partnership. One express purpose, he said, would be,

"To secure additional channels of communication throughout the continental limits of the United States that can be used in time of an emergency such that the land lines, both telephone and telegraph are seriously damaged or destroyed by flood, fire, tornado, earthquake, ice or from other causes."

There was also the firm intention of building a pool of trained operators who could be immediately mobilized in the event of need. MARS' mission remains largely the same today.

The great 1928 September Hurricane that racked southern Florida, claiming an estimated 2500 to 3000 lives in Palm Beach County thrust the fledgling A-ARS into the national headlines when pioneer members Ralph Hollis, 4AFC, and Forrest Dana, 4GR, remained on the air at Palm Beach for over four days without a break, as the only means of long-distance communications, to get appeals for help through to Washington.

Figure 24.1 — Checking out the new MARS on December 30, 1948 are Major General Francis L. Ankenbrandt (left), chief of the Air Communications Group at USAF HQ in the Pentagon, and Major General Spencer B. Akim (right), the Army's Chief Signal Officer. [*AIR FORCE COMMUNICATIONS COMMAND: 1938-1991 — AN ILLUSTRATED HISTORY*, THOMAS S. SNYDER, GENERAL EDITOR]

Figure 24.2 — This 1955 photo shows an Air Force MARS station in Taegu, Korea. Stations like this one enabled servicemen to contact their loved ones at a low cost.

"By this means," *QST* recorded, "the first reports to the American Red Cross were made, starting the relief machinery of that organization, and by the same channel the word was conveyed that brought Army blankets, cots and supplies for the stricken area from Atlanta." Belated recognition of Hollis and Dana finally came with their posthumous inclusion in the *CQ* Magazine Amateur Hall of fame in 2010, which was the 85th anniversary of the original A-ARS.

In 1941, the A-ARS won strong praise for providing Texas authorities with emergency communications during another September Hurricane. But only 2 1/2 months later — on the day after Japanese planes attacked Pearl Harbor — the Signal Corps shut the system down and referred future emergency backup requests to a wartime civil agency. Not until 1948 was the service reactivated under the new MARS acronym, with two independent branches: Army and Air Force (**Figure 24.1**). The third MARS service, Navy-Marine Corps MARS, followed in 1962, just as the US troop and fleet buildup got underway for the Vietnam War.

Two notable AF MARS members, Senator Barry Goldwater, K7UGA / AFA7UG (1909–1990) and General Curtis LeMay, K0GRL / K4RFA / W6EZV (1906–1990) were also posthumously honored by *CQ* Magazine in its initial Amateur Hall of Fame in 2001. Goldwater was Amateur Radio's ardent champion in Washington. His station handled thousands of MARSGRAMS during the Vietnam War. General LeMay remained active on the ham bands while serving as chief of the Strategic Air Command and, later, as Air Force Chief of Staff.

In Vietnam as in Korea, MARS had refocused its resources from domestic disaster to morale-building for service personnel abroad and at sea. Members provided free connections to family and friends back home. No count could be kept, but there must have been millions of MARSGRAMS and phone connections exchanged (**Figure 24.2**).

MARS was better prepared when US military involvement in Vietnam sharply escalated during the early 1960s. By the war's end, the Signal Corps had 47 stations in Vietnam, and the Navy had MARS stations on most large ships. In 1970, these stations were completing an average of 42,000 phone patches a month. This was a time of great pride for MARS members as families and war-fighters alike showered the organization with thanks.

But MARS work wasn't for the faint-hearted. Radio stations of any kind were an important military target, and, as the photo of the NMCMARS station (**Figure 24.3**) shows, N0EFA, the 3rd Marine Division's HQ MARS station just south of the DMZ (Demilitarized Zone) frequently came under artillery fire from the North. "By early 1968," the N0EFA website at **www.marinecorpsmars.com/usn-mc_house/ N0EFA/n0efa.htm** records, "the old Alpha [station] was badly damaged…wounded operators and destroyed equipment were taking their toll." Navy-Marine Corps MARS had its own network of two dozen stations in Vietnam manned entirely by licensed hams; the Corps had no job title for MARS operators.

By the first Gulf War (1990–91), e-mail and satellite phone service had begun replacing the morale message trafficking

Figure 24.3 — Navy-Marine Corps MARS station under fire in Vietnam. [COURTESY JIM ELSHOFF]

role played by the Military Affiliate Radio System, but there was still plenty of need for the traditional MARSGRAMS and "morale & welfare" phone patches during Desert Storm.

RESUMING THE DISASTER MISSION

In 1993, Army MARS conferred with an officer from the Army's Directorate of Military Support (DOMS) to update the role of volunteer operators in home-front disasters. Within weeks, southern California suffered the catastrophic Northridge earthquake of 1994. All communication was wiped out, and freeways, railroads, and airport runways sustained crippling damage. DOMS requested MARS members to survey what facilities might be immediately usable for getting relief supplies into the San Fernando Valley. Mobilized via HF radio, southern California MARS members provided the requested "ground truth."

The operational message form now known as the Incident Notification / Essential Elements of Information (IN / EEI) grew out of the Northridge response. This is a standardized checklist of data required for organizing immediate response from outside. All three MARS services adopted its use.

By order of Department of Defense Directive (DoDD) 4650.2 on January 26, 1998, the Department of Defense (DoD) established a policy of actively supporting the MARS mission "to provide emergency communications on a local, national, or international basis as an alternate communications capability." DoDD 4650.2 tasked various senior officials within the Office of the Secretary and Defense, along with the Secretaries of the Military Departments, with organizing and managing the MARS program. DoDD 4650.2 was superseded by DoD Instruction (DODI) 4650.02, on December 23, 2009, which renamed MARS the Military "Auxiliary" Radio System, upgrading its status to more closely resemble military auxiliaries like the Civil Air Patrol and Coast Guard Auxiliary; broadened the MARS mission; and expanded benefits for MARS members.

RECENT EXAMPLES OF MARS EMCOMM ACTIVITIES

Hurricane Katrina, August 2005

In August and September 2005, Air Force MARS operated with Army and Navy/Marine Corps MARS as well as with other Amateur Radio-based emergency response organizations to support state and local emergency management agencies and Non-Governmental Organizations. MARS disaster relief efforts were concentrated in Louisiana, Mississippi, Alabama, and Texas in response to this federally declared disaster. Broad working relationships with the National Guard and other military entities had not been established at that time; MARS operations were typically carried out using mobile or portable equipment powered by battery or generator with field-expedient antennas.

Hurricane Dolly, July 2008 / Hurricane Gustav, September 2008

Hurricane Dolly hit southern Texas in July 2008. From July 22–24, MARS members worked with Texas National Guard communications teams to create communications nodes where information could be passed between Texas National Guard and MARS nets, to assist in handling the heavy volume of coordinating traffic between the State Operations center and National Guard response units. This case illustrates how MARS members affiliated with the state EOC may, in an actual emergency, provide contingency communications using MARS capabilities until regular communications can be established.

From September 1–4 2008, in response to Hurricane Gustav, deployment teams from Army MARS provided backup voice and digital communications at several evacuation centers in Mississippi and Louisiana, and National Guard refueling points in Texas. Throughout the emergency, some 850 Army MARS volunteers in FEMA regions four and six were on standby to relay critical message traffic from their home stations, a goodly number of them ready to respond with portable Emergency Communications rigs if needed. Fellow hams from the Air Force and Navy-Marine Corps branches of MARS shared net operations during the emergency in a carefully-prepared demonstration of inter-operability.

New England and Midwest Ice Storms, December 2008

A combination of strong coastal storm and weather systems over the US and Canada resulted in ice accumulations of 1.5 inches or more, flooding, wind damage, power and communications outages, with an estimated one million people without power and heat for one week. This regional disaster affected all of New England and extended as far south as Kentucky (see **Figure 24.4**) and Tennessee. MARS worked from December 11-18 with ARES, RACES, and

Figure 24.4 — A December 2008 ice storm that affected all of New England had far-reaching effects, made evident by this photo taken in Kentucky. [COURTESY TENNESSEE EMA]

SKYWARN to provide vital communications to Emergency Operations Centers, evacuation shelters, and the National Weather Service offices.

Presidential Inauguration, January 2009

The US Department of Homeland Security's request for a nationwide activation for the Presidential Inauguration resulted in the participation of 134 USAF MARS stations and establishment of a continuous on-the-air liaison with the FEMA National Emergency Coordination Net, National Communications System SHARES network, US Army MARS C2 nets, and the Ft. Monmouth Joint MARS Communications Center for the purpose of disseminating information on events during the January 20, 2009 Inauguration and to pass emergency traffic, if required.

MARS TODAY

As of April 2011, over 5,600 MARS members (2,900+ Army, 1,600+ Navy-Marine Corps, and 1,100+ Air Force) hold their stations in readiness to provide contingency communications support, should they be called upon to do so, to emergency responders in the event of disasters, including to military, federal, state, and NGO/civilian entities such as the American Red Cross, when normal communications are unavailable. MARS members continue to train and equip themselves, at their own expense, to maintain a pool of skilled radio operators who are ready to assist. MARS members supply their own equipment and volunteer their time and skill, and the DoD supplies the frequencies and defines the MARS missions.

Today, MARS is focused on longer-distance HF traffic-handling, particularly for national or federal agencies. Each of the three Armed Forces branches — Army, Air Force, and Navy-Marine Corps — maintain independent MARS organizations, with rigorous and continual training and operating requirements. MARS services operate worldwide, in many places where United States armed forces serve, as well as within the United States. MARS stations operating outside the continental United States (OCONUS) do so with the approval of the host nation, based on agreements negotiated between the DoD service spectrum manager and the host nation's radio communications frequency management authority. Interoperability between the three independent MARS services facilitates the MARS mission, as each MARS service has unique and complementary capabilities.

MARS trains members for specialization in net control, administration, digital operation, liaison with civil agencies and military units, and the responsibilities of leadership. (All field leader and staff positions are held by volunteer members except for the small headquarters contingents at the three services' branches).

MARS provides, in order of priority, contingency radio communication support to:

- The Department of Defense (DoD) Operational Requirements
- Federal Agencies supporting DoD Operational Requirements
- Military Units acting in Defense Support to Civil Authorities
- Federal Agencies supporting the National Response Framework
- State Agencies supporting the National Response Framework

Because the government personnel that MARS supports are professionals, MARS members are expected to apply similar standards in their behavior and dress, appropriate for their mission and circumstances, and to present a professional appearance. Since some MARS members may also be acting in support of other agencies, such agencies may dictate appropriate dress and other requirements, such as identification.

Members are expected to become familiar with new digital modes such as Automatic Link Establishment (ALE), which facilitates inter-agency calling on HF. "Interoperability" has become a one-word mission statement, with the three MARS branches partnering with each other, with ARES and the other ham organizations, as well as with active duty, reserve, and National Guard units.

MARS nets run seven days a week, 52 weeks a year. State and regional nets provide training in both voice and digital procedures, and regular exercises. MARS traffic nets

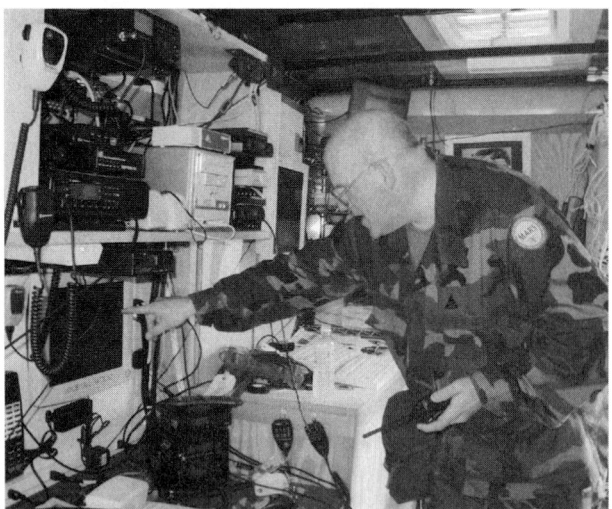

Figure 24.5 — Delaware Army MARS member John Scoggins operating from the EmComm van during the 2004 "Grecian Firebolt," an annual two-week-long signal exercise. Scoggins "dedicated hundreds of hours to the Grecian Firebolt mission, problem solving, trouble shooting and testing new equipment and technology," according to the *Army Communicator*, Vol. 29, No. 3, Fall 2004. [COURTESY US ARMY]

function much like the ARRL's National Traffic System, except that official government traffic is primary. Although this may occur infrequently, especially during "normal" non-emergency situations, MARS members are trained for this contingency. MARS member training is generally based on the Allied Communications Protocols (ACP), with some customized procedures that are specific to each MARS service, but which generally conform to the radio communications procedure used by their parent military service.

In addition, over 500 MARS members are also members of the SHAred RESources (SHARES) program, which is part of the US Government's National Communications System. The SHARES program is charged with promoting interoperability between High Frequency radio systems used by the Federal departments and agencies. SHARES members participate in regular HF radio nets and provide additional HF radio communications capabilities. MARS members are not automatically licensed in SHARES, but some MARS members may qualify for licensing in the SHARES program.

Other tasks range from providing trained operators for civil emergency operations centers to assisting National Guard and Reserve units with contingency communications during maneuvers/exercises, emergency exercises, and actual emergencies (see **Figure 24.5**).

Figure 24.6 — Members of the Connecticut National Guard Joint Incident Site Communication Capability (JISCC) team stand by as Lt. Col. Timothy J. Symonds, administrative officer for the 103rd Communication Squadron, Connecticut Air National Guard, takes part in a video teleconference with the Idaho Air National Guard's JISCC team. [STAFF SGT. NICHOLAS A. MCCORKLE; US AIR FORCE]

HOW TO JOIN MARS

To join Air Force MARS you must meet the following minimum requirements:
- Be a US citizen or resident alien
- Be at least 18 years old
- Possess a valid FCC Amateur Radio license
- Have digital capability
- Have the capability to transmit on MARS HF frequencies (2.0–30 MHz)

Visit **www.marsregionone.org/App/Join-New.html** for more information and to apply.

To join Army MARS you must meet the following minimum requirements:
- Be 17 years of age or older (Signature of parent or legal guardian is required when an applicant is under 18 years of age)
- Be a United States citizen or resident alien
- Possess a valid Amateur Radio license issued by the Federal Communications Commission, and upgrade to General Class or higher within 12 months of joining MARS
- Possess a station capable of operating on MARS frequencies, (2.0–30 MHz) and a valid e-mail account
- Agree to operate a minimum of 15 hours per calendar quarter, with nine hours being on VHF and/or HF networks; serve as NCS at least once per quarter, and participate in at least two three-hour exercises each year; members must agree to complete FEMA ICS courses IS-100, 200, 700, and 800 within one year of joining
- Successfully complete Army MARS on-air basic training course within six months of joining

Visit **www.netcom.army.mil/mars/join.aspx** for more information and to apply.

To join Navy-Marine Corps MARS you must meet the following minimum requirements:
- Be 18 years of age or older
- Be a United States citizen or resident alien
- Possess a valid Amateur Radio license issued by the FCC or other competent US authority
- Possess a station capable of operation on MARS HF frequencies (2.0–30.0 MHz)

Visit **www.navymars.org/national/join.html** for more information and to apply.

The Army National Guard subsequently created a portable system, the Joint Incident Site Communications Capability system (JISCC), to provide HF/VHF/UHF voice and data interoperability for domestic emergency communications (see **Figure 24.6**).

In 2004, Army MARS (ARMARS) negotiated rights to Automatic Link Establishment (ALE) sound card software developed by a British IT specialist, Charles Brain, G4GUO. An all-volunteer team headed by Carlos Santiago, WB2FOZ (ex-AAV2AS) and Steve Hajducek, N2CKH / NNNØWWL began updating the program (*MARS-ALE*) to MARS specifications. Navy-Marine Corps MARS subsequently assumed responsibility for the ALE project, which is ongoing. *MARS-ALE* is now a highly sophisticated sound card software program implementing most ALE modes.

In 2011, Navy-Marine Corps MARS (NMCMARS) initiated programming of "data modem terminal" software to allow MARS members to operate using the MIL-STD-188-110A PSK mode, to facilitate interoperability with military and government units equipped with this digital communication mode. In addition, NMCMARS is developing additional digital modes of communications.

One goal for MARS is to be able to operate an "infrastructure-independent," "radio-only" communication system that will continue to operate even with widespread outages of conventional communications modes such as telephone and Internet services. It is anticipated that there will be continuing developments in MARS radio communications capabilities, particularly in sound card and "software modem" modes, with the overall goal of increasing interoperability with military and government agencies to increase redundancy and survivability of communications in case of a widespread communications outage. Many MARS members actively participate in the development of these systems as "alpha" and "beta" testers, providing feedback to developers.

Air Force MARS (AFMARS) operates a transcontinental continental United States (CONUS) HF network, both voice and data, and provides unique phone patch capability from aircraft both CONUS and OCONUS. See **www.marsphonepatch.net/** for more information.

Because contingency communications support (the Space Support Net) to the space shuttle program is winding down with the end of the shuttle program, the Space Support Net has now been superseded by a "Mission Support Net" intended to provide contingency communications support in both voice and digital modes to contemporary military operations, homeland security and homeland defense activities, as well as CONUS and OCONUS disaster relief missions.

In addition, Air Force MARS has Base Support Teams (BST) at USAF, USAF Reserve, and Air National Guard Bases. The function of a BST is not only to provide contingency back-up HF and VHF voice point-to-point, phone patch, and *Winlink* e-mail communications support to the host military unit, but also to provide a resource of civilian volunteer communicators to assist with projects when requested by the Installation MARS Director.

Figure 24.7 — David Little, KD4NUE / AAA4GA demonstrating his self-contained portable *Winlink 2000* station powered by auxiliary battery. [DON WELLONS, AAM4RT / N4CMA]

Just as AFMARS maintains communication with aircraft, NMCMARS provides contingency voice and data HF radio communications to US Navy and Coast Guard vessels worldwide. During the Haiti earthquake, HQ Navy-Marine Corps MARS had a complete HF radio station with a PACTOR III modem flown in to provide *Winlink 2000* HF e-mail at the scene. In addition, VHF radios were used to provide voice communications between both the University of Miami Medical Center and Nassau University Medical Center clinics and the USNS *Comfort*, allowing medical staff at these clinics to talk directly to Navy medical staff on the USNS *Comfort*.

MARS members staff both ends of the communications link. They might deploy to incident locations as members of NGOs or state or local emergency communications teams, and they might operate the "home base" stations to take the calls coming from the affected area. MARS members provide critical contingency radio communications infrastructure, such as HF phone-patch and *Winlink 2000* HF e-mail Remote Message Servers (RMS) that operate when normal means of communication are not available, because they are relatively "infrastructure independent." MARS operators are needed at both ends! For those MARS members deploying to the field in support of civil authorities, "fly-away kits," or "go-kits" are critical when they need to provide stand-alone HF or VHF communications capability (see **Figure 24.7**).

The Pentagon MARS Station (**Figure 24.8**) operates under multiple call signs, including the Air Force MARS call

Figure 24.8 — AFMARS operator Ken Heitner, WB4AKK / AFA3PB, operating the Pentagon ARC ALE / *Winlink 2000* voice and digital station. Ken is the AFMARS National Deputy for ALE with the billet call sign AFD3LE. Ken has published articles on antennas in the ARRL's *Antenna Compendium* (Volumes 1 and 2) and on computer reception of RTTY in ARRL's *QEX*.

Figure 24.9 — The ARRL's Army MARS station, AAN1ARL.

sign AGA3DC, Army MARS call sign AAN3PNT, and Navy-Marine Corps MARS call sign NNNØPNT. When operating as a Department of Defense activity, the historic Army call sign WAR is used.

Not to be outdone, the ARRL received an Army MARS station license, AAN1ARL, in September 2007, and has constructed a MARS operating position at its venerable W1AW station at ARRL's Newington, Connecticut headquarters (see **Figure 24.9**). This MARS station provides ARRL's Emergency Preparedness staff with yet another means of emergency communications.

IN SUMMARY

MARS members are a pool of trained radio communicators with particular expertise in HF radio communications, who continuously train and equip themselves to provide contingency radio communications support to US government operations, including to their parent military services and civil agencies in fulfillment of DoD responsibilities. Just as importantly, MARS members are trained radio communicators who can provide to civil authorities high-quality radio communications skills, and, with appropriate authorization from their MARS service, access to MARS capabilities that can supplement and complement existing agency communications capabilities.

Chapter 25
Handihams and Emergency Communications: A Winning Team

When you think of someone who could rescue you if you were in trouble in a natural disaster or other emergency, what do you imagine that person would look like? In your mind, perhaps he or she looks like the people shown in military recruiting posters, those able-bodied men and women protecting our country. After all, television shows usually portray first responders as hale and hearty, so it seems logical for us to envision them that way. But that image is based on a stereotype, and like all stereotypes, it frequently breaks down in real life. This is especially true of the responders from the ranks of Amateur Radio who often participate in emergency communications and in disaster response. Amateurs with disabilities can and do find themselves intimately involved in public service and emergency communications operations. Their expertise keeps people and property safe. It even saves lives.

An important aspect of Amateur Radio has always been to offer assistance to one's community. Amateur Radio operators who train for, and participate in, emergency communications bring a variety of skills, experiences, and abilities to bear. Unfortunately, amateurs with disabilities must often overcome stereotypes and assumptions in proving that they are able to participate in organized emergency communications programs or public service events, despite the fact that they may be just as qualified for many roles as their "differently abled" comrades. An important resource for amateurs with disabilities is the Courage Center Handiham System.

THE COURAGE CENTER HANDIHAM SYSTEM

The Courage Center Handiham System, also known as "Handihams," is one of the primary resources through which people with disabilities can learn Amateur Radio and technology skills, and earn their Amateur Radio licenses. Handihams serves those with physical disabilities and/or sensory impairments, such as blindness, and also provides services to people who have other disabilities, such as dyslexia.

The worldwide Amateur Radio community is full of opportunities for all sorts of life-enhancing activities. One can make friends on the air; stay in touch with other Handiham members who might use similar assistive technology, such as blind-friendly computing systems and radios; and take part in competitions throughout the year. Handiham members and participants quickly learn about new technologies, including assistive technologies, that will help them in all aspects of their lives — not just Amateur Radio.

The mission of the Handiham parent organization, Courage Center, is "To empower people with disabilities to reach their full potential in every aspect of life." The Handiham program takes members through a process that builds confidence, achievement, planning, friendships, and volunteerism in service to others. While these values have always been a part of Amateur Radio, the Handiham program realizes that they transfer to all aspects of a successful and happy life.

Handiham History

The Courage Center Handiham System was established in Rochester, Minnesota, in 1967. The system was the brainchild of Ned Carman, WØZSW. Carman worked for a clinic, and, in the course of his work, would visit people with severe physical disabilities, who often had few opportunities to leave their homes. Carman realized that Amateur Radio would be the perfect hobby for them.

In April 1967, Carman enlisted the help of some local nuns from the Sisters of St. Francis. Several of the nuns received their licenses. Among them was Sister Alverna O'Laughlin, WAØSGJ, former Educational Coordinator for the Handiham System, who is now retired. The first Handiham was Edna (Eddy) Thorson, NØYL, who took her General Class license exam in December 1967.

Soon the Rochester Amateur Radio Club, followed shortly by the PICONET network of ham operators in southeastern Minnesota, took up the torch of service that Carman had lighted. Word of the Handiham System spread rapidly throughout southern Minnesota and northern Iowa.

By 1969 it was evident that the Handiham services could not expand further without substantial financial support. This support came from the non-profit Minnesota Society for Crippled Children and Adults (MiSCCA), whose name would later change to "Courage Center." MiSCCA granted full affiliate status to the fledgling Handiham System and helped with money and equipment. The Minnesota Handiham System merged with Courage Center in 1975 and became the Courage Center Handiham System. As a fully integrated service of Courage Center, the present Handiham System is able to call on the resources of its parent organization, from accounting and counseling to rehabilitation medicine and physical therapy, in order to better serve its members and participants wherever there is a need.

Handiham stations and remote base stations controllable via the Internet are now in operation at Camp Courage, Maple Lake, Minnesota, and at Courage North, Lake George, Minnesota. These remote base stations are designed to help Handiham members whose living situations do not allow them to have outdoor antennas.

Sessions of the week-long annual Radio Camp and other Handiham services benefit members throughout the United States and around the world. Members learn radio, electronic theory, and computing, but they also learn that they can accomplish what they set out to do.

STATION CONSIDERATIONS FOR PEOPLE WITH DISABILITIES

In order for a Handiham member to get the most satisfaction out of their experiences in Amateur Radio, some modifications may need to be made to station equipment to enhance ease of operability. Here are just a few potential solutions.

Radio Features for Blind and Low Vision Operators

Blind and low vision users need some way to read the radio's frequency display. Some models have large, high-contrast digital displays that will suffice for those who have some eyesight but require large print. Operators who are blind will need speech frequency readout, which is sometimes an optional feature. The availability of speech frequency readout varies by manufacturer and model, so it is important to shop carefully.

Speech frequency readout is actually a very limited accessibility feature. In its most basic form, it only provides readout of what is on the frequency display of the radio, though often audio feedback of the mode of operation and the S meter reading (signal strength) will also be offered. Some models come with these features built in; others require the end user to add a special module — actually a small circuit board — inside the radio. Installing the module generally requires nothing more than small screwdrivers, but there can be a fair amount of manual dexterity involved, and it's definitely not something for a person who has a visual impairment to tackle alone, unless he or she already has experience doing such installations.

Kenwood USA offers several radios with enhanced speech feedback for menu items, in addition to the spoken frequency output. This gives a blind person far greater access to the equipment's capabilities, as well as independence in operating it. Blind users who want this feature need to shop carefully to make sure the radio supports it, and it is well worth any extra cost to have the dealer install the speech module before shipping the radio.

Used equipment may or may not have speech frequency readout installed. Speech frequency output was not available at all on some older models, and available only as an option on others, so do your homework before making a purchase. It will not be practical to retrofit old equipment for speech frequency readout unless it was originally designed to accept an optional speech module. Even if it was designed as such, sometimes it can be difficult to find a module on the used market. The secret to success is to be sure you know what you're getting when you buy a used radio, especially if you must have speech frequency readout.

An Amateur Radio operator may prefer another brand of radio, regardless of the availability of an enhanced speech feedback system. Indeed, there are more features to consider; simplicity and convenience may be even more important than speech frequency readout. A person with mobility impairments may find a front panel keypad with larger, easier-to-use controls essential. Some radios, like the ICOM IC-7200 HF transceiver (which ships equipped with basic speech frequency readout), offer a front panel keypad frequency entry system and large, easy-to-use front panel buttons and controls. The layout of the front panel should be logical and easy to remember. A blind user will have to commit the locations of the various buttons and knobs to memory, and a simple, straightforward layout will be easier to remember than a large, complicated one. A special consideration for the frequency entry keypad is that it be laid out in the same general way as a standard telephone keypad. This universal design will help avoid errors. It seems like a simple thing, but you cannot assume that every piece of equipment will have this design.

No discussion of adaptations for ham radio operators with vision impairment would be complete without mentioning the instruction manual. Until fairly recently, a blind operator could not read a manual, traditionally a paper document, without using a computer scanner and optical character recognition. Fortunately, that is not necessary for newer radios, because manufacturers have put instruction manuals online in accessible PDF format that can be read by screen reading software program on a *Windows*, Mac, or Linux computer. PDF manuals are searchable, too, making it easy for a blind user to look up something in the manual.

The Courage Center Handiham System has some audio tutorials online at its website and on tape or other media for popular radios like the Kenwood TH-F6A. Another helpful website for blind operators is **www.icanworkthisthing.com**, which has some good resources on ham radio equipment.

Computers and Computer Control

The personal computer is one of the greatest accessibility advancements ever for people with disabilities. It is also one of the most common pieces of equipment found in ham radio shacks today, and it offers excellent opportunities for people with disabilities to use the many Amateur Radio online resources as well as to control their radio equipment.

Most Amateur Radio software applications work on the Microsoft *Windows*® operating system. There are a few appli-

REAL-LIFE STORIES

Here are the real-life stories of three special people who, like many other disabled amateurs, have worked hard to overcome stereotypes and to serve their communities.

Matt Arthur, KAØPQW

Matt Arthur, KAØPQW, of Ellendale, Minnesota, became a key communications resource for the town of Rushford in the summer of 2007. The small town in Fillmore County in southeastern Minnesota was flooded when Rush Creek overtopped the dikes in August. Matt, who is blind, was monitoring a local repeater when he learned via his local county ARRL Emergency Coordinator of a desperate need for communications in Fillmore County. He and two other amateurs from his county were transported to the affected area. Matt was assigned to an emergency communications trailer parked, ironically, on a bridge over the threatening river.

Matt was initially paired with another amateur, one who seemed completely disinterested in involving him in the effort. Eventually, the other amateur became fatigued and decided to leave. Matt explained, "He told me, 'Will you take this over? I'm going home.' I was given no instructions, so I had to figure it out as I went along."

And figure it out, he did. KAØPQW was net control station for the simplex net for most of the day. He transmitted health and welfare traffic with the nearby town of Winona. He recalled, "I was the only comms in and out of Rushford until the National Guard arrived later with their satellite communications van."

As one might expect, the local amateurs on scene were at first skeptical that Matt was up to the task.

"One guy was a little bit surprised that I could get things done. But once he figured out I could take care of myself, he was fine," Matt said. He also recalls that mistakes were made. But he and the other amateurs involved made a positive impression in the end. "Everyone was very happy with what we did," he said, adding, "The Mayor didn't want us [amateurs] in. He said he would take care of everything himself. But I can tell you, he was very happy afterwards."

Matt has these words of advice for other amateurs with disabilities who find themselves in the thick of public service or emergency communications situations: "Don't be afraid. Just do the best you can. If somebody questions you, prove that you have the skills to take the job. If you don't have the skills, don't do it."

Katherine "Kitty" Hevener, WB8TDA

Katherine "Kitty" Hevener, WB8TDA (**Figure 25.1**), has been a licensed radio amateur for nearly 40 years. Born and raised in West Virginia, she has lived in many different places in the United States, including both east and west coasts.

Kitty became involved in public service and emergency communications as a natural extension of her participation in traffic handling and the National Traffic System. As her proficiency grew, she rose through the ranks of section, region, and eventually, area nets. This traffic handling experience led her to volunteer as an Emergency Coordinator for Pendleton County, West Virginia. "I volunteered to be the EC, but I didn't know what the job entailed, nor did I receive any mentoring," Kitty said. Despite the experience of on-the-job training with little formal introduction, she rose to the occasion and effectively served Pendleton County for years.

Kitty left her native West Virginia and moved to Nashville, Tennessee, to attend college. While a student, she helped out with numerous weather-related emergencies, generally operating as net control station. She later moved to Connecticut, where she continued her amateur pursuits. In 1985 Hurricane Gloria, a massive storm that caused extensive damage and was responsible for eight fatalities, struck the U.S. eastern seaboard. Kitty and other members of the Newington (CT) Amateur Radio League were called upon to as-

Figure 25.1 — Katherine "Kitty" Hevener, WB8TDA, is seen here with a BrailleNote device, on which she can send and receive e-mail, send documents that can be opened in Microsoft *Word*, read Microsoft *Word* e-mail attachments in braille, and more.

sist with emergency communications. Kitty was assigned to a net control position at the Newington Police Department.

She remembered, "I was scared, because I'd never done anything like that, and the stakes were high. But I knew I had the skills to be an emergency communicator." She also recalled a police officer remarking to her, "You obviously know what you're doing, or they wouldn't have sent you here." The officer's assessment proved correct. In Kitty's words, "The training and day-to-day experience from my NTS activities prepared me well." She added, "While my colleagues were in the field radioing in reports of downed power lines and other threats to life and property, I was doing the job of running an orderly net."

Kitty says that not every experience during a public service operation or emergency communications event was positive. She recalls one incident in which another amateur heard an important message come into the net while she was acting as net control. He grabbed the microphone from Kitty and said, "Let me handle this."

She says she was disappointed with this attitude. "Often, people get caught up in stereotypes, especially what they think people can and cannot do. Instead of thinking this, why not just ask the person with the disability? Just ask!"

To public service and emergency communications organizers and leaders, Kitty has this to say: "It's very important to find out the strengths of all of your potential volunteers and make assignments accordingly. People think, 'you're disabled, so we'll put you at net control.' This may or may not be a good fit. For example, that individual might be a good candidate for assignment in a vehicle."

She is also a strong advocate for emergency communications groups purchasing radios that are "disability-friendly." This includes transceivers that contain large, easy-to-move knobs and switches, large displays, and voice readouts.

So what would Kitty say to amateurs with disabilities who wish to participate in emergency communications and public service events?

Be patient, but also be persistent.

Really practice doing public service or the traffic nets.

Remember that as a person with a disability, you end up having to prove yourself and your value, but once you have done so, the question won't be "Can you do it?" but "Who can do it better than you?"

Lucinda Moody, AB8WF

Lucinda Moody, AB8WF, has been a licensed amateur since 2006. She lives in Middleville, Michigan. At a very early age, her parents instilled in Lucinda a spirit of giving and community service.

"My parents were missionaries in Togo," she explained. "My mother was an operating room nurse. It was only natural that I accompanied her to medical procedures at a very early age." When she was 17, Lucinda took over duties at the hospital pharmacy.

Lucinda suffers from neurological problems that have required her to undergo numerous brain surgeries. As a result, her balance and mobility have been severely affected. She walks with a walker or forearm crutches. Unfortunately, Lucinda has also experienced prejudice and stereotyping while serving others. "People look at me and sometimes they decide I can't do anything," Lucinda said. She's also experienced the frustration of being assigned, as a ham volunteer, to a position on a racecourse where there was little or no reason to have an amateur present. And there was the time she was assigned with another ham who felt threatened or insecure around her.

"She had to repeat every transmission I made, even though I was doing just fine." At the 2010 Grand Rapids Mud Run, a challenging 5k run in Kentwood, Michigan, Lucinda was given a net control assignment — a first for her. It turned out to be quite a challenge, but Lucinda was equal to it.

"One woman jumped off a six-foot wall and broke her leg," she recalled. "Another man tried to dive head first into a mud pit and lacerated his head." In the end, the net handled two ambulance requests for transport, as well as numerous calls for treatment on scene. Lucinda credits her work with the Salvation Army Team Emergency Radio Network (SATERN) in helping hone her communications skills. "I run a SATERN net every week. That's a great learning experience — it helps me handle the chaos coming at me."

Lucinda has a tremendous respect for her fellow SATERN volunteers. "They know my limitations, and they respect what I can do," she said.

Lucinda has this advice for other amateurs who want to become involved in public service and emergency communications:

"Get out there and do something. Even someone who is stuck in a wheelchair has something to contribute. Look around and figure out what you can do to help. I don't ask; I just pitch in. Of course, I don't do something that is going to cause a problem. I just try and see where I can fit in, and then I help."

Lucinda explains that a good sense of humor can often go a long way, especially around people who are nervous when around those with disabilities. "I may not be able to stand up, but when I'm on the ground, I can do great CPR."

cations available for Apple and Linux operating systems, but they are somewhat limited by comparison. Blind users will want screen-reading software installed on their computers. Two popular programs on the commercial market for *Windows* are *JAWS*®, by Freedom Scientific, and *Window-Eyes*®, by GW Micro. A free, open source screen reader for *Windows* is *NonVisual Desktop Access* (*NVDA*); new Apple computers come with a built-in screen reader called *VoiceOver*. Some blind users access computer output using refreshable braille displays.

Screen reading capability in the ham shack can give a blind operator access to Internet call sign database lookups, Amateur Radio news and information pages, audio streams related to Amateur Radio, and computer programs like *EchoLink* that allow the user to connect to Amateur Radio repeater systems worldwide through the Internet. The Courage Center Handiham System operates two Internet remote base stations for HF operation. These stations can be controlled via the Internet from a *Windows* computer running remote control software, and they allow people whose disability requires them to live in a place that prohibits outside antennas to run a real Amateur Radio station, right from their personal computers.

People with other disabilities can make use of voice-controlled computers to access the many available Amateur Radio applications. For those with disabilities that prevent them from using a keyboard, voice input computing software, such as Nuance Communications' *Dragon NaturallySpeaking*, can be used to control the computer simply by speaking into a microphone. And, of course, even a computer without special software can be made to show large print and high contrast to make it easier to see what is on the screen.

The open-source software program *Ham Radio Deluxe* works on the *Windows* operating system and is free for any Amateur Radio operator to download and use. With *Ham Radio Deluxe* and a simple, inexpensive interface between the devices, a personal computer can be used to control a radio. This software has some accessibility features built in, including voice frequency announcement, which works even if you are using a radio without a speech module installed.

Antennas and Tuners

There are many different types of antennas and they vary in size and design, depending on the desired frequency band and operating preferences. Some antennas offer very straightforward designs, will fit nearly anywhere, and require little maintenance, making them likely the most suitable for operators with disabilities. Generally speaking, these are vertical antennas. The simplest, smallest ones are tuned for the VHF and UHF bands. They need little more than a place to be mounted (such as an outdoor pole or tower, or even an indoor platform in some cases) and a feed line that connects directly to the radio. For the HF bands, the antennas can be quite large but can be easily ground mounted outdoors and fed with coaxial feed line buried underground. Directional antennas, such as beams, require more robust mounting support (such as a tower that can support the rotor) and an electric motor that turns the antenna as directed by a control head in the ham shack.

A directional antenna system adds complication and expense to the station; however, directional antennas do have some great advantages over omnidirectional antennas like verticals. Blind operators who opt for a beam may want to choose a rotator system that will allow them to remove the protective plastic bezel from the control head so they can feel the position of the directional indicator needle. This is a fairly low-tech solution, but it does allow the operator to know which direction the antenna is facing for the best transmission and reception.

Since most modern Amateur Radio transceivers operate on multiple frequency bands, a typical ham station will include multiple antennas for different bands or some kind of single antenna that operates on all desired bands. A multiband antenna will often require an antenna tuner, either a stand-alone device connected between the radio and the antenna somewhere along the feed line, or the built-in one that comes with many modern HF transceivers. Having a built-in tuner can be an important consideration for people with disabilities who may not be able to see the meters on a stand-alone tuner or may not have the fine motor skills necessary to manually tune one (built-in antenna tuners work automatically or at the press of a button). A visually- or motor-impaired user only needs to know the position of the button on the front panel of the radio. That said, there are, however, some excellent stand-alone antenna tuners that operate automatically as soon as they sense RF energy from the transmitter. These can have a wider tuning range and may be a better choice for some antenna systems. Since they require little or no user intervention in normal use, they are an excellent choice for the Amateur Radio operator with a disability.

Ergonomics

Ergonomics plays a critical role in the design of a good ham radio station. For instance, someone using a wheelchair will need to be able to position the chair in such a way that the he or she can easily reach and control all the equipment. This may mean adjusting the table height and positioning equipment on the surface of the operating desk in a way that works best for the user. Having electrical outlets and lighting switches within reach will also be an important consideration.

There should be plenty of good lighting focused on areas where it is needed. Since everyone's needs are unique, a certain amount of experimentation will be required to get everything set up for maximum comfort and efficiency. Failing to get this right could lead to neck pain if an operator has to strain to see a computer screen or frequency readout, or to back pain if seating is at the wrong height or in the wrong spot.

A special consideration for disabled Amateur Radio operators is that the layout of the station, once decided upon, should be respected and other family members should not move anything around or place unrelated items on the operating desk. A blind Amateur Radio operator will learn where everything is located on the operating desk and will be comfortable maintaining that station layout unless there is some compelling reason to change it. At worst, a thoughtlessly put-together ham shack, or a careless change made by another, can be an accident waiting to happen. At the very least, it can be inconvenient for the person with the disability to have to call for someone else to help adjust a piece of equipment that is out of reach.

Editor's Note — Every ARES organization should assign someone to be an assistant EC for Handihams. Make a special effort to bring all hams into our fold.

RESOURCES

The following resources can help people with disabilities learn more about getting involved in Amateur Radio or improve their existing skills.

INFORMATION

Courage Center Handiham System

Specialty services for people with disabilities, on-line rig tutorials, and ham radio training in spoken word audio, residential radio camps, remote base HF stations, and Echolink nets.
Courage Center Handiham System
3915 Golden Valley Road
Golden Valley, MN 55422
866-426-3442 (toll free)
www.handiham.org

Handiham text file of *WorldRadio* article, "Accessibility Roundup," featuring accessible radios and equipment:
http://handiham.org/local/downloads/accessibility_roundup_2010.txt

ARRL

Articles, local radio clubs, accessible VE sessions, disability resources related to ham radio.
ARRL
225 Main St.
Newington, CT 06111
860-594-0200
www.arrl.org

List of ARRL *QST* articles related to equipment, adaptations, and operating for the blind: **www.arrl.org/access-to-amateur-radio-for-the-blind**

National Library Service for the Blind and Physically Handicapped (NLS)

Free talking book access to *QST*, *CQ*, and many other publications of interest for Amateur Radio operators who cannot read regular print.
National Library Service for the Blind and Physically Handicapped (NLS)
The Library of Congress
888-NLS-READ (888-657-7323 toll free)

Blind-Hams E-mail Reflector

A very active discussion list with a searchable database of posts. To subscribe, send mail to listserv@listserv.icors.org with the command SUBSCRIBE BLIND-HAMS.

SOFTWARE

CW Skimmer

By Alex Shovkoplyas, VE3NEA. Software for copying CW, for deaf and hard of hearing operators:
www.dxatlas.com/CwSkimmer

JAWS

JAWS screen reader and other low-vision-related products:
Freedom Scientific
11800 31st Court North
St. Petersburg, FL 33716-1805
800-444-4443 (toll free)
www.freedomscientific.com

Window-Eyes

Window-Eyes screen reader and other low-vision-related products:
GW Micro
725 Airport North Office Park
Fort Wayne, IN 46825
260-489-3671
www.gwmicro.com

NonVisual Desktop Access

Free, open source screen reading software for *Windows* computers:
www.nvda-project.org

VoiceOver

Built-in screen reader offered on new Apple computers:
www.apple.com/accessibility/voiceover/

Dragon NaturallySpeaking

Voice input computing software and similar products:
Nuance Communications
1 Wayside Road
Burlington, MA 01803
781-565-5000
www.nuance.com

Ham Radio Deluxe

Rig control software with built-in speech frequency announcements from Simon Brown, HB9DRV, and the HRD team of volunteers: **www.ham-radio-deluxe.com**

INSTRUCTION MANUALS

Equipment instruction manuals and other helpful audio resources in blind-accessible formats
www.icanworkthisthing.com

Chapter 26
An International Perspective on Emergency Communications

The use of the word "perspective" in the title of this chapter is deliberate, because the definition of a disaster depends on the viewer's perspective. To use a variation on a Mel Brooks quotation: "If you fall down the stairs, that's hilarious; if I get a paper cut, that's a disaster." It follows that the definition of a disaster varies significantly between countries, and so does the requirement for Amateur Radio to provide emergency communications. In this chapter we will highlight some of the differences in approach as well as some of the great similarities found within emergency communications activities in the International Amateur Radio Union (IARU) Region 1, which encompasses Europe, Africa, the Middle East, and Northern Asia.

WHAT IS A DISASTER?

From an outsider's viewpoint, North America could be seen as a disaster hotspot, with its wildfires, floods, hurricanes, tornadoes, and even the occasional volcano. This simple view is caused by a lack of awareness in many cases of the work done by ARES groups, which offer communications support, such as to local hospitals and municipalities in case of communication system failure, or during nuclear plant drills.

A similar simplified view may be taken of Africa; perhaps as a continent with little technological infrastructure, but beset by famine; or of highly networked Europe as experiencing few natural disasters. Yet Africa has a growing technology-based economy, and Europe suffers frequent floods along its great rivers, such as the Danube. As society becomes more dependent on technology, disruption of electrical networks and the Internet may present as much of an emergency as a natural disaster, as critical networks underpinning an economy's financial services or utilities could take weeks or months to rebuild.

Our perceptions are greatly affected by what we see in the media, especially with the expansion of the Internet and the 24/7 news culture. Within the context of emergency communications itself, we have Amateur Radio news services looking for any sign of our involvement in an event to promote the hobby, and that in turn makes additional demands on radio amateurs in disaster areas.

National governments and radio regulators have their own ideas on what qualifies as a disaster and how, or even if, Amateur Radio should be involved in a disaster response. This, of course, leads to some major differences among countries in the various approaches to providing emergency communications.

WHY ARE INTERNATIONAL REPONSES DIFFERENT?

That is rather a double-edged question. In one respect, although a natural disaster like a flood cannot be predicted accurately, it is relatively easy to understand and affects populations similarly. Since all disasters reach us at a basic, personal level, it is sometimes hard to see why a flood in one country will produce a major response from Amateur Radio, but one in another country would not initiate an amateur response at all. The difference is in each country's ability to react to its unique disaster scenario and its resilience after the fact.

Each country decides how it will use volunteers to assist in disaster response. A decision to employ them may be driven by a need for their specialized skills or because professional resources be too expensive to maintain on standby; a decision not to do so may be based on some bad experiences with volunteer groups in the past.

From an amateur viewpoint, this can create a challenge as we strive not impose our own model of how a disaster response should unfold when it occurs in someone else's country. To illustrate this point, let's look at three different arrangements for emergency communications within Europe.

The United Kingdom

The Radio Amateurs Emergency Network (RAYNET) was set up in 1953 following severe flooding caused by a North Sea storm that knocked out maritime coastal radio stations. Prior to this, there had been a government monopoly on telecommunications and the value of a voluntary emergency communications group was not understood. Afterward, however, Amateur Radio license terms were modified to allow Third Party messages to be passed when requested by a

RAYNET MESSAGE FORM

NUMBER	PRECEDENCE * ROUTINE PRIORITY IMMEDIATE EMERGENCY	STATION OF ORIGIN	CHECK **	PLACE OF ORIGIN	FILING TIME HH:MM	FILING DATE MMM-DD

TO:

FROM: (name /role in BLOCK letters) :

REC'D	FROM	DATE (MMM-DD)	TIME (HH:MM)	SENT	TO	DATE (MMM-DD)	TIME (HH:MM)

* Delete where not applicable ** The word count in the heavy outline box

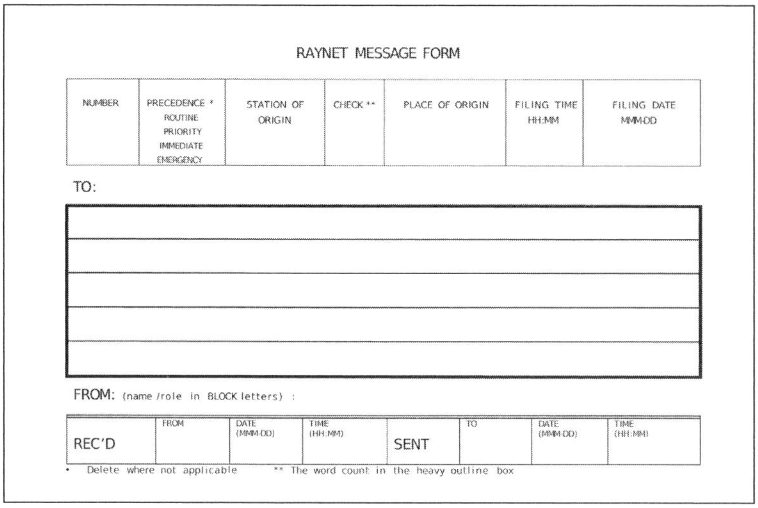

Figure 26.1 — A RAYNET message form. A standardized form allows a volunteer Amateur Radio group like RAYNET to more effectively pass written messages on behalf of its user services.

select group of emergency responders, such as RAYNET (see **Figure 26.1**).

At the time of this writing there are approximately 2,600 RAYNET members in the UK. All are volunteers who fund their own activities. RAYNET has no official role in Government contingency arrangements, but it is highly regarded and heavily referenced in government guidance documents for Emergency Planners. While most operation takes place in the Amateur Radio bands, there is some use of business radio frequencies to enable volunteers to work with services like the Red Cross and because 5 MHz channels are the easiest way to interoperate with UK military communications units.

Spain

Amateur Radio licenses granted in Spain require stations to cooperate in providing emergency communications when instructed to do so by the authorities. If an operator receives a distress communication it is compulsory for it to be relayed to the authorities as soon as possible.

In practice, emergency communications are handled by REMER (Red Radio de Emergencia), an official organization of the Spanish Directorate of Civil Protection and Emergencies. REMER comprises approximately 4,000 Amateur Radio operators who are authorized to operate on some frequencies outside the amateur bands. Amateur band operation is only contemplated for major emergencies. The REMER volunteers use their own equipment and vehicles, but have their insurance covered by the Spanish government.

Italy

A single government department is responsible for all emergencies, which are categorized into local, regional, and national levels. Volunteers of any description are well respected in Italy and there are over one million people involved in the volunteer sector, so a database is used to record the capabilities of each group and what kind of response it can provide. While they are active responding to an emergency, individuals in registered groups are protected by having their jobs held open, their normal salaries paid, insurance provided, and funds made available for training, equipment, and the like (see **Figure 26.2**).

Many groups have also set up their own communications units within larger organizations, such as the Red Cross and an aid group known as the Misericordie. These units may comprise radio amateurs who only coordinate within their own organization and not between groups. There are also reportedly over 100 groups of radio amateurs who are separately registered as responders in the country. This can make the task of providing a coordinated and interoperable response quite challenging.

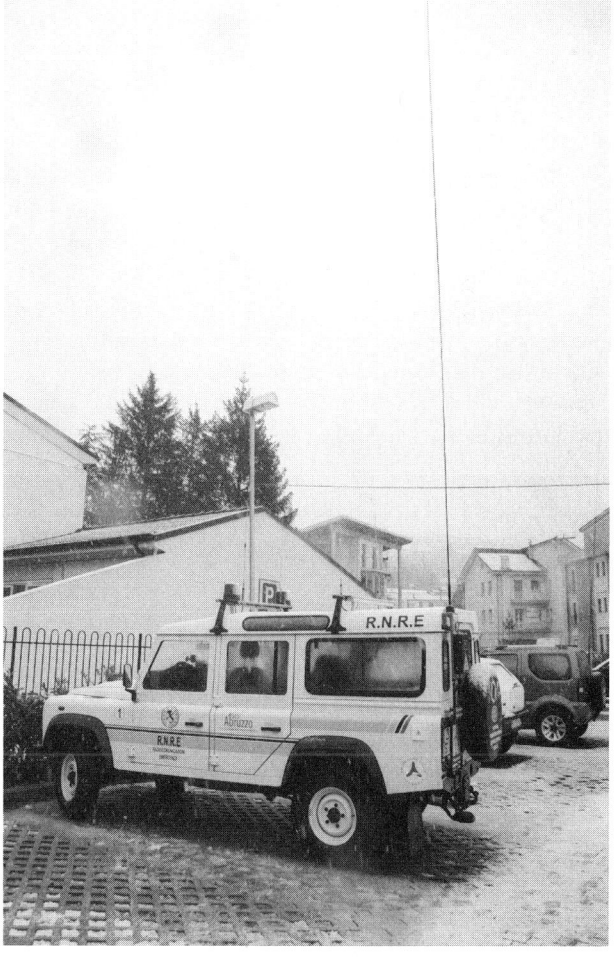

Figure 26.2 — A communications vehicle funded by the Italian Government for one of the voluntary amateur groups in that country's RNRE (Raggruppamento Nazionale Radiocommunicazioni Emergenza).

A CASE FOR STANDARDS AMID DIVERSITY

This brief glimpse of European groups may indicate that, other than comprising radio amateur volunteers, there is little similarity among them. What we as responders cannot avoid, however, given world population distributions and the interconnectedness of electrical and communications networks, is that a disaster will eventually cross borders.

The communications requirements of each country are laid down by their served agencies, which is how it should be. However, if HF were used for communications in a wide-area disaster, then wouldn't working through a foreign relay station become a necessity? Wouldn't a common message format help bridge the language barrier? Obviously, the answer to both is, yes. With so many tongues and dialects spoken in relatively close proximity, knowing that a message is composed of a strict number of elements always transmitted in the same order helps break down these natural communications barriers.

This applies to any language-rich area of the world. But keeping our focus on Europe for a moment, which message format should be used? NATO standards may not be welcome in former Eastern Bloc countries, and the West might not appreciate theirs. Since the Amateur Radio community has a Memorandum of Understanding with the International Red Cross to work together, the adoption of any military style format may not be acceptable in some scenarios.

We are relatively lucky in that a common message format for Amateur Radio has existed for many years, although it suffered from a lack of promotion until 2009. While the point may be contentious to some North American readers, the IARU's HF emergency message format (see **Figure 26.3**) is the same as the ARRL's National Traffic System (NTS) radiogram, although it lacks the HX (handling instructions) field as licensing conditions around the world rarely allow the same degree of third-party traffic handling as seen in North America.

To promote this standard, an event known as the Global Simulated Emergency Test (GlobalSET) has grown. The GlobalSET started as the "EmComm Party on the Air" in IARU Region 1 in 2006 with the intent of bringing volunteer groups within different countries together in a non-competitive environment to meet the following objectives:

- Increase the common interest in emergency communications
- Test how usable the Emergency Centre of Activity frequencies are across ITU regions
- Create practices for international emergency communication
- Practice the relaying of messages using all modes

The name of the event changed to eliminate the use of the word "party," and institute a name that would make the event identifiable as an emergency training exercise. The event asks stations to pass a simple message to a designated amateur station using the IARU message format. Such an amateur-to-amateur message circumvents third-party traffic restrictions in some countries, but allows the exchange of messages across country and even language barriers, providing valuable training opportunities.

When served agencies are apparently clamoring to pass large quantities of information, why train with what seems to some an outdated and limiting format? There are several reasons.

The first reason is that this format is a basic building block on which other messaging skills/networks can be built. For instance, in the world of computer networking, the TCP/IP protocol suite has been the backbone of the modern Internet, transferring packets of data between networks in an orderly manner. Simply put, the data packets are broken into "headers" to ensure correct routing and payload data. Packets are kept short enough that errors can be corrected in a

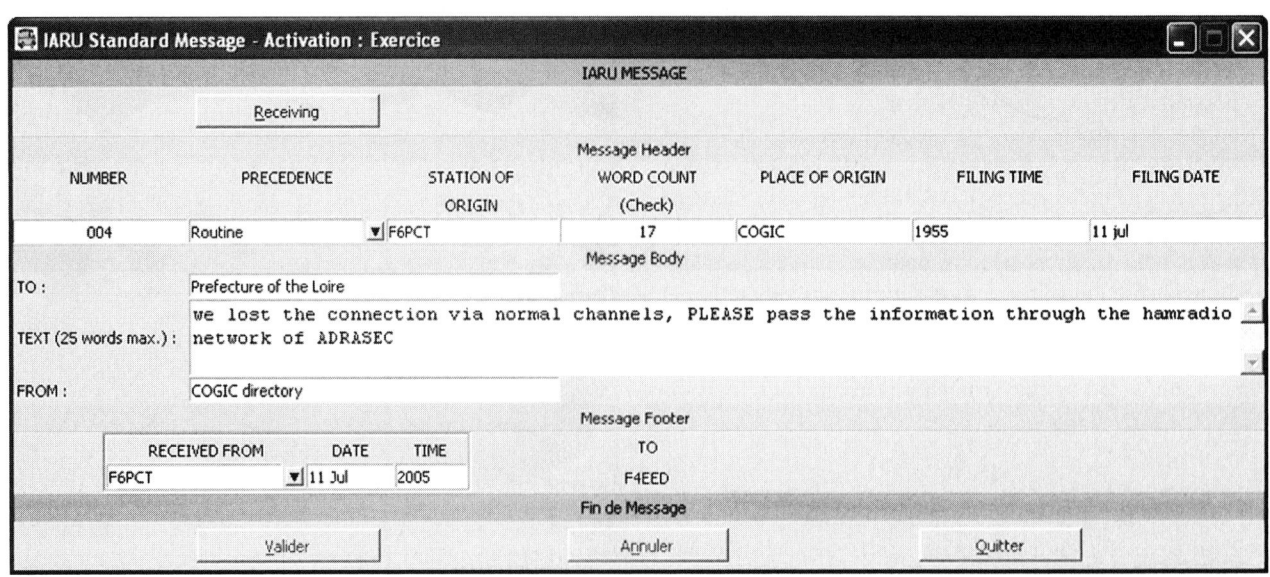

Figure 26.3 — IARU message displayed in the GesADRA program.

timely manner. The only difference between the Amateur Radio concept of a message packet and a data packet is that we have not dealt with error detection and correction.

This brings us to another reason for using a human-based, rather than machine-based, communications protocol. Groups establishing an emergency communications network must instill in their members the discipline required for proper message passing and the development of an audit trail so there is accountability at all stages of the relay chain. None of this stops innovation with groups in Europe developing and building wide-area High-Speed Multimedia (HSMM) networks in the 2.4 and 5.7 GHz bands, but we must remember that any relay chain is only as strong as its weakest link. Messages may start from an evacuation center as a simple speech message and be relayed by CW before it gets to a data network. It is important to ensure that the message is relayed correctly at all points in the chain. If we become dependent on just one network, then we are no better in some cases than the systems we may be called upon to replace. Diversity of available modes is our best asset.

A third reason for using a basic communications protocol is that served agencies must realize that their emergency communications are being routed through a scarce and bandwidth-limited resource. A recent national exercise in the United Kingdom to practice for a response in the event of flooding realized the problems in concise communications and the interim report stated:

"In the post-exercise debriefs communications personnel also said that if people had the discipline of responding in 140 characters — as one has to do on Twitter — this would sharpen correspondence in general and speed up the entire communication process between responders."

This interim report is no longer available on the Internet, and the focus on communications was reduced in the final version. Sadly it will take time to persuade emergency responders — who are used to the immediacy of e-mail and the "bells and whistles" inherent in complex spreadsheets and presentation formats — of the value of efficient, basic text communications, but it is something we must consider to ensure communications are passed as efficiently as possible.

THE ELEPHANT IN THE ROOM

Now to an uncomfortable issue that will not get solved by itself: Emergency communicators still need to use data modes. With the widely varying requirements and licensing restrictions of different countries, getting a common agreement on a data mode is going to be difficult, but ignoring the need would be bad practice for resilient communications.

We must focus on a common interface for operators, one that supports the common message format. Solutions, both well-known and relatively obscure, do exist to meet this need. W1HKJ's *Fldigi* modem software (**www.w1hkj.com/**) is one such solution. This English-language program offers a companion package, *Flmsg*, which allows the sending of message templates over data modes and even prints out a correctly formatted version for sending on CW. A French-language alternative comes from that country's emergency group, FNRASEC. The group's *GesADRA* package (**www.gesadra.fr/**) runs in the Python programming language on most compatible operating systems and includes the IARU message format alongside local French requirements.

Having a common data format lets us exchange messages intelligently despite language barriers, and allows us to detect whether the message may have been corrupted in transmission. The IARU format means that at any point in the chain, an operator can look at the received message and recognize the priority (or any other part of the message header) just because of its position in the message, even if the originating operator has entered the text in a form in his or her own language. The addressee and sender are plain because of their position in the message and the content should be sent as accurately as possible. If it is not in your own language, then send it as phonetic letter groups. Either way, if the word count is incorrect, then something has gone wrong.

The ideal of a single data format is still some way away; switching messages between software will remain necessary for some time, but while a human-readable format remains, we can still offer a message service that is robust and resilient.

ONE GLOBAL CONSTANT

Although the role of Amateur Radio is different the world's many countries, we still promote our services internationally to bodies such as the ITU and the Red Cross. At any moment, somewhere in the world, a radio amateur is involved in planning, exercising, or actually providing emergency communications to his or her community. Sadly, this activity frequently goes unreported. What does this mean for emergency communications?

For one thing, it means we lose out on opportunities to learn from the experience of others. In countries where emergency events are rare, there may be no relevant lessons local amateurs can build upon, or no known successes with which they can go to their government agencies to say, "we can do that!" For another, emergency communications appears to interest a minority of amateur hobbyists. A survey of emergency communications coordinators within IARU Region 1 showed that over 10,000 radio amateurs were involved in emergency communication groups in 34 countries where about 140,000 are members of an IARU National Society. While that is a significant number, it is a disparate assemblage that does not speak with one voice, making recognition inside and outside the amateur community much harder to achieve.

Unfortunately, while we are a "communications" group, we are not that good at communicating among or about ourselves, and this is something we need to change. If we don't, then media interests may unintentionally paint a distorted picture of our activities, or not put a brush to the canvas at all. We also need to maintain our relevance to the agencies we serve, so we must be able to show recent examples of the services we provide.

Since we do not always share the same values or licensing conditions, we have to be especially cautious that we do not do anything to alienate our served agencies. The Tampere Convention, which came into force in 2005, was hailed for simplifying the provision of emergency communications to disaster areas by removing many of the regulatory barriers normally in place such as licensing requirements, customs restrictions and use of frequencies. It carries, however, one important proviso: Assistance is only covered by the convention if the disaster-affected country actually requests help. Also, the convention only applies to the 44 countries (at the time of this writing) that have ratified it. The adage, "It's easier to ask for forgiveness than it is to get permission" does not work in some countries and could harm the image of Amateur Radio in those areas that are still suspicious of our activities.

ARE WE REALLY THAT DIFFERENT?

None of the issues highlighted here as "international" are really that different from local ones. Standardization is always difficult to achieve and is driven by perceived local needs. Getting help across internal borders between states and counties also can be difficult, perhaps because of local politics or volunteers coming in without invitation. Standard message forms may have been "adjusted" by local authorities to meet their own requirements, leading to confusion among operators coming from outside the area.

Although perhaps our greatest distinction as radio amateurs is that we can operate off the grid and replace the high-tech communications systems that are so fragile in the face of disaster, emergency communications groups all over the world have to weigh where and how to incorporate data modes. The goal here again is to improve our efficiency, and one of the best ways to do that, anywhere, is by standardizing methods among groups.

Amateur Radio groups everywhere need to introduce, explain, and promote the hobby and the invaluable skills radio operators offer, especially in emergencies. Emergency communications volunteers must also ensure that their served agencies, relevant governing authorities, and the media remain informed about their capabilities and activities during actual emergency response efforts, no matter where they occur.

THE NEXT STEP

Can we improve upon our effectiveness as we move forward? Of course, but how much depends on us all sharing more information so we can learn about experiences in different areas. Even if what we learn can't be used in our own groups, it can stimulate discussion about how we can improve what we do.

We must not allow ourselves to be faced with a "Betamax moment," the turning point in the competition between the competing VHS and Betamax home video formats, when users flocked to one system without fully considering the technical merits of each. We have many "competing" data modes and methods of linking to the Internet. We have operators still debating the merits of Morse code over voice for traffic handling. In an emergency communications context, this diversity should be celebrated as our unique selling point.

Whatever natural or manmade disaster hits our communities, we have a method to cope with it and to help in the recovery process. If one mode one gets knocked out, there is another one to take its place. If help is required from across borders, those who are tasked with providing it will know what is required get the message through, even if it's falling back on a method of communications they share with the affected area.

After all, that is what all our communities expect from us, because "When all else fails..."

Appendices

Appendix A
Synopsis and Extract from
TOPOFF 3 Exercise, Connecticut

by Chuck Rexroad, AB1CR, Wayne Gronlund, N1CLV, and Betsey Doane, K1EIC

TOPOFF 3 was the largest single disaster drill ever to take place in the United States. TOPOFF 3 was held simultaneously in New Jersey and Connecticut and involved multiple federal and state agencies as well as the United Kingdom and Canada. This full-scale exercise took place from April 4–8, 2005 and involved more than 10,000 participants representing more than 200 federal, state, local, tribal, private sector, and international agencies, organizations, and volunteer groups. The exercise offered agencies and jurisdictions a way to exercise a coordinated national and international response to a large-scale, multipoint terrorist attack. It allowed participants to test plans and skills in a real-time, realistic environment and gain the in-depth knowledge that only experience can provide.

During TOPOFF 3, ARES® was tasked to provide communications support for the American Red Cross, and to provide backup communications to the state Office of Emergency Management (OEM). While none of us asked for this opportunity, everyone was determined to show the value of Amateur Radio communications in a large-scale disaster, according to the District Emergency Coordinator (DEC) for the State OEM Area involved.

The ARES and Section Leadership teams played important roles in many ways. A drill of this nature provides certain advantages over a real event, but also has some severe disadvantages. On the plus side is the ability to pre-schedule and pre-plan to some extent. On the negative side, the drill was a weekday event and employers were not likely to allow employees time off to participate in a drill. This meant some people who wanted to participate simply could not, due to job demands. Others were required to take vacation time to participate. The drill went more than 36 hours straight in the beginning, which also provided a number of challenges.

The after-action report was broken into five sections:

1. Preparation — What we did to prepare for the event
2. Planning — How and why we planned our response
3. Operations — What actually happened
4. Lessons Learned — Analysis of what could have been done better, and suggestions for how to improve
5. Appendices — E-mails to members and leaders, communication and frequency plans, etc.

PREPARATION

ARES had the benefit of a great deal of preparation time before the TOPOFF 3 drill. We conducted an ARES symposium-style training session, a Recon Rally that allowed people to become familiar with the operating area, and other preparation activities. These activities were recognized as providing a great deal of value to ARES members who participated.

Training in preparation for TOPOFF 3

We conducted a training session in a day-long symposium-style set of seminars. The agenda covered: Go-Kits and Deployment Preparation, Directed and Tactical Nets, Formal Message Handling, and the National Incident Management System (NIMS).

Recon Rally

This "scavenger hunt" preparation exercise was conducted on a Saturday four weeks before the TOPOFF event, at the primary ARES Marshaling Center (a local firehouse). The day started off with a *PowerPoint* briefing that provided the 30 rally participants with an overview of the area. The briefing outlined all of the significant points of interest in the OEM Area that might either be "targets" for TOPOFF events or resources to be utilized during the exercise. These locations included defense installations (U. S. Naval Submarine Base, General Dynamics Electric Boat, National Guard Camp), transportation infrastructure (Interstate Highway Bridge, Railroad Station and Railroad Bridge, several Ferry Terminals), commercial establishments (State Pier, Pfizer, Hess Oil terminal), nuclear power stations (Units 1, 2, 3), local colleges (U. S. Coast Guard Academy, Connecticut College), and exercise-specific venues (Red Cross New London, Ocean Beach State Park, Fort Trumbull). **Figure 1** presents the Connecticut ARES Areas and the scope of Area 4.

The participants then moved on to a scavenger hunt activity wherein the Rapid Response Teams (comprised of two or three ARES members each) had to drive through the area, following a series of 12 clues and maps provided in a sealed envelope. During the hunt, each team had to locate and gather

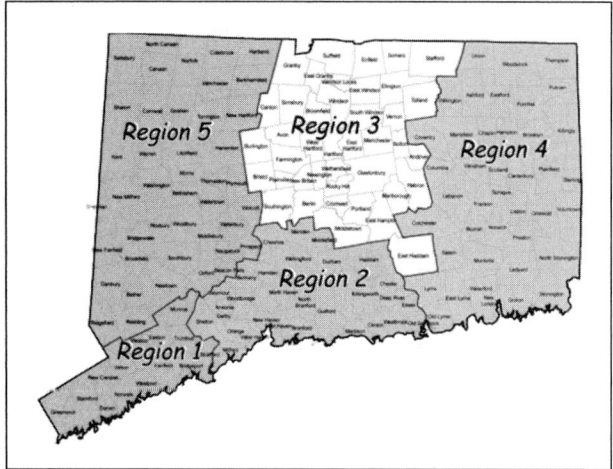

Figure 1 — The state of Connecticut, with ARES Areas delineated. TOPOFF 3 took place in Area 4.

a piece of information at each of several key locations in the OEM Area. The teams also had to originate and send an NTS-formatted message back to the Marshaling Center. The required use of several local repeaters helped to test the exercise frequency plan. Teams were also required to contact a local municipal EOC on a simplex frequency to pass informal traffic. All teams had two hours to complete the course, and all the teams but one finished within the allotted time. A point-based reward scheme recognized those who used proper NTS procedure, found answers to the clue questions, etc. The event not only helped to orient and familiarize ARES members who were from outside of the area, it also helped us all get to know each other better.

ARES provided a significant number of pre-built base, temporary base, mobile, and portable communications sites for TOPOFF. To this end, Rapid Response Teams (RRTs) from around the state were needed, but since the exercise was being held during the work week, the turnout was smaller than in the case of a real disaster. It was suggested that ARES members meet somewhere in their home area and check in, then travel as a group, perhaps at staggered intervals, to ease congestion at the Marshaling Center at this end.

Speaking at Club Meetings and ARRL Headquarters

The Section Emergency Coordinator (SEC) spoke at numerous club meetings, and was invited to ARRL Headquarters to brief ARRL on our plans as well as to recruit volunteers. The presentation at HQ was well received and numerous volunteers signed up after these sessions. Additionally, the live question-and-answer discussions provided more information to those attending the meeting, and alerted the SEC to additional areas that needed to be considered in planning for the TOPOFF 3 event. The Section Manager (SM) reported that the training the SEC provided was very well received; it was the first of its kind given to ARRL staff by volunteers from the Section.

Personal Preparation

Throughout our preparations as a team we emphasized the need for personal preparedness in terms of volunteers bringing items to support their well-being. In our meetings and discussions as well as in the TOPOFF weekly bulletins, we frequently emphasized personal self-sufficiency, not just radio equipment. Though this event was taking place in a metropolitan area, there was no way to determine exactly when someone's shift would end, or where he or she might be at meal times. For this reason, everyone was encouraged to bring food, water, medications, and other essential items and to keep these items with them at all times.

PLANNING

The Section Manager and the ARES Leadership team spent a great deal of time in planning for TOPOFF 3. Planning included recruiting and sign-up, scheduling and assignments, communications with the team, repeater usage, contingency planning, and other aspects.

Recruiting and Sign-Up

Recruitment was done extensively from ARES members, neighboring jurisdictions, and volunteers from areas as far away as Pennsylvania. The Section Manager's e-letters and posts on the ARES Discussion List were especially helpful with recruitment. We gathered a significant amount of information in the sign-up process and asked follow-up questions where necessary. Our sign-up form, which was simply copied into an e-mail and sent to the SEC, is shown in **Figure 2**.

Recruiting of additional EmComm volunteers was actively performed by the SEC and SM from the neighboring Section. One of the major lessons learned during the ARES operation in response to the September 11th attack was the value of Mutual Assistance Teams.

Scheduling/Assignments

The Section Emergency Coordinator handled all scheduling and assignments, serving as a central clearinghouse for all sign-ups. An *Excel* spreadsheet was used to keep track of Red Cross needs, OEM needs, reserve staffing, and leadership staffing. This spreadsheet was provided to the entire ARES leadership team periodically for their review and input.

Communicating With the Team

Internet e-mail was the primary means of communication with team members, individuals, all TOPOFF volunteers, and all ARES members. Numerous conference calls were held with the ARES leadership team as well.

Sign-up Form

Please provide the following information:
- Full Legal Name:
- Call Sign:
- Home Phone:
- Work Phone:
- Cell Phone:
- Address:
- What types of positions you can cover (based on the list below):
- When available:
  - April 4 —
  - April 5 —
  - April 6 —
  - April 7 —
  - April 8 —

(Note: If you say are available from 6:00 am to midnight, rest assured we will not give you an 18-hour shift. It would be helpful if you can tell us the maximum shift duration you are comfortable with, if that is longer than your availability.)

Other information:

The four types of positions are:

1 — Permanent Fixed locations where the radios and antennas are already set up and we just need operators.

2 — Temporary Fixed Locations where the communicator will need to bring at least a 25-watt radio, power supply, backup battery, and portable antenna with mast.

3 — Shadow stations that will stay with key people and will need a handheld radio and plenty of battery (either AA battery backup or small gel cell battery backup).

4 — Mobile in Red Cross or other vehicles will need at least a 25-watt radio, mag mount, and the ability to take power from a cigarette lighter jack or straight from the vehicle battery.

Figure 2 — Recruits for TOPOFF 3 filled out this sign-up form.

Repeater Usage

All of the local repeaters were made available to ARES by the repeater trustees. Additionally, there was a UHF repeater established in the local area specifically for the drill. A linked-repeater system was established that covered most of the Section on VHF/UHF. All of these repeaters provided exceptional service. Although one repeater in the linked system had to be replaced during the drill, the UHF repeater collocated with it stayed on the air and provided service during the absence of the VHF repeater.

Use of Magnetic Mount Antennas on Fiberglass Vehicles

Eighteen-by-twenty-four-inch steel plates and gaffer's tape were provided to hold the plates to the fiberglass roofs without damaging the roofs. These plates allowed the placement of magnetic mount antennas on the fiberglass vehicles.

Marshaling Center

The local Fire Department was kind enough to open their doors to ARES, giving us access to the firehouse facilities for both the Recon Rally and the TOPOFF 3 drill. We used several rooms, including the bunk rooms (for resting), a room we designated for communications, and a room that we used as our check-in area, refreshment area, and relaxation area. The facilities enabled ARES to have a Marshaling Center near enough to the event to be there within a few minutes, but far enough away to be clear of most hazards in the "hot zone."

Radio/Antenna Plan

The Marshaling Center was equipped to handle all ARES communication needs, even though the Tactical Net Control Station (NCS) was planned to be located at a nearby ARES Leader's home. In fact, the Tactical Net was run from the Marshaling Center on occasions when the net was slow.

A three-element beam was located at the front of the fire station. The APRS and local Resource net antennas were on the side of the building running low power. The 47.42 MHz antenna was located on the back of the building along with a two-meter antenna. The HF dipole was strung from the siren tower (using a pulley and rope installed by the Fire Department) and a rope over a tree placed using a Wrist Rocket® and fishing reel.

Two radios were placed at the front of the room for the Tactical Net and the State-Wide Resource Net. The rest of the radios were placed along the side wall.

Emergency Power

The firehouse had an automatic-start emergency power generator. In fact, we held our Recon Rally during their weekly generator test and we were completely unaware that we were actually operating on generator power until we were notified that it was time to switch back to commercial power and that there might be a momentary power interruption.

OPERATIONS

Marshaling Center Check-In and Assignment. Ample parking facilities were available at the Marshaling Center. Signs were placed at the Center entrance directing personnel to the check-in desk, at which they were given a packet that included the following:

- An ARES-type Application in which they listed their call sign and license type, name, address, various telephone numbers, e-mail address, Driver's License number, and verification of current Amateur Radio license.
- An Emergency Reference Sheet containing emergency contact data, medical data and the like, which was placed in an envelope and sealed.
- A Volunteer Assignment Information Sheet that included the ARES Team Leader's name, the volunteer's call sign, the assigned Tactical call sign (if appropriate), the assigned served agency's name, the agency's point of contact's name, and the location. (Additionally, this same sheet contained the contact numbers for all the ARES officials, frequency information for the various nets, and a list of served agencies. On the back side of the sheet was a check list of items to take upon arrival at the served agency, instructions for setting up the station; instructions on what to do after the station is on the air, and a quick review of how to serve the agency.)
- An Exercise Frequency Plan that translated all of the available frequencies in use during the drill to such names as "Alpha," "Bravo," "Charlie," etc. (This simplified the transmission of frequency assignments over the air without actually divulging the frequencies, and made the plan consistent with ICS frequency format.)
- Finally, a four-page document containing "briefing" information (what to have with you before leaving for your assignment, safety considerations, dress codes, expected behavior, etc.) was included.

Upon filling out the ARES Application and getting their licenses verified, volunteers were asked to go to the assignment desk. Here, their emergency data envelope was accepted and safeguarded. Their application was examined and their name cross-matched to a duty roster that had been prepared ahead of time. On their Volunteer Assignment Information Sheet, their team leader, call sign, tactical call sign, Served Agency, the Agency's point-of-contact's name, and the location to which the volunteer was to report were all entered, and a copy was given to the volunteer, along with an ID badge. Volunteers assigned to a Red Cross location were additionally issued an ARES vest. All pertinent volunteer data (name, call sign, time of arrival, TAC call, vest number, time of deployment, etc.) was entered on a log sheet to provide a ready reference for others when trying to determine the status of any volunteer. Some volunteers had to borrow a radio in order to fulfill their assignment, so the radio types and serial numbers were also tracked at the assignment desk.

A magnetic whiteboard drawn into a grid-like arrangement was made available to officers at the Marshaling Center, along with magnetic cards that had been preprinted with each volunteer's name and call sign. Each section of the grid represented a different assignment. As volunteers reached their assignment, their magnetic cards were placed in the appropriate section of the grid. Those in transit to their assignments were grouped together in a corner of the board, indicating that they had been processed through the Center.

Upon returning to the center, individuals were logged back in and their vests and/or badges reclaimed. These volunteers were either placed back on stand-by status or, if they had fulfilled all their assignments, they were free to return to their homes.

American Red Cross Check-In and Assignment. After checking in at the ARES Marshaling Center, all Amateur Radio operators who filled any assignment "inside" the exercise area (e.g., Emergency Response Vehicles, Unified Command Center, etc.) had to check in again at the local American Red Cross chapter office in the exercise area, where execution of an American Red Cross Local Disaster Volunteer Staff Registration (ARC Form 1492A) was required. Upon completion of processing, participants were issued red neck lanyards and Red Cross stick-on name tags to indicate they were officially part of the exercise. At that point, they stood by for transportation or proceeded to their assignment as appropriate.

On-Air Operations

Tactical Net. The Tactical Net served as the primary voice communications network for TOPOFF 3 Amateur Radio operations. The primary Net Control Station (NCS) for this net was selected because the ARES Leader who owned the station had the best overall knowledge of the area, concept of operations, resources available, and experience. Prior to the

---

MARSHALING CENTER RADIO CONFIGURATION

The radios set up at the Marshaling Center included:
- 2-meter radio with magnetic mount antenna for Tactical Net
- 2-meter radio with three-element beam for statewide linked system Resource Tracking Net
- 2-meter radio with magnetic mount antenna for local Resource Net
- 2-meter radio with TNC and laptop for APRS, which was used to track shuttle vans
- ICOM IC-706 connected to 1/2-wave dipole for 3.965 MHz (NTS and ARES Nets)
- 47.42 MHz commercial with appropriate Ringo for Red Cross liaison
- 800 MHz scanner for listening to ITAC/ICALL 800 MHz interoperability communications

exercise, the decision was made to have the Tactical NCS operate from a location separate from the Marshaling Center whenever possible, because the expected activity and noise level at the Marshaling Center was likely to make operating as NCS difficult. In fact, Net Control operated from this gentleman's home station since it had redundant VHF/UHF radios and antennas; APRS, packet, and HF capabilities; cellular and landline telephone service; broadband Internet access; and was located less than five miles from the Marshaling Center. In order to provide operator relief for NCS, personnel at the Marshaling Center served as alternate NCS on an as-needed basis.

Command Net. An "unpublished" 440 MHz repeater was setup at a local repeater site and made available for use as a dedicated ARES Command Net. This repeater/frequency provided a means for senior ARES leadership personnel (SEC, DECs, ADECs, SM, NCSs) to coordinate operations and discuss issues "semi-privately" without interfering with the normal Tactical and Resource Net operations. However, all particularly sensitive discussions were conducted "off the air" — either in person or via telephone (cellular or landline).

Statewide Resource Tracking Net. Operators participating in the exercise were asked to check into the Statewide Resource Tracking Net upon leaving home. This net was conducted using a statewide linked-repeater system. A Net Control Station at the Marshaling Center kept track of those stations while they were en route, to allow for any needed scheduling adjustments.

Local Resource Net. When incoming exercise participants got within range of the local VHF repeater, they were asked to check in with the Local Resource Net to provide Marshaling Center officials with an updated estimated time of arrival. This net provided a means to assist any operators from out of the area with final directions to the Marshaling Center as needed. This net also tracked operators after they departed the Marshaling Center en route to the Red Cross shuttle area.

Packet. Use of VHF Packet was expected, and the capability was available at the NCS and Marshaling Center operating positions. Plans called for using the local digipeaters and their respective bulletin board systems (BBS). Since most of our efforts during the exercise were ultimately focused on tactical voice communications between Red Cross vehicles, the food service location, and Red Cross headquarters, very little actual use was made of packet communications.

Automatic Position Reporting System (APRS). Several exercise participants had APRS capability. APRS map plots were maintained by the Tactical NCS, as well as by operators at the Marshaling Center and the Unified Command Center (UCC). APRS was used to track some of the incoming ARESMAT personnel from other OEM Areas. Use of APRS trackers on Red Cross vehicles not staffed by full-time Amateur Radio operators also provided a means of tracking some of our resources. All of the Red Cross Emergency Response Vehicles (ERVs) had operators assigned, but the personnel shuttle vehicles (usually small vans) were of limited capacity — using a seat for an operator would have reduced their capacity even more. For operational security, the APRS unproto paths were set to "NOGATE" to prevent RF signals from being gated to the Internet.

HF. A full-size resonant 80-meter dipole was installed at the Marshaling Center. The HF antenna was used to pass NTS traffic on 3.965 MHz until interference with Fire Department radios was noted. Due to other outlets/frequencies being available for the traffic (and also due to a lack of time), troubleshooting was not done to attempt filtering that would alleviate the interference, and the HF radio was not used further during the exercise.

Red Cross 47.42 MHz. We were able to install a radio operating on the national Red Cross frequency of 47.42 MHz. This radio was connected to a 6-meter Ringo which had been previously re-tuned to 47.42. For reasons that we never understood, no radio traffic was heard on this frequency. The Red Cross New London Chapter did use 47.42 MHz to talk to some of the Red Cross mobile units, but we were unable to receive their base station or any of the mobile units. Local terrain may have been a factor.

Contingency Planning for NCS and Repeaters. In order to provide contingencies in case of the actual or simulated loss of our primary repeaters, the ARES TOPOFF Frequency Plan designated both a secondary/backup repeater and a tertiary/simplex frequency for each of the nets or functions. These designations were made based on the area of effective coverage of the respective repeaters and whether they had emergency power available. This area was reasonably "repeater rich" and suitable alternative repeaters were readily available. As part of prior contingency planning, a simplex site had been identified that would allow the TOPOFF NCS to use a mobile or portable station to cover the necessary operational area without the use of repeaters. The site was conveniently close to the Marshaling Center (less than 3 miles direct and 5 miles by road) and could be reached by a mobile unit within 10 minutes if needed. The utility of this site was initially determined by use of topographic maps to evaluate terrain considerations. The site was located on a broad hilltop about 600 feet above sea level and had a little-used open parking lot area available for setting up a portable station. (The fact that the telephone company had long ago located several microwave relay towers in the area served to validate the preliminary conclusion.) After selecting the site, operational tests were conducted during the regularly scheduled ARES training nets prior to TOPOFF to confirm the expected coverage.

Use of Unified Command for NCS Overnight. ARES had an operating position in the Unified Command Center, on behalf of the American Red Cross. The station was equipped with APRS as well as VHF/UHF voice. During the overnight shift, this station became the net control for all Amateur Radio communication when the Marshaling Center and normal NCS locations were shut down. Communication was maintained with the SEC at the firehouse throughout the night, and approximately six ARES members also spent the night at the firehouse as a ready-reserve in case any needs were encountered overnight.

Unified Command Emergency Power. Automatic switching to battery backup was put in place at the Unified Command Center (UCC) as power appeared to be a potential source of difficulty. There were sporadic outages and a simulated power failure at the Unified Command. West Mountain Radio's DCtoGO, along with a RIGrunner, PowerGate, and a 12-volt 1-amp sealed lead-acid automatic charger, were put in place to run the radio and APRS system at Unified Command. The radio was a Kenwood TM-D700A APRS mobile radio, used on VHF/UHF voice and 2-meter APRS simultaneously. The laptop ran on ac power until manually switched to a power inverter. This system allowed the ARES UCC station, including APRS, to stay on the air during simulated and real power outages. We did have a deep cycle battery (with charger) powering this radio, so it was the only radio that did not go down during the outages.

Public Relations. Reporters from many of the major news agencies gathered in the media area, where they received frequent official briefings on the status of the drill. ARRL Media and Public Relations Manager Allen Pitts, W1AGP, and ARRL Chief Development Officer Mary Hobart, K1MMH, were both on hand to speak directly with reporters and give them a packet of information relating to ARES and our operation at TOPOFF. This packet included a modified version of the CT ARES brochure, along with contact information for the SEC and the SM who were both coordinating information to reporters from the Marshaling Center. A separate cell phone line was used for this purpose.

Operators in the field were also given instructions to direct reporters to the Marshaling Center for information. It was felt that operators should not be disturbed while performing their duties because, among other reasons, the Amateur Radio effort (as well as the entire TOPOFF event) was being assessed by government agency evaluators. Some calls were received at the Marshaling Center, but not nearly as many as expected primarily because Allen and Mary were on hand to give them good information. However, several interviews with reporters were conducted and news articles written. See the chapter on PIO for more information on the information dissemination process.

LESSONS LEARNED

Transportation. Transportation from the ARES Marshaling Center to the Red Cross facility for sign-in, and then on to specific assignments was the single largest difficulty encountered in the ARES response to TOPOFF. As a result, the ARES Leadership Team decided that, in any future drill or actual deployment, ARES will be responsible and will provide for transportation of ARES members. We feel that we can perform this function more efficiently than another organization because we know our own needs and have the communication capabilities to coordinate this activity.

Sign-In Process and Equipment Checkout. The sign-in process for ARES members needed to include an equipment checkout phase. A service monitor or at least a frequency counter should be in place, along with a radio programmed to act like each of the repeaters. This equipment will afford us the ability to ensure that ARES members have their radios properly programmed and in good working order before they are deployed. With the use of Anderson PowerPoles for all 12 V dc connections, it would be straightforward to provide power for mobile radios for checkout.

Scheduling/Assignments. In retrospect, it would have been more appropriate to have someone other than the SEC handle scheduling and assignments. There was a significant amount of administrative work that could have been delegated, which would have relieved the SEC from a number of competing priorities in the last few days before the drill. The situation was exacerbated by the fact that the entire plan and schedule for the drill was changed on the Friday afternoon immediately before the drill, which ran Monday through Friday of the next week. Keep in mind, this was a statewide (entire Section) response. We recruited statewide for volunteer assistance, knowing that we would need about 100 operators to fill all the designated time slots.

HF Interference with Fire Station Radios. We tested for interference between our radios at the ARES Marshaling Center prior to the drill, but during the drill we found out that our HF radio on 3.965 MHz interfered with the Fire Department station radio (on low-band VHF) to the extent that we had to terminate HF operations. It turned out this was not a significant problem, but there are parts of CT in which the terrain limits all reliable communications except HF. While we could easily relay through our Tactical Net Control, located a couple of miles away in the home of an ARES leader, or we could have relayed through relay stations, it is important to check the operation of any local two-way or other radios as part of equipment testing. Amateurs with a technical knowledge are a plus in this kind of situation.

Logistics with the Red Cross. Logistics coordination with the Red Cross was somewhat hampered by their use of Amateur Radio operators who were affiliated with and known to the Red Cross chapter personnel, but who were not as familiar with directed net and other ARES procedures. Some of their operational issues also resulted from the radio operator not having anyone available at the Red Cross chapter to coordinate and to answer the specific questions being asked by the Red Cross units conducting field operations (ERVs, etc.). At opportune times the Tactical Net Control Station was able to do some limited "on-air" training with the Red Cross operators to improve the overall flow of tactical traffic. Future operations will include an assignment of an ARES liaison operator, even if other Red Cross-affiliated Amateur Radio operators are present.

Message Prioritization on a Tactical Net. As the drill came to a close, we realized that there was one pivotal station in our efforts to support the Red Cross Mass Care: Feeding effort, which was the kitchen preparing the meals and which was staffed by a key Red Cross person. With the value of hindsight it is easy to see that having this station listen on one of the alternate 2-meter repeaters would have enabled the Tactical Net Control station to simply direct people to that alternate frequency and continue to handle other traffic. This would have placed all the high-priority traffic on one channel, away from lower-priority as well as possible emergency traffic.

Debriefing Volunteers on Finishing Assignment. Debriefing of volunteers did not occur as thoroughly and as formally as originally planned. This will be a high priority in future drills and deployment. While we did get information anecdotally from ARES members upon their return, we certainly lost some information by doing the debriefing informally.

TOPOFF ARES leaders needed to communicate in some fashion. We had the means but didn't perform well as we might have. We had a significant number of ARES leaders, and the Section Manager at the ARES Marshaling Center, but did not have scheduled team meetings or discussions. In retrospect, it would be good to have such meetings during the inevitable quiet period that occurred when a group of ARES members had been sent out. Such meetings need to occur at least twice a day — four times a day would be even better, to assure all leaders that they have current information and that all pertinent discussion takes place with the entire leadership team, including local ECs.

Marshaling Center Check-In and Assignment. The check-in process must be more thorough. Many volunteers were admitted without their admittance sheet showing their Drivers' License or government-issued ID Number, or the fact that their Amateur License had actually been presented. This was a requirement due to ARES and club liability exposure. By having additional entries on the form showing the expiration dates for these documents and having the assignment desk double-check and initial that they had checked the form, we would alleviate the possibility of unauthorized personnel arriving on the scene.

The assignment desk needs more thorough documentation on given assignments. There were instances in which a volunteer would walk up and ask who he would be reporting to and where he was expected to go. Although the assignment forms had an entry field for the person to report to, none of this was known to the assignment person, so the field was either left blank or the Marshaling Center's on-site manager's name was entered. As an after-action discussion, we realized that in many cases, we did not know what the operator was going to do. They got that information when they reported to their assigned location in the exercise zone.

Many volunteers asked for directions to pick up the shuttle bus. Although there were printed route maps available, some road construction and detours near the shuttle site created confusion for operators not familiar with the area. When possible, each person being told to drive to an off-site location to pick up transportation should be given a handout showing turn-by-turn directions for getting to that location. Those volunteers possessing GPS units would be appreciative of actual address data, consisting of number and street.

There was no pre-determined routine in place to track whether the people due at the Center were actually going to be there on time. This meant that schedules had to be shuffled to help fill assignments that were due but still left unassigned. A better communications system between the Enroute Tracking station and the Intake or Assignment Desk would have helped keep assignments from being missed. Additionally, instructions to the volunteers prior to their departures from home should stress the importance of being on time.

The use of assigned tactical calls was important when making assignments. There were both "long" tactical calls (e.g., "Tactical Net Control") and "short" tactical calls (e.g., "TacNet"). These tactical calls allowed the Tactical Net to flow smoothly and with continuity even when actual operators changed. (Keep in mind that one still has to identify with one's own FCC call sign, per Part 97.) When personnel were en route to and from, use of their FCC call sign made it clear that they were operating as an individual (necessary during shuttle transportation, etc.). Instructing personnel not currently assigned to check in with their status at least hourly would have helped to keep all assignments filled. There were almost always volunteers standing by at the Center to help fill in assignments that had been lost or changed.

The routine for checking people back in was, at times, confusing. There were ARES vests distributed that required tracking, as well as radios loaned and badges that needed to be returned. Although most people promptly returned their materials to the Center, some kept it for the three days they were on duty. We also felt that ARES identification badges should have been developed for our personnel, regardless of whether they would have been "accepted" by field personnel as ID. At the very least, ARES badges would have made it known that ARES was there!

A suggestion was made that the assignment, tracking of personnel, deployment status and the personnel database be automated to the extent possible in a user-friendly, Windows-type environment in which a laptop would make the assignments based on the availability of volunteers, track their deployment to their assignments, and track their release from these assignments to return to the Center. Such a system networked at the Center at critical posts would have kept all the data current and available.

IN CLOSING

The TOPOFF 3 exercise was the largest multi-agency, multi-organization emergency preparedness exercise ever held in Connecticut or, for that matter, New England. ARES was fortunate to be recognized by the public safety and emergency management officials in Connecticut as an organization necessary for participation in the exercise. ARES did the organization proud in its performance.

Appendix B
Amateur Radio Emergency Service® Material

Amateur Radio Emergency Service® Manual

Chapter One: Amateur Radio Emergency Service ®

(ARES ®)

The Amateur Radio Emergency Service ® (ARES ®) consists of licensed amateurs who have voluntarily registered their qualifications and equipment for communications duty in the public service when disaster strikes. Every licensed amateur, regardless of membership in ARRL or any other local or national organization is eligible to apply for membership in the ARES. Training may be required or desired to participate fully in ARES. Please inquire at the local level for specific information. Because ARES is an Amateur Radio service, only licensed radio amateurs are eligible for membership. The possession of emergency-powered equipment is desirable, but is not a requirement for membership.

1.1 ARES Organization

There are four levels of ARES organization--national, section, district and local. National emergency coordination at ARRL Headquarters is under the supervision of the ARRL Membership and Volunteer Programs Manager, who is responsible for advising all ARES officials regarding their problems, maintaining contact with federal government and other national officials concerned with amateur emergency communications potential, and in general with carrying out the League's policies regarding emergency communications.

1.2 Section Level

At the section level, the Section Emergency Coordinator is appointed by the Section Manager (who is elected by the ARRL members in his or her section) and works under his/her supervision. In most sections, the SM delegates to the SEC the administration of the section emergency plan and the authority to appoint District and local ECs, Assistant SECs and Assistant DECs. Some of the ARRL sections with capable SECs are well-organized. A few have scarcely any organization at all. It depends almost entirely on who the section members have put into office as SM and whom he/she has appointed as SEC.

1.3 Local Level

It is at the local level where most of the real emergency organizing gets accomplished, because this is the level at which most emergencies occur and the level at which ARES leaders make direct contact with the ARES member-volunteers and with officials of the agencies to be served. The local EC is therefore the key contact in the ARES. The EC is appointed by the SEC, usually on the recommendation of the DEC. Depending on how the SEC has set up the section for administrative purposes, the EC may have jurisdiction over a small community or a large city, an entire county or even a group of counties. Whatever jurisdiction is assigned, the EC is in charge of all ARES activities in his area, not just one interest group, one agency, one club or one band.

1.4 District Level

In the large sections, the local groups could proliferate to the point where simply keeping track of them would be more than a full-time chore, not to mention the idea of trying to coordinate them in an actual emergency. To this end, SECs have the option of grouping their EC jurisdictions into logical units or "districts" and appointing a District EC to coordinate the activities of the local ECs in the district. In some cases, the districts may conform to the boundaries of governmental planning or emergency-operations districts, while in others they are simply based on repeater coverage or geographical boundaries. **Figure 2** depicts the typical section ARES structure.

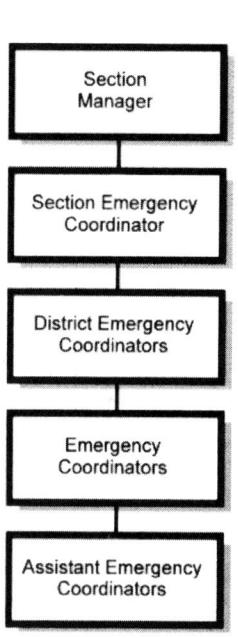

1.5 Assistant ECs

Special-interest groups are headed up by Assistant Emergency Coordinators, designated by the EC to supervise activities of groups operating in certain bands, especially those groups which play an important role at the local level, but they may be designated in any manner the EC deems appropriate.

1.6 Planning Committee

These assistants, with the EC as chairman, constitute the local ARES planning committee and they meet together from time to time to discuss problems and plan projects to keep the ARES group active and well-trained.

There are any number of different situations and circumstances that might confront an EC, and his/her ARES unit should be organized in anticipation of them. An EC for a small town might find that the licensed amateur group is so small that appointing assistants is unnecessary or undesirable. On the other hand, an EC for a large city may find that even his assistants need assistants and that sometimes it is necessary to set up a special sub-organization to handle it. There is no specific point at which organization ceases and operation commences. Both phases must be concurrent because a living organization is a changing one, and the operations of a changing organization must change with the organization.

1.7 Operation and Flexibility

We have discussed how a typical ARES unit may be organized. Just what shape the plan in your locality will take depends on what your EC has to work with. He/she uses what he/she has, and leaves provision in the plan for what he/she hopes, wants and is trying to get. Flexibility is the keynote. The personnel, equipment and facilities available today may not be available tomorrow; conversely, what is lacking today may be available tomorrow. In any case, bear in mind that organizing and planning are not a one-person task. The EC is simply the leader, or, as the title indicates, the coordinator. His/her effectiveness inevitably will depend on what kind of a group he/she has to work with. Make yourself available to your EC as a member of his planning committee, or in any capacity for which you think you are qualified.

Local ARES operation will usually take the form of nets -- HF nets, VHF (repeater) nets, even RTTY, packet or other special-mode nets, depending on need and resources available. Your EC should know where your particular interests lie, so that you can be worked in where your special talents will do the most good.

It is not always possible to use the services of all ARES members. While it is general policy that no ARES member must belong to any particular club or organization to participate in the program, local practical considerations may be such that you cannot be used. This is a matter that has to be decided by your EC. In some cases, even personality conflicts can cause difficulties; for example, the EC may decide that he cannot work with a particular person, and that the local ARES would be better served by excluding that person. This is a judgment that the EC would have to make; while personality conflicts should be avoided, they do arise, more often than we would prefer. The EC on the job must take the responsibility for making such subjective evaluations, just as the SEC and DEC must evaluate the effectiveness of the job being done by the EC.

1.8 ARES Operation During Emergencies and Disasters

Operation in an emergency net is little different from operation in any other net, requires preparation and training. This includes training in handling of written messages--that is, what is generally known as "traffic handling." Handling traffic is covered in detail in the ARRL *Operating Manual*. This is required reading for all ARES members--in fact, for all amateurs aspiring to participate in disaster communications.

The specifications of an effective communication service depend on the nature of the information which must be communicated. Pre-disaster plans and arrangements for disaster communications include:

- Identification of clients who will need Amateur Radio communication services.
- Discussion with these clients to learn the nature of the information which they will need to communicate, and the people they will need to communicate with.
- Specification, development and testing of pertinent services.

While much amateur-to-amateur communicating in an emergency is of a procedural or tactical nature, the real meat of communicating is formal written traffic for the record. Formal written traffic is important for:

- A record of what has happened--frequent status review, critique and evaluation. Completeness which minimizes omission of vital information.
- Conciseness, which when used correctly actually takes less time than passing informal traffic.
- Easier copy--receiving operators know the sequence of the information, resulting in fewer errors and repeats.

When relays are likely to be involved, standard ARRL message format should be used. The record should show, wherever possible:

1. A message number for reference purposes.
2. A precedence indicating the importance of the message.
3. A station of origin so any reply or handling inquiries can be referred to that station.
4. A check (count of the number of words in the message text) so receiving stations will know whether any words were missed.
5. A place of origin, so the recipient will know where the message came from (not necessarily the location of the station of origin).
6. Filing time, ordinarily optional but of great importance in an emergency message.
7. Date of origin.

The address should be complete and include a telephone number if known. The text should be short and to the point, and the signature should contain not only the name of the person sending the message but his title or connection, if any.

Point-to-point services for direct delivery of emergency and priority traffic do not involve relays. Indeed, the full ARRL format is often not needed to record written traffic. Shortened forms should be used to save time and effort. For example, the call sign of the originating station usually identifies the place of origin. Also, the addressee is usually known and close by at the receiving station, so full address and telephone number are often superfluous. In many cases, message blanks can be designed so that only key words, letters or numbers have to be filled in and communicated. In some cases, the message form also serves as a log of the operation. Not a net goes by that you don't hear an ARL Fifty or an ARL Sixty One. Unfortunately, "greetings by Amateur Radio" does not apply well during disaster situations. You may hear an ARL text being used for health and welfare traffic, but rarely during or after the actual disaster. Currently, no ARL text describes the wind speed and barometric pressure of a hurricane, medical terminology in a mass casualty incident or potassium iodide in a nuclear power plant drill. While no one is suggesting that an ARL text be developed for each and every situation, there is no reason why amateurs can't work with the local emergency management organizations and assist them with more efficient communications.

Amateurs are often trained and skilled communicators. The emergency management community recognizes these two key words when talking about the Amateur Radio Service. Amateurs must use their skills to help the agencies provide the information that needs to be passed, while at the same time showing their talents as trained communicators who know how to pass information quickly and efficiently. We are expected to pass the information accurately, even if we do not understand the terminology.

Traffic handlers and ARES members are resourceful individuals. Some have developed other forms or charts for passing information. Some hams involved with the SKYWARN program, for instance, go down a list and fill in the blanks, while others use grid squares to define a region.

Regardless of the agency that we are working with, we must use our traffic-handling skills to the utmost advantage. Sure, ARL messages are beneficial when we are passing health and welfare traffic. But are they ready to be implemented in times of need in your community? The traffic handler, working through the local ARES organizations, must develop a working relationship with those organizations who handle health and welfare inquiries. Prior planning and personal contact are the keys to allowing an existing National Traffic System to be put to its best use. If we don't interface with the agencies we serve, the resources of the Amateur Radio Service will go untapped.

Regardless of the format used, the appropriate procedures cannot be picked up solely by reading or studying. There is no substitute for actual practice. Your emergency net should practice regularly--much more often than it operates in a real or simulated emergency. Avoid complacency, the feeling that you will know how to operate when the time comes. You won't, unless you do it frequently, with other operators whose style of operating you get to know.

Chapter Two: Simulated Emergency Test (SET)

The ARRL Simulated Emergency Test is a nationwide exercise in emergency communications, administered by ARRL Emergency Coordinators and Net Managers. Both ARES and the National Traffic System (NTS) are involved. The SET weekend gives communicators the opportunity to focus on the emergency communications capability within their community while interacting with NTS nets. SET weekend is held in October, and is announced in *QST*.

2.1 Purpose of SET

1. To find out the strengths and weaknesses of ARES and NTS, the Radio Amateur Civil Emergency Service (RACES) and other groups in providing emergency communications.
2. To provide a public demonstration -- to served agencies such as Red Cross, Emergency Management and through the news media -- of the value to the public that Amateur Radio provides, particularly in time of need.
3. To help radio amateurs gain experience in communications using standard procedures and a variety of modes under simulated- emergency conditions.

2.2 SET Format

The scoring format reflects broad objectives and encourages recruitment of new hams and use of digital modes for handling high-volume traffic and point-to-point Welfare reports out of the affected simulated-disaster area. Participants will find SET an opportunity to strengthen the VHF-HF link at the local level, thereby ensuring that ARES and NTS are working in concert. The SET will give all levels of NTS the chance to handle exercise-related traffic. The guidelines also recognize tactical traffic on behalf of served agencies.

ARES units and other groups are free to conduct their SETs anytime during September 1 and November 30 if an alternative date is preferred. The activity period should not exceed 48 hours. The deadline for receipt of all reports is January 31. A complete array of reporting forms will be sent to affected Field Organization appointees.

2.3 Preparing for SET

Emergency Coordinators sign up all available amateurs in their area and work them into the SET plans. They make special efforts to attract new Technicians as outlined earlier.

A meeting of all ARES members and prospective members is called to briefly outline (no details!) SET activities, and give general instructions. ECs contact served agencies and explain the intent and overall purpose of the SET, offer to send test messages to other branches of their agencies, and invite officials to ARES meetings and SET operating sites. Publicity is arranged in consultation with an ARRL Public Information Officer in local newspapers and radio/TV stations.

2.4 During the SET

The "emergency" situation is announced and the emergency net is activated. Stations are dispatched to served agencies. Designated stations originate messages on behalf of served agencies. Test messages may be sent simulating requests for supplies. Simulated emergency messages (just like real emergency messages) should be signed by an authorized official. Tactical communications for served agencies is emphasized.

At least one session -- or substantial segment of a session -- of the local net should be conducted on emergency-only basis. Or, if a repeater is on emergency power, only emergency-powered stations should be allowed to operate through the repeater for a certain time period.

2.5 After the SET

An important post-SET activity is a critique session to discuss the test results. All ARES (and RACES) members should be invited to the meeting to review good points and weaknesses apparent in the drill.

2.6 NTS in SET

The main function of NTS in an emergency situation is to tie together all of the various local activities and to provide a means by which all traffic destined outside of a local area, section or region can be systematically relayed to the addressee.

The interface between NTS and ARES lies in the liaison function between local nets and other NTS nets, particularly at the section level. Responsibility for representation of the local network on the section net lies with the local net manager who may or may not be the EC.

At least one net session or substantial segment of a session should be conducted on emergency power. A surprise session or two should be conducted.

2.7 Summary

One of the first steps on the way to a successful SET is to try to get as many people involved as possible, especially new hams. In a real emergency, we find amateurs with all sorts of varied interests coming out of the woodwork. Get them involved in SET so they will know more about how emergency communications should be handled. Promote SET on nets and repeaters, and sign up new, enthusiastic Technicians.

Chapter Three: ARES Mutual Assistance Team (ARESMAT) Concept

The ARESMAT concept recognizes that a neighboring section's ARES resources can be quickly overwhelmed in a large-scale disaster. ARES members in the affected areas may be preoccupied with mitigation of their own personal situations and therefore not be able to respond in local ARES operations.

Accordingly, communications support must come from ARES personnel outside the affected areas. This is when help may be requested from neighboring sections' ARESMAT teams.

To affect inter-sectional support mechanisms, each Section Emergency Coordinator (SEC) should consider adopting the following principles in their ARES planning:

- Pre-disaster planning with other sections in the Division, and adjoining sections outside the Division. Planning should be conducted through written memoranda, and in-person at conventions and director-called cabinet meetings. An ARESMAT inter-sectional emergency response plan should be drafted.
- Development of a roster of ARESMAT members able, willing and trained to travel to neighboring sections to provide communication support inside the disaster area.
- Inter-sectional communication/coordination during and immediately following the onslaught of the disaster.
- Post-event evaluation and subsequent revision/updating of the inter-sectional emergency response plan.

When developing ARESMAT functions, ARES leadership should include the following basic action elements:

3.1 Pre-Departure Functions

Team leaders should provide ARESMAT members with notification of activation/assignment. Credentials should be provided for recognition by local authorities. They should provide a general and technical briefing on information drawn principally from the requesting authority, supplemented by reports from Amateur Radio, commercial radio, W1AW bulletins and ARRL officials. The briefing should include an overview of equipment and communication needs, ARESMAT leadership contacts and conditions in the disaster area.

The host SEC's invitation, transportation (including routes in disaster area) and accommodations considerations, and expected length of deployment should all also be reviewed with the team members.

3.2 In-Travel Functions

Before and while in travel to the affected areas, team leaders should review the situation's status with the team: job assignments, checklists, affected area profile, mission disaster relief plan, strengths and weaknesses of previous and current responses, maps, technical documents, contact lists, tactical operation procedures and response team requirements.

3.3 Arrival Functions

Upon arrival, team leaders should check with host ARES officials and obtain information about frequencies in use, current actions, available personnel, communication and computer equipment, and support facilities that could be used by the team to support the relief effort. The host's ARES plan in effect for the disaster should be obtained. A priority upon arrival should be the establishment of an initial intra-team communication network and an HF or VHF channel back to the home section for morale traffic.

Team leaders should meet with served agencies, Amateur Radio clubs' communications staff, local ARRL communications authority, and others as needed to obtain information and coordinate the use of frequencies. Communication site selections should take into account team requirements and local constraints.

3.4 In-situ Functions

Team leaders should make an initial assessment of functioning communication facilities, and monitor host ARES officials' communications, and other response team relief efforts to coordinate operations and reduce duplication of effort. Team members should be monitored and their capabilities to perform their duties evaluated. Proper safety practices and procedures must be followed. A daily critique of communication effectiveness with served units and communication personnel should be conducted.

3.5 Pre-Demobilization and Demobilization Functions

An extraction procedure for ham communicators should be negotiated with served agencies and host ARES officials before it is needed. To get volunteers' commitment to travel and participate, they must be assured that there will be an end to their commitment. Open-ended commitments of volunteers are undesirable, partly because they make potential volunteers hesitate to become involved.

Leaders must coordinate with the host ARES officials and served agencies, and other functions to determine when equipment and personnel are no longer needed. A demobilization plan should be in effect.

A team critique, begun on the trip home, should be conducted. Individual performance evaluations on team members should be prepared. Copies of critiques should be sent to both the home SEC and in-disaster SEC. Problems stemming from personality conflicts should be addressed and/or resolved outside of formal reports, as they only provide distractions to the reports. Equipment should be accounted for.

A post-event evaluation meeting should always be conducted, and a final report prepared so that an update to the inter-sectional ARESMAT plan can be made.

3.6 ARESMAT Member Qualifications

The individual filling the role of ARESMAT member must have high performance standards, qualifications, experience, and the ability to work with a diverse group of team members that will be required to provide relief to the affected areas. He or she must be able to work efficiently in a disaster relief operation under the most adverse conditions.

Additionally, a member should have demonstrated ability to be an effective team player, in crisis situations, a strong personal desire and strong interpersonal communication skills. A knowledge of how ARRL, Red Cross and other agencies function at both the national and local levels is helpful. A working knowledge of the incident command system is useful as many events are managed under this system.

Members should be respected and recognized by officials and peers as competent communicators and should understand a broad range of disaster response organizations' capabilities and communication requirements.

Important: Members must be available with the consent of their employer to participate!

They should be physically fit to perform arduous work under adverse environmental conditions.

3.7 Summary

It should be noted that there is a fine balance of authority over a deployed ARESMAT. The in-disaster SEC (or delegated authority) should be able to make decisions as to use and deployment of an incoming team. Therefore, an incoming team should be prepared to submit themselves to such authority; this is evidenced by the fact that any team, internal or external, has only a limited view of the overall operation. The supervising authorities will have a better overview of the whole situation.

In turn, however, the in-disaster authority should be discouraged from abusing the resources of incoming teams. Should a team no longer be required, or a situation de-escalate, the team should be released at the earliest possible time, so that they may return home to their own lives.

The ARESMAT tool should be one of last resort. Whenever possible, amateurs from the affected section should be used for support. It is a lot to ask of a volunteer to travel far from home, family and job for extended periods of arduous and potentially dangerous work.

Chapter Four: ARES and RACES

After World War II, it became evident that the international situation was destined to be tense and the need for some civil-defense measures became apparent. Successive government agencies designated to head up such a program called on amateur representatives to participate.

In the discussions that followed, amateurs were interested in getting two points across: First, that Amateur Radio had a potential for and capability of playing a major role in this program; and second, that our participation should be in our own name, as an Amateur Radio Service, even if and after war should break out. These principles were included into the planning by the formulation of regulations creating a new branch of the amateur service, the Radio Amateur Civil Emergency Service, RACES.

Recognition of the role of Amateur Radio as a public service means responsibility. RACES regulations are printed in full in the ARRL publication, *The FCC Rules and Regulations for the Amateur Radio Service*, along with the rest of the amateur regulations. Every amateur should study closely and become familiar with these rules; civil preparedness, now a major function, may become our only on-the-air function if we are plunged into war.

4.1 What is RACES?

RACES, administered by local, county and state emergency management agencies, and supported by the Federal Emergency Management Agency (FEMA) of the United States government. It is a part of the Amateur Radio Service that provides radio communications for civil-preparedness purposes *only*, during periods of local, regional or national civil emergencies. These emergencies are not limited to war-related activities, but can include natural disasters such as fires, floods and earthquakes.

As defined in the rules, RACES is a radiocommunication service, conducted by volunteer licensed amateurs, designed to provide emergency communications to local or state civil-preparedness agencies. It is important to note that RACES operation is authorized by emergency management officials only, and this operation is strictly limited to official civil-preparedness activity in the event of an emergency-communications situation.

4.2 Operating Procedure

Amateurs operating in a local RACES organization must be officially enrolled in the local civil-preparedness agency having jurisdiction. RACES operation is conducted by amateurs using their own primary station licenses and by existing RACES stations.

The FCC no longer issues new RACES (WC prefix) station call signs. Operator privileges in RACES are dependent upon, and identical to, those for the class of license held in the Amateur Radio Service. All of the authorized frequencies and emissions allocated to the Amateur Radio Service are also available to RACES on a shared basis.

While RACES was originally based on potential use for wartime, it has evolved over the years, as has the meaning of civil defense (which is also called civil preparedness), to encompass all types of emergencies.

While operating in a RACES capacity, RACES stations and amateurs registered in the local RACES organization may not communicate with amateurs not operating in a RACES capacity. Such restrictions do not apply when such stations are operating in a non-RACES--such as ARES--amateur capacity. Only civil-preparedness communications can be transmitted.

Test and drills are permitted only for a maximum of one hour per week. All test and drill messages must be clearly so identified. With the approval of the chief officer for emergency planning and applicable state, Commonwealth, district or territory, however, such tests and drills may be conducted for a period not to exceed 72 hours no more than twice in any calendar year.

4.3 ARES and RACES

Although RACES and ARES are separate entities, the ARRL advocates dual membership and cooperative efforts between both groups whenever possible for an ARES group whose members are all enrolled in and certified by RACES to operate in an emergency with great flexibility. Using the same operators and the same frequencies, an ARES group also enrolled as RACES can "switch hats" from ARES to RACES and RACES to ARES to meet the requirements of the situation as it develops. For example, during a "non-declared emergency," ARES can operate under ARES, but when an emergency or disaster is officially declared by a state or federal authority, the operation can become RACES with no change in personnel or frequencies.

This situation is still not well understood and accepted throughout the United States; both ARES and RACES still exist, separately, in many areas. League officials will have to determine the situation in their own area.

Where there is currently no RACES, it would be a simple matter for an ARES group to enroll in that capacity, after a presentation to the civil-preparedness authorities. In cases where both ARES and RACES exist, it is possible to join both or to be involved in either. As time progresses, the goal would be the merger into one strong organization, with coordination between ARES and RACES officials using the same groups of amateurs. In some sections of the U.S. today, the ARES structure has also been accepted as the RACES structure.

4.4 Other Amateur Facilities

There are a number of other Amateur Radio facilities, not sponsored or directly affiliated with the League, which are nevertheless an integral part of our public service effort. Some of these organizations are the monitoring services, MARS, independent nets -- both international and domestic -- and other similar activities. While naturally we want you to participate in organizations sponsored by your League, it's better to participate in a non-League sponsored public service organization than not to participate at all. In this booklet we cannot give details of the operation of these other organizations because there are too many of them, and their operations change too rapidly.

Chapter Five: ARES Principles of Disaster Communication

It is impossible to state exact rules that will cover every situation that arises. The good amateur faced with a disaster situation may, however, benefit greatly from certain rules of thumb. These rules are, or should be, part of his/her training in his/her ARES group. They are presented here and should be reviewed by all amateurs, even those not active in disaster communications preparation.

1. Keep the QRM level down. In a disaster, many of the most crucial stations will be weak in signal strength. It is most essential that all other stations remain silent unless they are called upon. If you're not sure you should transmit -- don't. Our amateur bands are very congested. If you want to help, study the situation by listening. Don't transmit unless you are sure you can help by doing so. Don't ever break into a disaster net just to inform the control station you are there if needed.

2. Monitor established disaster frequencies. Many localities and some geographical areas have established disaster frequencies where someone is always (or nearly always) monitoring for possible calls. When you are not otherwise engaged, it is helpful simply to sit and listen on such frequencies, some of which are used for general ragchewing as well as disaster preparedness drilling. On CW, SOS is universally recognized, but has some legal aspects that should be considered where the need is not truly crucial. On voice, one can use "Mayday" (universal, the phone equivalent of SOS) or, to break into a net or conversation with the word "emergency."

3. Avoid spreading rumors. During and after a disaster situation, especially on the phone bands, you may hear almost anything. Unfortunately, much misinformation is transmitted. Rumors are started by expansion, deletion, amplification or modification of words, exaggeration or interpretation. All addressed transmissions should be officially authenticated as to their source. These transmissions should be repeated word for word, if at all, and only when specifically authorized. In a disaster emergency situation, with everyone's nerves on edge, it is little short of criminal to make a statement on the air without foundation in authenticated fact.

4. Authenticate all messages. Every message which purports to be of an official nature should be written and signed. Whenever possible, amateurs should avoid initiating disaster or emergency traffic themselves. We do the communicating; the agency officials we serve supply the content of the communications.

5. Strive for efficiency. Whatever happens in an emergency, you will find hysteria and some amateurs who are activated by the thought that they must be "sleepless heroes." Instead of operating your own station full time at the expense of your health and efficiency, it is much better to serve a shift at one of the best-located and best-equipped stations. This station will be suitable for the work at hand, and manned by relief shifts of the best-qualified operators. This reduces interference and secures well-operated stations.

Appendix B

6. Select the mode and band to suit the need. It is a characteristic of all amateurs to believe that their favorite mode and band is superior to all others. For certain specific purposes and distances, this may be true. However, the merits of a particular band or mode in a communications emergency should be evaluated impartially with a view to the appropriate use of bands and modes. There is, of course, no alternative to using what happens to be available, but there are ways to optimize available communications.

Long experience has developed the following advantages:

CW Mode

- 1. Less QRM in most amateur bands.
- 2. Secrecy of communications--contents of communications are much less likely to be intercepted by the general public to start rumors or undue concern.
- 3. Simpler transmitting equipment.
- 4. Greater accuracy in record communications.
- 5. Longer range for a given amount of power.

Voice Mode

- 1. More practical for portable and mobile work.
- 2. More widespread availability of operators.
- 3. Faster communication for tactical or "command" purposes.
- 4. More readily appreciated and understood by the public.
- 5. Official-to-official and phone-patch communication.

Digital Modes

(1) The first two advantages of CW, the (2) second advantage of voice mode, plust (3) greater speed in record communication than some of the other modes. In most of these modes, (4) error detection. In addition, (5) digital modes offer the potential for message store and forward capability of "digipeating" messages from point A to point Z via numerous automatically-controlled middle points.

The well-balanced disaster organization will have CW, phone, and digital mode capabilities available in order to utilize all of the advantages. Of course, one must make the best use of whatever is available, but a great deal of efficiency is lost when there is lack of coordination between the different types of operation in an emergency. Absolute impartiality and a willingness to let performance speak for itself are prime requisites if we are to realize the best possible results.

7. Use all communications channels intelligently. While the prime object of emergency communications is to save lives and property, Amateur Radio is a secondary communications means; normal channels are primary and should be used if available. Emergency channels other than amateur which are available in the absence of amateur channels should be utilized without fear of favoritism in the interest of getting the message through.

8. Don't *broadcast*. Some amateur stations in an emergency situation have a tendency to emulate *broadcast* techniques. While it is true that the general public may be listening, our transmissions are not and should not be made for that purpose. Broadcast stations are well equipped to perform any such service. Our job is to communicate *for*, not *with* the general public.

9. Communication support. Within the disaster area itself, the ARES is primarily responsible for communications support. When disaster strikes, the first priority of those NTS operators who live in or near the disaster area is to make their expertise available to their Emergency Coordinator where and when they are needed. For timely and effective response, this means that NTS operators need to talk to their ECs before the time of need so that they will know how to best respond.

Chapter Six: Working with Public Safety Officials

Public service communications performed by ARES members are based on a number of requirements. Specifically, we must be accepted by public-safety officials. Once accepted, our continued ability to contribute in times of disaster is based on the efficiency and effectiveness of our performance. While acceptance, image, efficiency and effectiveness are all important to the ongoing working relationships between amateurs and officials, it is the initial acceptance that is often difficult to achieve.

Police and fire officials tend to be very cautious and skeptical concerning those who are not members of the public-safety professions. This posture is based primarily on experiences in which well-intended but somewhat overzealous volunteers have complicated, and in some cases jeopardized, efforts in emergencies. The amateur operator or other volunteer who wishes to be of assistance must be aware of this perception.

The police have generally had their fill of "groupies" or "hangers on." They can ill afford to tolerate frustrated individuals who have always wanted to be police officers or firefighters, but for one reason or another have never reached that objective. There seems to be an abundance of people, especially during a crisis, who will quickly overstep the limits of their authority and responsibility if they are given any opportunity to assist in an official capacity. In their zeal, such persons often inhibit the actions of trained personnel. Worse yet, they can make an already dangerous situation even more so by their getting in the way. With rare exception, Amateur Radio operators do not fall into this category. The problem is, however, that police officers in the midst of stressful operations may have extreme difficulty in distinguishing between those volunteers who are problem solvers and those who are problem makers.

Those very few hams who behave emotionally, are overzealous in offering their services or in describing their abilities or who abuse the established limits of their authority are doing the amateur fraternity a real disservice. The typical police officer or firefighter, like the typical civilian, does not understand the vast differences among various radio services, the types of licensing involved or the high level or expertise and discipline that is characteristic of the Amateur Radio Service.

When an amateur arrives at a scene and jumps out of a vehicle with a hand-held in each fist and two more clipped to the belt, all squawking at once, officials simply don't know how to respond. They are either overwhelmed by equipment they don't understand, or so awe-struck that they try to avoid what they perceive as threatening.

How Amateur Radio volunteers are accepted depends on their establishing a track record of competent performance in important activities. This begins with convincing officials that amateurs offer a cost-effective (otherwise known as free) substitute for functions previously paid for by the taxpayer. Local radio amateurs also must demonstrate that they are organized, disciplined and reliable, and have a sincere interest in public service.

The most effective way to accomplish this is for you, as head of your communications group, to initiate the contact with public safety agencies in an official capacity. This is better than having individual amateurs, particularly outside an organized structure, making uncoordinated and poorly prepared contacts that often result in an impression that your group is disorganized.

Approach that first meeting well-prepared, and give a concise presentation of Amateur Radio's capabilities. Illustrate accomplishments with newspaper clippings, *QST* articles, etc., highlighting Amateur Radio public service. Discuss the existing Amateur Radio structure, emphasizing that a certain number of qualified operators will be able to respond to the public's needs.

Demonstrate the reliability and clarity of amateur gear. Nothing is more impressive than asking for a roll call on a 2-meter repeater using a hand-held radio in the police or fire chief's office and having amateurs respond with full-quieting signals from locations where municipal radios are normally ineffective. Such a demonstration several years ago convinced officials in Laguna Beach, California to ask for the assistance of the South Orange County ARES. The wisdom of this decision became evident a short time later when that seaside resort community was hit by a series of local emergencies.

Suggest specific ways in which amateurs can be of assistance. Indicate you are aware that police and fire radio frequencies are usually saturated with tactical or operational traffic in emergencies, and offer to provide an administrative frequency for use in overall management and coordination of the relief effort. More importantly, offer to demonstrate what you are capable of doing by supplying a demonstration of your communications capabilities. It is of tremendous importance that you emphasize that the services supplied by your group will free public-safety officers for other duties.

Demonstrate how easily amateurs and their equipment can interface with public-safety efforts. A perfect way to do this is to demonstrate equipment that can be made operational quickly inside the headquarters building, in a mobile command post or in field units.

Express your group's willingness to meet the needs of the sponsor or agency you are dealing with. Show a readiness to provide training to your membership. Offer public-safety officials the opportunity to have their own representatives appear before your group and provide orientation and training they feel is essential.

Finally, be realistic and objective in terms of what your group promises to provide. Be fully prepared to keep all promises you make. Remember to be organized and competent. Once you have implemented these suggestions, be patient. The requests for your services will be forthcoming, perhaps in a volume you had not anticipated!

Grass-roots action is the name of the game when it comes to achieving effective liaison. With the proper ground work accomplished in advance, recognition among those sponsors and agencies having communications needs can be dramatically increased. It's symbiotic; these people need us, and we want to help. Now that all the necessary introductions have been made, the rest is easy, for we are indeed the experts in meeting communications requirements of every sort.

The Amateur Radio Public Service Handbook

Chapter Seven: On Serving "Served" Agencies

Meeting the communications needs of served agencies is a challenging, and often daunting proposition in today's complex disaster/emergency relief arena. With the proliferation of emergency relief organizations, increasingly sophisticated needs, all competing for that scarce resource -- the volunteer -- coupled with the emergence of other non-ARES amateur providers, it's enough to make an ARES member's head spin. As more of the population moves to disaster-prone areas, and less government funding is available, more pressure is consequently placed on agencies to appropriately use the volunteer sector for support of their missions in disaster mitigation.

The League's formal relationships with served agencies are vitally important and valuable to radio amateurs. They provide us with the opportunity to contribute meaningfully to the relief of suffering among our fellow human beings. Another substantial benefit not to be overlooked is that the relationships lend legitimacy and credibility for Amateur Radio's public service capability, and that is important when it comes time to defend our frequencies and privileges before the FCC and Congress. So, ARRL's relationships with the emergency/disaster relief world need to be nurtured.

7.1 What to Do?

First, it is imperative that a detailed local operational plan be developed with agency managers in the jurisdiction that set forth precisely what each organization's expectations are during a disaster operation. ARES and agency officials must work jointly to establish protocols for mutual trust and respect. Make sure they know who the principle ARES official is in the jurisdiction. All matters involving recruitment and utilization of ARES volunteers are directed by him/her, in response to the needs assessed by the agency involved.

Make sure ARES counterparts in these agencies are aware of ARES policies, capabilities and perhaps most importantly, resource limitations. Let them know that ARES may have other obligations to fulfill with other agencies, too. Technical issues involving message format, security of message transmission, disaster welfare inquiry policies and others should be reviewed and expounded upon in the detailed local operations plans.

7.2 Pulled Every Which Way But Loose

Another challenge ARES faces is the number of agencies that demand communications support during a disaster. A local ARES unit only has so much to go around, and it can't possibly meet every agency's needs.

While the League maintains several formal Memoranda of Understanding (MOU) with disaster and emergency response agencies including the Federal Emergency Management Agency (FEMA), National Weather Service, Salvation Army, National Communications System and Associated Public- Safety Communications Officials - International. These documents merely set forth a framework for possible cooperation at the local level. While they are designed to encourage mutual recognition, cooperation and coordination, they should not be interpreted as to commit, obligate or mandate in any way that an ARES unit *must* serve a particular agency, or meet *all* of its needs, in a jurisdiction. MOUs are "door openers," to help you get your foot in the door. It's up to you to decide whether or not to pursue a local operational plan with an agency, a decision that will be based on a number of factors including the local needs of the agency and the resources you have available to support those needs, given that you may have other prioritized commitments as well.

To address this, sit down with your fellow ARES members, EC and SEC, and determine what agencies are active in your area, evaluate each of their needs, and which ones you are capable of meeting. Then prioritize these agencies and theri needs. After you're all in agreement, sit down with your counterparts in each of the agencies and execute local, detailed operational plans and agreements in light of your priority list based on the above.

Given the above, however, you should also be working for growth in your ARES program, making it a stronger, more valuable resource and hence able to meet more of the agencies' local needs. A stronger ARES means a better ability to serve your communities in times of need and a greater sense of pride for Amateur Radio by both amateurs and the public. That's good for all of us.

7.3 Another Kind of Competition

With a strong ARES program, and a capability of substantially meeting most of the local served agencies' needs, you might avoid another problem that is cropping up in some parts of the country -- competition with emerging amateur groups providing similar communications services outside of ARES. Some of these groups may feel that their local ARES doesn't do the job, or personality conflicts and egos get in the way, so they set up shop for themselves, working directly with agency officials, and usurping ARES' traditional role. Some agencies have been receptive to their assistance.

There continues to be "RACES versus ARES" polarization in some areas. And some agencies, including at least one with statewide jurisdiction, are forming their own auxiliary communications groups, and recruiting their own hams, some away from ARES.

There's not much you can do about this, except to work to find your ARES program's niche and provide the best services you can as outlined above. Strive for growth and enhancement of ARES members' abilities, and make sure you present a "professional" face to potential "served" agencies and your opportunities will grow. Make your program better than the next guy's, and agencies will be more attracted to you.

If possible, setting egos and personalities aside, seek out these other groups and take the initiative to try to establish a rapport, and the fact that "we're all in this *together*," for the good of the public and Amateur Radio. With good communication, mutual respect and understanding between you and the other groups, at the least, you should be able to coordinate your program's missions with theirs (i.e., divide up the pie, or who will do what for which agency) to foster an efficient and effective Amateur Radio response overall. At best, you may find other groups willing to fold their tents and join your camp! Try it.

Minute 20 of the ARRL Board of Directors Meeting; ARES® Groups and Use of Registered Trademarks

I. The Board Motion

The following Motion was passed by the ARRL Board of Directors at its Second Annual Meeting in July of 2004. The motion has left some ARRL members with misconceptions about the intention of the Board of Directors concerning the ARES program. This paper will address those concerns.

> After discussion about the Amateur Radio Emergency Service, led by Mr. Walstrom, the Board was in recess for luncheon from 11:56 AM until 1:20 PM with current meeting participants present except Mr. Day. On motion of Mr. Walstrom, seconded by Mr. Goddard, the following resolution was unanimously ADOPTED:
>
> WHEREAS, for security reasons and for purposes of establishing formal relationships with served agencies, ARES® groups within an ARRL section are increasingly in need of affirmative recognition; and
>
> WHEREAS, ARES® (Amateur Radio Emergency Service®) is a program of, and both logos are registered trademarks of the American Radio Relay League, Incorporated;
>
> NOW THEREFORE, it is ARRL policy that ARES® groups, and any group using the ARES® logo, shall acknowledge the nature of the ARES® program as an ARRL program, and abide by the guidelines of the program established by ARRL as amended from time to time, according to the following principles:
>
> 1. Each ARES® group will, when using the term "ARES" or "Amateur Radio Emergency Service," utilize the ® symbol in any printed matter or in any electronic media, and will acknowledge that both logos are registered marks of the American Radio Relay League, Incorporated and are used by permission.
>
> 2. In bylaws or other organizational documents, or by amendment of any existing bylaws or organizational documents, ARES® groups will acknowledge that ARES® is a program of the American Radio Relay League, and that entity will abide by the Rules and Regulations of the ARRL's Field Organization, as they may be amended from time to time, and by ARRL policies, rules, and guidelines contained in ARRL publications.
>
> 3. All ARES® records, membership rosters, and other data pertaining to the ARES® program wherever located are the property of the American Radio Relay League, Incorporated.

II. What was the purpose of the Board motion?

There were several reasons why this motion was necessary. First of all, numerous ARES groups around the country, concerned about liability and wishing to operate in a manner that might encourage the donation of equipment for emergency communications purposes, expressed to ARRL the desire to incorporate. Two benefits of incorporation are the protection against individual liability of persons operating under the "umbrella" of a corporation, and the possibility of obtaining tax exemption as a Section 501(c) organization. Section 501(c)(3) charitable, educational or scientific organizations provide tax deductions for certain donors of equipment. It was felt by some that incorporation of ARES groups might better protect participants against liability and might encourage grants or donations to further the ARES purposes. While these are worthy and beneficial goals, they are somewhat at odds with the nature of ARES as part of the ARRL's field organization. ARES is an ARRL program, and for ARRL to permit ARES groups to incorporate would, without some specific reservation of rights, allow the ARES group to operate independently of the ARRL, and perhaps at odds with the rules of the Field Organization. ARRL wishes for ARES to remain an ARRL program, and for ARES participants to be a part of the ARRL Field Organization. ARES is widely recognized by State and Federal emergency services agencies and other served agencies as a national program which is under the auspices of the ARRL. An ARES corporation, in order to be considered an ARES entity, should be considered a part of this nationwide program.

Some years ago, ARRL registered its servicemark rights in the marks "ARES" and "Amateur Radio Emergency Service." ARRL has established the exclusive right to use those marks as representative of a nationwide emergency communications organization. ARRL has, and will continue to authorize ARES groups to use the mark. There are some clubs and groups which are already incorporated and which use the mark in the name of the club or group. This is acceptable, as long as the group or club which uses the logo operates in accordance with the rules and policies established by ARRL for the ARES program.

To those ARES groups which stated a desire to incorporate, the ARRL Board wished to allow them the greatest flexibility to do so, consistent with maintaining ARES as an ARRL program. However, permitting the formation of a separate corporation which is not a subsidiary of the ARRL and which operates as an independent entity, but which uses the name of the ARRL program creates the possibility of dividing the ARES program into small, separate groups without consistency. Enactment of the three requirements in the Board motion seemed to ARRL to be the minimum practical means of insuring that an incorporated ARES group remains a part of the ARRL field organization.

Second, there have been rare, but occasional instances in the past in which ARES field appointees who are replaced by a new Section Manager or Section Emergency Coordinator are dissatisfied with the termination of their voluntary service, and have refused to assist in the transition. In several instances, the membership rosters of these groups were retained and the terminated appointees refused to provide them to the incoming volunteers. This made it very difficult to continue or reconstitute the ARES groups and continue emergency communications planning and service. Whether or not an ARES group chooses to incorporate, the tools necessary for continuing ARES operation in a given community following changes in field appointments have to be protected. Hence, the affirmation that membership rosters and data are the property of the ARRL is a necessary component of any ARES group.

Finally, in a very few instances, the ARES logo has been misappropriated by certain individuals who have no association with any ARES group at all, or who refuse to abide by the rules of the ARRL Field Organization. This is a misappropriation of the mark and the program itself. Such misappropriation would be discouraged by compliance with the three obligations adopted in the Board motion.

III. Concerns and Questions from ARES Participants.

Since the Board Motion was adopted, some ARES participants asked reasonable questions about the policy. These are generally summarized and responded to as follows:

1. How does this apply to combined ARES/RACES organizations? Separate membership rosters should be maintained, even if the same individuals are members of both ARES and RACES in a given area. A local Office of Emergency Management might consider RACES rosters to be its property, rather than that of ARRL. However, ARES membership rosters are in fact the property of ARRL and should be maintained independently. No one is suggesting taking anything away from an OEM staff person. However, ARES groups and their successors have to also have access to that information.

2. Does the registered trademark "®" symbol need to be used each and every time that a person types the letters ARES in an e-mail and other written correspondence, on web pages, etc.? The requirement generally is to use the symbol in published documents and where printed for public display or distribution, and on web sites, for example. It is not necessary to use it each time in private written or e-mail correspondence.

3. Can the symbol be used once, at the beginning of a published document, and not thereafter? Yes. Normally, a footnote is used stating, for example, that "ARES" and "Amateur Radio Emergency Service" are registered servicemarks of the American Radio Relay League, Incorporated and are used by permission.

4. Under what circumstances must ARRL be asked for permission to use the ARES and Amateur Radio Emergency Service logos? ARRL offers ARES group participants the right to use the logo without specific written permission for all ARES functions and while engaged in ARES operations in accordance with the rules and regulations of the ARRL field organization. Any other use, or any use by a person who is not a member of an ARES group, requires written permission of ARRL. Where written permission is necessary, e-mail correspondence to **permission@arrl.org** or in writing to ARRL HQ is sufficient.

5. Does the ARRL intend to claim ownership or rights to property donated to or held by an ARES group, such as titles to motor vehicles, communications equipment, antennas, or related hardware? <u>Absolutely not</u>. Nothing in the Board motion states or implies such. ARRL neither has nor claims any property interest in anything except the ARES and Amateur Radio Emergency Service logos; membership rosters and data showing who the ARES participants are and how to communicate with them; and written information with emergency communications plans and protocols developed for ARES use and operation. Communications vans, equipment, hardware, or anything else that is owned or used by an ARES group remains the property of individuals or any association or entity that is formed. ARRL has no claim to those items of property. Likewise, ARRL has no responsibility or liability for maintenance or use of those vehicles or equipment.

6. What should an appointee such as an EC, DEC, or SEC do with membership rosters, communications and operations plans after that person is replaced? Quite simply, the information should be delivered to the person's successor, or as instructed by the SEC or SM if there is a vacancy in the appointment for any significant period. The successor will be under the same obligation as the predecessor to maintain confidentiality of the membership rosters, contact information, and the like. There will be no instances where the membership rosters, etc. will be made public or disclosed to third parties.

7. If the membership rosters, etc. have been maintained for many years, how far back must they be maintained? Principally, the only information that must be passed on to successors is that which is of current interest in maintaining and continuing the ARES program. A good faith effort to provide current information to successors is all that is called for.

It is hoped that this addresses the majority of questions and concerns about the ARRL policy. Please address any specific questions to **sewald@arrl.org**.

Appendix C
Examples of Net Preambles

THE ARIZONA TRAFFIC AND EMERGENCY NET

Calling the Arizona Traffic and Emergency Net. Calling the Arizona Traffic and Emergency Net, This is [*call sign*], Net Control station for [*date with month, day, and year*], my name is [*name*], and I am located in [*location*].

This is a Section Net, and is organized to handle traffic in times of emergency, and meets for drill on 3986 kHz at 5:30 PM Mountain Standard Time, or 0030Z daily.

The purpose for this net is the passing of formal written NTS Radiogram traffic into and out of our coverage area, and to wherever third party traffic is allowed by law.

This is a directed net. Unless you have an emergency, or are checking back into the net with your suffix, please do not transmit unless directed by the Net Control Station.

Visitors and late members can check in after roll call.

Is there any priority traffic?

Are there any announcements for the net?

Calling the following Liaison Stations: WB6OTS, W7JSW, K0LQB, KC7ZZ, KC5ZGG, K7OAH.

Is there any outbound traffic?

Is there any local traffic?

[*At this point, the NCS would assign traffic to stations off frequency, telling them what frequency to go to; to handle the traffic unless conditions are poor, then handle traffic on the net frequency.*]

Stand by for roll call, this is [*call sign*], Net Control Station for the Arizona Traffic and Emergency Net.

[*NCS calls the roll of listed members, pausing now and then for stations trying to check back into the net.*]

Are there any late ATEN members wishing to check into the net?

Are there any visitors wishing to check into the net?

[*List and welcome visitors.*]

Is there any further business for the net before I close tonight's session?

The Arizona Traffic and Emergency Net meets for drill on 3986 kHz at 5:30 PM MST 0030Z Daily. All amateur stations licensed to operate on this frequency are welcome to check in. This Net is an affiliated member of the ARRL National Traffic System, maintaining liaison with the 12 region and the Pacific Area Net. For further information, our website is at atenaz.net. This is [*call sign*] closing the net at [*time*] MST. Thanks for your help tonight. Good to hear everyone, 73, and a very good evening to all. [*Name/call sign*] is clear.

THE ARIZONA, NEVADA, NEW MEXICO MERCURY NET

Calling the Arizona, Nevada, New Mexico Mercury Net. Calling the Arizona, Nevada, New Mexico Mercury Net. Net Control station today is [*name/call sign*], located in [*city/state*]. This net meets each Saturday morning at 0700 hrs. AZ time on 3.990 MHz.

The purpose of this net is to help prepare amateurs for communications in the event of a communication emergency and to provide those with a common interest the opportunity to meet. This is a directed net, so please transmit as requested by Net Control. Stations checking in are asked to remain on the frequency until net business has been completed. Should it become necessary to leave the net, please check out with Net Control.

Any licensed amateur is welcome to check in to this net. A station checking in twice in a calendar month will be placed on the roster and remain until absent from the net for a long time.

Are there any announcements for the net?

Is there any traffic that needs to be handled at this time?

Are there any stations that need to check in early?

Relays are appreciated if a responding station is not acknowledged by Net Control.

Roll call follows: [*Remember to ID every 8–10 minutes during roll call.*]

Are there any late or missed member stations?

Are there any visitors who would like to check in now?

Is there any other business or comments for the net at this time?

We will now close the Arizona, Nevada, and New Mexico Mercury Net at [*time*] MST. Thank you for being here today and participating in the net. We appreciate you sharing your morning with us. This is [*name/call sign*], clear.

The Amateur Radio Public Service Handbook

Appendix D
Go-Kit Checklists

For 2-hour, 12-hour, and extended activations [COURTESY OF SANTA CLARA COUNTY ARES/RACES]

Legend:
X = Required (must have in kit at all times)
R = Recommended (likely useful on many assignments)
O = Optional (useful on some assignments)

2-Hour Carry Kit

Purpose: To be kept nearby at all times for immediate (within minutes) communication of damage reports during Resource Net Level 1 ops. Also used to remain in contact with Resource Net Level 2 while returning home to retrieve 12-hour Go-Kit.

Items:
- X 2m/70cm dual-band radio
  - HT recommended (min. 5 W on 12 V/2.5 W on batt)
  - Mobile 25 W optional (if vehicle will not be far away)
  - Programmed with Resource Net frequencies
- X Charged batteries for 2-3 hours operation
- X 2 m/70 cm dual-band mobile antenna (mag mount, window mount or existing mobile antenna)
- X Modified Mercalli (Mike-Mike) scale
- X Notepad/pens
- R Cigarette lighter adapter
- R Emergency county and city telephone contact list
- R Cell phone
- R Water (16 oz.)

12-Hour Go-Kit

Purpose: For fully independent operation; unknown environment (heat, cold, wind, rain); unknown time (day, night, up to 12 hours). Return home to retrieve.

Equipment
Portable Radio:
- X 2m/70cm dual-band handie-talkie (HT)
  - minimum 5 W on 12 V/2.5 W on batteries [Note 1]
  - dual-receive recommended
- X Radio user manual or cheat sheet
- X Earbud or headphones minimum; headset, earbud/mic, or speaker/mic/earbud, or similar recommended
- R Small backpack, vest, chest harness or other similar method for carrying HT while operating portable

Power Source:
- X Charged batteries for 12 hours (min. 3000 mAH) [Note 2]
- X Power cord adapters — connect to various power sources:
  - Powerpoles
  - Cigarette lighter socket
  - Vehicle battery terminals
- X Spare fuses
- R Powerpole splitter or fused distribution panel
- R Extension cord, 3-wire, 3-6 ft., multi-outlet
- O Extension cord, 3-wire, 50-100 ft.
- O Power inverter

Antennas:
- X Coax adapters: connect HT to coax, coax to the following:
  - BNC plug (male) & BNC socket (female)
  - UHF plug (PL-259) & UHF socket (SO-239)
  - N-type plug (male) and N-type socket (female)
- X Min. 25 feet of 50 ohm coaxial cable
- X 2 m/70 cm dual-band magnetic or window mount antenna
- R 2 m/70 cm high gain HT antenna
- R 2 m/70 cm dual-band portable base antenna (e.g. roll-up J-pole or other)
- R Portable mast (elevates antenna min. 10 ft.)
- R Tripod or self-supporting base for mast
- R Window clip antenna mount (for non-metallic vehicles)

Other Communications Gear:
- R Cell phone & charger and/or cigarette lighter adapt.
- O FRS/GMRS Radio
- O Satellite phone

Tools:
- R Duct tape
- R Electrical tape
- R Nylon tie-wraps/wire ties
- R Utility knife
- R Small multi-tool or tool kit
- O Volt-Ohm meter
- O SWR/Power meter

Operating Position:
- X Sign(s) for operating position
- R Lighting for operating position
- R Rope or Dacron cord (50')
- R Folding chair
- O Magnetic sign for car
- O Folding table
- O Pop-up Canopy
- O Tarp (8' by 8' or larger)
- O Folding cart
- O Safety strobes or flares
- O Caution/flagging tape (for marking cables, antennas, etc.)

Documentation
Identification:
- X Driver's license or state-issued ID card
- X Amateur Radio license
- X County Emerg. Resp. ID card
- X If issued: county ID badge, city badge

Maps:
- X County map

X	Compass or GPS
R	Maps of antenna locations (if available)
R	City, county, or other detail maps

Forms and Documentation:

X	Modified Mercalli (Mike-Mike) scale
X	ICS 205-SCCo – Communications Plan (min. 5)
X	ICS 211A-SCCo – Communications Check-In (min. 5)
X	ICS 213-SCCo – Message (min. 10)
X	ICS 214-SCCo – Unit Activity Log (min. 5)
X	ICS 309-SCCo – Communications Log (min. 5)
X	ICS 314-SCCo – Windshield Survey (min. 5)
X	Phone message pad (two-part style recommended)
R	County Performance Standards

Logging / Note taking:

X	Clipboard (covered type recommended)
X	Notepads (standard or waterproof)
X	At least two pens/pencils
O	Highlighters/felt-tip pens

Contact Lists:

X	Voice and packet frequency lists
X	DEC/ADEC and city EC telephone contact list
X	Police/Fire direct dial phone numbers
O	Repeater directory

Personal Gear

Vehicle:

X	Reliable operating condition
X	Fueled — minimum ½ full at all times
R	Jumper cables

General Items:

X	Money (paper and coin) – in case ATMs are down
X	Watch or clock
R	Trash bags

Personal Safety Gear:

X	Flashlight or headlamp and spare batteries for 12 hours
X	Safety vest, ANSI standard (lime yellow recommended)
R	First Aid kit
R	Whistle
R	Work gloves
R	Sunglasses
R	Sunscreen lotion
R	Insect Repellent
R	Safety glasses
R	Mask (NIOSH-certified N95 or better)
O	Hearing protection (e.g. foam ear plugs)
O	Hard hat (lime yellow recommended)
O	Chemical light sticks

Clothing:

X	Sturdy, closed-toe shoes (no sandals)
X	Long pants (no shorts)
X	Hat (broad brim recommended)
X	Seasonal jacket / rain gear

Food & Water:

X	Food for 12 hours (make your own list)
X	Water for 12 hours (3-4 quarts recommended)
R	Small cooler or ice chest

Toiletries:

R	Hand soap and/or sanitizer
R	Toilet paper
O	Pain reliever
O	Antacid tablets

As Needed / Appropriate:

- Prescription medication
- List of medication used
- Eyeglasses & spare

Miscellaneous (as needed)

- Portable AM radio and spare batteries
- Binoculars
- Small plastic bags to seal/protect items
- Shovel
- Fire extinguisher
- Disposable camera

Mobile Radio Kit (as needed)

- 2 m/70 cm mobile radio
  - 25 W minimum
  - Dual-receive, cross-band repeat recommended
- Radio user manual or cheat sheet
- Headset (stereo recommended for VFO per ear)
- Battery for 12 hours of operation (20 AH min.; 26 AH rec.)
- Battery charger
- Power cord adapters — connect to various power sources:
  - Powerpoles
  - Cigarette lighter socket
  - Vehicle battery terminals
- Coax adapters: connect mobile to coax, coax to following:
  - BNC plug (male) & BNC socket (female)
  - UHF plug (PL-259) & UHF socket (SO-239)
  - N-type plug (male) and N-type socket (female)

Packet Equipment (as needed)

- Laptop with Outpost and PacFORMS installed
- USB flash drive (i.e. USB key)
- TNC (may be hardware, software or built into radio)
- Cables: TNC to radio; TNC to PC
- Shade cover for display
- Portable printer
- Entire station can operate for min. 1 hr on battery

The Amateur Radio Public Service Handbook

Extended Go-Kit

Purpose: Additional items for fully independent operation over an extended period of time. Used in situations where returning home after shift is not possible or not ideal.

As Needed

Power Source:
- Regulated DC power supply
- Battery charger
- Spare batteries (for charging while operating)
- Portable generator and fuel
- DC distribution panel & cables (Powerpoles recommended)

Clothing:
- Rain gear
- Jacket
- Warm clothing (preferably in layers)
- Under garments (3 sets)
- Socks (3 sets)
- Pants (3)
- Belt
- Shirts (3)
- Alternate boots or shoes
- Sleepwear
- Cold water laundry soap (e.g. Woolite)

Food and Water:
- MREs (self-heating) or other non-perishable meals
- Water (1 gal/day recommended, depending on conditions)
- Water purification tablets or devices
- Can opener
- Cooler or ice chest
- Bowl and eating utensils
- Coffee cup

Shower Items:
- Washcloth and towel
- Soap and shampoo
- Razor and shaving cream
- Toothbrush and toothpaste
- Comb and/or brush
- Deodorant/antiperspirant
- Wash basin (in case of no sink)

Shelter:
- Sleeping pad
- Sleeping bag/blanket
- Pillow
- Blanket
- Tent
- Alarm clock

Personal Go-Kit Items/Notes:

Notes:

1. Most recently manufactured handheld radios are capable of 5 W output when 12–13.8 V dc is connected to the dc-IN jack and at least 2.5 W output power using rechargeable battery packs. Check your radio's user manual to be sure your radio outputs at least 2.5 W on rechargeable batteries. However, most handheld radios are not capable of producing a minimum of 2.5 W output power using AA batteries. Some known exceptions are the Kenwood TH-D7 and the Yaesu FT-60. For all other radios, rechargeable battery packs will be needed unless the radio can be shown to have a minimum of 2.5 W output on AA batteries (check user manual or test with power meter).

2. A review of the most popular handheld radios was conducted. Receive current, transmit current, and rechargeable battery pack capacity were reviewed. 3000 mAH was determined to be the minimum capacity needed for 12 hours of operation. (Some radios may require a little more). Depending on the make and model, this translates to 2 or 3 rechargeable battery packs. This minimum requirement correlates well with real-world experience in drills and real incidents such as Katrina.

Appendix E
MoU Documents

Memorandum of Understanding

between

The Salvation Army in the United States of America

and

ARRL, the national association for Amateur Radio

Purpose

The purpose of the agreement between the American Radio Relay League, Incorporated and The Salvation Army in the United States of America is to establish a framework for cooperation between the two organizations for relief of disaster victims. It is intended that coordination of facilities, equipment and personnel of the two organizations may provide better service of victims of natural or man-made disasters.

Responsibilities

The American Radio Relay League, since its inception in 1914 up to the present, has observed a self-imposed responsibility for the welfare and conduct of the Amateur Radio Service as regulated by Part 97 of FCC's Rules and Regulations. Principal in that responsibility has been the rendition of public service and communication through the handling of third party communications for the general public, and communications in time of emergency when normal communications are not available. Using amateur radio operators in the amateur bands, the American Radio Relay League has been in the forefront of this activity in serving the general public directly and through government and welfare agencies, and continues to do so. To that end, in 1935, the Amateur Radio Emergency Corps was organized; and in 1949, the National Traffic System was established.

The Salvation Army has, for many years, provided emergency services to individuals and groups in time of disaster. This service has received public recognition. The Congress of the United States of America enacted the Disaster Relief Act of 1970, which, as amended by the Disaster Relief Act of 1974, Public Law 93-288, officially recognized the capabilities of The Salvation Army.

Since that time, The Salvation Army has entered into specific agreements with other agencies concerned with emergency and disaster relief services both public and private.

Recognition

The Salvation Army recognizes that the American Radio Relay League, because of its organized emergency communications facilities, can be of invaluable assistance in providing communications during emergencies and disasters when normal lines of communication are disrupted.

The American Radio Relay League, Incorporated, recognizes The Salvation Army as an agency whose corporate charter merits sanction by the Federal government to provide community aid in times of disaster. It further recognizes The Salvation Army as a channel for voluntary service during such time.

The Amateur Radio Public Service Handbook

Organizations of the American Radio Relay League and The Salvation Army

The American Radio Relay League (ARRL) is the principal organization representing the interests of U.S. Radio Amateurs. It is governed by a Board of fifteen directors elected by the membership. For more than 80 years, ARRL has been the standard-bearer in amateur radio affairs throughout the U.S.

For emergency communications, ARRL sponsors the Amateur Radio Emergency Service (ARES), a division of its over-all public service organization. The ARES is organized under local emergency coordinators, with local plans coordinated through section (usually state) emergency coordinators and a public service coordinator located at ARRL's Newington, Connecticut international headquarters. The National Traffic System (NTS) functions daily in handling medium and long haul message traffic, and is ready at all times to function in an emergency situation.

The Salvation Army in the United States of America has its national headquarters in Alexandria, Virginia, and is incorporated under the laws of New Jersey. For administrative purposes, the United States is divided into four territories, each having its own headquarters and corporate structure. These territories and headquarters are:

Reporting to each territorial office are from nine (9) to eleven (11) divisional administrative centers, strategically located in the territories. Salvation Army personnel in these centers direct activities in from one (1) to four (4) states. Reporting to divisional centers are local corps community centers (churches) and social service institutions of other types; also reporting to divisional centers are numerous local volunteer committees operating in smaller communities.

Principles of Cooperation

In order that dependable communications might be maintained and that relief operations might be quickly expedited, the American Radio Relay League, Incorporated and The Salvation Army agree that:

A. Each organization will, through channels to its local units, encourage ongoing liaison with the other, urging both staff and volunteers to create and maintain adequate communication and effective relationships at all levels.

B. Each organization will participate in cooperative pre-disaster planning and training programs at local, regional and national levels.

C. Each organization will, in times of disaster, cooperate to meet the needs of disaster victims, and of the agencies and organizations attempting to serve them. Each will make its facilities, resources, and capabilities accessible to the other, in accordance with established plans and procedures for cooperative service.

D. Each organization will work through its own lines of authority and respect the lines of authority of the other.

E. Each organization will distribute copies of this agreement through channels to its own field units, and to other organizations, both public and private, which may have an active interest in emergency and disaster relief.

revised January 1996

MEMORANDUM OF UNDERSTANDING BETWEEN THE NATIONAL WEATHER SERVICE AND THE AMERICAN RADIO RELAY LEAGUE, INC.

I. PURPOSE

The purpose of this document is to state the terms of a mutual agreement (Memorandum of Understanding) between National Oceanic and Atmospheric Administration's (NOAA) National Weather Service (NWS) and the American Radio Relay League, Inc. (ARRL), that will serve as a framework within which volunteers of the ARRL may coordinate their services, facilities, and equipment with NWS in support of nationwide, state, and local early weather warning and emergency communications functions. It is intended, through joint coordination and exercise of the resources of ARRL, NWS, and Federal, State and local governments, to enhance the nationwide posture of early weather warning and readiness for any conceivable weather emergency.

II. RECOGNITION

The National Weather Service recognizes that the ARRL is the principal organization representing the interests of more than 690,000 U.S. radio amateurs. Because of its field organization of trained and experienced communications experts, Amateur Radio Service volunteers can be of valuable assistance in early severe weather warning and tornado spotting.

ARRL recognizes the National Weather Service's statutory responsibility to provide the following meteorological services for the people of the United States:

> 1. NOAA's National Weather Service provides weather, hydrologic, and climate forecasts and warnings for the United States, its territories, adjacent waters and ocean areas, for the protection of life and property and the enhancement of the national economy; and,

> 2. NWS data and products form a national information database and infrastructure which can be used by other governmental agencies, the private sector, the public, and the global community.

III. ORGANIZATION OF THE AMERICAN RADIO RELAY LEAGUE

ARRL is a noncommercial membership organization of radio amateurs, organized for the promotion of interest in Amateur Radio communication and experimentation, for the establishment of networks to provide communications in the event of disasters or other emergencies, for the advancement of the radio art and of the public welfare, for the representation of the radio amateur in legislative matters, and the maintenance of fraternalism and a high standard of conduct. A primary responsibility of the Amateur Radio Service, as established by the Federal Communications Commission, is the rendering of public service communications for the general public, particularly in times of emergency. Using Amateur Radio operators in the amateur frequency bands, the ARRL has been serving the public, both directly and through government and relief agencies, for more than ninety years. To that end, the League created the Amateur Radio

Emergency Service ® (ARES) ® and the National Traffic System (NTS). The League's Field Organization consists of seventy-one administrative sections managed by elected Section Managers. A Section is a League-created political boundary roughly equivalent to states (or portions thereof). The Section Manager appoints expert assistants to administer the various emergency communications and public service programs in the section. Each section has a vast cadre of volunteer appointees to perform the work of Amateur Radio at the local level, under the supervision of the Section Manager and his/her assistants.

IV. ORGANIZATION OF THE NATIONAL WEATHER SERVICE

National Oceanic and Atmospheric Administration's (NOAA) National Weather Service consists of 122 weather forecast offices, 13 river forecast centers, 9 national centers, and other support offices. NWS scientists provide weather, water, and climate forecasts and warnings for the United States for the protection of life and property, and the enhancement of the national economy. The NWS' national headquarters is located in Washington, D.C., and there are six regional headquarters: Eastern, Southern, Central, Western, Alaska, and Pacific.

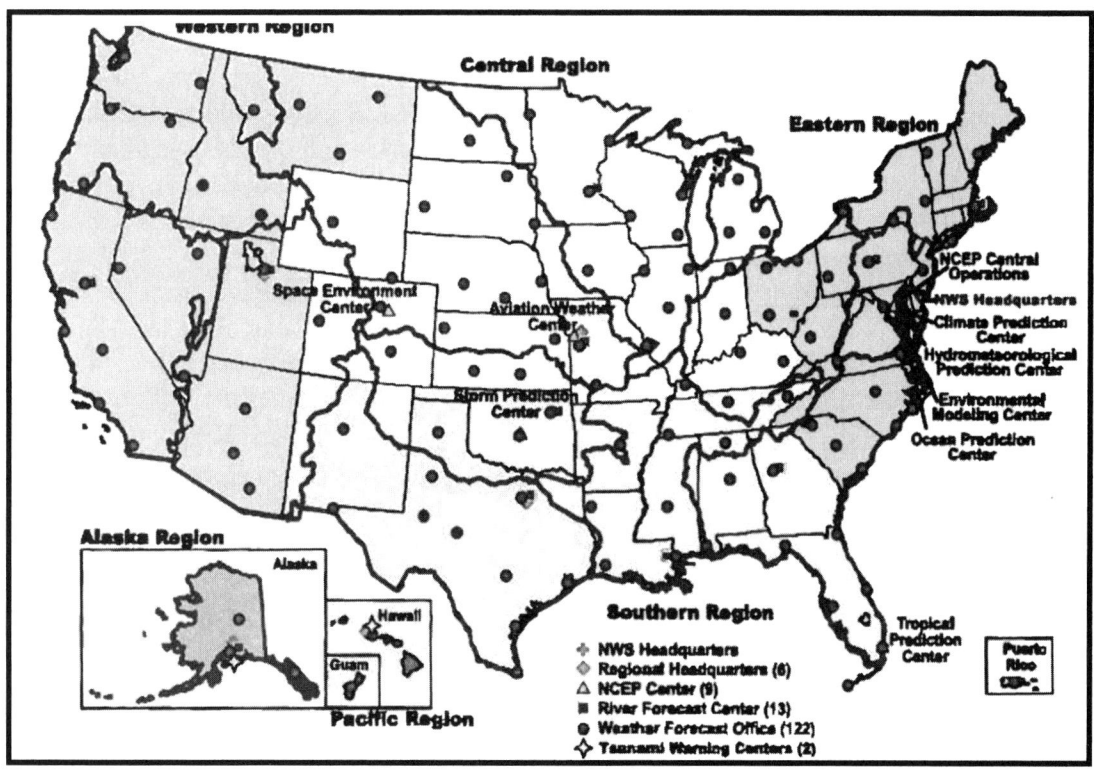

Skywarn® is the National Weather Service's severe weather spotting program. Radio amateurs have assisted as communicators and spotters since the program's inception in the late 1960s. In areas where tornadoes and other severe weather have been known to threaten, NWS recruits volunteers, and trains them in proper weather spotting procedures. These dedicated citizens help keep their local community safe by conveying severe weather reports to their local NWS Forecast Office. Skywarn spotters are integral to the success of our Nation's severe weather warning system.

Warning Coordination Meteorologists (WCMs) serve as the NWS' principal liaison with its customers and partners in the evaluation and improvement of its products and services. WCMs are responsible for maintaining the working partnership with the local ARRL Skywarn organizations. There are 132 NWS Warning Coordination Meteorologists (WCMs) located throughout the country: 122 Weather Forecast Offices, 6 Regional Headquarters, National Headquarters, the Storm Prediction Center, the National Hurricane Center, and the National Aviation Weather Center.

V. PRINCIPLES OF COOPERATION

A. ARRL agrees to encourage its volunteer Field Organization appointees, especially the Amateur Radio Emergency Service, to contact and cooperate with National Weather Service Warning Coordination Meteorologists for the purpose of establishing organized Skywarn networks with radio amateurs serving as communicators and spotters, consistent with rules and regulations of the Federal Communications Commission.

B. ARRL further agrees to encourage its Section management teams to provide specialized communications and observation support on an as-needed basis for NWS offices in other weather emergencies such as hurricanes, snow and heavy rain storms, and other severe weather situations.

C. The National Weather Service agrees to work with ARRL Section Amateur Radio Emergency Service volunteers to establish Skywarn networks, and/or other weather emergency alert and relief systems. The principal point of contact between the ARRL Section and local NWS offices are the Warning Coordination Meteorologists. Local Warning Coordination Meteorologist contact information is available at: www.stormready.noaa.gov/contact.htm. Contact information for ARRL Section volunteer leaders is available at www.arrl.org/sections. The national contact for ARRL is the Emergency Preparedness Manager at ARRL Headquarters, Newington, CT 06111. The national contact for NWS Warning Coordination Meteorologists is the Office of Climate, Weather and Water Services, WCM Program, 1325 East-West Highway, Silver Spring, MD 20910.

Kay Craigie, N3KN
President, American Radio Relay League, inc.

Date May 12, 2011

Date 6/9/2011

Printed Name David B. Caldwell

Title OCWWS Director

NOAA National Weather Service

Memorandum of Understanding

between

The American National Red Cross

and

ARRL, the national association for Amateur Radio

American Red Cross

I. Purpose

The purpose of the Memorandum of Understanding (MOU) is to document the relationship between the American National Red Cross (the "Red Cross") and the ARRL, the national association for Amateur Radio (the "ARRL"). This MOU provides a broad framework for cooperation between the two organizations in preparing for and responding to disaster relief situations at all levels in rendering assistance and service to victims of disaster, as well as other services for which cooperation may be mutually beneficial.

II. Independence of Operations

Each party to this MOU will maintain its own identity in providing services. Each organization is separately responsible for establishing its own policies and financing its own activities.

III. Organization Descriptions

The American Red Cross is a humanitarian organization led by volunteers and guided by its Congressional Charter and the Fundamental Principles of the International Red Cross and Red Crescent Movement. The Red Cross provides relief to victims of disasters and helps people prevent, prepare for and respond to emergencies. The Red Cross provides services to those in need regardless of citizenship, race, religion, age, sex, national origin, disability, sexual orientation, veteran status or political affiliation.

The ARRL is the national membership association for Amateur Radio operators. The ARRL is a not-for-profit organization that engages in the promotion of interest in Amateur Radio communication and experimentation; the establishment of Amateur Radio networks to provide electronic communications in the event of disasters or other emergencies; the furtherance of the public welfare; the advancement of the radio art; the fostering and promotion of noncommercial intercommunication by electronic means throughout the world; the fostering of education in the field of electronic communication; the promotion and conduct of research and development to further the development of electronic communication; the dissemination of technical, educational and scientific information relating to electronic communication; and the printing and publishing of documents, books, magazines, newspapers and pamphlets necessary or incidental to any of the above purposes.

The Amateur Radio Public Service Handbook

IV. Methods of Cooperation

The Red Cross and ARRL desire to expand their mutually-beneficial relationship to enhance community disaster preparedness and coordinate disaster planning and response activities as follows:

Relationship building

- **Open Communications:** Each organization will share current appropriate data regarding disasters, disaster declarations, and changes in regulations, technology and legislation related to communications. The same interaction and liaison will be encouraged at all levels of both organizations, to include all Red Cross chapters, ARRL sections and subordinate levels.
- **Local partnerships:** Each organization will encourage its local units to communicate with the other organization's corresponding local unit to explore opportunities for collaboration. These units may perform cooperative efforts such as disaster planning and preparedness, first aid, cardiopulmonary resuscitation (CPR), health courses, communications training and licensing, and community disaster education. Cooperative efforts could include participation in predisaster planning or any other of the methods of cooperation listed here or as listed in the sample local agreement found in *Attachment C, Sample Statement of Cooperation for local organizational units*. *Attachment C* may be modified or updated by joint agreement of each organization's points of contact (listed in *Attachment A, Organization Contact Information*) without requiring a resigning of this MOU.
- **Shared members**: Each organization will encourage interested volunteers to become members and participate in the activities of the other organization. Such volunteers shall meet the standards, have the responsibilities and be entitled to the privileges of each organization.
- **ARRL volunteers supporting the Red Cross:** The ARRL may provide volunteers to assist the American Red Cross with communications in support of disaster relief roles as may be mutually agreed upon at the local and national levels.
The Red Cross requires the completion of a criminal background check to participate in Red Cross activities. A criminal background check may be performed through the Red Cross process at no cost to the volunteer, or by State or local law enforcement agency at the volunteer's own initiative and expense. The Red Cross is only responsible for the costs of background checks conducted through their processes.
The ARRL accepts the requirement of a criminal background check for volunteers but prefers that such checks be performed by law-enforcement entities. The Red Cross agrees that ARRL volunteers shall not be asked or required to consent to credit checks, mode of living investigations, or investigative consumer reports in order to provide a communications function.
- **Red Cross members supporting the ARRL:** Red Cross volunteers affiliated with a local Chapter that hold a valid Federal Communications Commission (FCC) Amateur Radio License are encouraged to participate in the Amateur Radio Emergency Service (ARES®) program to develop emergency communications skills, cross-train in local disaster drills and exercises, and integrate Chapter communications resources into the local emergency management structure.

Assumptions

- **Radio station operations:** It is understood and agreed that amateur radio operators, being licensed and regulated by the Federal Communications Commission (FCC), shall at all times exercise sole and exclusive control over the operation of their radio stations. Such control cannot be surrendered or delegated, in accordance with Federal law.
- **Radio operators:** It is understood and agreed that radio operators have skills that extend beyond amateur radio frequencies and equipment. These skills may be applied to operate on Red Cross frequencies and equipment.
- **FCC Licenses:** The Red Cross is responsible for any licensing arrangements necessary for Red Cross operations that occur outside amateur radio licenses, or any amateur radio licenses established by American Red Cross Amateur Radio Club Stations. Individual amateur radio operators are responsible for the maintenance and renewal of their personal licenses.

Activities

- **Training**: The Red Cross recognizes the leadership and expertise of the ARRL in the area of amateur radio communications. Where appropriate, the Red Cross will rely on materials created by the ARRL to train radio communicators. Additionally, the ARRL offers training in Amateur Radio emergency communications that is mutually beneficial to the ARRL and to the American Red Cross. Volunteers holding valid ARRL Emergency Communications certificates of completion will be recognized for this knowledge.
- **Joint exercises:** Chapters, Sections and subordinate units of each organization will be encouraged to engage in joint training exercises.
- **ARRL Field Day:** The Red Cross will encourage all chapters to participate in ARRL Field Day, the Simulated Emergency Test (SET) and other emergency exercises. Participation may take many forms, including Red Cross officials visiting and touring sites to better understand the capabilities of local ARRL volunteers and ARES® units, or the joint use of Red Cross equipment such as vehicles or trailers.
- **Planning:** Planning needs will be identified, tasked and completed to address issues beneficial to both organizations in responding to events. Such issues can be, but are not limited to pre-staging communications equipment, coordination of Mass Care and Damage Assessment support activities, and catastrophic disaster plans for high risk areas of the United States.

During disasters

- **On-scene cooperation:** Both ARRL volunteers and American Red Cross workers will work cooperatively at the scene of a disaster and in the disaster recovery, within the scope of their respective roles and duties as recommended in *Attachment D, ARRL Roles on Red Cross Disaster Relief Operations.*
- **National HQ coordination:** Operational coordination between Red Cross HQ and ARRL HQ will occur through the primary points of contact as shown in *Attachment A, Organization Contact Information* or other officially designated staff. Reports and data that are mutually beneficial to each organization's operations and mission assignments

will be exchanged.
- **Communications:** Whenever there is a disaster requiring the use of amateur radio communications resources and/or facilities, the local Red Cross Chapter may request the assistance of the local ARES organization responsible for the jurisdiction of the scene of the disaster. This assistance may include: alert and mobilization of ARRL ARES® personnel in accordance with a prearranged plan; establishment and maintenance of fixed, mobile, and portable station emergency communication facilities for local radio coverage; point-to-point contact between Red Cross personnel and locations; and the maintenance of the continuity of communications for the duration of the emergency period until normal communications channels are substantially restored, or until radio communications are no longer necessary in support of the response to the disaster.
- **Equipment sharing:** Each organization may request equipment for temporary use to support operations. The specifics of responsibility and liability of the loaned equipment will be developed as part of plans and procedures, in writing, and are separate from this agreement.
- **Health and Welfare Messages:** The Red Cross processes general welfare messages through the Red Cross Safe & Well web site. ARRL volunteers are encouraged to assist in registering people on the Safe & Well website by passing the required information from a point in the disaster area to someone outside the disaster area who can enter the information on the Safe & Well website. No special training or pre-defined agreements are necessary for ARRL volunteers to do this. The Safe and Well website is located on www.redcross.org.

V. General

a. The Red Cross and ARRL will use or display the name, emblem, or trademarks of the other organization only in the case of defined projects and only with the prior, express, written consent of the other organization.
b. The Red Cross and ARRL will keep the public informed of their cooperative efforts through their public information offices during the time of disaster.
c. The Red Cross and ARRL will widely distribute this MOU within the respective departments, administrative offices and subordinate levels of each organization and urge full cooperation.
d. The Red Cross and ARRL will allocate responsibility for any shared expenses in writing in advance of any commitment.
e. Local units of the Red Cross and subordinate levels in the ARRL Field Organization that desire a localized MOU to meet specific needs and conditions will utilize a format as shown in *Attachment C, Sample Statement of Cooperation for local organizational units.*
f. ARRL agrees to adhere to *Attachment B - the Code of Conduct for the International Red Cross and Red Crescent Movement and NGOs in Disaster Relief* as it applies to disaster-caused situations in the USA. Attachment B will not be changed without a resigning of the MOU by both parties.

VI. Periodic Review and Analysis

Representatives of the Red Cross and ARRL will, on an annual basis on or around the anniversary date of this MOU, jointly evaluate their progress in implementing this MOU and revise and develop new plans or goals as appropriate.

VII. Term and Termination

This MOU is effective as of the date of the last signature below and expires on March 24, 2015, five years from the signature date. The parties may extend this MOU for an additional period not exceeding five years, and if so shall confirm this in a signed writing. It may be terminated by written notice from either party to the other at any time.

VIII. Miscellaneous

Neither party to this MOU has the authority to act on behalf of the other party or bind the other party to any obligation. This MOU is not intended to be enforceable in any court of law or dispute resolution forum. The sole remedy for non-performance under this MOU shall be termination, with no damages or penalty.

IX. Signatures

American Red Cross

By: _[signature]_
Name: Joseph C. Becker
Title: Senior Vice President, Disaster Services
Date: March 25, 2010

ARRL

By: _[signature]_ Kay Craigie
Name: Kay Craigie
Title: President, ARRL
Date: March 25, 2010

ATTACHMENT A – Organization Contact Information

Primary Points of Contact

The primary points of contact in each organization will be responsible for the implementation of the MOU in their respective organizations, coordinating activities between organizations, and responding to questions regarding this MOU. In the event that the primary point of contact is no longer able to serve, a new contact will be designated and the other organization informed of the change. Contact changes do not require any renegotiation of this MOU.

Relationship Manager* and Operational Contact**

American Red Cross		ARRL	
Contact	Keith Robertory	Contact	Michael P. Corey
Title	Manager, Disaster Technology	Title	Manager, Emergency Preparedness and Response
Office phone	202-303-8628	Office phone	860-594-0222
24x7 Contact	202-303-4126	Mobile	860-597-8643
e-mail	robertoryk@usa.redcross.org or dst@usa.redcross.org	e-mail	W5mpc@arrl.org

*The Relationship Manager is the person that works with the partner organization in developing and executing the MOU.

**The Operational Contact is the person each organization will call to initiate the disaster response activities as defined in the MOU.

Organization Information

American Red Cross		ARRL	
Department	Disaster Services Technology	Department	ARRL
Address	2025 E Street, NW Washington, DC 20006	Address	225 Main Street Newington, CT 06111-1494
e-mail	dst@usa.redcross.org	e-mail	info@arrl.org
Website	http://www.redcross.org/	Website	www.arrl.org

ATTACHMENT B

Code of Conduct for
The International Red Cross and Red Crescent Movement
and
NGOs in Disaster Relief

Principle Commitments:

1. The Humanitarian imperative comes first.

2. Aid is given regardless of the race, creed or nationality of the recipients and without adverse distinction of any kind. Aid priorities are calculated on the basis of need alone.

3. Aid will not be used to further a particular political or religious standpoint.

4. We shall endeavor not to act as instruments of government foreign policy.

5. We shall respect culture and custom.

6. We shall attempt to build disaster response on local capacities.

7. Ways shall be found to involve program beneficiaries in the management of relief aid.

8. Relief aid must strive to reduce future vulnerabilities to disaster as well as meeting basic needs.

9. We hold ourselves accountable to both those we seek to assist and those from whom we accept resources.

10. In our information, publicity and advertising activities, we shall recognize disaster victims as dignified human beings, not hopeless objects.

More information about the code of conduct can be found at http://www.ifrc.org/publicat/conduct/

The Code Register
The International Federation is keeping a public record of all those NGOs who register their commitment to the Code. The full text of the Code including a registration form is published by the International Federation and is available upon request. (Telephone +41 22 7304222, Fax +41 22 7330395).

Non-governmental Organizations who would like to register their support for this Code and their willingness to incorporate its principles into their work should fill in and return the registration form.

The Amateur Radio Public Service Handbook

ATTACHMENT C – Sample Statement of Cooperation for local organizational units

American Red Cross XXX Chapter and <<XXX>>Cooperative Agreement

The purpose of this Statement of Cooperation is to document the relationship between the American Red Cross XXXXX Chapter and the <<XXX (insert ARRL Section, ARES® unit or local radio club)>> for the purposes of disaster planning and response. This Statement of Cooperation provides the methods of cooperation between the two organizations in rendering assistance and service to victims of disaster, as well as other services for which cooperation may be mutually beneficial. This Statement of Cooperation incorporates by reference the details and limitations contained in the national MOU between the American Red Cross and the ARRL, the national association for Amateur Radio (the "ARRL"). Each organization retains its own identity in providing services, and each is responsible for establishing its own policies and financing its own activities.

Concept of Cooperation

The American Red Cross XXXXX Chapter and <<XXXX>> agree to the methods of cooperation listed in the American Red Cross and ARRL national MOU. In addition, they agree to the following specific local methods of cooperation.

The American Red Cross XXXXX Chapter will:
- Incorporate <<XXX>> in its response plans (EXAMPLE)
- Provide preparedness training opportunities (EXAMPLE)
- Provide shelter training (EXAMPLE)

<<XXX>> will:
- Provide personnel to assist with communications in support of disaster relief roles as agreed upon (EXAMPLE)
- Expand their communications support to other activities within the disaster response system (Disaster Assessment, ERV driving)
- Add another action as needed (EXAMPLE)

This Statement of Cooperation is effective as of the date of the last signature below and expires on _____. It may be terminated by written notice from either party to the other at any time.

Neither party to this Statement of Cooperation has the authority to act on behalf of the other party or bind the other party to any obligation. This Statement of Cooperation is not intended to be enforceable in any court of law or dispute resolution forum. The sole remedy for non-performance under this Statement of Cooperation shall be termination, with no damages or penalty.

The primary points of contact are:

American Red Cross XXXXX Chapter Contact: e-mail: Office: Mobile:	<<XXX>> Contact: e-mail: Office: Mobile:

Signature American Red Cross XXXXX   Signature <<XXX>>

Print Name: _____   Print Name: _____

Date: _____   Date: _____

Review Date (after one year): _____

ATTACHMENT D – ARRL Roles on Red Cross Disaster Relief Operations

During a Red Cross Disaster Relief Operation (DRO), ARRL volunteers may perform in any of the following roles. These are examples of actual roles; they may or may not actually be included in all operations depending on the needs of the operation. It is possible that one person can support multiple roles or one role may require support from several people. This is not an exhaustive list and ARRL volunteers who have taken Red Cross Disaster Services training can participate in other roles. ARRL volunteers who are assigned roles by the Red Cross during a DRO will be provided with Red Cross credentials as required by the role, consistent with Red Cross policy.

Amateur Radio Liaison: This role is for a person who is familiar with both Red Cross and local amateur radio operations. This role would establish contact with the local ARES unit, amateur radio club and repeater owners to provide a single technical-level point of contact for the DRO. If local agreements already exist, this role could be pre-designated. It would be expected that this role would be linked to a similar role in the partner organization.

Communication Equipment Operator: This is a standard radio operator role for someone who would operate a two-way radio or other communication device at a fixed facility or mobile/portable location to support the DRO. They would pass messages from point to point either directly or through a message relay. Operators may use DRO-issued equipment or personally-owned equipment, and they may be on amateur radio frequencies or frequencies coordinated or licensed by the Red Cross.

Communication Equipment Installation / Repair: This is a more technically hands-on role than the Operator. In this role, the person would be asked to temporarily install two-way radio equipment into a facility or vehicle that is under Red Cross authority through ownership, lease or rental. The equipment could include base-station radios, mobile radios and appropriate antennas. Equipment may also require field repairs, such as the radios installed into Red Cross ERVs.

Disaster Assessment: Individuals who have taken the necessary training with the Red Cross can assess the damage caused by a disaster, and use their radio skills to relay that information back to a central point that will use the information to develop a complete picture of the event.

MINOR DOCUMENT REVISIONS

November 1, 2010
- Corrected sentence structure in <u>Section IV, *During Disasters*, Communication</u> to read properly
- Updated ARRL Contact Information

Appendix F
Emergency Power Projects and Information from *QST*

Establishing a reliable source of power in an emergency is one of the most vexing tasks you'll face. After all, your electronic equipment is utterly useless if you can't find the energy to run it.

On the pages that follow, you'll find a collection of helpful articles selected from the pages of *QST*. The topics are diverse and include battery chargers and accessories, a solar-power charge controller, and a homebrew gasoline generator (one of the most popular emergency power projects ever published in *QST*), as well as detailed advice on selecting batteries, generators and more.

QST ARTICLES REPRODUCED IN APPENDIX F

"A Long-Haul H-T Battery System"	Thurman Smithey, N6QX
"A Low-Voltage Disconnect"	Mike Bryce, WB8VGE
"Honey, They've Shrunk the Batteries"	Ken Stuart, W3VVN
"The Micro M+ Charge Controller"	Mike Bryce, WB8VGE
"An Automatic Sealed-Lead-Acid Battery Charger"	Bob Lewis, AA4PB
"Modern Portable Power Generators-Small, Sleek and Super Stable"	Kirk Kleinschmidt, NTØZ
"The 12 Volt Pup: A DC Generator You Can Build"	Yaniko Palis, VE2NYP

A Long-Haul H-T Battery System

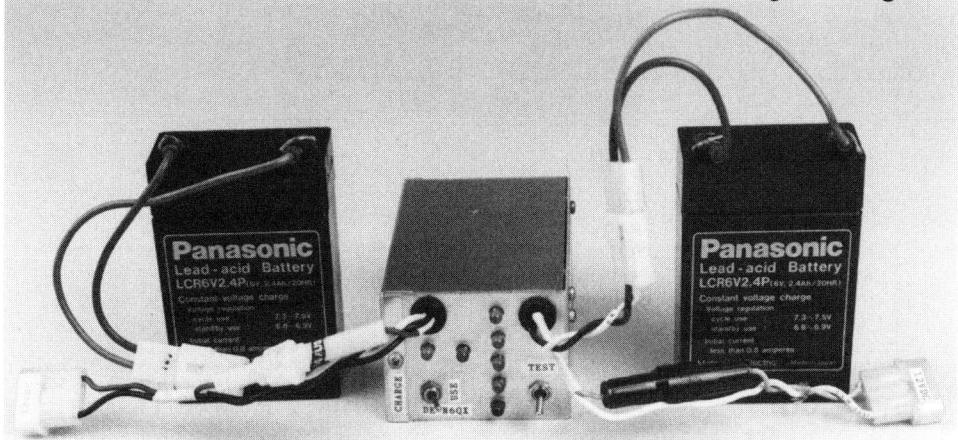

It's inexpensive, portable and you can build it yourself!

By Thurman Smithey, N6QX
56 Center St
Chula Vista, CA 91910

It seems that almost every ham owns a VHF and/or UHF hand-held transceiver (H-T), undoubtedly because they serve so many purposes. Sometimes, however, an H-T's utility is limited by its standard-issue, (usually) short-duration batteries. If you're providing public-service communications for an all-day event, for example, you may find that your battery has died long before your stint is over. There

---

Reconditioning Small Lead-Acid Batteries

Small lead-acid batteries are available for very little money at surplus outlets, swap meets and hamfests. I have learned a few things about these batteries that I feel are worth passing along.

Most of the used batteries I have found are completely dead—showing no open circuit voltage at the terminals. A battery in this condition can still be returned to a portion of its original capacity, but it takes a bit of doing and I'm not sure it's worth the effort.

When you first place the battery on charge, it appears for all intents and purposes to be an insulator. Check it with a milliammeter, though, and you find a small current is flowing, which increases with time. If you have the facilities, put a higher voltage on it (I have used 50 volts on a 6-volt battery to get the current started). Be warned: I have also nearly melted a battery or two by not connecting a suitable resistor in the charging circuit to prevent excessive current if the battery came "alive" when I wasn't around.

The application of a higher voltage may, in some cases, not be enough to get the current flow started. I've been successful in moving the process along by applying the charging voltage in *reverse* for about 30 seconds, allowing no more than 0.5 ampere of current to flow. Strange as it may seem, this procedure is often recommended by the manufacturers of these types of batteries. The rationale is that when the battery is inactive for a long time, one of the electrodes becomes surrounded by a film of distilled water, which prevents current flow. Charging in reverse for a brief time has the effect of stirring up the juices and mixing some ions with the distilled water.

Once current flow is started, it can be increased by repeated charging and discharging until the battery begins to act very much like a normal battery. So far, however, I have not been successful in restoring more than about 60% of the original capacity of a battery that has been resurrected in this manner.

When shopping for a lead-acid battery, bring along a small load, such as a small 12- or 6-volt lamp, and use it to test the battery. If the battery lights the lamp, chances are reasonably good that you have a winner. If the battery is completely flat, you have your work cut out for you and may wind up with a mediocre at best. For example, I have one set of used 2.5 Ah batteries that did not require reconditioning. That set puts out as much or more power than the best of three different sets of 4.0 Ah batteries that *did* require reconditioning.

You'll find D-sized cells to be quite popular in the surplus market. They can usually be purchased as individual cells, or as packaged assemblies. I bought one 12-volt assembly (six cells) which I then split to make two 6-volt batteries. The assembly had a decent charge when I purchased it, and made two good 6-volt batteries.

On other occasions, I haven't been so lucky. I recently purchased 20 individual D cells (the price was right), all of which were showing 2.0 volts or greater at the terminals. When I started checking them for use in this project, I kept discarding substandard cells until only eight good ones were left. Except for the time involved, I still wound up with one good battery set for very little money.—*N6QX*

The Amateur Radio Public Service Handbook

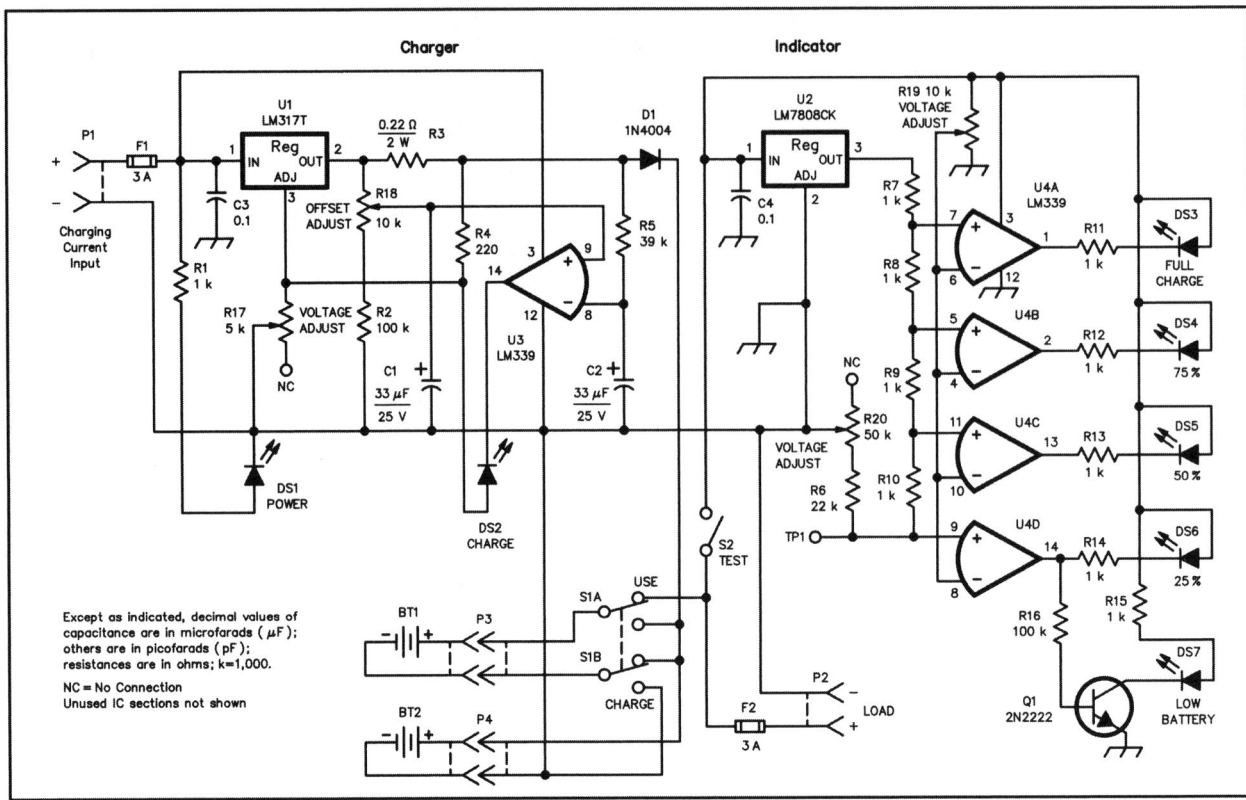

Fig 1—Schematic of the long-haul H-T battery system. Resistors are ¼-watt, 5%-tolerance carbon-composition or film except as noted below.

BT1, BT2—Panasonic LCR6V2.4P (Digi-Key Corp, 701 Brooks Ave South, PO Box 677, Thief River Falls, MN 56701-0677, tel 800-344-4539. Digi-Key p/n P262), or Radio Shack 23-181.
C1, C2—33-µF, 25-V electrolytic.
C3, C4—0.1-µF ceramic disc.
D1—1N4004.
DS1—T-1¾ yellow.
DS2, DS3, DS4 DS5, DS6—T-1¾ green.
DS7—T-1¾ red.
Q1—2N2222.
R3—0.22Ω, 2 W (Ocean State RM2-0.22).
R17—5 kΩ, ½ W, linear taper, 15 turn (Digi-Key 3006P-502-ND).
R18, R19—10 kΩ, ½ W, linear taper, 15 turn (Digi-Key 3006P-103-ND).
R20—50 kΩ, ½ W, linear taper, 15 turn (Digi-Key 3006P-503-ND).
S1—Miniature double-pole, double-throw toggle switch.
S2—Miniature single-pole, single-throw momentary normally open switch.
U1—LM317T adjustable voltage regulator (Radio Shack 276-1778).
U2—LM7808CK voltage regulator (Ocean State 7808).
U3, U4—LM339 quad comparator (Radio Shack 276-1712).

are many other situations, including emergencies of all kinds, where a portable, heavy-duty H-T power source would prove advantageous.

Having been caught with a dead battery a time or two, I decided to develop a long-endurance battery system for H-Ts—one that could be carried comfortably in a "fanny pack," or a similar-sized bag slung from a shoulder strap. Here I'll describe the system I developed, and tell you how to build one yourself. Parts cost, including the cost of new batteries, is probably less than the list price of one H-T replacement battery. Purchasing surplus batteries can reduce the cost by approximately half.

And how well does the system work? It runs my H-T (mostly in receive mode, admittedly) *continuously* for 2½ days. Charge time for the 2.5 ampere-hour (Ah) battery is only 6 to 8 hours.

System Description

My long-endurance H-T battery system requirements included:

• **Battery Charging**: The battery must be chargeable from any 10- to 15-volt dc source.

• **Automatic shut off**: The charger must shut off automatically when the battery is completely charged. An indicator must be provided to signal when charging is complete.

• **Discharge level indication**: There must be an accurate means to indicate the discharge level of the battery *as it is being used*.

• **Output regulation**: The battery output voltage must be regulated to suit the requirements of any H-T that can't be operated directly from its 12-volt output.

The Batteries

Battery choice is very important. I selected sealed, past-electrolyte,

Appendix F

lead-acid types. They hold their charge better than NiCds and they're readily available at reasonable prices. I chose two 6-volt batteries which are paralleled for charging, then connected in series to provide a 12-volt source for powering H-Ts. An added benefit of this switchable series/parallel approach is that it allows the use of either battery to supply 6-volt loads (video cameras, video lights, portable electric lanterns and so on).

In addition to the battery, the other three parts of the system are the charger, the battery-condition indicator and the output regulator.

Charger Circuit Description

A sealed, 12-volt lead-acid battery (2 to 4 ampere hours [Ah] capacity) is fully charged when its terminal voltage reaches about 15 volts and the charge current has dropped from its initial value to about 0.25 amperes. This assumes that the charging source maintains a constant voltage at the end of the charge cycle. The charger shut-off circuitry uses this current drop to define the full-charge condition.

The batteries are connected through connectors P3 and P4. S1 is placed in the **CHARGE** position to connect the batteries to the charger circuit. Charge current is supplied through connector P1 and applied to the input (pin 1) of the LM317T regulator, U1. A yellow LED (DS1) lights to indicate the application of charging power. R17 sets the output of U1 to 8.5 volts. R18 is adjusted so that 100 mV appears between its wiper and the junction of R3 and pin 2 of U1. With power applied to the circuit and no current flowing in R3, this offset voltage appears between pins 8 and 9 of U3, an LM339 quad comparator. In this state, the voltage at pin 9, the noninverting terminal, is 100 mV less than the voltage on pin 8, the inverting terminal. Therefore, the comparator output at pin 14 is *low*.

With batteries connected to the charger, however, the initial charging current flowing through R3 and

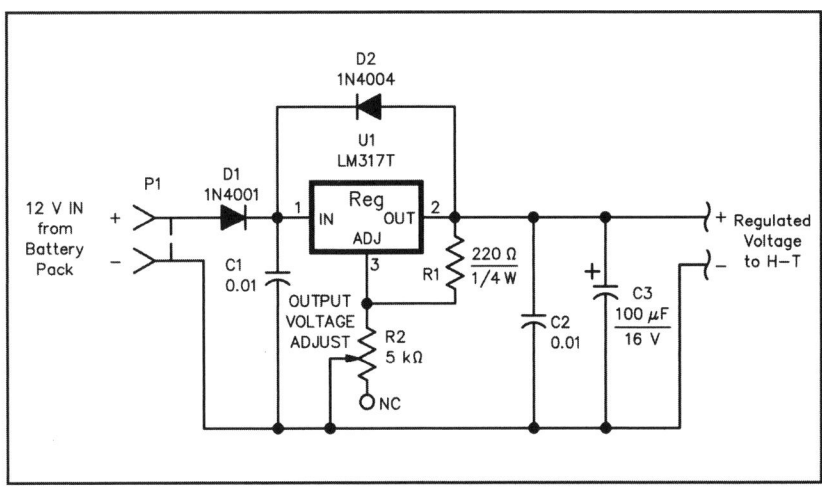

Fig 2—Voltage regulator schematic. Resistors are $^1/_4$-watt, 5%-tolerance carbon-composition or film except as noted below.
C1, C2—0.01-µF ceramic disc.
C3—100-µF, 15V electrolytic.
D1—1N4001.
D2—1N4004.
R2—5 kΩ, $^1/_2$ W, linear taper, 15 turn (Digi-Key 3006P-502-ND).
U1—LM317T adjustable voltage regulator (Radio Shack 276-1778).

D1 is approximately 0.8 amp, resulting in a voltage drop across R3 of about 176 mV. The inverting terminal of comparator U3 is now negative with respect to the noninverting terminal by 76 mV. As a consequence, the comparator output switches to *high*. C1 and C2 prevent a racing condition that might otherwise cause the comparator to change state before the charging current is established. At the outset, the output of U1 is less than 8.5 volts because R3 is used in the voltage-determining circuit in a negative feedback (current limiting) mode. U1's output voltage rises as the battery voltage increases, until it reaches about 8.4 volts. The charging current, and the voltage across R3, remain nearly constant at less than their initial values for most of the charge cycle. When the batteries approach their full-charge condition, their voltage rises. This decreases the charging current through R3, which results in decreased voltage across R3. The voltages appearing at pins 8 and 9 of U3 become equal when the voltage drop across R3 is reduced to the amount of the offset (100 mV). This equals a charging current of 455 mA, shared between two batteries, or about 227 mA for each battery.

When the voltage across R3 becomes *less* than 100 mV, comparator in U3 changes state and pin 14 goes low. This draws current through R4 and lights DS2. R4 is in the voltage-determining circuit of U1. The additional current drawn through R4 by U3 reduces the voltage at pin 3 of U1, dropping the output voltage lower than the battery voltage. D1 prevents current from the batteries from flowing backward in the circuit, so there is essentially no current through R3.

With no current flowing through R3, pin 9 of U3 is lower than pin 8 by 100 mV. The comparator output remains low and no additional charging takes place. The lighting of DS2 signals that the charge cycle is complete. At that point, the charging power source is disconnected, S1 is switched to the **USE** position and the batteries are available to power whatever device is connected to P2.

Battery-Condition Indicator Circuit Description

In Fig 1, U2, an LM7808CK 8-

volt regulator, provides a stable reference voltage when S2 is closed. The string of equal-value resistors (R7 through R10) functions as a four-way voltage divider. Since the resistor values are equal, the resulting voltage drops across each resistor are equal. Even so, the voltage drops can be increased or decreased (as a group) by adjusting R20. A simple computation and a voltage measurement at TP1 determines the R20 adjustment—as we'll see later.

The reference voltages are applied to the noninverting terminals of the four comparators of U4, another LM339. The inverting terminals are all connected to a common voltage which is referenced to the battery voltage. (The ratio is adjusted by R19.) Four green LEDs (DS3 through DS6) are connected to the comparators.

When S2 is closed (placing power on the indicator), the battery voltage is applied to pin 1 of U2 as well as the LEDs. If the referenced battery voltage is greater than a given comparator's reference voltage, that comparator's output is low and its LED glows. If the referenced battery voltage is less than a given comparator's reference voltage, that comparator's output goes high and its LED does not illuminate.

If the reference voltage is correct and R19 and R20 are properly adjusted, all four LEDs glow when the battery is fully charged. The first LED (DS3) switches off when 75% of the charge remains. The second LED is extinguished when a 50% charge remains. The third LED winks out at 25% and the fourth switches off when the low-voltage condition is reached (when the battery should be taken out of service and recharged). As a safeguard against the problems of negative indications, a fifth LED (DS7) was added (with Q1 and U4D acting as a switch) to provide a constant **LOW BATTERY** indication. This red LED lights when the fourth green LED is extinguished.

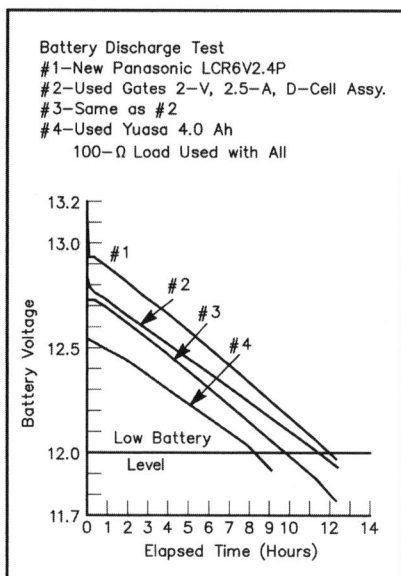

Fig 3—Battery discharge test results.

Output Regulator Circuit Description

Many H-Ts are designed to be operated on 12 volts and thus don't require this regulator circuit. For this reason, the regulator is not included on the same board with the charger/indicator. (The regulator schematic is shown in Fig 2.) If your H-T requires less than 12 volts, the regulator can be set to provide the required voltage. U1, an LM317T adjustable regulator, is at the heart of this simple circuit. R2, a 5-kΩ potentiometer, adjusts the output of U1 to suit your H-T.

You'll need to provide a connector to fit your H-T. I made an adapter for my Kenwood TR-2500 using the case of a defunct battery pack. This is a good way of making the power connection, since it is reliable and attractive. In addition, the empty case provides space for the voltage regulator.

Construction

The charger is built on a 2¼- × 3½-inch circuit board. The prototype was constructed on perf board, but I highly recommend that you use a printed-circuit board. You can make your own or order the PC board from FAR Circuits.[1]

The circuitry is housed in a 4 × 2 × 2¾-inch aluminum box. The circuit board is supported inside the box on three angle brackets made of stiff steel wire bent to shape. The circuit board is mounted flush with the end of the box that is farthest from the indicator. This leaves space between the box and the board edge at the indicator for wires to pass to and from the switches, LEDs and grommets. The board edge on which U1 is mounted must also be flush with the side of the box so that U1 can use the box as a heat sink. The LEDs are cemented into their ³⁄₁₆-inch holes using epoxy adhesive. U1 is mounted near the board edge abutting the side of the box and is bolted to the box using an insulating kit.

Four connectors are used—one for each battery, one for the charging source, and one for the device to be powered by the batteries. Select connectors that fit your requirements.

I recommend you make a power cable with an in-line connector and fuse. Use connectors that mate with whatever dc-power source you intend to use to charge the batteries. I use a cigar-lighter plug which allows me to charge my batteries from an automobile electrical system.

The cable connecting the batteries to your H-T or other device should also include an in-line fuse. If you intend to build the voltage regulator circuit, be advised that the voltage regulator IC should be mounted on a small heat sink. This regulator can deliver up to half an amp or so on transmit.

All parts used in the project, with the exception of the new batteries, are

---

[1] A PC board and part overlay are available from FAR Circuits, 18N640 Field Court, Dundee, IL 60118; price $4.50 plus $1.50 shipping and handling per order. Check or money order only; credit cards not accepted. The PC-board template and part overlay are available free of charge from the ARRL Technical Department Secretary. With your request for the SMITHEY LONG-HAUL H-T BATTERY SYSTEM PC BOARD TEMPLATE PACKAGE, send a #10 SASE.

common parts which can be found in any electronics parts store or catalog. A source for the batteries is listed in the parts list (see Fig 1 caption).

Calibrating the Charger

After you've completed construction and checked your work, connect a 12-volt power source to the charger input (P3) *before* you install U3 and connect the batteries. DS1 should glow. Measure the voltage at pin 2 of U1 (referenced to ground) and adjust R17 until U1's output is 8.5 volts. Disconnect the power source and install U3. Reconnect the power source and measure the voltage difference between pins 8 and 9 of U3, adjusting R18 until pin 9 is 100 mV less than pin 8.

Calibrating the Battery-Condition Indicator

I have determined that no single calibration of the indicator unit is truly accurate with several different battery types. Fig 3 shows the discharge characteristics of four different batteries, all with the same 100-ohm load, all having just been charged using the charger. The variations are great enough to significantly affect the accuracy of the indicator.

If you buy new batteries of the type shown in the parts list, you can be confident using the discharge characteristics of battery #1 in Fig 3 to calibrate your indicator. If you have elected to use batteries that have seen previous service, I recommend that you run a simple discharge test on the batteries before performing final calibration of the indicator. (Charge the battery using the calibrated charger, then attach a 100-ohm load and plot the discharge characteristic as was done for Fig 3.)

If you have more than one set of batteries with different discharge characteristics, I recommend that you calibrate for the best set, and take into account the difference when you read the charge remaining in the weaker set(s). That way, the charge remaining at any time will be nearly equal for the different batteries.

Given a small, constant load—such as an H-T in the receive mode—the discharge voltage curve over time is nearly linear until the battery voltage drops to about 12 volts. A battery should be taken out of service and recharged when its voltage under load drops to 12 or less. This rule seems to apply regardless of the battery brand.

Fully charged battery voltage can vary considerably (see Fig 3). Depending on the brand, battery potential at the beginning of the discharge cycle can vary from 13 to 12.55 volts. Fortunately, the provisions for adjustment of the indicator are flexible enough to accommodate any batteries you're likely to find.

Calibrate the indicator based on the assumptions we've just discussed. In other words, assume that its voltage-versus-time curve will be linear with a constant, small load, and that it will need recharging when the voltage decreases to 12 volts. Compute the voltage you want to see at TP1 as follows:

$$V_{TPI} = V_{REF} \times \frac{V_{LOW}}{V_{FULL}}$$

V_{Ref} is the voltage output of U2 at pin 3, in this case 8.0 volts. V_{Low} is the voltage selected for **LOW BATTERY** indication (12 volts). V_{Full} is the battery voltage at full charge.

To calibrate the indicator, you'll need a variable-voltage power supply with a range of 11.5 to 13 volts. After determining the voltage that you want at TP1, apply 12 volts to pin 1 of U2 and adjust R20 until the voltage at TP1 reaches the desired level.

Now reduce the voltage to the V_{Low} value and adjust R19 until DS6 goes out and DS7 comes on. That's all there is to it! DS3 should now go out at a battery voltage of 12.75, DS4 at 12.50, DS5 at 12.25 and DS6 at 12.00. When DS6 turns off, DS7 lights to tell you your battery needs to be charged.

While you have your variable-voltage supply connected, check that all four green LEDs do in fact go on and off at the correct voltages. If not, you may want to try calibrating using the transition of DS5 and a calibrating potential of 12.25 volts. Then, recheck for accuracy on all four LED set points. You should be able to get them all transitioning within 50 mV of the stated voltages.

For maximum accuracy, V_{Ref} should be measured with the same voltmeter you used to set the voltage at TP1. To calibrate the indicator for another battery with different characteristics, merely substitute the appropriate numbers in the equation above. To further improve accuracy, I measured all the 1-kΩ resistors used in the project and selected the four that were most nearly equal for R7 through R10.

Additional Thoughts

I use a common fanny pack to house the batteries and the charger/indicator when I want to carry the system around. In the one I bought, there's plenty of room for the equipment and accessory cables. Although I elected to discontinue using large batteries, I was able to get a pair of 4.0-Ah batteries in the pack and the weight wasn't too objectionable. The box housing the charger/indicator gets quite warm while batteries are being charged, so it should not be left in the pack when charging is in progress.

Summary

I've had a great deal of enjoyment in developing this low-tech project, and even more enjoyment out of using the long-haul battery system with my own H-T. Try one yourself and I'm sure you'll like it as much as I do.

Thurman Smithey, N6QX, was introduced to Amateur Radio in the late 1930s as a high school student. It wasn't until his retirement from the Navy, 30 years later, that he finally obtained his license. He was first licensed to the General class as WA6FUY in 1968. Thurman acquired a sailboat the same year and has enjoyed operating maritime mobile while doing some blue water sailing. Thurman holds a Master of Science degree in Engineering Electronics.

A Low-Voltage Disconnect

Whether you're operating a repeater, operating on emergency power or just watching TV in your RV, this little gadget protects your batteries from damage.

By Michael Bryce, WB8VGE
2225 Mayflower NW
Massillon, Ohio 44647

Many Amateur Radio stations use batteries to power their radio equipment during commercial electric power outages. Some of us use battery power all the time in the shack. Keeping an eye on the battery's charge sometimes can't be done (or is forgotten altogether) until you are unexpectedly—and unwillingly—off the air!

What you need is a battery watchdog—something to keep track of the battery and disconnect loads when the battery just about goes kaput. No matter what your use of battery power—whether you own a camper, RV, or just fish on the lake beside your cottage—this contraption does the battery monitoring for you and protects your battery from severe discharge as well. Repeater owners and operators may find the device an ideal way to extend operating time while on emergency power.

What's It Do?

The low-voltage disconnect (LVD) automatically disconnects a load from the battery before damage is done to the battery or the load. The potential across a discharging battery's terminals depends on the battery's state of charge and the load's discharge rate. This circuit monitors the battery terminal voltage, and when it reaches a preset level, a relay is de-energized disconnecting the load from the battery. There's an approximate 5-second delay before the device senses the low-voltage set point and the relay drops.

Take a look at some of the features of this watchdog:
- User-adjustable turn-off voltage.
- User-adjustable reset voltage.
- Built-in delay to ignore temporary low-voltage conditions.
- 30-A-capacity relay contacts.
- Low current consumption.
- Easy construction using a readily available PC board and components.

The load can be connected directly to the LVD's relay, or its relay can drive an off-board, heavy-duty power relay, if need be. You can also use the PC-board-mounted relay to control a 120-V ac load (I'll talk about that later). The LVD can also supply logic to a repeater controller, too.

Circuit Description

To see how the circuit works, refer to Fig 1. U5, an LM317LZ 100-mA adjustable-voltage regulator, creates a reference voltage for the comparators. This voltage (4.00) is set by R30 (**REFERENCE ADJ**), a 1-kΩ potentiometer. R27 places a 4-mA load on the regulator's output and improves regulator stability. U2C buffers the reference voltage. From here, the reference voltage goes to the voltage comparators. D1, a 1N4001 diode, protects U5 from reversed power-supply voltage polarity.

R1 and R2 halve the battery terminal voltage. R3 and C9 help filter out battery-line noise. U2B acts as a buffer between the voltage divider and the battery sense line.

Two set points are needed to control the LVD. One turns on the LVD. That causes K1 to drop out, disconnecting the load. The other set point turns off the LVD, closing the relay contacts and reconnecting the load. If it weren't for the two different set points, the LVD would constantly switch on and off. The difference in the set-point voltages allows the battery to recover before the load is reconnected. (This assumes you have some means of recharging the battery after the load has been disconnected.)

The voltage comparators are nearly identical. The battery voltage, now divided by two, is applied to two voltage dividers. Let's first look at the trip comparator.

The battery's output voltage is divided in half and applied to comparators U1C and U1B through U2B. If the battery was discharged to 10 volts,[1] the trip comparator (U1C) input would see less than 5 volts. The reference voltage at pin 10 is 4 volts. By adjusting R5 (**LVD TRIP**), we can set the comparator to switch states when the input voltage at U1C pin 9 equals the reference voltage. D5 and R12 provide a bit of hysteresis to keep U1C from oscillating. Because U1C may not provide the needed high-to-low positive switching action, a second op-amp section (U1D) is used. U1D provides the switch-like on/off state needed by the delay circuit composed of D4, R16 and C8, which provides a delay of about 5 seconds. Again, to provide the required logic levels, U2A is used. Its output goes to the SET point of the R/S latch, U4A and B.

The **ON RESET** circuit (U1B) works similarly, but has no delay circuit. U1B's output is routed through U1A and goes to the RESET of the R/S latch, U4A.

When the battery voltage is above the LVD trip point, the output of the R/S latch (U4A pin 3) is high. This turns on Q1, a TIP-120 NPN Darlington power transistor, which energizes K1. A 1-W, 47-Ω resistor (R18) limits relay current. This minimizes the overall current demand of the LVD. D8 protects Q1 from the back EMF produced when the relay coil's magnetic field collapses.

As the battery discharges, its terminal voltage falls. When the LVD turn-on voltage is reached, there's a 5-second delay, then the output of U2A goes low, setting the R/S latch. This removes Q1's base drive, causing K1 to de-energize. The output of U4 pin 4 connects to U4 pin 8 to allow the oscillator (U2D) to output through U4C to U3A, which turns on the **LOW-VOLTAGE** LED, DS1. DS1 flashes at a rate determined by R25, R26 and C7. With the values shown, the on-time is about 1/20th of a second. Otherwise, DS1 remains dark. This arrangement reduces the circuit current drain during a low-voltage battery condition.

Construction

There's nothing critical here: perfboard, wire-wrap, dead-bug or PC-board construction are all suitable. A PC board is available[2] as well as a complete kit of parts.[3] Using a PC board speeds construction and makes troubleshooting easier.

You can buy most of the parts from a well-stocked Radio Shack store. Mouser Electronics[4] can supply the parts "the Shack" does not carry. In both cases, the single exception is the relay. Obviously, the one specified fits the PC board. Secondly, it has a hefty contact rating (30 A) and is inexpensive (less than $4). This relay is available from Digi-Key.[5] Certainly, you can use a different relay, but you'll probably have to mount it off-board. Also, you may have to change R18's value, as it's dependent on the relay's coil current. The value shown is calculated for the relay identified in the caption. (By the way, mount R18 with a ¼- to ½-inch clearance between the resistor body and the PC board. This allows air to circulate around the resistor and prevents the dissipated heat from discoloring the PC board.)

Component values aren't critical. Use equivalent parts you have on hand. If you don't have a 300-kΩ resistor, a 270-kΩ will work just fine. (The ±10% rule of thumb can be safely applied.) Use sockets for the ICs and be careful when handling U4: It's subject to damage by static discharges, so use a wrist strap. Install all the parts—*except for C8*. After you've assembled the PC board, check your work to ensure the diodes and capacitors are properly polarized. Ensure you have the ICs oriented properly before you apply power to the board.

Set Up and Test

You'll need a digital VOM[6] and a variable-voltage power supply to adjust the LVD. Connect your power supply to the terminal block at the battery terminals.

Set the power supply output to 14 volts. Apply power to the board. Probe pin 10 of U2. Verify the presence of the reference voltage; in all probability, it's not 4.00 volts—yet. Move the probe to U2 pin 8 and verify the reference voltage is there, too. Now, adjust R30 (**REFERENCE ADJ**) until the voltage at U2 pin 8 is 4.00 volts. Next, set your power supply to 10.5 volts. Measure the voltage at pin 7 of U2. It should be 5.25 volts (half of 10.5 volts).

Adjustment Method

First, *remove C8* (in case you forgot the earlier warning and soldered it in) from the PC board. This defeats the delay circuit. Turn both R7 and R5 fully counterclockwise. Set your power supply output to 13.5 volts. Connect the power supply to the LVD, then turn it on. The **LOW VOLTAGE DISCONNECT** LED (DS1) should be blinking. Slowly adjust R7 (**ON RESET**) until the relay closes and DS1 goes dark. Reduce the power supply output to 10.5 volts. Slowly turn R5 (**LVD TRIP**) until K1 drops out and DS1 begins to flash. Verify the two set points by raising the power supply output voltage to 13.5 volts. K1 should energize. Reduce the voltage to 10.5 volts and K1 drops out. *Now* you can install C8! You're done!

Troubleshooting

If you can't get the circuit to

The Amateur Radio Public Service Handbook

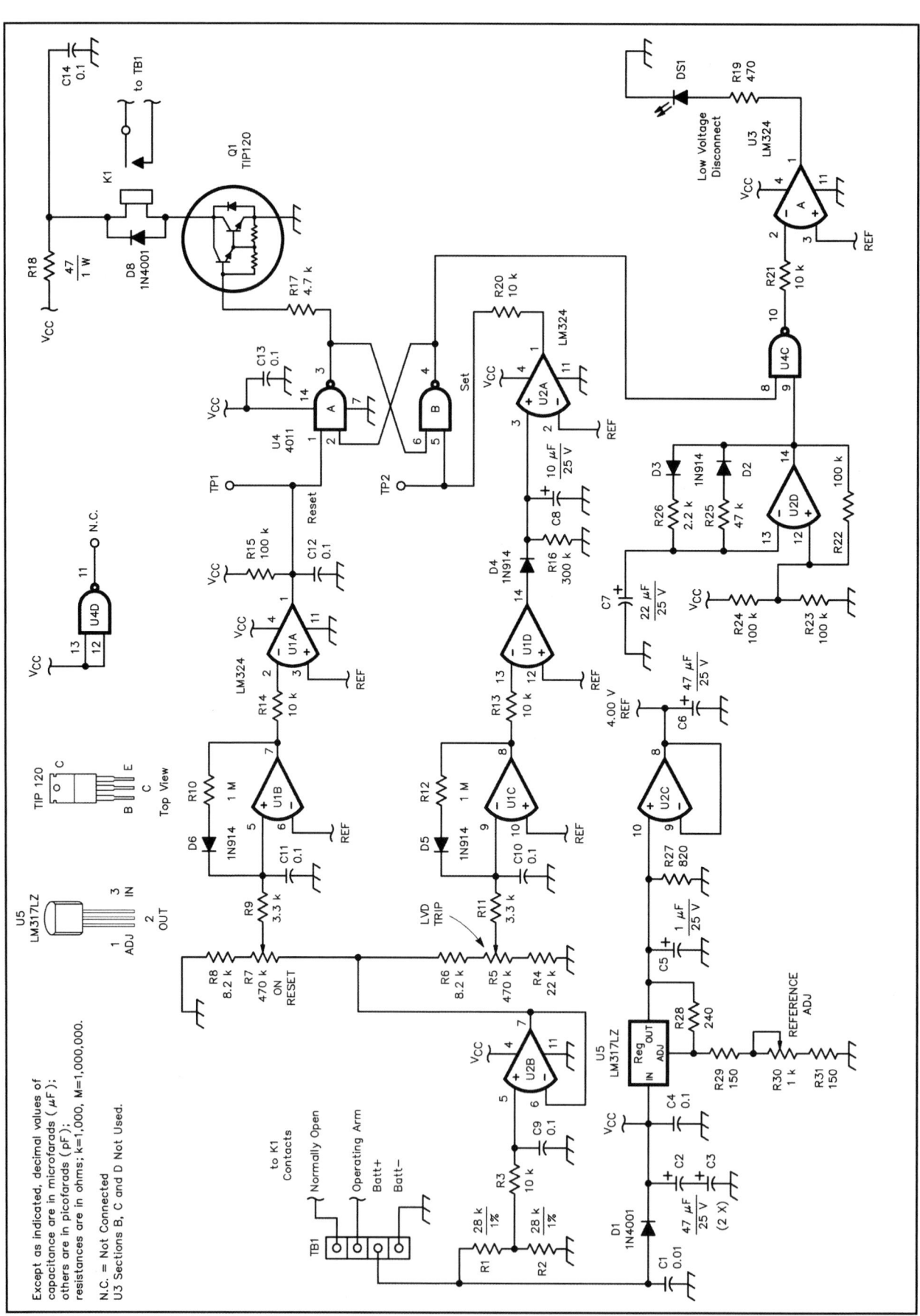

Appendix F

Fig 1—Schematic of the low-voltage disconnect circuit. Equivalent parts can be substituted. Unless otherwise specified, resistors are ¼-W, 5%-tolerance carbon-composition or film units.

U1-3—LM324 quad op amp (Mouser 511-LM324; Digi-Key LM324N).
U4—4011 quad gate (Mouser 511-4011; Digi-Key 4011CD).
U5—LM317LZ 5-V, 1-A adjustable regulator (Mouser LM317LZ; Digi-Key LM317LZ).
K1—Potter and Brumfield T-90 series; 12-V dc, 155-Ω coil, SPST, 30-A normally open contacts (Digi-Key PB110-ND).
TB1—Terminal block (Mouser 506-8PCV-04).
R5, R7—470-kΩ trimmer potentiometer (Mouser 531-PT15D-470K).
R30—1-kΩ trimmer potentiometer (Mouser 531-PT15D-1K).
Q1—TIP-120 Darlington power transistor (Mouser 511-TIP-120; Digi-Key TIP120PH-ND).

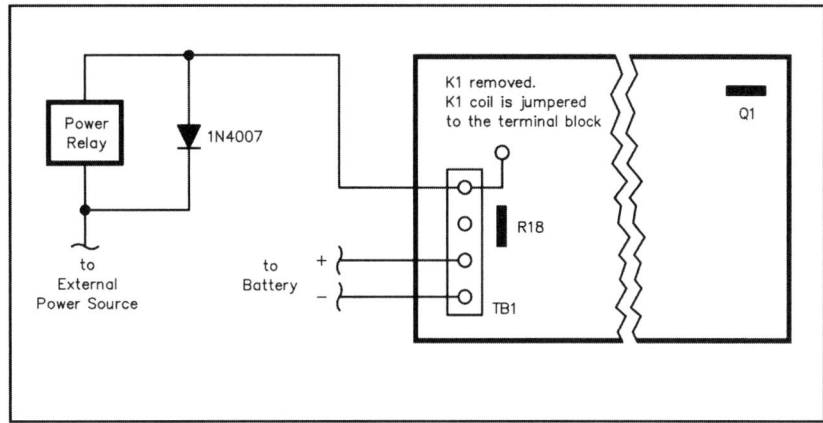

Fig 2—An off-board solid-state relay provides an excellent means of controlling 120-V-ac operated loads. Here, the relay derives its operating current and voltage from the battery being monitored. The resistance value and power rating of RX is chosen for proper relay operating current.

work, first check for the presence of the reference voltage. Without it, you'll be dead in the water from the start. Check for 4.00 volts on pins 3, 6, 10 and 12 of U1, and at U2 pin 2 and U3 pin 3.

If the comparators won't switch (and you have the proper reference voltage) check the battery sense line by checking the output voltage at U2 pin 7. (This voltage should be one-half of the power-supply voltage applied to the battery sense line.) As you can see, replacing U1 replaces *all* the battery sense comparators.

When the battery voltage is at 10.5, you'll be able to see the delay action by probing U2 pin 3. At this pin, you'll see the voltage slowly drop during the 5-second delay period.

If you used a relay other than the one specified, the value of the current-limiting resistor (R18) may be too high to allow the relay to energize. Try reducing R18's value (or short it out).

Hooking Up the LVD

With only four wires, hook-up is a breeze! Simply connect the battery you intend to monitor to TB1's battery terminals. K1's contacts are completely isolated from the battery. By connecting a jumper from the +12-volt battery terminal to one of the relay contacts, you can deliver battery power through the relay contacts to your load.

Remember, although K1's contacts can carry 30 amperes, voltage drop caused by long wire runs can have an effect on the load. If you need to control heavy current loads, use K1 to control a power relay located right at the load. Install a protective diode across the power-relay's coil terminals to prevent inductive kick-back.

You can use K1's contacts to control logic levels to a repeater. Connect K1's contacts to ground or +12 V via a current-limiting resistor. If you have a repeater controller and it requires a logic input, this is one way to go.

Controlling a 120-V AC Load

Although K1's contacts easily handle a 120-volt load, having an exposed 120-volt line connected to TB1 would keep me up at night! A safer way to control such a load is to use an off-board solid-state relay (see Fig 2).[7] (A solid-state relay is an op- tically coupled device that provides excellent isolation between the load and the driving source. In this case, between the 120-V ac mains and your battery.) Solid-state relays with various control voltages are readily available. RX, an external resistor, serves to limit the current flowing to the solid-state relay. By properly altering the value of RX, you can use a 5- or 12-volt control line. As mentioned earlier, the battery can supply power to operate the solid-state relay via K1's relay contacts.

Life with an LVD

While the LVD certainly protects your battery from deep discharge, it's not perfect. (What is?) Every LVD consumes *some* power from the battery it's trying to protect. In this case, when the LVD has the relay pulled in, it draws about 90 mA. If you run the LVD 24 hours a day, you have a 2.16-Ah load just for the LVD. Even with the relay off, and the battery at 10.5 volts, the LVD draws about 12 mA. So, if your battery is being charged by a solar array, be sure to include the LVD load requirements when performing your sizing calculations. To save power when you're not using the load, turning off the LVD automatically turns off the load connected to the relay.

If your battery is charged from a

120-V ac charger, you'll not have to worry about the extra LVD load. Repeaters operators normally have a battery back-up system constantly being charged. When the grid power fails, the battery takes over. When the battery discharges to the point that the LVD trips, the LVD can then take the power amplifier off line to extend battery operation until the grid power comes back on.

The LVD load should not be your main load. *Shedding* loads is the main job of the LVD. You don't want to have it shut down *everything*, but disconnect what you can live without. For instance, in an RV, you may want to connect the running lights to the LVD and leave your TV bypassed. When the battery becomes so low as to trip the LVD, the running lights will be disconnected. (Given the nature of what's on TV these days, it's probably a better idea to take the TV off line and keep the running lights on!)

Here's another example of choice: You have a sailboat docked in the lake. A bilge pump is connected to the LVD. If too much water leaks into the boat and the pump is running all the time, the LVD will disconnect the pump from the battery protecting the battery from damage. With the pump disconnected from the battery, the pump won't work any more and before you know it, your sailboat has become a submarine!

A much better way to prevent your sail-boat from sinking is to have the LVD warn you of the discharged battery. The warning could be as simple as a flashing light or a buzzer. If you really want to go to the extreme, you could combine the LVD and the METCON II[8] for telemetry. The LVD keeps a constant eye on your batteries, while you work the world on your radio, or just fish in the lake by your cottage.

Notes

[1] A lead-acid battery is generally considered dead when the terminal voltage is 10.5 under load.

[2] A PC board for this project is available from FAR Circuits, 18N640 Field Ct, Dundee, IL 60118-9269. Price: $12, plus $1.50 shipping. A PC-board template package is available free from the ARRL. Address your request for the LOW VOLTAGE DISCONNECT TEMPLATE to: Technical Department Secretary, ARRL, 225 Main St, Newington, CT 06111. Please enclose a business-size SASE.

[3] A complete kit of parts, including the PC board and relay, is available from SunLight Energy Systems, 2225 Mayflower NW, Massillon, OH 44647. Price: $55 plus $3 shipping.

[4] Mouser Electronics, 2401 Hwy 287 N, Mansfield, TX 76062; tel 800-346-6873, 817-483-4422, fax: 817-483-0931.

[5] Digi-Key Corp, 701 Brooks Ave S, PO Box 677, Thief River Falls, MN 56701-0677, tel 800-344-4539, 218-681-6674, fax 218-681-3880.

[6] You can use an analog meter to calibrate the LVD, but a digital voltmeter provides better resolution.

[7] Available from All Electronics Corp, PO Box 567, Van Nuys, CA 91408-0567, tel 800-826-5432, 818-997-1806, fax 818-781-2653.

[8] P. Newland, "Introducing METCON, a New Remote control and Telemetry System," QST, Jan 1993, pp 41-47.

By Ken Stuart, W3VVN

Honey, They've Shrunk the Batteries!

Microminiaturization of electronic components has taken a giant leap forward in the past few decades. Combining these components into integrated circuits has also resulted in squeezing more and more circuitry into smaller and smaller space to the extent that couldn't have been imagined some 40 or 50 years ago. For example, one prophet of the electronics industry went on record around 1950 as saying that he foresaw computers in the year 2000 as weighing less than one-and-a half tons! Well, he was right, and then some. Today's laptop has a capability that vastly exceeds the computer of his day, which took up rooms of space and gobbled up electricity at a rate that would supply several of today's homes.

With the ensuing emergence of electronic miniaturization has come a formidable market of personal electronics products such as the cellular and portable phone; laptop computers and pocket organizers; H-Ts; compact disk, cassette and MP3 players; pocketable GPS navigation units; and many others. The development of these mini electronic gadgets has brought about the need for smaller and smaller batteries with greater stored energy.

Driven by the development of these consumer products with their smaller power sources, battery technology has taken giant leaps forward in the past decade. Consumer items like the laptop computer, cellular phone and similar portable products have forced technology to produce lighter, smaller cells with increased energy storage. As a result, new battery chemistries—for example, nickel metal hydride (NiMH) and lithium—are rapidly replacing older technologies such as nickel cadmium. As in any technology, however, one doesn't get something for nothing, and there are few miracles. It pays to know the tradeoffs before making a switch in battery types. This article is an overview of some of the new varieties of battery cell now available, and will provide comparisons between new and old technologies.

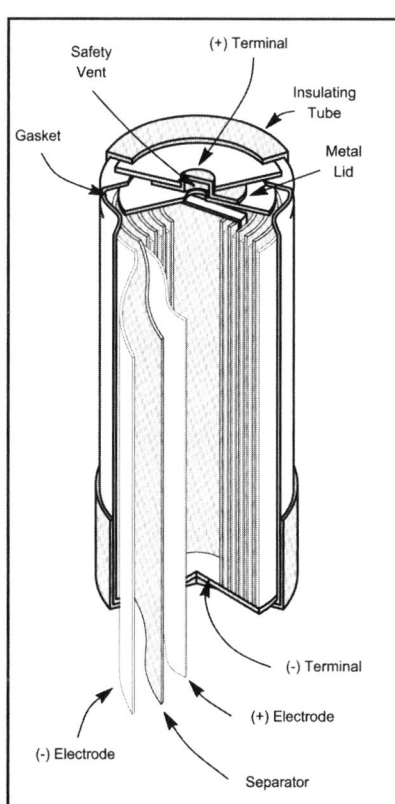

Figure 1—Cutaway drawing of a NiMH cell. Except for the cathode, which is made of hydrogen-storage metal instead of cadmium, the construction is similar to that of the NiCd cell.

Nickel Cadmium (NiCd): The Old Standby

For many years the reigning king of the miniature rechargeable battery, the NiCd cell still has a number of things to recommend it. Although largely superseded by the nickel metal hydride, the NiCd cell still leads the field in the number of charge/discharge cycles, easily reaching to a thousand or more for larger volume cells like the "C" and "D" sizes. It is also the undisputed champ in providing extremely high output currents for its cell size, which is why it is still the predominant battery in Amateur Radio hand-held transceivers (HTs).

Recently, the environmental impact of landfill disposal of NiCds has become an issue. Cadmium metal is used to form the anode of the cell (the *negative* terminal of a cell, since the anode of a device is defined as the terminal into which current flows). Cadmium is one of the most toxic of metals, and disposal of these cells has become a serious problem. In fact, there are recycling programs in several areas of the country for used NiCds, and we should take advantage of them. The Rechargeable Battery Recycling Corp (1000 Parkwood Cir, Ste 450, Atlanta, GA 30339, tel 678-419-9990, **www.rbrc.org**) is a nonprofit organization that provides recycling assistance in the US. With the improvement in the capability of NiMH cells, the thrust now is to change to nickel metal hydride, which eliminates the cadmium problem.

Memory

NiCds have gotten a lot of bad press because of the "memory" effect, which is now often referred to as "voltage depression." It's not nearly as prevalent a problem as it sounds, and if it does occur, it is completely reversible. We now know that this effect is caused by crystallization of the nickel electrode which reduces the area of the active material available for chemical action. Nickel metal hydride (NiMH) cells also have nickel anodes, however. Therefore, contrary to popular belief, the NiMH cell is also subject to anode crystallization and therefore memory. The memory effect is not seen in NiMH cells as often as in NiCd, simply because the NiMH cell does not have the longevity of NiCd.

The memory effect can be avoided and reversed by subjecting the battery to a hefty discharge and full recharge once every couple of months. Note that this does *not* mean discharging the battery to zero, since full discharge can cause cell reversal which can shorten cell life.

Nickel Metal Hydride (NiMH): Serious Contender

The nickel metal hydride (NiMH) cell has been evolving for a number of years, and until recently it was not a serious contender for the throne occupied by the NiCd. Figure 1 is a cutaway view of an NiMH cell. Recently, the state-of-the-art of the NiMH cell has advanced to where it can often be considered as a one-for-one replacement, although it has only about one-third the number of charge-discharge cycles (cyclic lifetime) and higher internal resistance than nickel cadmium. Figure 2 shows a NiCd (left) and NiMH battery pack for a popular H-T, while Figure 3 shows three AA cells—from the left, an older NiCd, a higher-rated NiCd and an NiMH.

Advantages

NiMH batteries have approximately the same electrical characteristics as NiCd with one exception—they have about 30 to 50 percent more energy capacity per cell.

• Per-cell cost is now competitive with NiCd.

• Readily available in single cells or in ready-to-go battery packs for popular ham transceivers.

• Can be recharged using the same charger as the NiCd battery that it replaced; no new charger is needed.

Disadvantages

• NiMH cells have fewer charge-discharge cycles than NiCd. Typical cyclic lifetimes are around 500 charge/discharge cycles as compared to about 1500 for NiCd. For most hams, this is not a significant problem.

• Internal resistance is about twice that of NiCd, which means that NiMH will not provide as much output power in higher power H-Ts as NiCds.

• NiMH self-discharge is greater than NiCd—about 30% per month, compared with NiCd's 20%.

As the NiMH technology continues to improve and prices drop, the NiMH cell will, in all likelihood, supplant nickel cadmium.

Lithium-ion (Li-ion): Up and Coming

In the past decade, Lithium battery technology has made Li-ion batteries an up-and-coming contender in the portable battery field. Already, rechargeable lithium cells are making their way into cellular phones, where their superior energy storage capability provides increased talk and standby time.

Lithium primary (non-rechargeable) cells were the first of the family to evolve. Starting about a dozen years ago, the lithium button cell first appeared in electronic watches where it gave years of operating life before its power was consumed. Its main advantages are that it has an energy storage capability of about twice that of alkaline cells by volume, and about four times by weight, as well as an extremely long storage life. Lithium primary cells are now available in the popular "flashlight" sizes, but they are expensive and not readily obtainable. These cells are commonly seen in such applications as key-chain flashlights, wristwatches and memory backups in computers and ham rigs.

Lithium secondary, or rechargeable, cells are becoming popular for cell phone and laptop battery packages. Their light weight and high energy storage capacity provides longer life while not burdening the consumer with a heavy power pack. Although many lithium rechargeable technologies have been developed, the most popular is the lithium-ion (Li-ion). See Figure 4.

Rechargeable lithium cells, however, have had some problems. In its pure state, lithium metal is extremely reactive, and any contact with water results in the liberation of hydrogen and possible fire or explosion. In lithium cells, the lithium is normally in the form of a salt, which makes it non-reactive. Certain battery manufacturers, however, have stated that with overcharging, lithium metal can be extracted from the salt inside the cell casing, and that reactions have taken place causing rupturing of cell cases, and damage to the equipment in which they were installed. Therefore, charging of lithium batteries is usually handled by special protective balancing and charging circuits built into the battery package. These circuits carefully regulate the state of charge, and terminate charging before overcharge can occur.

Early lithium batteries had a relatively high internal impedance that was about three times higher than that of NiCds. Recently, this internal impedance has been lowered as a result of improved manufacturing techniques and research. Although the impedance is still not as low as NiCds, it is sufficiently low for application in some Amateur Radio H-Ts. Yaesu's VX-5R is a 5-W unit with a small Li-ion battery as standard equipment (see Figure 5).

One negative point is that Li-ion batteries do not have a particularly long lifetime even if they are not used. One manufacturer has stated that the lithium rechargeable cell can last only about two to three years after manufacture. On the good side, their rate of self-discharge is very low—only about 10 percent per month—which means they are excellent for standby equipment applications.

Sealed Lead-Acid (SLA): Old Standby

Although the rechargeable sealed lead-acid (SLA) cell is heavy and bulky, and doesn't hold a lot of energy for its size and weight, it has the advantage of a very low self-discharge rate. In addition it is relatively inexpensive and very reliable. Complete batteries are readily available at electronics dealers. One can come across batteries that have been routinely pulled from emergency lighting systems that still have lots of life left in them. One manufacturer, Quantum, used to provide a battery pack specifically made for ham radio use; although they have discontinued this model, they have a higher power alternate with state-of-charge indication and included charger.

Larger SLA battery packages, which can power a desktop or automotive transceiver for hours on end, can also be found at automotive accessory dealers, discount buyer's "clubs," and so forth. These are sold as emergency automobile starting units, complete with jumper cables and cigarette lighter outlets. The au-

Figure 2—Two 13.8-V battery packs for my ICOM IC-2GAT H-T. The pack on the left is an older NiCd unit having a rating of 1200 mAh. The one on the right is an NiMH with a rating of 2700 mAh. They have identical case sizes.

Figure 3—Three different types of AA-size cells. At left, an older NiCd cell rated at about 600 mAh; in the middle, a newer NiCd cell with an 1100 mAh rating; at the right, an NiMH cell rated at 1600 mAh. Note that 1300 mAh is a "comfortable" rating for an NiMH cell of this size; higher capacity cells are readily available.

Figure 4—Removable Li-ion battery back for ICOM H-Ts.

Figure 5—Small Li-ion battery pack for the Yaesu VX-5R H-T.

thor has used one of these units over the past few years to power a Kenwood TM-V7A for an entire day of communications at Scouting events. Figure 6 shows different sizes of gel-cell lead acid batteries.

Reusable Alkaline

Before nickel cadmium cells were readily available, it was fashionable to "recharge" flashlight batteries by passing a very small current into them for a day or two. Devices to perform this recharging function were sold at novelty and specialty stores, and under certain circumstances a certain amount of energy could be restored to a discharged cell. Unfortunately, the amount of energy that could be recovered was nowhere near what a new cell from the dealer's shelf could deliver, and continued recharging could result in leakage of electrolyte into a flashlight or radio.

With the thrust to provide a cheaper cell than NiCd and the desire to give the consumer a cell that would provide the higher terminal voltage of the alkaline cell, the idea arose to return to the old flashlight battery recharger, and the reusable alkaline was born.

Reusable alkalines do not have a high cyclic life. Testing performed on these cells showed that after one initial discharge and recharge, the energy capability was down to only about 60 percent of the original capacity. Cyclic life is also highly dependent upon the depth of discharge. Only about 10 charge/discharge cycles can be expected if the cell is repeatedly discharged to depletion, more if the cell is only slightly discharged and then recharged.

A rechargeable alkaline's internal resistance is also higher than an equivalent regular alkaline cell, which limits the reusable cell's capability for high discharge current applications. This all but eliminates the reusable alkaline for most ham radio applications.

Self discharge, however, is excellent for these cells and is only about 0.3 percent per month. This makes them a good choice for emergency flashlights that are used for home power outages and other occasional purposes.

Comparison of Rechargeable Cell Types

Table 1 is a quick comparison of the capabilities of the most popular rechargeable cell types. Included in this chart is the popular "Gel Cell," a sealed lead-acid type.

Figure 7 is a graph showing the ability of cell types to provide high levels of discharge current versus the energy storage capacity of each cell. As can be noticed, although Li-ion is rated to have lots of capacity, this is not the case under high discharge conditions, such as during transmit mode in an H-T. Note that only NiCd, NiMH and Li-ion are depicted on the graph. Lead-acid and rechargeable alkaline are in classes by themselves.

Making an Intelligent Choice

Handheld Transceiver

The first step is to decide what is important to you. Do you want minimum battery weight and lots of power regardless of the cost, and are willing to sacrifice battery life? Or perhaps you are located in northern

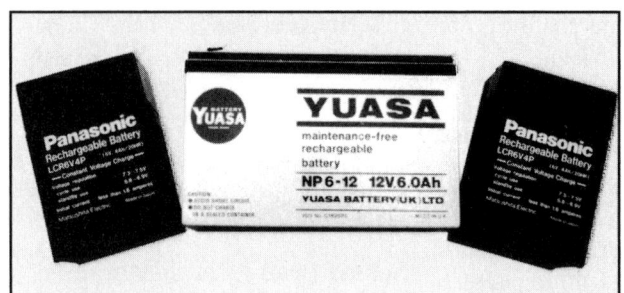

Figure 6— These gel-cell lead-acid batteries are rechargeable and won't leak.

Table 1

Comparison of Types of Cell Chemistries

	Nickel Cadmium	Nickel Metal Hydride	Sealed Lead Acid (Gel Cell)	Lithium-Ion	Reusable Alkaline
Energy density (watt-hours per kilogram)	40-60	60-80	37	100	80 (initial)
Cycle life	1500	500	200-300	500-1000	10
Self discharge, % per month	20	30	5	10	0.2
Maximum load current	Greater than 2C	0.5-1C	0.2C	Less than 1C	0.2C

Table Glossary
Energy density—Stored energy versus weight. The higher the number, the more total energy available.
Cycle life—The approximate theoretical number of charge/discharge cycles which the cell can sustain before its energy storage capacity degrades to a specific level (about 60%). Many factors influence this figure, including the depth of discharge, average temperature, etc.
Self discharge—The amount of stored energy lost per month with the cell lying unused.
Maximum load current—The amount of discharge current that the cell can provide without significant terminal voltage drop. This is an indicator of the cell's internal resistance. Note: "C" is the cell's ampere-hour rating which is stated by the manufacturer as a 10 hour discharge rate.

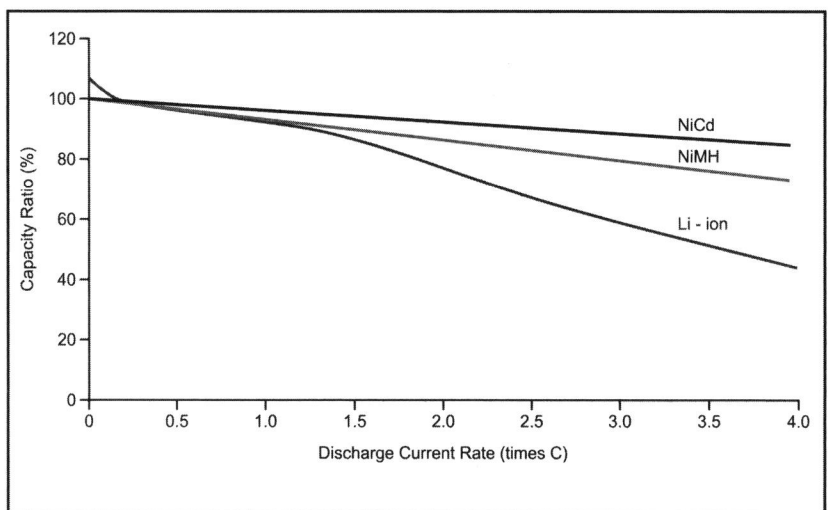

Figure 7—The graph shows capacity ratio percentage versus discharge current rate.

climates where you need a battery that will still pump out the watts even if the temperature is sub-zero. Maybe you need a battery that has the capability to sit in a ready state for many months, that you can just "grab and go." Or how about a battery that is a good compromise? Let's look at the options.

If you are an avid hiker, biker or camper, you will appreciate something which gives minimum size and weight without compromising performance. Lithium-ion is what you will want. With the highest energy density of all types of rechargeables, it will be comfortable to carry or pack. Expect to pay a higher price, however, for replacement batteries and a shortened lifetime. Lithium is also top choice for emergency standby use since it has a low self-discharge rate, and is therefore going to have more energy available when called upon for action after a lengthy period of sitting idle.

Nickel cadmium has the edge for cold weather operation as well as having the lowest internal resistance. It allows you to put out the maximum RF watts in the coldest extremes. Cost is also relatively low, and it has the highest charge/discharge cycle capability and lifetime of all types, meaning that it is a battery which will stay with you for a long time.

Emergency Shack Power

The leader here is still the lead acid. Whether you opt for the classic top-cap battery or the sealed gel cell type, the charging and maintenance is similar. If the battery is going to be inside the house, the sealed unit is the optimum choice due to its cleanliness and minimal gas evolution. Also, if it gets knocked over, there is no safety issue from spilled acid. Expect to pay more at the time of purchase, however. If the shack is basement or garage located, a deep cycle variety of marine or golf cart battery is possible. These are cheaper than the sealed variety and easier to find (see Figure 8).

Whichever type you choose, don't make the mistake of using a cheap automotive charger. Make sure that the charger is of the automatic variety, preferably one that has two or three charging states such as bulk charge, current limited, and float (by the way, a small 7-A power supply from Astron or similar manufacturer is good substitute for maintaining a charged battery, but it will not bring it to a fully charged state).

Getting the Most Life from your Battery

Nickel Cadmium and Nickel Metal Hydride

These cell types are so similar in chemistry that they can be considered together.

First, remember that both of these cell types tend to lose their stored energy quickly with time. NiCds should be recharged about once every two months, and NiMH cells about once every 4-6 weeks if they are to be kept in a ready state. Another thing is that these chemical powerhouses are like human muscles—both need exercise to retain their capability. If you use your H-T a lot, like every day or two, the battery is getting all the exercise it needs; but if your radio sits on the shelf unused, the battery can get lax and weak. If this is the case, fire up your H-T once a month and give it a day or two of good usage followed by a generous recharge afterwards. It's like a shot of vitamins.

Don't, however, allow the battery to fully discharge. To do so means that one or more cells will discharge first and will be pushed into a reverse charged state. When that happens, the cell can generate gas from the breakdown of electrolyte, which will vent into the air. The cell is robbed of some of its capability as a result and will be even more likely to reverse charge again. A basic rule of thumb is *never* let the battery discharge to the point where the "battery low" indicator comes on, since this indicates that one or more cells have already been subjected to reverse charging.

Sealed Lead Acid

The SLA cell has requirements that are different from the NiCd and NiMH cells. Whereas the NiCd and NiMH cells don't mind being in a partially charged state (or even fully discharged, as long as they haven't been reverse charged), the SLA must be kept in a near full charge condition continuously for best life and energy content. These batteries should be recharged frequently or kept on a float charge (note the reference to a small voltage regulated power supply in a preceding paragraph). The problem of plate sulphation, capable of destroying the unit, can occur if the battery becomes fully discharged for a length of time. Maintain your battery near full charge for the best life and service.

Lithium-ion

The lithium-ion is the closest of the group to being a no maintenance cell. The only real concern is overcharging, and that is usually prevented by a charge maintenance system either built into the battery pack or contained externally. Self discharge is also less than the other types, and a recharge once every two or three months should suffice.

I hope this article has provided a little insight into what's going on in that little package of power in your hand or on the shack floor. If you follow the tips on charging and maintenance, your portable equipment will be ready to serve you fully on a moment's notice.

Figure 8—This type of marine deep-cycle battery can be discharged hundreds of times.

A power systems design engineer for the last 40 years, Ken Stuart, W3VVN, has developed equipment for spacecraft and deep ocean environments as well as airborne and shipboard. He has served as ARRL Technical Advisor and lecturer on power supplies and batteries since 1980, and has held a ham license continuously since 1953. Ken presently works for Lockheed Martin in Baltimore. You can reach Ken at 1235 Hillcreek Rd, Pasadena, MD 21122, w3vvn@arrl.net.

By Mike Bryce, WB8VGE

The Micro M+ Charge Controller

Current capacity of up to 4 A, positive line switching so all grounds tie together, standby current of less than 1 mA and more features make the Micro M+ the ideal photovoltaic charge controller for use at home or in the field. It's an easy-to-build, one-evening project that just about anyone can master.

he Micro M proved a very popular project.[1] It seems hams really do like to operate their rigs from solar power while in the outback. Many hams find solar power to be very addictive. I had dozens of requests for information on how to increase the current capacity of the original Micro M controller. The original Micro

[1]Notes appear at end of article.

M would handle up to 2 A of current. The PC board traces and blocking diode limited the design to this current capacity. I also wanted to improve the performance of the Micro M while I was at it. Because the Micro M switched the negative lead of the solar panel on and off, the negative lead of the solar panel had to be insulated from the system ground. While that's not a problem with portable use, it may cause trouble with

a home station where all the grounds should be connected. Here's what I wanted to do:
• Reduce the standby current at night
• Increase current handling capacity to 4 A
• Change the charging scheme to high (positive) side switching
• Improve the charging algorithm
• Keep the size as small as possible, but large enough to build.

The Micro M+

I called the end result the Micro M+. You can assemble one in about an hour. Everything mounts on one double-sided PC board. It's small enough to mount inside your rig yet large enough so you won't misplace it. You can stuff four of them in your shirt pocket! And, you need not worry about RFI being generated by the Micro M+. It's completely silent and makes absolutely zero RFI!

The Micro M+ will handle up to 4 A of current from a solar panel. That's equal to a 75-W solar panel.[2] I've reduced the standby current to

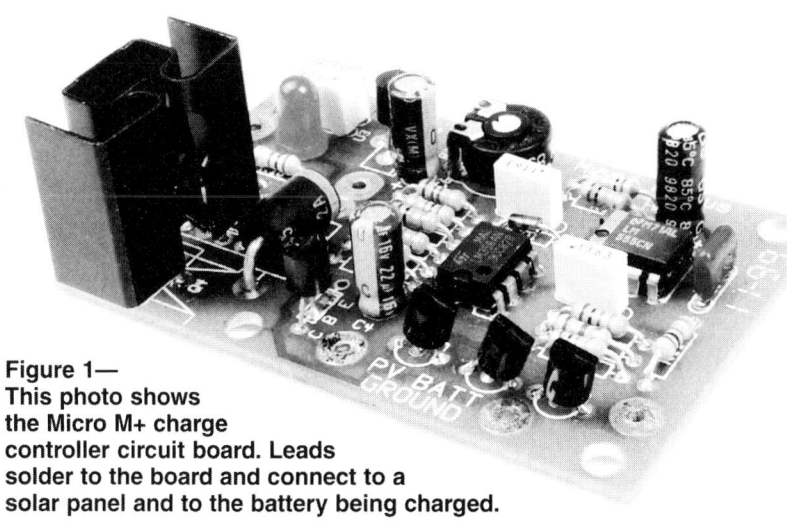

Figure 1—
This photo shows the Micro M+ charge controller circuit board. Leads solder to the board and connect to a solar panel and to the battery being charged.

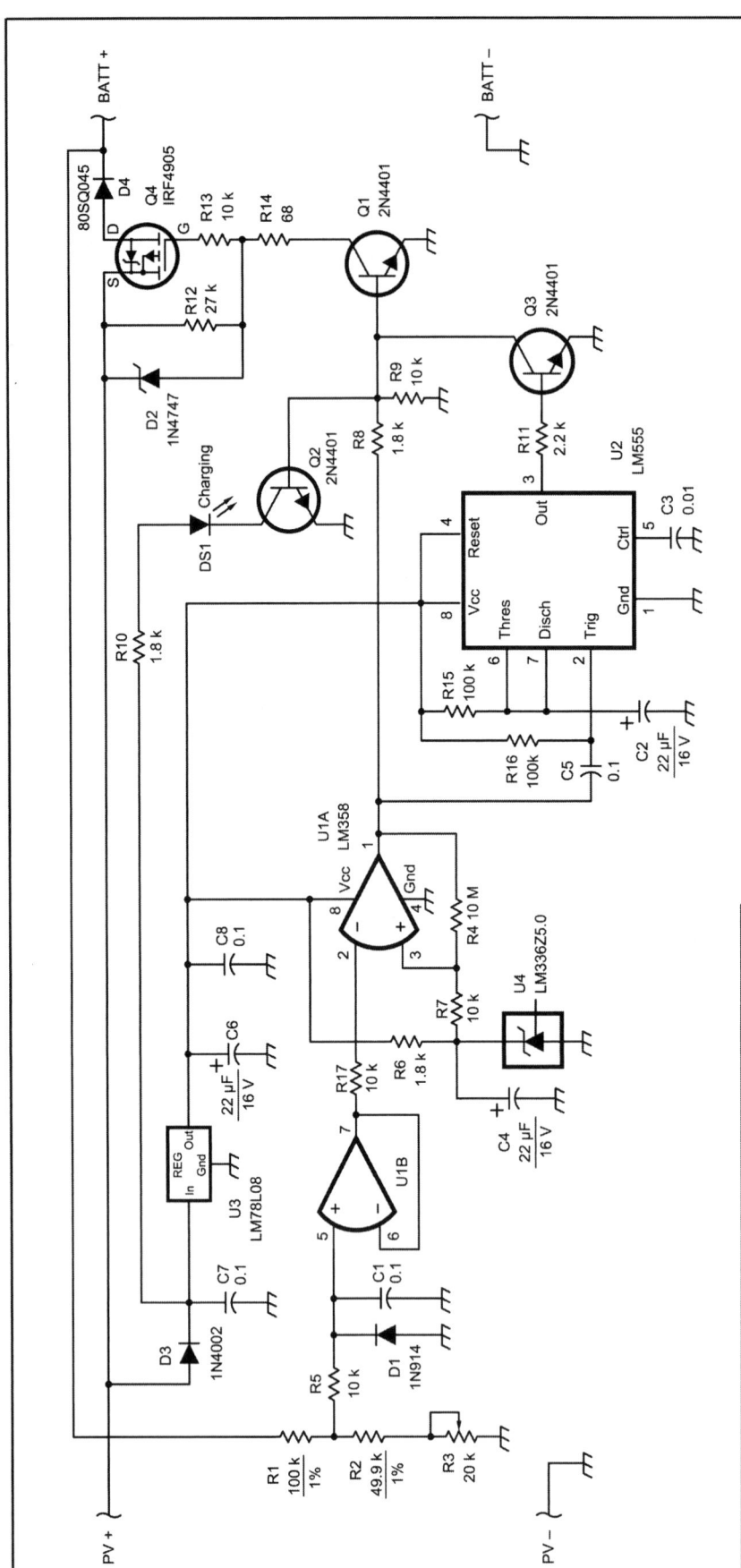

Figure 2—The schematic diagram of the Micro M+ charge controller.
C1, C5, C7, C8—0.1 µF.
C2, C4, C6—22-µF, 16-V electrolytic.
C3—0.01 µF.
D1—1N914, small signal silicon switching diode.
D2—1N4747, 20-V, 1-W Zener.
D3—1N4002, silicon rectifier diode.
D4—80SQ045, 45-V, 8-A Schottky diode.
DS1—LED, junkbox variety.
Q1, Q2, Q3—2N4401 NPN small-signal transistor (2N2222 or 2N3904 will also work).
Q4—IRF4905 P-channel MOSFET in TO-220 case. You will also need a small clip-on heat sink for this case.
R1—100 kΩ, 1%.
R2—49.9 kΩ, 1%.
R3—20-kΩ trimmer.
U1—LM358AN, dual op-amp.
U2—LM555AN timer.
U3—LM78L08, 8-V regulator.
U4—LM336Z-5.0, 5.0-V Zener diode in TO-92 case. The adjust terminal allows control of the temperature coefficient and voltage over a range. The adjust terminal is not used for the Micro M+.

less than 1 mA. I've also introduced a brand new charging algorithm to the Micro M+. All the current switching is done on the positive side. Now, you can connect the photovoltaic (PV) array, battery and load grounds together.

A complete kit of parts is available as well as just the PC board. The complete kit, including the PC board and all parts is $30.[3] The Micro M+ is easy to build, making it a perfect first-time project.

Here's How it Works

Figure 1 shows the complete Micro M+, while Figure 2 shows the schematic diagram. Let's begin with the current handling part of the Micro M+. Current from the solar panel is controlled by a power MOSFET. Instead of using a common N-channel MOSFET, however, the

Micro M+ uses an International Rectifier IRF4905 P-channel MOSFET. This P-channel FET has a current rating of 64 A with an RDS_{on} of 0.02 Ω. It comes in a TO-220 case. Current from the solar panel is routed directly to the MOSFET source lead.

N-channel power MOSFETs have very low RDS_{on} and even lower prices. To switch current on and off in a high side application, the gate of an N-channel MOSFET must be at least 10 volts higher than the rail it is switching. In a typical 12-volt system, the gate voltage must be at least 22 volts to ensure the MOSFET is turned completely on. If the gate voltage is less than that required to fully enhance the MOSFET, it will be almost on and somewhat off (the MOSFET is operating in its linear region). The device will be destroyed at high current.

To produce this higher gate voltage, some sort of oscillator typically is used to charge up a capacitor via a voltage doubler. This charge pump generates harmonics that may ride on the dc flowing into the battery under charge. Normally, this would not cause any problem, and in most cases, a filter or two on the dc bus will eliminate most of the harmonics generated. Even the best filter won't get rid of all the harmonics, however. To compound the problem, long wire runs to and from the solar panels and batteries act like antennas.

The P-channel MOSFET eliminates the need for a charge pump altogether. To turn on a P-channel MOSFET, all we have to do is pull the gate lead to ground! Since the Micro M+ does not have a charge pump, it generates *no RFI*!

Now, you may be wondering, if the P-channel MOSFET is so great, why have you not seen them in applications like this before? The answer is twofold. First, the RDS_{on} of a P-channel MOSFET has always been much higher than its N-channel cousin. Several years ago, a P-channel MOSFET with an RDS_{on} of 0.12 ohms was considered very low. At that time an N-channel MOSFET had an RDS_{on} of 0.009 ohms. Suppose you want to control 10 A of current from your solar panel. Using the N-channel MOSFET above we find the MOSFET will dissipate less than a watt of power. On the other hand, the P-channel MOSFET will dissipate 12 W of power! Current generated by our solar panels is way too expensive to have 12 W of it go up as heat from the charge controller.

The second factor was price. The P-channel MOSFET I described above would have sold for $19 each. The N-channel would have been a few dollars.

The Micro M+ never draws current from the battery. The solar panel provides all the power the micro M+ needs.

In the last year or so the RDS_{on} of the P-channel MOSFET has fallen to 0.028 ohms. The price, while still a bit on the steep side, has dropped to about $8 each.

With the P-channel MOSFET controlling the current, diode D4—an 80SQ045 Schottky—prevents current from the battery from flowing into the solar panel at night. This diode also provides reverse polarity protection to the battery in the event you connect the solar panel backwards. This protects the expensive P-channel MOSFET.

Zener diode D2, a 1N4747, protects the gate from damage due to spikes on the PV line. Resistor R12 pulls the gate up, ensuring the power MOSFET is off when it is supposed to be.

The Micro M+ Likes to Sleep

The Micro M+ never draws current from the battery. The solar panel provides all the power the Micro M+ needs. At night, the Micro M+ goes to sleep. When the sun rises, the Micro M+ will start up again. As soon as the solar panel is producing enough current and voltage to start charging the battery, the Micro M+ will pass current into the battery.

To reduce the amount of standby current, diode D3 passes current from the solar panel to U3, the voltage regulator. U3, an LM78L08 regulator, provides a steady + 8 V to the Micro M+ controller. Bypass capacitors C6, C7 and C8 are used to keep everything happy. As long as the solar panel is producing power, the Micro M+ will be awake. At sundown, the Micro M+ will go to sleep. Sleep current is on the order of less than 1 mA.

Using the Micro M+ with the Yaesu FT-817

With the introduction of the new Yaesu FT-817 all mode, all band QRP transceiver, more and more of us will be using solar power in the field. The Micro M+ was designed to use a 12-V solar panel to charge a 12-V battery. The Yaesu FT-817 can operate from 12 V supplied externally or from an internal 9.6-V NiCd battery. The NiCd battery may be charged when the battery is installed in the radio. Or, if you want, it can be charged separately from the 817 via a solar panel and the Micro M+ controller.

To use the Micro M+ to charge this NiCd pack, you'll have to change the value of resistor R2 from 49.9 kW 1% to 82.5 kW 1%. This will allow the logic to switch correctly at 11.6 V, the voltage of a fully charged 9.6-V NiCd battery. This assumes you use the standard of 1.45 V per NiCd cell. With the new value for R2, there's plenty of adjustment in the state-of-charge trimmer to allow you to fine-tune the state-of-charge.

Since the NiCd battery is rated at only 9.6 V, this throws the power point of the solar panel in the trash. A typical 5-W solar panel is rated at 290 mA at 17.1 V. Because of the lower battery voltage, there will be more than the 290 mA of current flowing. However, if the panel is designed to produce 5 W, that's all it will do. As the voltage goes down, the current will increase, up to the Isc (current short circuit) of the panel. The panel will not produce any more current than it was designed for.

Battery Sensing

The battery terminal voltage is divided down to a more usable level by resistors R1, R2 and R3. Resistor R3, a 20-kΩ trimmer, sets the state-of-charge for the Micro M+. A filter consisting of R5 and C1 helps keep the input clean from noise picked up by the wires to and from the solar panel. Diode D1 protects the input of the op-amp in the event the battery sense line were connected backward.

An LM358 dual op-amp is used in the Micro M+. One section, U1B, buffers the divided battery voltage before passing it along to the voltage comparator, U1A. Here the battery sense voltage is compared to the reference voltage supplied by U4. U4 is an LM336Z-5.0 precision diode. To prevent U1A from oscillating, a 10-MW resistor is used to eliminate any hysteresis.

As long as the battery under charge is below the reference point, the output of U1A will be high. This saturates transistors Q1 and Q2. Transistor Q2 conducts and lights LED DS1, our CHARGING LED. Q1, also fully saturated, pulls the gate of the P-channel MOSFET to ground. This effectively turns on the FET and current flows from the solar panel into the battery via D4.

As the battery begins to take up the charge, its terminal voltage will increase. When the battery reaches the state-of-charge set point, the output of U1A goes low. With Q1 and Q2 now off, the P-channel MOSFET is turned off, stopping all current into the battery. With Q2 off, the CHARGING LED goes dark.

Since we have basically eliminated any hysteresis in U1A, as soon as the current stops, the output of U1A pops back up high again. Why? Because the battery terminal voltage will fall back down as the charging current is removed. If left like this, the Micro M+ would sit and oscillate at the state-of-charge set point.

To prevent that from happening, an LM555 timer chip, U2, monitors the output of U1A. As soon as the output of U1A goes low, this low trips U2. The output of U2 goes high, fully saturating transistor Q3. With Q3 turned on, it pulls the base of Q1 and Q2 low. Since both Q2 and Q1 are now deprived of base current, they remain off.

With the values shown for R15 and C2, charging current is stopped for about four seconds after the state-of-charge has been reached.

After the four second delay, Q1 and Q2 are allowed to have base drive from U1A. This lights up the charging LED and allows Q4 to pass current once more to the battery.

As soon as the battery hits the state-of-charge once more, the process is repeated. As the battery becomes fully charged, the "on" time will shorten up while the "off" time will always remain the same four seconds. In effect, a pulse of current will be sent to the battery that will shorten over time. I call this charging algorithm "Pulse Time Modulation."

As a side benefit of the pulse time modulation, the Micro M+ won't go nuts if you put a large solar panel onto a small battery. The charging algorithm will always keep the off time at four seconds allowing the battery time to rest before being hit by higher current than normal for its capacity.

Building Your Own Micro M+

There's nothing special about the circuit. The use of a PC board makes the assembly of the Micro M+ quick and easy. It also makes it much easier if you need to troubleshoot the circuit. You can build the entire circuit on a piece of perf-board if you want.

The power MOSFET must be protected against static discharges. A dash of common sense and standard MOSFET handling procedures will work best. Don't handle the MOSFET until you need to install it in the circuit. A wrist strap would be a good idea to prevent static damage. Once installed in the PC board, the device is quite robust.

A small clip-on heat sink is used for the power MOSFET. If you desired, the MOSFET could be mounted to a metal chassis. If you do this, make sure you insulate the MOSFET tab from the chassis.

If you plan on using the Micro M+ outside, then consider soldering the IC directly onto the board. I've found that cheap solder-plated IC sockets corrode. If you want to use an IC socket, use one with gold-plated contacts.

Feel free to substitute part values. There's nothing really critical. I do suggest you stick with 1% resistors for both R1 and R2. This isn't so much for the close tolerance, but for the 50-PPM temperature compensation they have. You can use standard off-the-shelf parts for either or both R1 and R2, but the entire circuit should then be located in an environment with a stable temperature.

Adjustments

You'll need a good digital voltmeter and a variable power supply. Set the power supply to 14.3 V. Connect the battery negative and power supply negative leads together at a circuit-board ground point. Connect the PV positive and battery positive lead, and the power supply positive leads together. The charging LED should be on. If not, adjust trimmer R3 until it comes on. Check for +8 V at the V_{cc} pins of the LM358 and the LM555. You should also see + 5 V from the LM336Z5.0 diode.

Quickly move the trimmer from one end of its travel to the other. At one point the LED will go dark. This is the switch point. To verify that the "off pulse" is working, as soon as the LED goes dark quickly reverse the direction of the trimmer. The LED should remain off for several seconds and then come back on. If everything seems to be working, it's time to set the state-of-charge trimmer.

Now, slowly adjust the trimmer until the LED goes dark. You might want to try this adjustment more than once as the closer you get the comparator to switch at exactly 14.3 V, the more accurate the Micro M+ will be. Here's a hint I've learned after adjusting hundreds of Micro M+ controllers. Set the power supply to slightly above the cutoff voltage you want. If you want 14.3 V, then set the supply to 14.5 V. I've found that in the time it takes to react to the LED going dark, you overshoot the cutoff point. Setting the supply higher takes this into account and usually you can get the trimmer set to exactly what you need in one try. That's all you need to do. Disconnect the supply from the Micro M+ and you're ready for the solar panel.

Odds and Ends

The 14.3-V terminal voltage will be correct for just about all sealed and flooded cell lead-acid batteries. You can change the state-of-charge set point if you want to recharge NiCds or captive sealed lead-acid batteries.

Keep the current from the solar panel within reason for the size of the battery you're going to be using. If you have a 7-amp hour battery, then don't use a 75-W solar panel. You'll get much better results and smoother operation.

The tab of the power MOSFET is electrically hot. If you plan on using the Micro M+ without a protective case, make sure you insulate the tab from the heat sink. A misplaced wire touching the heat sink could cause real damage to both the Micro M+ and your equipment. A small plastic box from RadioShack works great.

More Current?

Well yes, you can get the Micro M+ to handle more current. You must increase the capacity of the blocking diode and mount the power MOSFET on a larger heat sink. I've used an MBR2025 diode and a large heat sink for the MOSFET and can easily control 12 A of current.

Battery Charging Without a Solar Panel?

Yes, that's possible, too. The trick is to use a power supply for which you can limit the output current. A discharged lead acid battery will draw all the current it can from the charging source. In a solar panel setup, if the panel produces 3 A, that's all it will do. With an ac powered supply, the current can be excessive. To use the Micro M+ with an ac powered supply, set the voltage to 15.5 V. Then limit the current to 2 or 3 A.

No matter if you're camping in the outback, or storing photons just in case of an emergency, the Micro M+ will provide your battery with the fullest charge. The Micro M+ is simple to use and completely silent. Just like the sun!

Notes

[1]"The Micro M," Sep 1996 *QST*, p 41.
[2]A 75-W module produces 4.4 A at 17 V. The Micro M+ can easily handle the extra 400 mA.
[3]A complete kit of parts is available from SunLight Energy Systems, 955 Manchester Ave SW, North Lawrence, OH 44666. A complete kit including all parts and PC board is $30 plus $4 US Priority mail. Visa, MasterCard accepted. Tel 330-832-3114; **www.seslogic.com/**.

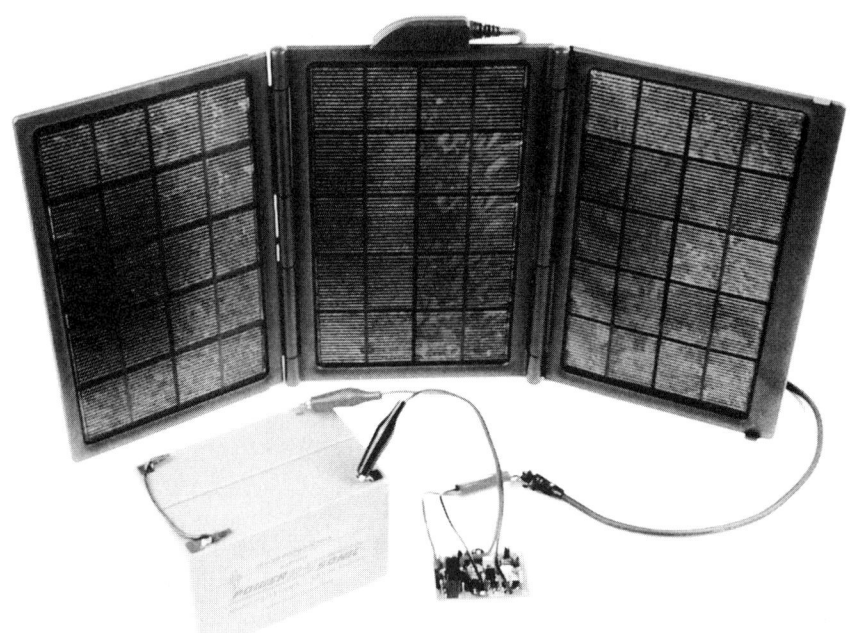

The Micro M+ Charge Controller board, small enough to mount inside your rig, is shown connected to a solar panel and a rechargeable battery.

By Bob Lewis, AA4PB

An Automatic Sealed-Lead-Acid Battery Charger

This nifty charger is just what you need to keep your SLA batteries up to snuff!

Photos by Joe Bottiglieri, AA1GW

After experiencing premature failure of the battery in my Elecraft K2 transceiver (most likely because I forgot to keep the battery on a regular charge schedule), I began searching for an *automatic* battery charger.[1,2] The K2 uses a Power-Sonic PS-1229A 12-V, 2.9-Ah sealed lead-acid (SLA) battery. SLAs are commonly called *gel-cells* because of their gelled electrolyte. As with all things, to obtain maximum service life from an SLA battery, it needs to be treated with a certain degree of care. SLA batteries must be recharged on a regular basis; they should not be undercharged or overcharged. If an SLA battery is left unused, it will gradually self-discharge.

Although my SLA battery experiences related here are linked to my K2 transceiver, you can think of the K2 simply as a load for the battery. The comments pertaining to the SLA batteries and chargers apply across the board and the charger described here can be used with any similar battery.

Using a Three-Mode Charger

My first attempt at keeping my K2's SLA battery healthy was to purchase an automatic three-mode charger. I soon discovered that most three-mode chargers work by sensing

[1] Notes appear at end of article.

current and were never intended to charge a battery under load.

Three-mode chargers begin the battery charging process by applying a voltage to the battery through a 500-mA current limiter. This stage is known as *bulk-mode* charging. As the battery charges, its voltage begins to climb. When the battery voltage reaches 14.6 V, the charger maintains the voltage at that level and monitors the battery charging current. This is known as the *absorption mode*, sometimes called the *overcharge mode*. By this time, the battery has achieved 85% to 95% of its full charge. As the battery continues to charge—with the voltage held constant at 14.6 V—the charging current begins to drop. When the charging current falls to 30 mA, the three-mode charger switches to *float mode* and lowers the applied voltage to 13.8 V. At 13.8 V, the battery becomes self-limiting, drawing only enough current to offset its normal

self-discharge rate. This works great æuntil you attach a light load to the battery, such as turning on the K2 receiver. The K2 receiver normally draws about 220 mA. When the charger detects a load current above 30 mA, it's fooled into thinking that the battery needs charging, so it reverts to the absorption mode, applying 14.6 V to the battery. If left in this condition, the battery is overcharged, shortening its service life.

UC3906-IC Chargers

Chargers using the UC3906 SLA charge-controller IC work just like the three-mode charger described earlier except that their return from float mode to absorption mode is based on voltage rather than current. Typically, once the charger is in float mode it won't return to absorption mode until the battery voltage drops to 10% of the float-mode voltage (or about 12.4 V). Although this is an improvement over the three-mode

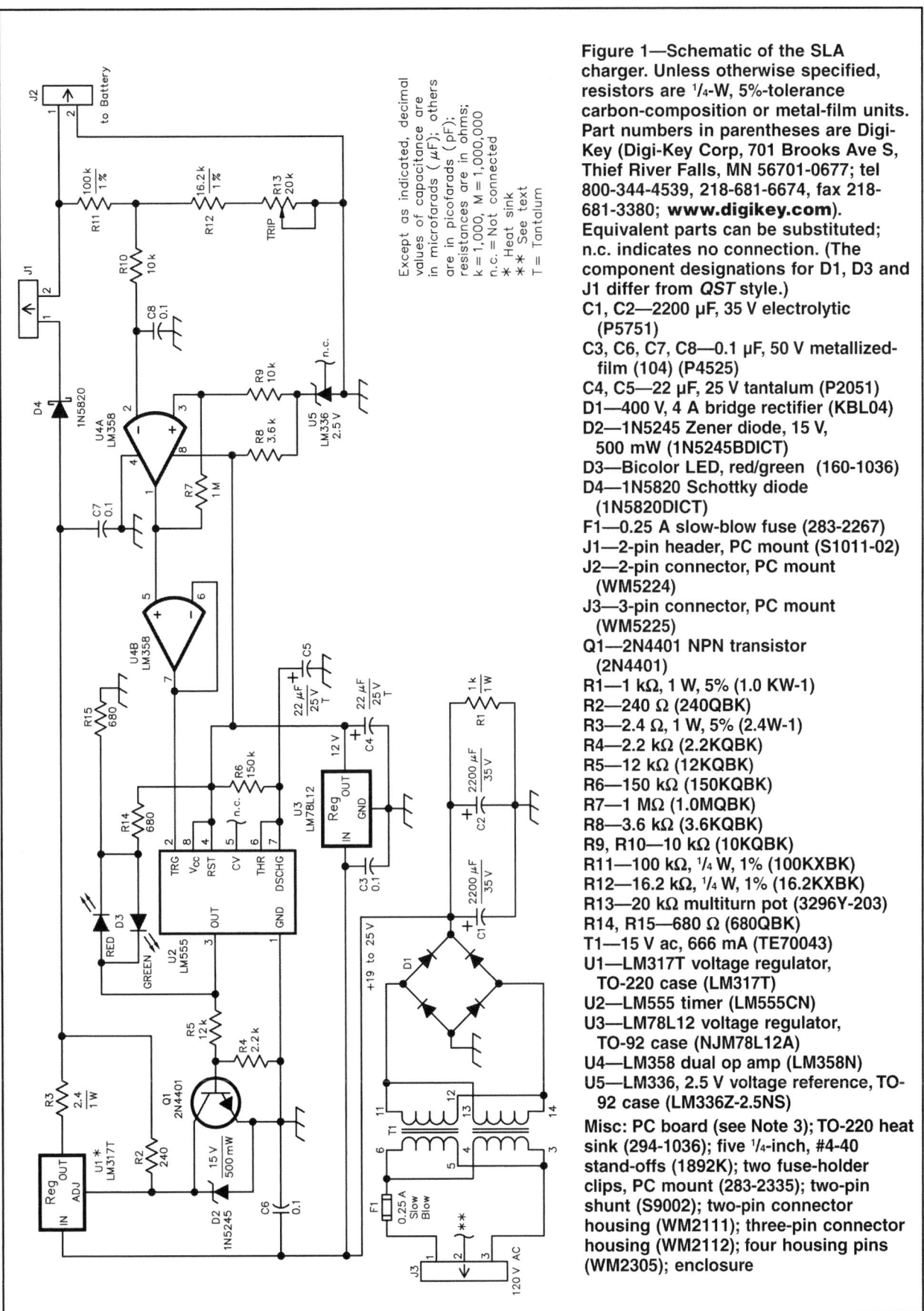

Figure 1—Schematic of the SLA charger. Unless otherwise specified, resistors are 1/4-W, 5%-tolerance carbon-composition or metal-film units. Part numbers in parentheses are Digi-Key (Digi-Key Corp, 701 Brooks Ave S, Thief River Falls, MN 56701-0677; tel 800-344-4539, 218-681-6674, fax 218-681-3380; www.digikey.com). Equivalent parts can be substituted; n.c. indicates no connection. (The component designations for D1, D3 and J1 differ from *QST* style.)

C1, C2—2200 µF, 35 V electrolytic (P5751)
C3, C6, C7, C8—0.1 µF, 50 V metallized-film (104) (P4525)
C4, C5—22 µF, 25 V tantalum (P2051)
D1—400 V, 4 A bridge rectifier (KBL04)
D2—1N5245 Zener diode, 15 V, 500 mW (1N5245BDICT)
D3—Bicolor LED, red/green (160-1036)
D4—1N5820 Schottky diode (1N5820DICT)
F1—0.25 A slow-blow fuse (283-2267)
J1—2-pin header, PC mount (S1011-02)
J2—2-pin connector, PC mount (WM5224)
J3—3-pin connector, PC mount (WM5225)
Q1—2N4401 NPN transistor (2N4401)
R1—1 kΩ, 1 W, 5% (1.0 KW-1)
R2—240 Ω (240QBK)
R3—2.4 Ω, 1 W, 5% (2.4W-1)
R4—2.2 kΩ (2.2KQBK)
R5—12 kΩ (12KQBK)
R6—150 kΩ (150KQBK)
R7—1 MΩ (1.0MQBK)
R8—3.6 kΩ (3.6KQBK)
R9, R10—10 kΩ (10KQBK)
R11—100 kΩ, 1/4 W, 1% (100KXBK)
R12—16.2 kΩ, 1/4 W, 1% (16.2KXBK)
R13—20 kΩ multiturn pot (3296Y-203)
R14, R15—680 Ω (680QBK)
T1—15 V ac, 666 mA (TE70043)
U1—LM317T voltage regulator, TO-220 case (LM317T)
U2—LM555 timer (LM555CN)
U3—LM78L12 voltage regulator, TO-92 case (NJM78L12A)
U4—LM358 dual op amp (LM358N)
U5—LM336, 2.5 V voltage reference, TO-92 case (LM336Z-2.5NS)

Misc: PC board (see Note 3); TO-220 heat sink (294-1036); five 1/4-inch, #4-40 stand-offs (1892K); two fuse-holder clips, PC mount (283-2335); two-pin shunt (S9002); two-pin connector housing (WM2111); three-pin connector housing (WM2112); four housing pins (WM2305); enclosure charger, it still has the potential for overcharging a battery to which a light load is attached.

First, let's look at the situation where a UC3906-controlled charger is in absorption mode and you turn on the K2 receiver, applying a load. The battery is fully charged, but because the load is drawing 220 mA, the charging current never drops to 30 mA and the charger remains in absorption mode, thinking that it is the battery that is asking for the current. As with the three-mode charger, the battery is subject to being overcharged.

If we remove the load by turning off the K2, the current demand drops below 30 mA and the charger switches to float mode (13.8 V). When the K2 is turned on again, because the charger is able to supply the 220 mA for the receiver, the battery voltage doesn't drop, so the charger stays in float mode and all is well. However, if the transmitter is keyed (increasing the current demand), the charger can't supply the required current, so it's taken from the battery and the battery voltage begins to drop. If we unkey the transmitter before the battery voltage reaches 12.4 V, the charger stays in float mode. Now it takes much longer for the charger to supply the battery with the power used during transmit than it would have if the charger had switched to absorption mode.

Let's key the transmitter again, but this time keep it keyed until the battery voltage drops below 12.4 V. At this point, the charger switches to the absorption mode. When we unkey the transmitter, we're back to the situation where the charger is locked in absorption mode until we turn off the receiver.

Why Worry?

So, why this concern about overcharging an SLA battery? At 13.8 V, the battery self-limits, drawing only enough current to offset its self-discharge rate (typically about 0.001 times the battery capacity, or 2.9 mA for a 2.9 Ah battery). An SLA battery can be left in this float-charge condition indefinitely without overcharging it. At 14.6 V, the battery takes more current than it needs to offset the self-discharge. Under this condition, oxygen and hydrogen are generated faster than they can be recombined, so pressure inside the battery increases. Plastic-cased SLA batteries such as the PS-1229A have a one-way vent that opens at a couple of pounds per square inch pressure (PSI) and release the gases into the atmosphere. This results in drying the gelled electrolyte and shortening the battery's service life. Both undercharging and overcharging need to be avoided if we want to get maximum service life from the battery.

Continuing to apply 14.6 V to a 12-V SLA battery represents a relatively minor amount of overcharge and results in a gradual deterioration of the battery. Applying a potential of 16 V or excessive bulk-charging current to a small SLA battery from an uncontrolled solar panel can result in serious overcharging. Under these conditions, the overcharging can cause the battery to overheat, which causes it to draw more current and result in thermal runaway, a condition that can warp electrodes and render a battery useless in a few hours. To prevent thermal runaway, the maximum current and the maximum voltage need to be limited to the battery manufacturer's specifications.

Design Decision

To avoid the potential of overcharging a battery with an automatic charger locked up by the load, I decided to design my own charger, one that senses battery voltage rather than current in order to select the proper charging rate. A 500-mA current limiter sets the maximum bulk rate charge to protect the battery and the charger's internal power supply. Like the three-mode chargers, when a battery with a low terminal voltage is first connected to the charger, a constant current of 500 mA flows to the battery. As the battery charges, its voltage begins to climb. When the battery voltage reaches 14.5 V, the charger switches off. With no charge current flowing to the battery, its voltage now begins to drop. When the current has been off for four seconds, the charger reads the battery voltage. If the potential is 13.8 V or less, the charger switches back on. If the voltage is still above 13.8 V, the charger waits until it drops to 13.8 V before turning on. The result is a series of 500-mA current pulses varying in width and duty cycle to provide an average current just high enough to maintain the battery in a fully charged condition. Because the repetition rate is very low (a maximum of one current pulse every four seconds) no RFI is generated that could be picked up by the K2 receiver. Because the K2's critical circuits are all well regulated, slowly cycling the battery voltage between 13.8 V and 14.5 V has no ill effects on the transmitted or received signals.

As the battery continues to charge, the pulses get narrower and the time between pulses increases (a lower duty cycle). Now when the K2 receiver is turned on and begins drawing 220 mA from the battery, the battery voltage drops more quickly so the pulses widen (the duty cycle increases) to supply a higher average current to the battery and make up for that taken by the receiver. When the K2 transmitter is keyed, it draws about 2 to 3 A from the battery. Because the charger is current limited to 500 mA, it is not able to keep up with the transmitter demands. The battery voltage drops and the charger supplies a constant 500 mA. The battery voltage continues to drop as it supplies the required transmit current. When the transmitter is unkeyed, the battery voltage again begins to rise as the charger replenishes the energy used during transmit. After a short time, (depending on how long the transmitter was keyed) the battery voltage reaches 14.5 V and the pulsing begins again. The charger is now fully automatic, maintaining the battery in a charged condition and adjusting to varying load conditions.

The great thing about this charging system is that during transmit the majority of the required 2 to 3 A is taken from the battery. When you switch back to receive, the charger is able to supply the 220 mA needed to run the receiver and deliver up to 280 mA to the battery to replenish what was used during transmit. This means that the power source need only supply the average energy used over time, rather than being required to supply the peak energy needed by the transmitter. (You don't need to carry a heavy 3-A regulated power supply with your K2.) As long as you don't transmit more than about 9% of the time, this system should be able to power a K2 indefinitely.

Have you ever noticed that sometimes when your H-T has a low battery and you drop it into its charger you hear hum on the received signals? This charger's power supply is well filtered to ensure that there is no ripple or ac hum to get into the K2 under low battery voltage conditions.

Circuit Description

The charger schematic is shown in Figure 1. I've dubbed the charger the PCR12-500A, short for Pulsed-Charge Regulator for 12-V SLA batteries with maximum bulk charge rates of 500 mA. U1, an LM317 three-terminal voltage regulator, is used as a current limiter, voltage regulator and charge-control switch. A 15-V Zener diode (D2) sets U1 to deliver a no-load output of 16.2 V. R3 sets U1 to limit the charging current to 500 mA. When Q1 is turned on by the LM555 timer (U2), the **ADJ** pin of U1 is pulled to ground, lowering its output voltage to 1.2 V. D4 effectively disconnects the battery by preventing battery current from flowing back into U1. A Schottky diode is used at D4 because of its low voltage drop (0.4 V).

An LM358 (U4A) operates as a voltage comparator. U5, an LM336, provides a 2.5-V reference to the positive input (pin 3) of U4. R11, R12 and R13 function as a voltage divider to supply a portion of the battery voltage to pin 2 of U4A. R13 is adjusted so that when the battery terminal voltage reaches 14.5 V, the negative input of U4A rises slightly above the 2.5-V reference and its output switches from +12 V to 0 V. When this happens, the 1-MW resistor (R7) causes the reference voltage to drop a little and provide some hysteresis. The battery voltage must now drop to approximately 13.8 V before U4A turns back on.

U4B is a voltage follower. It pulls the trigger input (pin 2) of U2 to 0 V, causing its output to go to 12 V. U4B's output remains at 12 V until C5 has charged through R6 (approximately four seconds) *and* the trigger has been released by U4A sensing the battery dropping to 13.8 V or less. While the output of U2 is at 12 V, emitter/base current for Q1 flows via R5 and Q1's collector pulls U1's **ADJ** pin to ground, turning off the charging current.

The output of U2 also provides either +12 V or 0 V to the bicolor LED, D3. R14 and R15 form a voltage divider to provide a reference voltage to D3 such that D3 glows red when U2's output is +12 V and green when U2's output is at 0 V. When ac power is applied but U1 is switched off and not supplying current to the battery, D3 glows red. When U1 is on and supplying current to the battery, D3 is green. As the battery reaches full charge, D3 blinks green at about a four-second rate. As the battery charge increases, the *on* time of the green LED decreases and the *off* time increases. A fully charged battery may show green pulses as short as a half-second and the time between pulses may be 60 seconds or more.

T1, D1, C1 and C2 form a standard full-wave-bridge power supply providing an unregulated 20 V dc at 500 mA. U3, an LM78L12 three-terminal regulator, provides a regulated 12-V source for the control circuits.

Note that the mounting tab on U1 is not at ground potential. U1 should be mounted to a heat sink with suitable electrically insulated but thermally conductive mounting hardware to avoid short circuits. Suitable mounting hardware is included with the PC board (see Note 4).

Other Bulk-Charge Rates

The maximum bulk-charge rate is set by the value of R3 in the series regulator circuit. The formula used to determine the value of this resistor is $R_{ohms} = 1200 / I_{mA}$. T1 must be capable of supplying the bulk charge current and U1 must be rated to handle this current. The LM317T used here is rated for a maximum current of 1.5 A *provided* it has a heat sink sufficiently large enough to dissipate the generated heat. If you increase the bulk-charge rate, you'll definitely need to increase the size of the on-board heat sink. Mounting U1 directly to the housing (be sure to use an insulator) may be a good option.

Transformer Substitution

I selected T1 because of its small size and PC-board mounting. You can substitute any transformer rated at 15 or 16 V ac (RMS) at 500 mA or more. You may find common frame transformers to be more readily available. You can mount such a transformer to an enclosure wall and route the transformer leads to the appropriate PC-board holes.

Construction

There is nothing critical about building this charger. You can assemble it on a prototyping board, but a PC board and heat sink are available.[4] The specially ordered heat sink supplied with the PC board is 1/4-inch higher than the one identified in the parts list and results in slightly cooler operation of U1. The remaining parts

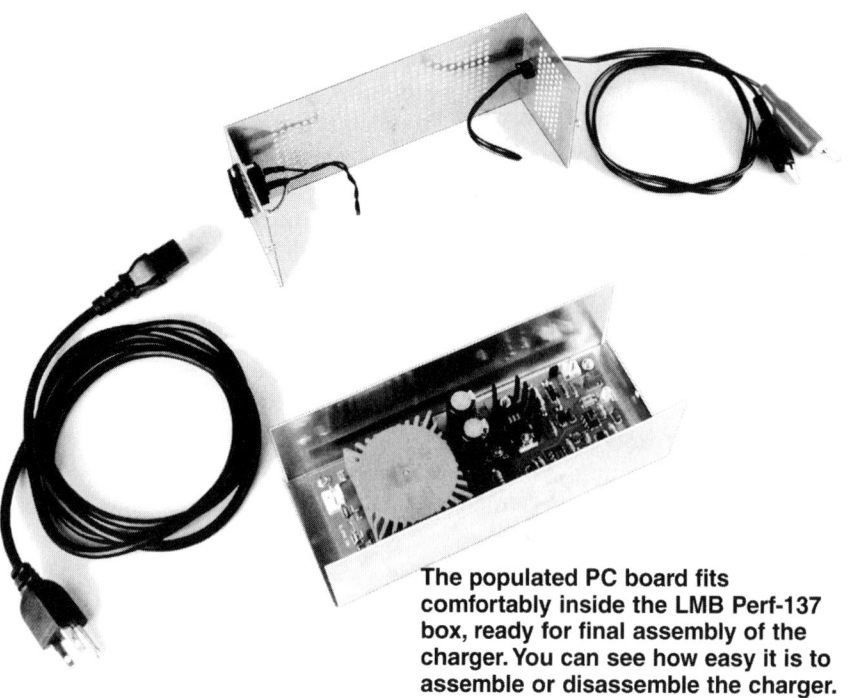

The populated PC board fits comfortably inside the LMB Perf-137 box, ready for final assembly of the charger. You can see how easy it is to assemble or disassemble the charger.

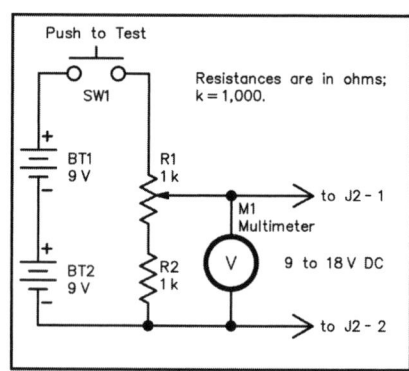

Figure 2—Test voltage source for the battery charger. (The component designation for the push-button switch differs from *QST* style.)

are available from Digi-Key.

Be sure to space R1 and R3 away from the board by $1/4$ inch or so to provide proper cooling. R13 can be a single-turn or a multiturn pot. You'll probably find a multiturn pot makes it easier to set the cutoff voltage to exactly 14.5 V.

R13 Adjustment

To check for proper operation and to set the trip point to 14.5 V dc, we need a test-voltage source variable from 12 to 15 V dc. A convenient means of obtaining this test voltage is to connect two 9-V transistor-radio batteries in series to supply 18 V as shown in Figure 2. Connect a 1-kΩ resistor (R2) in series with a 1-kΩ potentiometer (R1) and connect this series load across the series batteries with the fixed-value resistor to the negative lead. The voltage at the pot arm should now be adjustable from 9 to 18 V. During the following procedure, be sure to adjust the voltage *with the test supply connected to the charger at J2* because the charger loads the test-voltage supply and causes the voltage to drop a little when it's connected.

Remove the jumper at J1 and apply ac line voltage to the unit at J3. Turn R13 fully counterclockwise. D3 should glow green. Connect the test voltage to J2 and adjust R1 of Figure 2 for an output of 14.5 V. Slowly adjust R13 clockwise until D3 glows red. To test the circuit, wait at least four seconds, then gradually reduce the test voltage until D3 turns green. At that point, the test voltage should be approximately 13.8 V. Slowly increase the test voltage again until D3 turns red. The test voltage should now read 14.5 V. If it is not exactly 14.5 V, make a minor adjustment to R13 and try again. The aim of this adjustment is to have D3 glow red just as the test voltage reaches 14.5 V.

To test the timer functioning, remove the test voltage from J2 and set it for about 15 V. Momentarily apply the test voltage to J2. D3 should turn red for approximately four seconds, then turn green. The regulator is now calibrated and ready for operation. Remove the test voltage and ac power and install the jumper at J1.

A Suitable Enclosure

I used an 8×3×2.75-inch LMB Perf-137 box (Digi-Key L171-ND) to house the charger. An alternative enclosure is the Bud CU482A Convertabox, which measures 8×4×2 inches (available from Mouser). If you use the Convertabox, be sure to add some ventilation holes directly above the board-mounted heat sink. The LMB Perf box comes with a ventilated cover. If you are inclined to do some metal work, you could build your own enclosure using aluminum angle stock and sheet and probably reduce the size to perhaps 8×3×2 inches. If you use a PC-board-mounted power transformer, watch out for potential shorts between the transformer pins (especially the 120-V ac-line pins) and the case. If you use a metal enclosure, connect the safety ground (green) wire of the ac-line cord directly to the case.

Operation

It is very important that this charger be connected *directly* to the SLA battery with no diodes, resistors or other electronics in between the two. The charger works by reading the battery voltage, so any voltage drop across an external series component results in an incorrect reading and improper charging. For example, the Elecraft K2 has internal diodes in the power-input circuit, so it's necessary to add a charging jack to the transceiver that provides a direct connection to the battery. Now I can leave my K2 connected to the charger at all times and be assured that its internal battery is fully charged and ready to go at a moment's notice.

Notes

[1] Larry Wolfgang, WR1B, "Elecraft K2 HF Transceiver Kit," Product Review, *QST*, Mar 2000, pp 69-74.
[2] Although this charger was designed specifically for use with the Power-Sonic PS-1229A SLA battery used in the Elecraft K2 transceiver, its design concepts have wide ranging applications for battery operated QRP rigs of all types.
[3] Although it's labeled a 12-V battery, the terminal voltage is nominally 13.8 V with no load.
[4] A PC Board (double sided, plated through holes, solder masked and silk screened) and heat sink are available from Intelligent Software Solutions, PO Box 522, Garrisonville, VA 22463-0522. Price: $18 plus $1.50 shipping in the US and Canada.

Bob Lewis, AA4PB, became interested in Amateur Radio during junior high school in the late '50s. With the encouragement of his cousin, Al Krugler, K8DDX, Bob obtained his Technician license (K8KNI) and spent most of his time on 6-meter AM in the Detroit, Michigan area. His early interest in Amateur Radio resulted in a career in electronics, first as a radio mechanic in the air-transport industry, followed by ten years in the Navy as an aviation electronics technician. While in the Navy, Bob found 6-meter activity to be a bit sparse in the middle of the Atlantic Ocean, so he upgraded to General, then Advanced and finally, Extra class. He enjoys QRP, PSK31 and home-brewing. Bob is retired from Civil Service, currently working part-time for an electronics consulting firm. You can contact him at Box 522, Garrisonville, VA 22463; **rlewis@staffnet.com**.

The Amateur Radio Public Service Handbook

By Kirk A. Kleinschmidt, NTØZ

How to Choose and Use a Portable Power Generator

June—Field Day month—gives us a great excuse to operate from woods, mountain peaks or jungles (concrete or otherwise) in an age-old exercise to improve our emergency communication skills. But what about electrical power? If you're operating "off the grid," a portable generator may be just what you need. Here's some practical advice about choosing the right generator and using it safely.

Setting up a radio station at the local park and working stations here, there and everywhere with a bunch of your best radio buddies is what makes Field Day special. That, and the beer, hamburgers or whatever you're serving in your neck of the radio woods. In addition to a weekend of fun and camaraderie, we improve our ability to serve the public in times of need—and our operating skills probably inch up a notch or two as well.

If you're new to this game you may be wondering about how everything is powered. And even if your operating site has electrical power (campground, softball park, courthouse lawn, etc), you may not want to avail yourself of its convenience. Field Day is ostensibly about practicing for communication emergencies when the ac mains might be out of commission.

Common power sources include batteries, vehicle alternators, wind-powered alternators, solar panels and engine-driven portable power generators. And speaking of portable power generators (PPGs for short), we see them at golf courses, in hardware stores and on TV news reports during floods and ice storms—but we don't see much about them in Amateur Radio publications!

I hope this article gives you a leg up on choosing and using the right portable power generator for your applications, Field Day or otherwise. Pay special attention to the safety issues. Generators—like all engine-powered devices—can injure or even kill you if you don't respect them. And unlike your garden tractor, these powerhouses can *electrocute* you (or others). Don't be afraid—but do pay attention!

The most basic units have preset throttle/engine speeds that can be adjusted to match required loads. These are most useful for powering incandescent lights or small ac motors (saws, drills, etc), which can safely tolerate "cruddy power." Use them to power solid-state devices at your own risk!

Because there are no free lunches, PPGs that offer better regulation and greater output power cost more money. Units that have little or no automatic regulation and less capacity are more affordable—unless you're talking about the tiny "hand-held" units that weigh in at 25 to 60 pounds, which cost more than some higher-powered, beefier, standard models!

And as if that's not enough, in addition to power capacity and regulation, there are other factors to consider such as engine type, noise level, fuel options, fuel tank capacity, run time, size, weight, cost, connectors, miscellaneous bells and whistles, etc.

Buying a Portable Power Generator

Before you run down to your local hardware superstore and buy the first PPG that catches your eye, consider the following items in light of

Figure 1—The author tested these four portable power generators during the creation of this article. See Table 1 and the "Resources" sidebar for more information. (All photos by the author)

Appendix F

your personal requirements. Sure, you'll use the generator for Field Day, but don't forget to factor in other possible uses such as camping, power outages, and so on. Try to do some research of your own. Your exact requirements may vary, and you may need a solution that fits.

The generators we're discussing here are designed for consumers, contractors and farmers. They're designed for occasional use, not for continuous, long-term applications. Units designed for continuous service and ultra-reliability (for marine, medical and telecommunication systems) are available through specialty suppliers, but the prices are prohibitive for casual users.

PPGs powered by diesel, kerosene, propane and natural gas are also available (at similarly high prices), as are ultra-quiet, liquid cooled, and specially sized and shaped generators. The PPGs we'll be considering are air-cooled and powered by gasoline. The four PPGs I tested while writing this article are shown in Figure 1.

Capacity

To be useful, your generator must be able to safely power all of the devices that will be attached to it. On a basic level that's just common sense. Simply add up the power requirements of *all* the devices, add a reasonable safety margin (25 to 30%) and choose a suitably powerful generator that meets your other requirements.

When you read the fine print, however, things get tricky. Some devices—most notably motors—take a lot more power to start up than they do to keep running. For example, a motor that takes 1000 W to run may take 2000 to 3000 W to start. Light bulbs, soldering irons, space heaters and most radios don't require extra start-up power, but be sure to plan accordingly.

Size and Weight

PPG size and weight usually vary according to power output—low-power units are lightweight and physically small, while beefier models are larger and weigh more. See Table 2 for more details. Some models are wrapped in a large protective frame while others have less "air space" inside the "cage."

Tiny camper models (800 to 1000 W output) are amazingly small and lightweight, but some units lack sufficient regulation and may not be recommended for powering solid-state devices. On the other hand, some teeny gens can put out a whopping 70 A of 12-V dc for charging batteries. If your gear is battery-powered, you may still be in luck.

Engines and Fuel

Most portable generators are driven by small gasoline engines similar to those used to power lawnmowers or go-carts. Basic models are powered by standard side-valve engines. These often make more noise, need more-frequent servicing and often don't live very long. More expensive models have overhead-valve (OHV) engines, pressure lubrication, low-oil shutdown, cast-iron cylinder sleeves, oil filters and electronic ignition systems. These features may be overkill if your generator will be used only occasionally. But if your generator needs are more consistent, "upgrade" models may offer much better service.

Run Time

Let's face it: Filling the generator's gas tank every hour can be a hassle—especially if you do it safely by shutting off the engine and letting it cool briefly before carefully pouring in more gas.

As a rule, smaller PPGs have smaller gas tanks (and vice versa)—but that doesn't necessarily mean that they need more frequent refueling. Some small engines are more efficient than their larger counterparts and may run for half a day while powering small loads.

When you look at generator specs, remember that the run times for most units are shown for 50%

Figure 2—Prized by RVers, PowerWatch Technologies' *Good Governor* is a handy unit that visually indicates ac wiring faults and accurately displays the voltage and frequency of the ac line source it's plugged into. If you can't find one at your local RV dealership, contact the manufacturer at PO Box 22988, Denver, CO 80222.

loads. If you're running closer to max capacity, your run times may be seriously degraded. The opposite is also true. "Extended Run" models usually have more efficient engines and larger gas tanks. The generator unit, however, is usually unchanged.

Typical PPGs run from three to nine hours on a full tank of gas at a 50% load.

Noise

Subject to a few exceptions, generators are almost always too loud. That is, we'd always prefer them to be less obvious. If you're set up for Field Day way out in the woods, generator noise probably isn't a problem. If you're set up in a campground or other more-public space, however, PPGs can sound like a rock concert. Keeping the things quiet isn't always possible!

Noise levels for many models are stated right on the box, but because there's no set standard for measuring generator noise, take these with a grain of salt and try to test them yourself before buying.

Was the PPG three feet away

from the sound level meter, or was it 10 or 20? Was the muffler facing the test set, or was it hiding behind the unit's engine? Did the noise tests take place in an open field, or were buildings or other reflective structures nearby? You get the idea!

That said, some models are definitely quieter than others. Some gens *do* have quieter engines and muffler systems, but most of the noise is actually produced by rotating generator parts and vibrating sheet metal. If you take great pains to make the exhaust quieter—as some users attempt—you may be shocked to discover that your improved "stealth generator" is only marginally less noisy!

Water-cooled PPGs (rare and somewhat expensive) produce less noise, as do units designed to be housed in special compartments found in boats and RVs. They're not a free lunch, though. RV gens are expensive and heavy.

Regulation

As previously mentioned, voltage and frequency regulation—or lack thereof—may significantly influence your buying decision. The bottom line is that *any* PPG can safely power lightbulbs, heating elements and power saws, but when it comes to computers, TVs and expensive ham radios, units with mechanical or electronic regulation may be required, if only for peace of mind! (All of the gens I tested safely powered solid-state devices. Initial tests, however, were made with a small TV set I'd purchased for $5 at a garage sale, just to be sure!)

Unloaded generators may put out 130 V at 62-63 Hz. As loads increase, frequency and voltage decrease. Under full load, output values may fall as low as 105 V at 58-59 Hz. Normal operating conditions are somewhere in between.

To add an extra measure of safety, consider inserting an uninteruptable power supply (UPS) or a line conditioner between the generator and your sensitive gear. These devices are often used to maintain steady, clean ac power for computers and telecommunication equipment. As the mains voltage moves up and down, a line conditioner bucks or boosts accordingly. UPSs, with internal gel-cell batteries, provide power to the load if the ac mains (or your generator) go down.

If "electronic voltage regulation" isn't mentioned on the box, consider calling the manufacturer before you buy. And although you might get lucky, don't expect expert help from the salesperson at your local hardware store—they're used to helping contractors who want to power lights and saws. (To improve your odds of getting a unit with electronic regulation, consider buying a PPG intended for sale in Canada. Two manufacturers suggested that all Canadian PPGs must have electronic regulation.)

Dc Output

Some PPGs have 12-V dc outputs for charging batteries. These range from 2-A trickle chargers to 100-A powerhouses. Typical outputs run about 10 to 15 A. As with the ac outputs, be sure to test the dc outputs for voltage stability (under load if possible) and ripple. Batteries—especially when your car is stranded in a blizzard—aren't too fussy about a little ripple in the charging circuit, but your radio might not like it at all! It's better to be safe than sorry.

Miscellaneous

Other considerations include outlets (120 V ac, 240 V ac, 12 V dc, etc), circuit breakers (standard or GFCI), fuel-level gauges, handles (one or two), favorite brands, starters (pull or electric), engine operating speeds (faster means more noise, less weight and a shorter lifespan, and vice versa), wheels, handles or

Table 1
Measurements and Data from the PPGs Shown in Figure 1

Model	Output Surge & Cont	Run Time @ 50% load (hours/gals)	V/f @ 0 W 500 W 1000 W	Reg. Method	Engine Type & Size	Weight (lb)	Price (Street)	Notes
Coleman Vantage 3500	4375 3500	9/3	128/63 125/62 125/62	Elect.	OHV 5.5 hp	110	$1049	1,2,3,4 5,6,8,9
Coleman Pulse 1750	1750 1400	5/0.9	133/63 130/62 125/60	None	Std 3 hp	65	$499	1,2,5,7,9
Homelite LR2500	2500 2300	7.2/3	157/64 151/64 149/64	Mech	Std 5 hp	87	$479	2,5,9
Honda EZ2500A	2500 2300	2.6/1	124/63 122/61 119/60	Elect.	OHV 5.5 hp	81	$789	5,6,7,8,9

1—Has 15-A, 12-V dc output.
2—"Extended Run" model.
3—Unit intended for sale in Canada.
4—Engine automatically idles when no loads are attached.
5—Unit has low-oil shutdown.
6—Unit has electronic ignition.
7—Unit is physically compact.
8—Unit is noticeably quieter than typical units in its class.
9—Unit should be adjusted for best voltage and frequency before regular use.

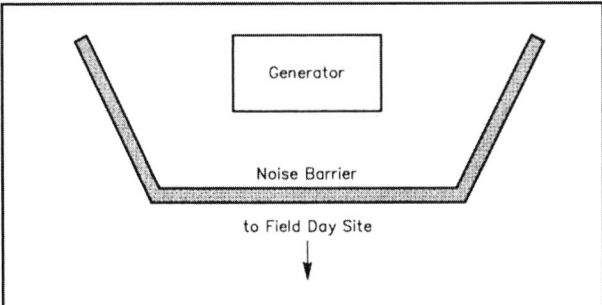

Figure 3—A two- or three-sided "room divider"-style noise shield is a handy, low-tech way to put a barrier between you and your Field Day generator. Use cardboard or carpet-covered plywood (with hinges, perhaps?), but don't put the generator in a covered box!

Resources

If you can't find a decent selection of PPGs at your local hardware superstore (with the Y2K craze, PPGs are in short supply everywhere), call Northern Hydraulics (800-533-5545) or Harbor Freight Tools (800-423-2567) and request catalogs. They're handy for other items, too!

For more information on the generators I mention in this article, surf or call: Honda generators, **www.honda-generators.com/generators/index.html**; Coleman generators, 800-445-1805; Home-lite generators, **www .homelite.com/homelite/products/**.

For information on UPSs, line conditioners and inverters, start with Statpower (**www.statpower.com/home.htm**) and American Power Conversion (APC) **www .apcc.com/**.

whatever you require.

Using Your Generator

Before we can connect "real" electrical loads in a Field Day situation we need to choose a grounding method—a real controversy among campers, RVers and home-power enthusiasts.

To complicate matters, almost all PPGs have ac generator grounds that are connected to the units' metal frames, but some units do not "bond" the ac neutral wires to the ac ground wires (as is done in typical house wiring). Although they might safely power your ham station all day long, units with "unbonded neutrals" may appear defective if tested with a standard outlet "polarity" tester.

Some users religiously drive copper ground rods into the ground or connect the metal frames of their generators to suitable existing grounds, while others vigorously oppose this method and let their gens float with respect to earth ground. Some user manuals insist on the ground connection, while others don't. The same is true for various electrical codes.

Follow the instructions in your user's manual and comply with local electrical codes. Grounding can also be a consideration with respect to lightning protection. See the ARRL Technical Information Service package on lightning-protection methods at **www.arrl.org/tis/info/lightnin.html**.

Regardless of the grounding method you choose, a few electrical safety rules remain the same. Your extension cords *must* have intact, waterproof insulation, three "prongs" and three wires, and must be sized according to loads and cable runs. Use 14-16 gauge, three-wire extension cords for low-wattage runs of 100 feet or less. For high-wattage loads, use heavier 12-gauge, three-wire cords designed for air compressors, air conditioners or RV service feeds. If you use long extension cords to power heavy loads, you may damage your generator and/or your radio gear. When it comes to power cords, think *big*. Try to position extension cords so they won't be tripped over or run over by vehicles. And don't run electrical cords through standing water or over wet, sloppy terrain.

During Field Day operations, try to let all operators know when the generator will be shut down for refueling so radio and computer gear can be shut down in a civilized manner. Keep the loads disconnected at the generator until the generator has been refueled and restarted. And keep a sharp eye out for late night ops who try to sneak space heaters, leg warmers or coffee makers into the tents. An extra 1500 W of power draw can crash the generator in a hurry!

Handy Tips

Ask any Generator Elmer and you'll get a flood of helpful hints—many learned the hard way. Here are a few:

Light Bulb Load Stabilizer

To keep generator output as stable as possible when switching loads on and off (keying a transmitter, for example), try keeping a small load (two light bulbs, for example) connected for the duration. The constant load can reduce power swings while the engine governor "hunts" to maintain proper shaft speeds.

Noise Reduction

According to many trial-and-error users, the best way to tame a noisy PPG during Field Day is to set it up in an out-of-the-way area and make a two- or three-sided sound shield from carpet-covered plywood or stiff cardboard (these look like small, folding room dividers). Keep the sound absorber/reflector between you and the gen. Do *not* make a four-sided shield or put the generator into any type of box. Gens need airflow to keep cool. See Figure 3.

Storage

When Field Day fun is over, don't just shove your generator into a dark corner of the garage. Follow the user manual's storage procedures and consider adding a small amount of gasoline stabilizer to keep the gasoline from oxidizing and gumming up the carburetor.

By Yaniko Palis, VE2NYP

The 12 Volt Pup: A DC Generator You Can Build

Grab a lawn-mower engine and an alternator to build a great 50 A power supply for Field Day or . . .

Field Day weekend is the best event of the year! I have always loved wilderness camping and almost any other adventure in the wide-open spaces. Coincidentally, my work often involves setting up all kinds of gear at remote locations for short periods of time—sort of a large-scale version of Field Day. Because of these two interests, Field Day has been my favorite event ever since I became a ham, six years ago. Now, thanks to what I have named "The 12 Volt Pup," I can easily generate enough power to operate a 100 W transceiver and plenty of accessories at almost any location I choose.

Generating power at remote locations is burdensome, in both equipment weight and cost. The Pup weighs about 45 pounds without a battery; so one person can handle it fairly easily. All told, expect to tote anywhere between 70 and 100 pounds, including batteries, fuel, oil and cables. If needed, you can easily disassemble the Pup into assemblies weighing less than 20 pounds each for backpacking.

The 12 V Pup combines a standard 3.5 horsepower lawn mower engine with an automotive alternator. These two components mount face downward onto two parallel, heavy duty, L-shaped steel rails, as shown in Figure 1. Spacers between the components and the rails precisely locate the pulleys and belt within the two steel rails. (See Figure 2.) Thus, the unit can rest on any appropriate flat surface. The engine takes a pulley for standard V belts, which makes it compatible with the alternator. Add a car battery and presto! You're in business. This design is amazingly simple.

An emergency version of this device could be jury rigged in an hour and a half. All you really need is a pulley for the engine, the right

[1]Notes appear at end of article.

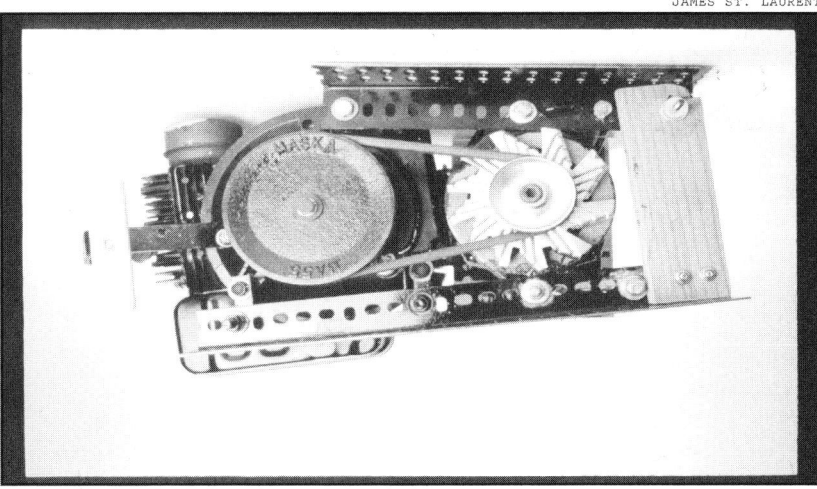

Figure 1—A bottom view without protective shields indicates the simplicity of the basic design. An engine bracket is visible at the left end of the lower (front) rail. The slot to mount the alternator (small pulley with fan) is in the upper (rear) rail. The Pup has wooden handles at each end for carrying.

Figure 2—One of the two small engine brackets is above the pipe-coupling spacer. The engine is at the upper right, the front rail at lower left. See Figure 4 for mechanical details.

Partial Parts List for the 12 V Pup:

Motor (1)—(See Figure 4.) After searching for a used engine, I bought a new, no frills lawn mower (for $99) and kept the engine. Recent models have a safety lever connected to a **KILL** switch on the engine that grounds a *neutralizing wire* to stop the engine. This neutralizing wire connects to the control box ignition switch and protection circuit.

Alternator (1)—(See Figure 3.) The one I used is modified as suggested by the folks at a large alternator-remanufacturing company. They rewound a standard alternator with fewer turns so that its internal regulator activates more often (50 A output). A modified unit should cost $65 to $85. Any standard internally regulated alternator with an internal charge controller should be fine, especially for charging automotive batteries. (A used alternator is worth $15 to $30.)

Motor Pulley (1)—Get one sized for standard **V** belts. Its rim diameter should be twice that of the alternator's pulley. This makes the alternator turn twice as fast as the engine. I used a $5^1/_2$-inch-diameter pulley. It's a big blessing that the engine shaft's dimensions are standard in every way. A common steel pulley fits right onto the engine's $^7/_8$-inch shaft and accepts a standard locking key ($^3/_{16}$ inch wide by $^1/_8$ inch deep).

V Belt to fit the pulleys, likely to be somewhere between 27 and 30 inches long; see text.

Storage Battery (1)—12 V lead-acid battery, 15 Ah or greater. Automotive or motorcycle batteries work, but a deep-discharge battery that tolerates fast charging is best. (Gel cells require a closely controlled charging regimen.)*

Steel rails (2) of **L**-shaped angle iron. This material is commonly used to support heavy-duty, industrial-grade storage shelves. It is perforated with rows of holes that ease assembly, provide ventilation and reduce its weight. The flanges should be at least $2^1/_4 \times 1^1/_2$ inches. The front rail is 18 inches long; the back rail is 14 inches long.

Motor Brackets (2)—Heavy-duty 1×1-inch angle iron. See Figure 4.

Hardware (Nuts, bolts and spacers—all of which may vary):
 (3) Engine-mount bolts, $^3/_8 \times 16 \times 2^1/_2$ inches long
 (3) Spacers, $^3/_4$-inch-diameter, $1^1/_4$-inches-long steel pipe couplings. These spacers place the engine pulley in the same plane as the alternator pulley. Buy longer couplings and/or shorten them as needed to accurately align the two pulleys.
 (2) Alternator mounting bolts to fit your alternator.

*The regimen is described in "A New Chip for Charging Gelled-Electrolyte Batteries," by Warren Dion, N1BBH, in *QST*, Jun 1987, pp 26-29.

---

size belt and two angle iron rails fitted with simple little mounts. Of course, you must also be willing to critically amputate your car and lawn mower! I decided to build a dedicated unit instead; it sports a control box and it cost me only $250 for all new parts. If you can scrounge up used parts, $125 should get you all the basic ingredients. My Pup took about four days to create. It's great to use the Pup with two or more deep-discharge lead-acid batteries. You can operate with power from one battery while charging the other. Because the Pup will probably charge a battery much faster than you would normally consume the stored energy, the generator may be switched off perhaps half of the time. This conserves fuel and reduces noise pollution.

You could also connect a load directly to the generator—as long as there's a battery connected across the load to stabilize the alternator's output. The engine's little governor works just fine, readily adapting the throttle to changing load conditions. While idling, the Pup provides about 6 A for normal battery charging. A 50% throttle setting produces about 30 A and ensures proper governor performance under varying loads.

Uses for the Pup go far beyond powering radios: I have inspired a friend to make one for his remote mountain cabin; it's a reliable supplement to his solar panels. A Pup

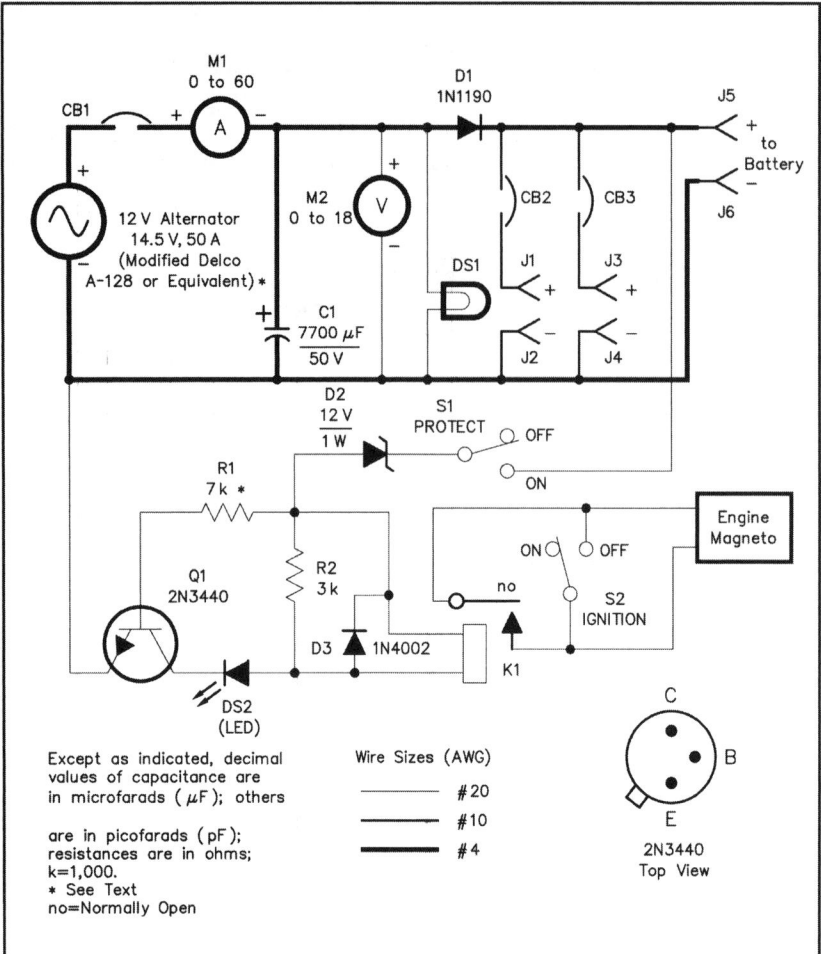

Figure 3—Control box schematic. Equivalent parts may be substituted for those shown. Many of the parts that carry large currents are not available from typical electronic-part suppliers. You'll have better luck at auto-part stores and local electrical-supply shops.

DS1—Automobile panel lamp, 12 V, 6 W, with socket and switch
C1—7700 µF, 50 V aluminum electrolytic
CB1—50 A automotive automatic-reset circuit breaker (from author's junk box; see Note 1)
CB2, CB3—30 A dc circuit breaker switches (65 V dc, 37.5 A trip, No. UPL1-1 from Philips Technologies Airpax Protector Group, 807 Woods Rd, Box 520, Cambridge, MD 21613-0520; tel 410-228-1500, fax 410-228-3456)
J1-J4—30 A terminals or connectors (builder's choice)
J5, J6—50 A terminals or connectors (builder's choice, look at your car's alternator connectors for ideas)

can charge vehicle batteries in the field. The Pup is also an excellent auxiliary power unit for an RV or at the race track, for deluxe golf carts and—my most ingenious use thus far—to charge batteries for electric trolling motors. "Ahoy, mateys! Let's visit a maritime mobile, haar!" I'm sure you'll find a use for a VE2NYP 12 Volt Pup.

Voltage Regulation

Cars do not run on 12 V, and regulated alternators are inherently unstable. Without some additional regulation, even a so-called "internally regulated" alternator will likely put out ugly inductive spikes at a dangerous 20 V, or more. Without other provisions to condition the output, a sizable lead-acid battery is essential; it should stabilize the output to a ripple-free 14.5 V.

The Control Circuit

The control box that I built is very simple. (See Figure 3.) The entire circuit is protected by an internal, auto-

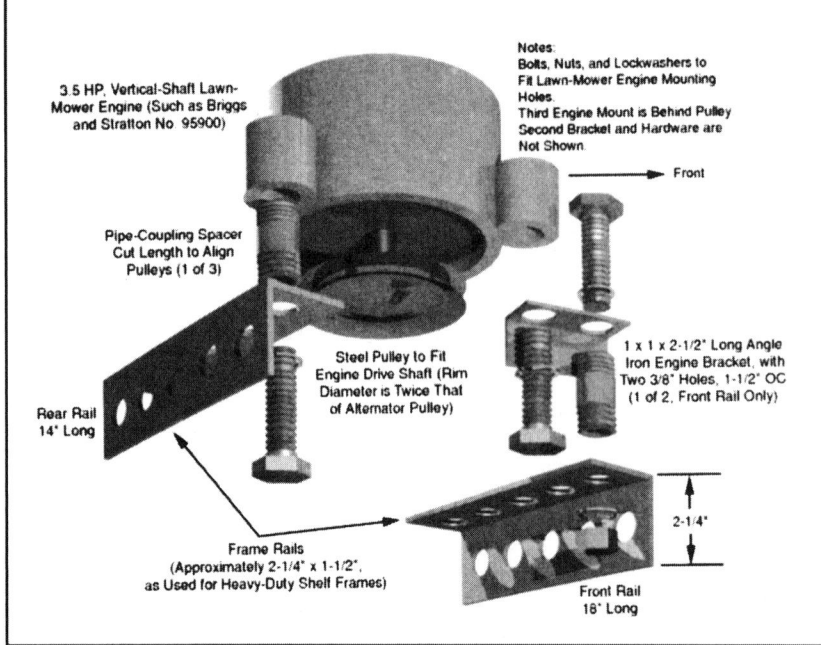

Figure 4—A pictorial of the engine mounting details.

Appendix F

motive, 50 A automatic-reset circuit breaker.[1] The two auxiliary outputs are each protected by 30 A breaker switches. Even with these breakers, this system is as hazardous as that of a car: Shorting the battery, alternator or internal wiring will cause a *big* explosive spark. (They might hear it in Calcutta, but we no longer send code like this!) Carefully avoid electrical shorts at all times—*especially* when handling the battery cables.

To filter the alternator's output, I installed a 7700 mF electrolytic capacitor across it. The capacitor absorbs the output spikes, leaving a rounded reverse-ramp wave as ripple at 0.40 V (a barely tolerable 3.5%). A 6-W panel lamp acts as a minimum load that protects the battery against overcharging. D1 is a high-current blocking diode. It prevents battery discharge through the lamp and reduces the voltage at the battery to about 13.8 V. I also built a very simple protection circuit that stops the engine should the output exceed 15.5 V (16.0 V peak ac).

During its brief life as a prototype, I have already received many good suggestions on how to improve my control box. For instance, one could stay on an automotive theme and use a ballast resistor, solenoid and an ignition relay to disconnect the battery. You could use a heavy-duty headlight switch with an internal circuit breaker for the power switch.[2] All this is to say, the control-box circuit that I show here is only one of many possibilities—you're welcome to improvise!

Finally, I recommend that you study the unit's output with an oscilloscope to be certain that your valuable equipment won't be damaged if the battery is disconnected while you are running the Pup. Also, some 12 V-only devices might be damaged by the 13.8 V dc that this device normally generates. [Most equipment built for automotive use is rated to +15 V.—*Ed.*]

Potential Hazards

There are mechanical dangers from the belt, pulleys and other moving parts. It is *your* responsibility to install adequate mechanical shields to prevent bodily harm. The photos show some metal shields and a plywood base that enclose the moving parts. Cut and fit similar shields to your Pup when the main construction is done. Keep fingers, hair, clothes, etc, completely away from all moving parts.

As with all combustion-powered generators, stray sparks may ignite the fuel. Stop the engine to refuel, and don't start it again until any spills have evaporated. Keep all cables, connectors, switches and relay contacts away from the fuel tank, and use this device only in well-ventilated areas. Closely follow *all* of the engine manufacturer's warnings.

Construction

The exact configuration of your Pup will depend on the actual engine and alternator pair that you acquire. That selection will determine the control-box size limitations. (I temporarily assembled the major parts several times to determine the final arrangement.) These notes may ease your construction. A socket set, wrenches and nut drivers turn this process into a breeze. So tune in your favorite listening frequency and enjoy the pleasures of being an insatiable tinkerer.

As you build, take measures against hazards: Prevent access to moving parts; tighten and seal connections against vibration; allow engine and alternator heat to escape; provide ventilation for cables and contacts carrying high currents; plan for exposure to the weather. Use plenty of grommets, wire ties, heat-shrink tubing, hot glue and strain reliefs to render all the connections Murphy proof.

Soldered connections may melt at the current levels found in this project. I crimped—and then soldered—heavy-duty lugs onto all the cable ends. For high-current connections, I bolted the lugs to the various components and jacks. Almost any circuit that shorts in the control box will likely melt. Finally, keep in mind that your Pup will probably operate in wet environments, so paint and seal its controls and connections against rain (and fuel vapors!).

Mechanical Assembly: Be an Iron Worker in your own Home

In the following assembly notes, I call the side of my engine with the fuel tank and carburetor on it the "front." The spark plug therefore sticks out of the right side and the crankcase is on the left. The alternator is to the left of the engine, beside the crankcase. This places the alternator on the cooler side of the engine (away from the cylinder). The control box is mounted atop the alternator.

Most lawn-mower engines seem to have the same three reinforced mounting holes on their base. (See Figure 4.) Two of the three holes line up with the front, so the long rail goes there. The third hole is at the "rear" and the shorter rail bolts to it. The engine mounts—via two angle-iron brackets, bolts and spacers—to the narrow flange of each main rail; the wide flanges become the vertical sides of the Pup's base. (Refer to Figure 2.)

Before you attach the rails, assemble the two engine brackets to the two front mounting holes on the engine. Position them to point away from the engine, toward the front. These brackets create plenty of elbow room for the engine's new pulley and permit easy access to the oil drain plug. They can swivel slightly, to easily mate with existing holes on the front rail.

Temporarily install the small mounting brackets to the engine, and measure the spacer length (Figures 2 and 4) required to perfectly align the two pulleys. Attach the two main rails so that they extend toward the left as far as possible. It is advantageous that the back rail has only one engine mount because the rail can pivot to accommodate alternators of any diameter.

My alternator did not require spacers because its two mounting holes are flush with the pulley side of its casting. The alternator's cooling fan blades scraped the edge of the rails so I trimmed the blade corners slightly. The threaded mounting hole of the alternator sits on the back rail and mounts to a slot you will cut out of the back rail later. The plain hole on the alternator casting pivots on the front rail, where it's attached. Check all clearances, and ensure once more that the two pulleys are in *perfect* alignment. Verify that the rails and spacers support the pulleys above the ground.

Now measure the arc that the alternator must swing along the back rail to accept a standard-length belt. A slot about 2 inches long allows for a $1\frac{1}{2}$-inch variation for belt size, eg, to accept *either* 28 *or* 29-inch belts. (I finished the unit before buying a belt—keep Murphy at bay, I say.) You can plan for standard-length belts during construction using the following formula:

$$BL = 1.57(D+d) + \sqrt{(D+d)^2 + 4C^2} \quad (Eq\ 1)$$

where

BL = Belt length (make all measurements in inches)
D = Diameter of large pulley
d = Diameter of small pulley
c = Distance between pulley centers

To use all available space, I installed the control box on simple rubber-damped mounts that I improvised. They poise the box about $1\frac{1}{2}$ inches above the alternator. This allows for air flow and protects the alternator from the rain. Once you have measured all the large internal components and cabling and have established the placement of the control box, pick a suitable cabinet and mark it for machining.

To finish, I picked a spot for a heavy-duty ground lug on the front rail. Thereafter, a few inches will remain open at the left end of the two rails. You can secure a small piece

Figure 5—A rear view clearly shows the largest mechanical shield in place and the carry handle—made from L brackets—that protects the spark plug from damage.

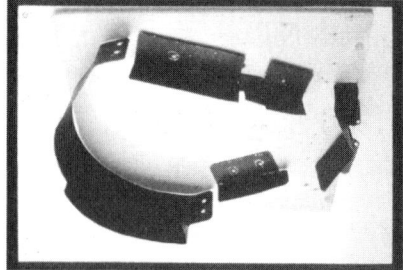

Figure 6—The protective shields, arranged on the plywood base to approximate their mounting positions.

of wood to them, to grasp when lifting the Pup by its left side.

Time to bend, bang, drill, flatten (bang some more), file and sand everything into its final shape. Polish all mechanical grounding points including the engine mounts. Cut the slot out of the back rail with a jigsaw. File off all sharp edges. When the relentless din of power tools, files, twisted blades and flying metal bits finally subsides, you will emerge victorious—and ready for subassembly and painting. Spray paint the mounts, rails and the control box with high-temperature engine enamel.

The protection circuit is built on a piece of perf board. When the output voltage exceeds 15.5 V, a heavy-duty, 5 V PC-board relay grounds the engine's magneto neutralizing wire to stop the engine.

The correct value for R2 depends on the relay's characteristics, so it must be set for each particular relay. To do so, install 10 kΩ pots in place of R1 and R2. Set both pots for maximum resistance. Connect an 18 V variable-voltage power supply across the circuit. (Connect the positive lead to D2's cathode and the negative lead to Q1's emitter.) Set the supply to your desired trigger voltage, and switch on the power. Adjust the R1 pot until the LED just lights. Then adjust the R2 pot until the relay just closes. The two adjustments may interact. Make a final adjustment of R1 when the Pup is complete with the control box installed and the battery disconnected. Finally, remove the pots, measure their values and replace them with combinations of fixed resistors.

Once my basic unit was tested, I added a pair of modified **L** brackets with a wood handle to the engine's right side. Together they span over the spark plug to protect it from being broken. (Do *not* loosen the cylinder head bolts to mount this!)

The protective mechanical shields that work well on my particular version are four custom-shaped pieces (cut from 22-gauge sheet metal stock, 7×24 inches). Machine screws hold them to the rails. (See Figures 5 and 6.) Attach the entire unit to a solid base (I used plywood) that blocks any access to the underside of the Pup. Editor Robert Schetgen, KU7G, suggests a lightweight hand cart as a base. Again, keep *all* the moving parts *completely* shielded!

You will love the 12 Volt Pup! It charges big batteries in a couple of hours. A gallon of gas lasts about four hours with a constant 20 A output. It usually loafs at low speed once a large battery has taken its initial charge. The gang at the Concordia University, VE2CUA, Field Day site was very interested in the Pup, and they first suggested that I write this article. Many members already have their own models churning in their minds. Richard Allix, VE2ARW, promises a miniature pup, to be born from a weed whacker and a motorcycle alternator. You are certainly welcome to write me with your comments and experiences. Good Health, Good Luck and Great DX from VE2NYP!

Notes

[1] I did not locate a suitable automatic-reset circuit breaker. Manual-reset breakers in that current range (*eg*, Potter & Brumfield W31X2M1G-50) cost about $20, or more. A large fuse would be less expensive. Automobile manufacturers use a fusible link to protect the alternator output.—*Ed.*

[2] According to E. P. Rolek, K9SQG's "A Source for High-Current Relays," in Hints and Kinks (p 73) Wal Mart may be a good source for such parts.

Yaniko "Nick" Palis first became interested in radio communications in his early teens. After some 20 years of SWLing, he finally decided to get on the air by becoming VE2NYP in 1990. Nick ran his college's broadcast radio station and designed many high-power laser light shows in their heyday (up to the early 1980s). He was a lighting director for films and television specials and would sometimes design custom electronic special effects for movies. He was a unit and location manager for many years. Yaniko is presently a supervising producer for feature films and television series in international distribution. Amateur Radio has revived all those previous technical interests and put them to good use again! You can reach Nick by mail at PO Box 61 station Place du Parc, Montreal, PQ H2W 2M9, Canada.

The Amateur Radio Public Service Handbook

Appendix G
Proformas

THE AMERICAN RADIO RELAY LEAGUE
RADIOGRAM
VIA AMATEUR RADIO

NUMBER	PRECEDENCE	HX	STATION OF ORIGIN	CHECK	PLACE OF ORIGIN	TIME FILED	DATE

TO

PHONE NUMBER
E-MAIL

THIS RADIO MESSAGE WAS RECEIVED AT
AMATEUR STATION _____ PHONE _____
NAME _____ E-MAIL _____
STREET _____
CITY, STATE, ZIP _____

FROM	DATE	TIME	TO	DATE	TIME

REC'D   SENT

THIS MESSAGE WAS HANDLED FREE OF CHARGE BY A LICENSED AMATEUR RADIO OPERATOR, WHOSE ADDRESS IS SHOWN IN THE BOX AT RIGHT ABOVE. AS SUCH MESSAGES ARE HANDLED SOLELY FOR THE PLEASURE OF OPERATING, NO COMPENSATION CAN BE ACCEPTED BY A "HAM" OPERATOR. A RETURN MESSAGE MAY BE FILED WITH THE "HAM" DELIVERING THIS MESSAGE TO YOU. FURTHER INFORMATION ON AMATEUR RADIO MAY BE OBTAINED FROM ARRL HEADQUARTERS, 225 MAIN STREET, NEWINGTON, CT 06111

THE AMERICAN RADIO RELAY LEAGUE, INC IS THE NATIONAL MEMBERSHIP SOCIETY OF LICENSED RADIO AMATEURS AND THE PUBLISHER OF *QST* MAGAZINE. ONE OF ITS FUNCTIONS IS PROMOTION OF PUBLIC SERVICE COMMUNICATION AMONG AMATEUR OPERATORS. TO THAT END, THE LEAGUE HAS ORGANIZED THE NATIONAL TRAFFIC SYSTEM FOR DAILY NATIONWIDE MESSAGE HANDLING.

PRINTED IN USA

ARRL Radiogram form (2011)

The Amateur Radio Public Service Handbook

GENERAL MESSAGE	
TO:	POSITION:
FROM:	POSITION:
SUBJECT:	DATE: TIME:

MESSAGE:

SIGNATURE: | **POSITION:**

REPLY:

DATE: | TIME: | SIGNATURE/POSITION:

ICS 213   NFES 1336

ICS-213 Form

Index

after-action reports (AAR) 37–41
Amateur Radio Emergency Service (ARES) 7–8, 10–19, 20, 22, 29, 30–31, 37, 40, 42, 51, 111, 115, 117, 125, 133, 142–146, 149, 168, 171, 184, 192–193, 203, 204, 209–215, 216–238
amplifier 80
ARRL 7–9, 10–19, 20, 25, 29, 31, 52, 54–55, 110, 114, 115, 119–120, 123, 142, 170–171, 190, 197, 206, 210, 214, 244–247, 248–250, 251–261, 300
Automatic Packet Reporting System (APRS) 31, 148, 186, 189, 212–214
aviation radio 131
batteries 99–100, 138, 263–267, 268–272, 273–278, 279–283, 284–289
Boston Marathon communications 154–159
Cabrillo log format 65–66
chargers 100–101, 279–283, 284–289
Citizens Band (CB) 130
Community Emergency Response Teams (CERT) 45, 51
contests, contesting 64–71
CW .... see Morse code
demodulation 79, 84
digital VOX soundcard interface 173–175
D-RATS 93, 187–188
D-STAR 31, 85, 93, 95, 149, 186–189
Echolink 23, 28, 31, 176–180, 202
Emergency Coordinator (EC) 10–19, 42, 215
Emergency Management Agency (EMA) 42–47
Emergency Operations Center (EOC) 42–45, 46–47, 50, 70, 96–97, 135–136, 145, 171, 184
emergency power projects 262–299
Emergency precedence 100–101, 121
Emergency Preparedness Program (EPP) 8
Family Radio Service (FRS) 130–131
FCC 7, 14–15, 21, 125, 130–131, 150, 215
Federal Emergency Management Agency (FEMA) 12, 16–17, 23, 35, 43, 52–54, 59, 63, 146, 188, 192
Field Day 64–71, 93
filter 79–82

Fldigi, Flmsg, Flwrp, etc. 160–172, 207
General Mobile Radio Service (GMRS) 131
generators 100–101, 138–139, 290–293, 294–299
go-kit 96–97, 102, 241–243
Handihams 198–203
handling instructions (HX) 100–101, 120, 123, 206
hazardous materials 61–63
health and welfare traffic 112–113
Incident Command System (ICS) 12, 51, 53, 120, 145, 169–170, 188, 212, 301
inverters 101
IRLP 28, 31
J-pole antenna 98
jump kit .... see go-kit
Log-Checking Report (LCR) 66–67
marine radio 131–132
marine VHF channels 131
message handling 115–125
Military Auxiliary Radio System (MARS) 16, 25, 26, 29, 182, 184–185, 190–197
mixer 80–81
modulation 77–79
Morse code 119–122, 140
multi-mode kit 85–95
Multi-Use Radio Service (MURS) 130
Narrow Band Emergency
 Messaging System (NBEMS) 160–175
National Fire Protection Association (NFPA) 62–63
National Hurricane Center 30–41
National Incident Management System (NIMS) 43, 51, 53, 152
National Traffic System (NTS) 8, 11, 15, 19, 26, 50, 55, 106–114, 159, 170–171, 188, 206
National Voluntary Organizations
 Active in Disaster (VOAD) 49
National Weather Service (NWS) 30–41, 248–250
Near Vertical Incidence Skyway (NVIS) 98–99, 103–105, 134
net, nets 17–19, 26–29, 30, 35, 106–114, 115–125, 136–137, 151, 192–194, 212–214, 239–240

net control, Net Control Station (NCS)............17–18, 22, 31, 35, 37–40, 43, 65, 97, 100, 108–109, 115–118, 124, 135, 148, 150, 157–158, 212–213, 239–240
network theory 126–132
oscillator..80
oscilloscope...77–81
PACTOR 85, 125, 134, 181–185, 196
Part 97 ....7, 14–15, 150, 215
Pelican case..85, 88, 91–92, 95
Personal Communications Service (PCS)..............................131
personal survival checklist...59
portable operation 96–105
precedence (of messages) 100–101, 120–121
Priority precedence (P) 100–101, 121
prosigns...119, 121–122
Public Information Officer (PIO)....................15, 142–146, 214
Public Safety Radio...131
public service events...147–153
QST articles............46–47, 102, 103–105, 173–175, 263–267, 268–272, 273–278, 279–283, 284–289, 290–293, 294–299
Radio Amateur Civil Emergency Service (RACES)..........15–16, 20, 22, 25, 29, 51, 117, 125, 192–193
Radio Amateurs Emergency Network (RAYNET).........204–205
Radio Emergency Associated Communications Teams (REACT)..25, 29, 131
Radiogram...................................120, 123, 170–171, 188, 300

rectifier 80–81
Red Cross, American Red Cross..................13, 20–24, 32, 40, 49, 115–116, 122, 125, 136, 155, 182, 193, 207, 209, 211–214, 251–261
Routine precedence (R)..100–101, 120
RSID tones 172
Safety 36, 58–63, 151–152
Salvation Army 25-29, 32, 40, 49, 201, 244-247
SATERN program...25–29, 38, 201
Served Agency(ies) 11, 20, 49, 54–55, 133
SHARES (Shared Resources)..........................27, 29, 193–194
Simulated Emergency Test (SET).............19, 51, 55, 189, 206
SKYWARN.......................................29, 30–41, 52, 134, 193
solar power...100–101
storm spotter..34
stress...137–138
tactical call signs...117, 155–156, 215
Tampere Convention 208
TOPOFF 3 exercise..209–215
training, training courses............................12, 16, 29, 42–43, 48–57, 64–71, 146
VoIP..28, 31, 35, 164
Welfare precedence (W)................................100–101, 120–121
Winlink, *Winlink 2000* 15, 17, 19, 93, 181–185, 188, 196

FEEDBACK

We're interested in hearing your comments on this book and what you'd like to see in future editions. Please email comments to us at **pubsfdbk@arrl.org**, including your name, call sign, email address and the title, edition and printing of this book. Or you can copy this form, fill it out, and mail to ARRL, 225 Main St, Newington, CT 06111-1494.

Name _____ ARRL member? ☐Yes ☐No

_____ Call Sign _____

Address _____

City, State/Province, ZIP/Postal Code _____

Daytime Phone (   ) _____ Age _____

If licensed, how long? _____

Other hobbies _____ E-mail _____

_____

Occupation _____

For ARRL use only	PSH
Edition	1 2 3 4 5 6 7 8 9 10 11 12
Printing	1 2 3 4 5 6 7 8 9 10 11 12